OBSERVING OUR ENVIRONMENT FROM SPACE
NEW SOLUTIONS FOR A NEW MILLENNIUM

PROCEEDINGS OF THE 21ST EARSeL SYMPOSIUM
PARIS / FRANCE / 14-16 MAY 2001

Observing our environment from space

New solutions for a new millennium

Edited by
GÉRARD BÉGNI
Medias-France / CNES, France

A.A. BALKEMA PUBLISHERS   LISSE / ABINGDON / EXTON (PA) / TOKYO

Published by: A.A. Balkema, a member of Swets & Zeitlinger Publishers
                www.balkema.nl and www.szp.swets.nl

ISBN 90 5809 254 2

Printed in the Netherlands

Observing our environment from Space: New solutions for a new millennium, Bégni (Ed.)
© 2002 Swets & Zeitlinger, Lisse, ISBN 90 5809 254 2

Table of contents

8 *Health and Risk Management Strategies*

9 *Monitoring*

Miscellaneous

Miscellaneous

Welcome Address

Observing our environment from Space: New solutions for a new millennium, Bégni (Ed.)
© 2002 Swets & Zeitlinger, Lisse, ISBN 90 5809 254 2

The Institut Géographique National: Training, research and production

J.Poulit
Institut Géographique National, Paris, France

ABSTRACT: Welcome to the training facilities of the Institut Geographique National in Marne la Vallee. The twin building where you are gathered hosts, since 1997, the Ecole Nationale des Sciences Geographiques (National School for Geographical Sciences), where all lGN staff are trained, and the Ecole Nationale des Ponts et Chaussees (National School for Public Works), which was founded in 1747 and happens to be the oldest engineering school in France. I am pleased to see that you are hosted in these convenient and still brand new facilities, as they witness how dynamic the Geographic Information world is, noticeably in relationship with space development and planning activities, and environmental monitoring. These facilities witness too that our Ministry, the Ministry for Equipment, Transportation and Housing is keen to modernise the education and training of its staff.

1 INTRODUCTION

The IGN France is among the major European producers of maps, digital geographic information and orthophotos. Our institute has therefore a strong interest in training, research and production in the field of remote sensing. Indeed, this domain of activity plays a major role in the production of geographic information. I would like to briefly mention our institute's activities in these three areas: training, research and production.

2 TRAINING

The IGN school (Ecole Nationale des Sciences Geographiques), which is one of IGN's directorates, hosts, for initial training, about 200 students of different categories. All of them spend some time, a short time or quite long periods of time, learning remote sensing. This school can take advantage of a wide array of image processing systems, as well as of dozens of analogue, analytical and digital stereo-plotting systems. In addition, the School for Geographical Sciences supervises the 10 months remote sensing training course that is provided by the "Groupement pour le développement de la télédétection aérospatiale" (Joint structure for the development of space remote sensing) based in Toulouse. Students who follow this course get a "Diplôme d'Etudes Supérieures Spécialisées"

(more or less a Master's level degree) jointly from the University of Paris VI, the University of Toulouse III and the Ecole Nationale des Sciences Geographiques.

The Ecole Nationale des Sciences Geographiques is also involved in a remote sensing course, with the Universities of Paris VI and VII, the University of Versailles -Saint Quentin en Yvelines, and the Ecole Nationale Supérieure des Télé-communications (National School for Telecom Sciences and Techniques). The students get a remote sensing degree "Diplôme d'Etudes Approfondies" (Master's level degree), which is a joint degree of the five schools.

Finally, starting in October this year, the school is to launch a new course, called a "mastère" (professional training at Master's level, as opposed to research training) for Digital photogrammetry and 3 dimension metric imagery. This 12 months course has got the Conférence des Grandes Ecoles' label. The students, who are assumed to have previously obtained an engineer's degree, will benefit from training that complements very well the previously mentioned courses. The Ecole Supérieure des Géomètres et Topographes (National School for Surveying Techniques) and the Ecole Nationale Supérieure des arts et industries de Strasbourg (Strasbourg's National Superior School for Arts and Industry) organise this "rnastère" course jointly with the Ecole Nationale des Sciences Geographiques.

3 RESEARCH

As for research, the IGN has chosen to spend a significant part of its research effort on image processing and remote sensing issues:

- Development of new sensors
- Image geometry and elevation computation
- Automated stereo plotting from digital images.

To be more specific, a good share of our large scale reference data still results from manual interpretation and stereo plotting from stereo pairs of aerial images. Any improvement that alleviates or diminishes the need for manual capture, hence reduces costs and time spent, is therefore crucial to us.

Our top priority is first to be able to process aerial pictures with a pixel size in the range of some decimetres (from 20 cm to 75 cm). We investigate interactive, assisted solutions, as well as fully automated solutions. Besides, we have an interest in looking at the potential of using already existing geographical data to help understand t~e images.

Research activities deal with image understanding, object detection and extraction, and update extraction. They aim at improving the production process, with respect to output accuracy, reliability and exhaustivity, but also performance, and ease of use for the human operator.

Another research priority consists in developing digital airborne cameras, which are going to replace the conventional analogue cameras we still use for aerial pictures. This is to be a major contribution to the evolution of our production process towards a fully digital process. Since 1999 we have begun using our first digital camera, truly operational for large-scale missions, and developed in house. We take advantage of it to tune our production processes to using such new digital image data.

I think it is worth noting that such digital cameras offer significant advantages when compared with conventional cameras:

- They provide a much better signal/noise ratio
- The images have a much larger dynamic range
- The radiometric response is fairly stable and linear.

They are therefore well suited for the production of orthophotos, and for feature extraction.

Research and development activities that address remote sensing issues take place in three units. Two are laboratories that are located in Saint-Mande: one laboratory for image processing and understanding, MATIS "Methodes d'Analyse et de Traitement d'Images pour la Stereo-restitution", and one laboratory for opto-electronics, LOEMI "Laboratoire d'Optique d'Electronique et de Micro-lnformatique". The last unit is the space department, IGN Espace,
located in Toulouse, which works in close relationship with the Centre National pour les Etudes Spatiales (French National Space Centre) located nearby.

4 PRODUCTION

Finally, as a producer, the IGN has been for a long time an important producer of space maps. Such productions routinely take place in the Toulouse space department IGN Espace. This department, where about 60 people (engineers, technicians and operators) work, outputs several hundreds of maps a year.

I may also mention the BD ORTHO© product, a nation-wide coverage with a 5 years update cycle, produced from aerial images, with a pixel size that ranges from 50 cm in rural areas to about 15 cm in town centres. Last but not least, you may know that our major topographical data set, the large scale BD TOPO© data base, is produced from stereo plotting of aerial images.

5 CONCLUSIONS

To summarise, the objectives of your symposium, the issues you will discuss have strong commonalities with the activities of the Institut Géographique National and the Ecole Nationale des Sciences Géographiques. Owing to the threats and constraints that surround our world, be they natural or man made, remote observation of our environment is, at the beginning of this new millennium, a stressing and unavoidable issue. Space remote sensing is for sure the only available technique for environmental assessment and monitoring at the global scale. This field was timely born in the last two decades, for the human being to properly identify the issues at stake, and areas at risk, and to be able to avoid ecological catastrophes that, unfortunately, tend to multiply.

This 21st symposium programme mirrors very well these priorities: soil and subsoil displacement studies, desert expansion, coastal and continental floods, forest fires, atmospheric risks (typhoons, hurricanes, pollution etc.), health risks and so forth. Besides, I understand that after the symposium itself, you will held an international workshop on remote sensing and GIS applications for forest fire prevention, detection and representation. This, again, addresses similar stressful issues.

These various approaches allow us to foresee that geographical information is to play a key role in our societies' activities. However, much work is still left to be done, for the various information sources to jointly contribute to operational applications. Institutes such as the IGN are mandated to make reference data available to users. I strongly believe that we must also help develop value added applications, be they developed by public entities, or by private partners. Such applications are to be developed, for the citizens to take full advantage of the available data and technologies.

Keynote lectures

Observing our environment from Space: New solutions for a new millennium, Bégni (Ed.)
© 2002 Swets & Zeitlinger, Lisse, ISBN 90 5809 254 2

The Kyoto Protocol: Legal statements, associated phenomena and potential impacts

G.Bégni & S.Darras
MEDIAS-France, Toulouse, France

A.Belward
JRC/SAI, Ispra, Italy

ABSTRACT: The United Nations Framework Convention on Climate Change of 1992 has the objective of "stabilisation of greenhouse gas concentrations in the atmosphere at a level that would prevent dangerous anthropogenic interference with the climate system." The Convention was ratified by the EU and entered into force in March 1994. Under the terms of this Convention the European Union must prepare inventories of anthropogenic emissions by sources and removals by sinks of various greenhouse gases (GHG). To achieve this all Parties are obliged to control, reduce or prevent anthropogenic emissions and conserve and enhance GHG sinks and reservoirs in terrestrial, coastal and marine ecosystems.

1 BACKGROUND

1.1 Description

The Kyoto Protocol to the UNFCCC is a key policy tool to be used to achieve the conditions stipulated in the UN 1992 Convention. The Protocol contains legally binding commitments to either reduce or limit the emissions of six GHG's: Carbon dioxide [CO_2], methane [CH_4], nitrous oxide [N_2O], hydrofluorocarbons [HFCs], perfluorocarbons [PFCs] and sulphur hexafluoride [SF_6]. There are agreed targets for the industrialised countries that collectively amount to a 5% reduction on 1990 emission levels of all six GHG (expressed in "equivalent CO_2"). The European Union collectively negotiated its commitment to reduce yearly emissions by 8% compared to 1990 levels within the commitment period 2008 – 2012. Significant achievements have to be evidenced by 2005. A number of Articles within the Protocol make provision for the use of biological sources and sinks to meet commitments.

A full monitoring system for terrestrial carbon must account for the carbon stored in terrestrial vegetation or soil (carbon stocks or pools) and the fluxes from or to the atmosphere and oceans through which it participates in the global carbon cycle. A carbon sink is where carbon is removed from the atmosphere. Thus a forest ecosystem may be a sink if its assimilation of carbon through photosynthesis

exceeds the carbon lost through harvest (or some other disturbance such as fire) and respiration.

The Kyoto Protocol actions dealing with biological sources and sinks centre on *Land Use, Land Use Change and Forestry (LULUCF) activities*. The Protocol also requires national systems for verification, reporting and accountability. Once ratified, signatories to the Protocol will be required to monitor measures promoting the protection and enhancement of greenhouse gas sinks and reservoirs and measure changes in carbon stocks resulting from human-induced land-use change and forestry activities. The resulting Carbon credits or debits are included in national reporting. Furthermore the Protocol makes provision for countries to trade these credits/debits via the so-called *flexible mechanisms*. These are governed by the Joint Implementation scheme in the industrialised nations, and by the Clean Development Mechanism (CDM) for the rest of the world. If carbon sink projects become acceptable activities within the CDM then world-wide assessment of changes in land use, especially afforestation, reforestation and deforestation becomes a requirement. CDM can be a win-win transaction for developing countries, provides that some equity principles are fulfilled. They can be a way to finance a more sustainable development at the national level and alleviate climate change, the impacts of which may endanger their national development. Adaptation activities within the

Global Environment Fund may include avoidance of deforestation and this will add further need. Collectively these raise the problem of accurate, reliable and verifiable monitoring of the terrestrial carbon sink.

1.2 Policy drivers and implementers

Monitoring of the terrestrial carbon sink is required at local, national, pan-European and the Global Scale. Sinks enhancements will be achieved by localised projects. National systems for the estimation of GHG sources and sinks are a Conventions commitment and a Protocol compliance requirement. The European Union has officially stated (Council Decision 1999/296/EC for a monitoring mechanism of Community CO_2 and other greenhouse gas emissions) that harmonised reporting across the Union is a goal and the European Commission as a Party to the Convention is also obliged to report as all other Parties.

Scientific and technical research to reduce uncertainties related to the climate system need global data and information. Strengthening endogenous capacities and capabilities to participate in international and intergovernmental efforts are stated objectives of both the Convention and the Protocol. Although local in scale, sinks projects may occur throughout the globe and will require consistent auditing. Finally the strategic value of independent assessments of the state of the globe's carbon sinks should not be underestimated. The relevant policies and owners at these various levels are listed below (global and European frameworks):

- Framework Convention on Climate Change (09/05/92, New York).
- Protocol to the Climate Change Convention (12.12.97, Kyoto).
- EC 1993. Council decision for a monitoring mechanism of community CO2 and other greenhouse gas emissions. Official Journal of the European Communities. L167: p.31ff.
- EC 1999. Council decision amending decision 93/389/EEC for a monitoring mechanism of community CO2 and other greenhouse gas emissions. Official Journal of the European Communities. L117: p.35ff.
- 6th Environmental Action Programme.

The issues related to the Kyoto protocol commitments monitoring is *one of the major items of the GMES (Global Monitoring for Environment and Security) European initiative*. The following

report is mainly based upon works by past or present ad hoc working groups from 1999 to 2001.

2 - SOME KEY SCIENTIFIC ISSUES

In 1999, a specific scientific working group was set up to characterise the information needs to monitor the UNFCCC and Kyoto Protocol commitments. The present chapter summarises some of its major conclusions.

2.1 Ecosystem monitoring

In order to take ecosystems into account, the following parameters should be monitored. This could most likely be achieved through space observations confirmed by in situ observations (at least for most of these parameters) when the Kyoto protocol comes into force.

1. *An index of vegetation cover*, that can be transformed into a fraction of the Photosynthetically Active Radiation (PAR) intercepted by the vegetation cover. Normalised vegetation indices observable from space at different temporal and spatial resolutions (LANDSAT, SPOT, IRS, NOAA-AVHRR, VEGETATION..) such as NDVI or SR are related to f_{PAR}, the fraction of PAR, by relations that do not seem to depend on the vegetation type.

The NPP (Net Primary Productivity) may then be computed in proportion to the absorbed radiation:

$$NPP = e\ f_{PAR}\ PAR$$

where e is the biological conversion efficiency of the biomass-absorbed PAR. In some models, it is considered as a characteristic of a vegetation type, in others it is calculated as:

$$e = e_{max}\ f(SWD)\ g(VPD)\ h(T)\ i(C_a)$$

where e_{max} is the maximum value of e and f, g, h are reducing factors (between 0 and 1) that take into account the effect of soil water deficit SWD, water vapour pressure deficit in the air VPD and non-optimal temperatures T. Function $i(C_a)$ depends upon the atmospheric CO_2 concentration C_a.

In more complex models e is computed according to a detailed model of land cover photosynthesis and plant respiration.

The NEP (Net Ecosystem Productivity) is the integrated net flux of CO_2 between an ecosystem and the atmosphere. It depends not only on NPP, but also on the heterotrophic respiration R_h:

$$NEP = NPP - R_h$$

R_h mainly depends on waste, soil organic matter, soil temperature and moisture, but is not directly measurable from satellite. It is usually calculated using a model of soil organic matter.

At the moment, NOAA-AVHRR and VEGETATION supply global NDVI information with a good temporal resolution (typically 1 week for a scene without clouds) and a spatial resolution of about 1 km. This is generally sufficient for global studies on carbon budget. However, regions such as Europe need to get information with a better spatial resolution while keeping the same temporal resolution, to allow crop monitoring for instance. The MSU-SK/RESURS Russian system, as well as the MODIS/TERRA system, the MERIS/ENVISAT future system and some medium term CNES projects intend to meet these requirements.

2. *An index of the nitrogen content of the vegetation cover*, which is strongly linked to the canopy photosynthetic capacity. There have been some attempts to derive it from reflectance in the MIR (1500 to 2000 nm range) but it may be difficult to get a general relation that holds for various vegetation types. It is also possible to first derive the chlorophyll content of the canopy from reflectance in the optical range (using a model of leaf optical properties) and to correlate it with the nitrogen content. This method could work but should be assessed with several types of vegetation cover.

3. *An index of the photosynthetic activity.* It is a good thing to know the vegetation cover, but it would be better to know whether it is under stress or not. This is especially true for forests that keep their leaves for a long time and respond to water stress by stomate closure long before they shed their leaves. Several solutions have been analysed with this aim. Temperature (thermal radiation) is not very sensitive for closed forests whose surface temperature does not increase by more than 3°C when stomates close. Considering chlorophyll fluorescence is a possible way, but it requires the use of Fraunhoffer lines and its feasibility has still to be demonstrated. This solution should nevertheless be explored. In the meantime, it is possible to use changes in reflectance in the 500-550 nm range that reflect changes in the carotenoid composition of leaves (the so-called xanthophyll cycle). This requires a sensor with several bands (10 nm ?) in this range, which will be available in the near future (MERIS/ENVISAT and MODIS/TERRA instruments).

4. *An index of biomass per ground area for forests.* Attempts to get this index from radar scattering in the P-band (about 0.6 m wavelength) have been successful for a given forest type and for biomass lower than about 150 t.ha^{-1}. However, we do not know if this relation can be applied generally. Another interesting approach is to measure vegetation height by Lidar. A US sensor (Vegetation Canopy Lidar, VCL) is scheduled to be launched in 2000. Biomass is related to height in a significant part of the forest growth cycle.

5. *Other variables* are rather interesting regarding the carbon cycle. The carbon and nitrogen content of the soil is a significant parameter, but cannot be provided through space measurements. Surface soil moisture is also quite important, and can be derived from radar measurements. Soil moisture influences its capacity to act as a sink of methane; on the other hand, global warming causes the permafrost to melt, which is an important source of methane. The SMOS project, developed by the CESBIO and selected as "opportunity mission" of the ESA programme "live planet", intends to reach such objectives. Its spatial resolution is not always accurate enough to meet all the requirements expressed here, particularly regarding approach (a), but it is a first important step in that direction.

The main sources of methane that could be determined through Earth observation techniques are rice paddies and wetlands (see figures 3 and 4). Wetland mapping has been undertaken under the IGBP-DIS programme. It is however difficult to determine quantitatively the actual methane fluxes, since complex biochemical processes are involved. Determination of fluxes related to other sources are closely linked to socio-economic parameters.

2.2 Land use and land cover changes

Together with industrial activities, land use changes are the main anthropogenic cause of GHG budget modification. They consequently require special attention (Adger et al, 1994; Estes *et al*, 1999;

Henderson-Sellers, 1994; Houghton, 1994; Lambin, 1997; Meyer et al, 1994; Turner II, 1994). Their understanding is a key issue of the LUCC programme (Turner II et al, 1995; Lambin *et al*, 1999). The information regarding land use and cover required in order to monitor the carbon dioxide budget are quite different from that required to monitor emissions of methane and nitrous oxide. In the first case (CO_2), what is required are mostly data on biomass changes for the forest ecosystems and for peat bogs. In the second case (CH_4 and secondarily N_2O), detailed information are needed on cultivation and livestock breeding practices (input use, manure application, irrigation, etc.).

There are four major information requirements related to land use and land cover:

- a baseline land cover map,
- a quantitative assessment of land cover conversions and modifications,
- the monitoring of landscape disturbances, particularly regarding forest ecosystems (e.g. fires, storms, diseases, etc.) and of the regeneration rate of the vegetation cover after these disturbances,
- information on land use practices (type of forest management, agricultural practices and input use, etc.).

Information on forestry practices is essential to understand the role of forests in the carbon cycle. For example, it would be worth getting information on forest conservation practices (forest protection, conservation and sustainable development; policies aiming at reducing the rate of deforestation and forest degradation), storage management practices (carbon storage in forests linked to the extension of the forest area, longer rotations, higher density of trees and better preservation of wood products) and substitute management practices (use of biofuels likely to be exploited in the long term). Such data cannot be estimated through remote sensing. They are usually not available as geo-referenced data.

As a consequence of these remarks, the highest relative uncertainty regarding the estimate of carbon dioxide emission is the "Land Use Change and Forestry" (LUCF) factor (see figure 4). Regarding the evolution survey of carbon sinks linked to land use change, the problem of estimating the situation in the base year 1990 needs to be solved. Earth observation is obviously an activity oriented to the future and aimed at ensuring a monitoring from 2000. It also meets the needs for archiving (LANDSAT, SPOT, ERS) in order to estimate the situation for the base years stated by the Kyoto protocol.

The uncertainty in the estimates of global methane emissions due to human activities is now estimated to be around 30%. Source estimates are uncertain, mainly due to the lack of representative emission factors. It is necessary to improve the data regarding LUCF activities, especially in developing countries. For example, the rate of methane emission from rice paddies is highly uncertain. For this activity, the required information consists of data on rice paddy areas and inundation periods. However, the highest uncertainty comes from an intricate mix of factors such as rice variety, soil properties, temperature, presence of microbes, local fertilising practices, etc...

In order to meet the requirements of approach (a), the required information must be spatially explicit and with an accurate resolution (dozens of meters). Information should also be quantitative, not only in terms of relevant areas but also in terms of biomass changes and fluxes involved. Only a combination of data from remote sensing-based systems, forest inventory data collected in the field, experimental data from sample plots, and detailed socio-economic data on management practices and technological factors will meet the information requirements regarding land use and land cover changes in the post-Kyoto period.

Some global models of land use change compute the rates of future land cover changes according to socio-economic factors and their interactions with biophysical processes. Whereas these models are useful to design global scenarios, they predictive value is limited since they are not yet actually reliable for long-term projections as many factors such as institutional, technological or political changes significantly interfere with the more predictable factors of land use change. Models of land use change are not yet fully operational either for predicting spatial patterns of changes or for taking into account spatial heterogeneity. However, a huge effort is currently dedicated to the design of robust models of land use changes.

2.3 *Socio-economic information*

As mentioned above, global monitoring directly requires socio-economic information for at least two reasons:
- First of all, some emissions and sinks can only be derived from national or local data. This concerns, for instance, industrial emissions and the consequences of the Kyoto protocol mechanisms on these emissions.

- Secondly, some information derived from satellite observations and/or *in situ* measurements must be coupled to socio-economic factors in order to deliver the required parameters. For instance, as developed above, data regarding cultivation practices or information on forest management must be coupled to data regarding the observation of land use and cover in order to enable the understanding of carbon fluxes.

To be more precise, the results of the four core projects of the IHDP research programme highlight the following needs:

I Land Use and Cover Change (LUCC). Many socio-economic issues are involved in this subject, described in §2.2. Here we will only recall the importance of ownership modes of land and resources, and the social issues related.

II Global Environmental Change and Human Securities (GECHS) & Disaster management. The societal demand may allow to better define the needs for a global environmental monitoring system, which may deliver extreme climate and weather forecasts in the medium term (10 days to 1.5 year). This is highly demanded for agriculture, water management, crop and extreme weather insurance. To perform such an analysis, it would be necessary to combine the physical aspect of risk assessment with detailed information on population and socio-economic geo-referenced information (e.g. vulnerability mapping).

III Institutional Dimensions of Global Environmental Change (IDGEC). This sensitive issue could generate problems likely to influence the monitoring strategy. The co-operative mechanisms that are part of the Kyoto protocol have a significant impact on the design and implementation of a monitoring network. In order to check its general compliance, the IPCC bottom-up/top-down workshop on GHG emissions considered essential to combine satellite data with local measurements and refined global dispersion models in order to monitor the local and regional emissions of CO_2, CH_4 and N_2O.

IV Industrial Transformation (IT). The ultimate goal of this programme is the understanding of the major factors of change in production and consumption processes (including service industries) and their implications regarding global environment. Four themes have been identified: Energy, Food, Cities and Communications.

1 *Energy*: In order to optimise energy production from renewable resources (which falls within the scope of CDMs), the following information are required: detailed maps of available Solar Energy (km scale); maps of the best locations for wind energy production (10-m scale); biomass availability & regrowth potential, burning intensity, CO/CO_2 ratio related to burning, tropospheric ozone formation, CO_2 emissions from forest and bush fires (area and physical dimension) coal burning (China), peat fires, waste dumps, gas pipelines. Such information can be taken into account within approach (a), in a specific way: they do not enable the observation or forecasting of GHG balance evolution, but are an important basis for the negotiations within the scope of the Kyoto protocol.

2 *Food*: Information on the protein chain: transformation from plant-food to plant-cattle-food.

3 *Cities*: In order to know better the GHG balance linked to direct anthropogenic effects, highly populated areas - and particularly megalopolis - have to be scrutinised. The information required include intensity of economic activities as well as water, energy and transport flows. Urban development, land use, biodiversity, direct GHG emissions, industrial waste and waste sites, as well as forecasts regarding population growth and urban development must be monitored, and suitable models developed accordingly.

4 *Communications:* Factors linked to communications may have a beneficial effect on the reduction of anthropogenic emissions. The use of available data should be enhanced in order to facilitate direct and operational decision making (can the stress suffered by environment be reduced if timely information are obtained regarding farming, biomass burning, wind and solar energy production?). Can relevant information improve the commitments of decision-makers? Telecommunication tools should be improved so as to reduce road and air traffic.

It should be noted that military activities, though a source of emissions, are not part of the protocol.

Some indirect aspects that have a deeper influence on data requirements. The two following points must be kept in mind when considering the socio-economic information required to implement the Kyoto protocol:

- First of all, the concern regarding the societal acceptance of the measures aiming at reducing the GHG budget should be associated with the "objective" measurements. One of the socio-economic issues is that the quantity of emissions is (in theory) a physically measurable parameter: however, the quality of emissions determines the societal acceptance of these measures. For instance, using wood instead of fuel for surviving (cooking, heating) is different from using wood for manufacturing newspapers. To be more precise, it is a matter of survival versus luxury. Still regarding societal acceptance, the Clean Development Mechanism (CDM) itself involves processes that favour investments aiming at reducing emissions in developing countries. But only the comparatively urban and wealthy fractions of these societies will gain by these measures, whereas the rural, poor, ethnic minorities or the women and children will not be concerned. A method only based on the physical study of these phenomena cannot cope with such matters and may even lead to counterproductive strategies.
- Secondly, the implementation of the Kyoto protocol requires an international consensus regarding the way qualitative and quantitative unquestionable and reliable information over several years will be operationally processed. As a consequence, proven concepts of Earth observation data collection and processing are required in order to monitor and assess the compliance of the protocol.

Lastly, the availability of operational systems is obviously a prerequisite for obtaining information. But the increased involvement of end-users is the key to a successful implementation of Earth observation data, which will enable to monitor the application of the protocol and the assessment of its compliance. Consequently, the development of software, methodologies, expertise, technology transfers and operational applications should be allotted a large part of the development effort and budgets.

Socio-economists begin to issue *global studies about the overall CDM market*. So, Zhong Xiang Zhang (2000) estimates the net value of the CDM market (gain to non-Annex-I countries) to lie in the range of US$ 254 to 2560 million. 75% of the total CDM flows are expected to go to China and India, while only a few percents (3%?) could go to Africa. Sokona *et al.* explain such a poor market by such elements as :
- Low contribution to GHG emissions
- Lack of institutional structures to address climate change policy issues
- Lack of monitoring capacities and skilled personnel to implement such projects
- Poor availability of a small private sector to properly identify CDM business opportunities
- Rigidity and instability of African economy in a difficult socio-economic and political context.

3 REQUIREMENTS STATEMENT

The present chapter presents the conclusions of the present GMES ad hoc working group on the monitoring Kyoto protocol commitments.

3.1 *Overarching considerations*

According to Article 4 (1) UNFCCC each Party - including the EU - has to develop, periodically update, publish and report to the Conference of the Parties (COP) national inventories of anthropogenic emissions by sources and removals by sinks of all greenhouse gases (GHG) not controlled by the Montreal Protocol. In the EU this has been implemented through the Council Decision for a Monitoring Mechanism of Community CO_2 and other greenhouse gas emissions (93/389/EEC) and its amendment (1999/296/EC). The GHG inventory required by the Convention includes changes in forest and other woody biomass stocks, forest and grassland conversion, abandonment of managed land and carbon emissions/removals from soil.

Under Article 3.3 of the Kyoto Protocol Parties must measure changes in carbon stock resulting from direct human-induced land use change and forestry activities, limited to afforestation, reforestation and deforestation (ARD) since 1990. Either within the first or follow on commitment periods Parties may be permitted to use other changes in land use as a means of gaining carbon credits. There is no agreement on which activities may be included (if any) and how they will be dealt with. If additional activities are included, then observation of all a Party's territory may become necessary, not just parts where ARD activities have occurred since 1990. This is much closer to the current state of reporting required by the Convention itself.

The Protocol's Clean Development Mechanism (CDM) aims to help non-industrialised countries achieve sustainable development whilst contributing to the ultimate objective of the convention. Sinks projects (such as new forest plantations) may become part of the CDM. If this happens then land use, land use change and forestry assessments will become a global requirement.

To calculate the benefits from a sink project the baseline must be established (i.e., conditions at the start of the project activities) and quantification of changes from this baseline measured. This is referred to as additionality.

Human induced or natural events (such as fire, wind throw, insect damage) can cause the subsequent release of carbon. Sinks projects will therefore require routine surveillance to determine or verify permanence. The projects will need monitoring over very long time frames. This is so that liability for any eventual release of stored carbon may be correctly assigned.

Sinks projects consume land, and in much of the developing world this is a scarce resource. As a consequence there is the need to closely monitor the immediate environs of any sinks project to track the problem of leakage. Leakage is the phenomena whereby carbon-releasing activities are displaced, rather than eliminated. Countries must be able to monitor undesirable phenomena, such as clearance of old-growth forest to liberate new land for sinks projects or to replace land consumed by a sinks project. Regular monitoring and surveillance should help countries to ensure that the benefits of any project are not outweighed by threats to existing agriculture, rural development programmes, sustainable forest management programmes and a region's biodiversity.

Land cover inventories and information on land cover changes (in terms of rates of change, land use changes, processes driving change and directions of change) can provide important information when considering the possible locations of sinks projects. Unfortunately this baseline information is often lacking in the developing world.

Article 10 (d) puts an obligation on Parties to contribute to global observing systems and support scientific and technical research concerning the climate system and climate change. Whilst there are no explicit measurement requirements, compliance with the spirit of this Article implies considerable effort in building global environmental databases.

The EU has established a Monitoring Mechanism for anthropogenic CO₂ and other GHG emissions in the Community. Each year Member States have to report their GHG emissions and removals to the Commission. The data have to be determined in accordance with the methodologies published in the Revised 1996 Intergovernmental Panel on Climate Change Guidelines. Although operational for a number of years the reporting by Member States on CO_2 sinks and sources from land use change and forestry follows no consistent methodology.

In Europe most countries use data from national forest Inventories. Independent of the quality of the underlying forest inventory, there remains the key problem that these inventories are performed for assessing marketable stem-wood and not for Carbon sinks/emission reporting. This results in a lack of transparency, consistency and harmonisation with regard to the different forest and other wooded land areas included and the expansion and conversion factors applied for transforming timber volume into overall CO_2-emissions and sinks. In addition inventories provide no means of assessing changes for a specific piece of land, hence current systems will not be appropriate for reporting on sinks projects.

Outside of forestry the situation is even worse and most Member States provide no data for categorise such as abandonment of managed land and carbon emissions/removals from soil. In response the EU's Monitoring Mechanisms Committee is looking for ways to improve the reporting situation concerning land use change and forestry chapters of the National GHG Inventories.

If the Kyoto Protocol is ratified there will be even *more pressure on Member States to build improved reporting structures* because the Protocol requires national systems for the estimation of GHG emissions and sinks before the year 2007 (Article 5 (1) KP). Such systems will be based on recommendations of the Special Report on LULUCF (IPCC 2000) and on decisions to be expected at future Conferences Of the Parties (COP).

Details concerning measurement requirements are still the subject of political debate. A summary based on current practices and interpretations is provided here. These are all required globally, though must be sufficiently spatially explicit to allow national, EU and project level reporting.

- Mapping and measurement of forest cover and forest cover conversion (area affected per year) at a resolution of 100 ha or better.
- Inventory, mapping and measurement of major disturbance features such as fire, wind-throw or disease (area affected per year) at a resolution of 100 ha or better.
- Inventory and mapping of forest type (down to species level) and stand age.

- Mapping and measurement of land use changes per year at a resolution of 100 ha or better.
- Monitoring changes in soil management techniques e.g. low till/standard plough/fallow.
- Measuring rates of soil carbon storage/loss, gC/ha/yr
- Measuring Net Ecosystem Productivity and Net Biome Productivity, gC/ha/yr

3.2 *Space Requirements and existing systems*

Use of Earth Observing satellites supports uniform, harmonised, globally consistent and comparable measurements. Earth Observation data provide georeferenced, spatially detailed information that permits verification of specific declarations. They provide unique measurements, and offer cost effective means of obtaining global, repetitive, long-term data sets. The following products have been developed:

- Land cover maps: These can currently be made in a consistent manner for the whole globe at a resolution of 100 ha. Improvements to 25 ha data sets are expected but are not currently guaranteed.
- Land cover change maps and statistics: Robust statistically based sampling methods for mapping and measuring change have been developed. These can report changes in the order of 0.5 ha.
- Active fire location and temporal distribution: Quasi operational global daily fire detection systems have already been demonstrated.
- Burnt area mapping: Global annual data sets at 100 ha resolution are under preparation. The statistical sampling methods used for land cover change may also be applied to burn scar mapping.
- Global fire history (from archived imagery). Currently only one year of global daily fire activity has been prepared (1992). Global archives dating back to 1981 are available, but much research is needed to turn these into a consistent product.
- Direct parameterisation of biophysical variables such as fraction of absorbed photosynthetically active radiation (fAPAR) and Leaf Area Index (LAI): Advanced algorithms allow consistent fAPAR data sets to be created every 10 days. LAI and vegetation structure can be measured, but

more research is needed to realise the full potential of new multi-angular satellite imagery.
- Leaf phenology and seasonal growing cycle: These can be inferred from time series of the biophysical parameters.
- Water stress: This is still in the research domain.
- Biomass : Research points to measurements possible below 250 T/ha. This may be improved with new technology.
- Digital Elevation Model (1 km resolution or better): 1 km global data sets are available, better resolution products are under development.

The main existing systems allowing to set up some of these products and information are :

- Very high resolution data, (IKONOS, IRS-1C/D, Corona) for mapping detailed changes
- High resolution data (SPOT - HRV, Landsat 7, ERS SAR, JERS SAR) for compiling land cover change statistics
- Moderate resolution (VGT, MISR, SeaWiFS, MERIS) for biophysical products and for trend detection (AVHRR, ATSR, MODIS for fire detection)
- Geostationary (METEOSAT) for land surface characterization for climate modelling

3.3 *Non-Space and in-situ observation requirements - assimilation techniques.*

The main *in-situ* measurement needs are :

- Meteorological / climate data
- Soil type and distribution
- Ecological variables such as stand age and allocation strategy
- Surface-atmosphere carbon, water and energy flux measurement
- Atmospheric CO_2, ^{13}C, O_2/N_2 concentration measurements
- Land use factors such as nitrogen fertilisation rates, forest management practices.

It should be emphasised that space data, non-space and *in-situ* observations are most often "assimilated" into more or less complex models in order to retrieve the useful information at the proper scale. One consequence of this is that improving the quality of models may be (at least) as important as

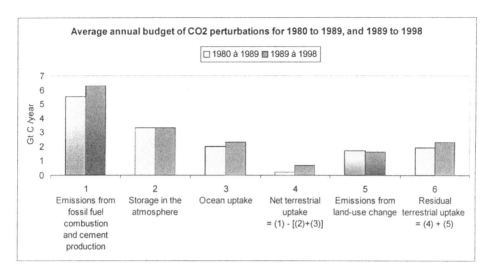

Figure 1 Global carbon sources and sinks distribution (IPCC 2001)

collecting new data. Another consequence is that any progress in modelling techniques may change the data requirements, in terms of nature, sampling and accuracy.

Besides setting up models, the hard point for the Kyoto Protocol monitoring is downscaling or upscaling techniques, since most of the models now available address either the global or the local scale, but are not adequate to the appropriate "mesoscale" (typically 100x100 km) at which the landscape heterogeneity has to be taken into account.

Some of the *in situ* systems allowing to get relevant information are:

- Fluxes of carbon, water and energy are continuously measured with sub-hourly time steps at 140 stations world-wide and analysed as synthesis products within the framework of FLUXNET
- Network of sampling sites maintained by various governments performing flask sampling in remote areas to capture variations in clean, background air far from local sources and sinks
- Network of meteorological stations and climate data from the ECMWF
- European Soil Bureau provides harmonised soils data base for all Europe, FAO for the world at coarser resolution

3.4 *Some Missing Elements and Future Requirements*

- On the global scale numerous activities are underway or planned that address terrestrial carbon observation in various geographic regions. These are not well co-ordinated at either the European or International scales, nor do they provide consistent information on carbon at the global scale. Systematic global observations are still lacking.
- The first progress report (November 2000 (COM(2000)749) on the EU's Monitoring Mechanisms Committee does not even include data on land use change and forestry in the tables and calculations because the data quality is so low. A Reference System for EU Greenhouse gas emissions and removals (sinks) can therefore be said to be missing. Notable improvements in the overall quality of the EU's GHG inventory could be achieved if the inventory is submitted to a pre-submission review. This is foreseen in the existing EU Monitoring Mechanism. In the longer term alternative methods for EU-wide GHG emission and sink estimates could be developed and brought into operation, especially where there are scientifically sound reasons to believe that a central estimate would provide higher quality EU inventory estimates. Alternative methods would also provide an independent check on the quality of the EU

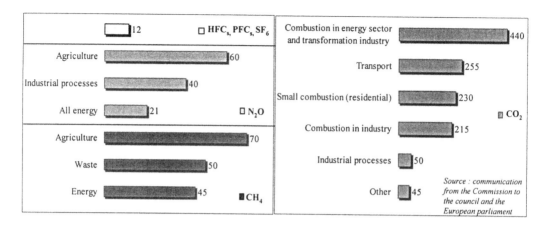

Figure 2 - The relative importance of the 6 greenhouse gases within the EU year 1990 in CO_2 equivalent

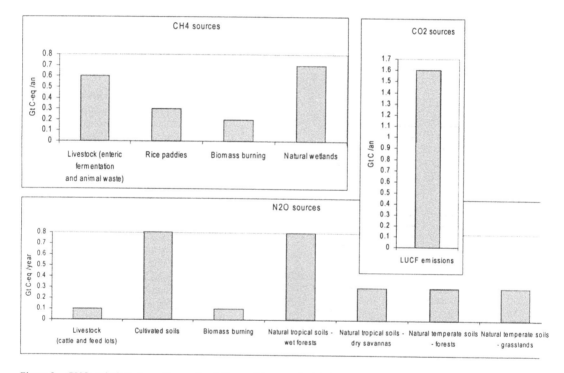

Figure 3 - GHG emissions depending on Land Use and Forestry (IPCC 2001)

inventory obtained by summing national inventories. Such a Reference System is proposed under the European Climate Change Programme.

In order to fill these gaps, future requirements might be:

- Institutional arrangements need to be made for product generation, standardisation and quality control.
- Continuity of calibrated, fine resolution optical data from both fixed-view and pointable sensors needs to be ensured.
- Continuity of moderate resolution and multi

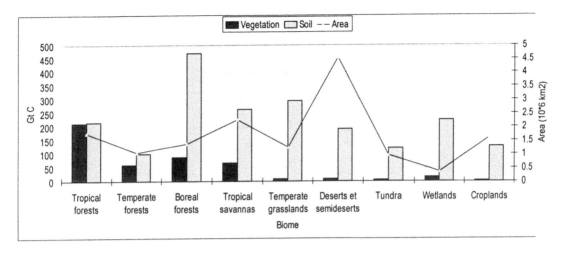

Figure 4 - Global carbon stocks in vegetation and top 1 m of soils (based on WBGU, 1998 - IPCC 2001)

angular optical sensor measurements must be ensured.

- Reprocessing of archived satellite data should be systematically addressed.
- Canopy structure measurements from satellite sensors such as lidars and advanced SARs should be ensured.
- Density of *in-situ* measurements of below ground biomass should be increased.
- Current network of flux measurement programmes should be maintained and new stations added in underrepresented regions (especially Africa and Asia), and in ecosystems undergoing major disturbances. Both are important as net carbon flux measurements are currently available only from a few dedicated, "healthy" sites.
- The long-term continuity and stability of the atmospheric sampling programme should be ensured.
- Appropriate mesoscale models have to be developed or improved to assimilate available data and information and generate the requested information with the relevant accuracy. A scientific consensus has to be found about their relevance and performances.
- Global level co-ordination and collaboration are needed to ensure international acceptance of any proposed terrestrial carbon sink measurement methods, and possibly to ensure such measurements are made in the first place.

5 CONCLUSION

The importance of terrestrial carbon sink monitoring cannot be overstated. Climate change is a major, if not the major environmental issue we face. Political solutions at the international level must be found if we are to safeguard the environment for future generations. Political solutions must be based on reliable information. Current capabilities and activities provide a sound basis on which to build reliable terrestrial carbon sink monitoring systems that satisfy both the European and International communities. These include satellite observations, in-situ observation networks, existing environmental databases and advanced models. But these lack a common focus and a co-ordinated and systematic mode of operation. The GMES can provide both the common focus and co-ordination. The GMES process too must ensure that gaps in the observing systems are filled and that the quality, continuity and consistency of our observing capabilities are guaranteed.

6 - ANNEX: SOME QUANTIFIED FIGURES

The figures may illustrate in a quantitative ways some of the parameters and trends described above.

7 BIBLIOGRAPHY

Adger, W.N. and Brown, K. 1994: *Land use and the causes of global warming*. Chichester: John Wiley & Sons.
Belward, A. S., 2000: A Policy context for global

environmental information systems. *Proceedings of the workshop on Remote sensing and long-term global data sets for climate studies*, Dundee 2-13 August 1999. (Springer Verlan : in press)

Belward, A. S., 2000: The Potential Use of Satellites and the Kyoto Protocol, Invited paper for presentation at the *28th International Symposium on Remote Sensing of Environment*, Cape Town, SA, 27-31 March 2000.

Bicheron, P., Leroy, M., 1998: A method of biophysical parameter retrieval at global scale by inversion of a vegetation reflectance model, accepted in *Remote Sensing of the Environment*, April 21, 1988.

Bricaud, A., 1998: SeaWIFS et la couleur de l'océan, *Lettre PIGB-PMRC France n° 8*, September 1998, 48-52.

Briffa, K.R., et al., 1995: Unusual twentieth century warmth in a 1000-year temperature record from Siberia. *Nature 376*, 156-159.

Commission des Communautés Européennes, Communication de la Commission au Conseil et au Parlement Européen, "Changement climatique - vers une stratégie communautaire post-Kyoto", Bruxelles, 03.06.1998, COM(1998) 353 final.

Commission of the European Communities, Communication from the Commission to the Council and the European parliament, "Preparing for Implementation of the Kyoto Protocol", Brussels, 19.05.1999, COM(1999) 230 final.

Coppin, P.R. and Bauer, M.E. 1996: Digital change detection in forest ecosystems with remote sensing imagery. *Remote Sensing Reviews*, in press.

Darras, S., Michou, M., Sarrat, C., 1999: IGBP-DIS Wetland data Initiative. A first step towards identifying a global delineation of wetlands, *IGBP-DIS Working paper # 19*, February 1999.

Dedieu, G., Lafont, S., Cayrol, P., Chehbouni, A., Kergoat, L., Maisongrande, P., Watts, C., Berthelot, B., Gérard J.C., François, L., Graetz, D., Ruimy, A., Moulin, S., Saugier, B. 2000: STEM-VGT: Satellite Measurements and terrestrial Ecosystem Modeling using VEGETATION instrument, *VEGETATION 2000 conference*, Belgirate, April 2000.

Dixon, R.K., Brown, S., Houghton, R.A., Solomon, A.M., Trexler, M.C. and Wisniewski, J. 1994: Carbon pools and flux of global forest ecosystems. *Science 263*, 185-190.

Downton, M.W. 1995: Measuring tropical deforestation: Development of the methods. *Environmental Conservation 22*, 229-240.

Estes, J. E., Belward, A. S., Loveland, T. R., Scepan, J., Strahler, A, Townshend, J. R. G., and Justice, C.O., 1999: Global Land Cover Mapping, The Way Forward, *Photogrammetric Engineering and Remote Sensing*, 65, 1089 - 1093.

Fontelle, J.P. et al. 1999: Inventaire des émissions de gaz à effet de serre en France au cours de la période 1990-1997. Format IPCC. CITEPA (Rapport au Ministère de l'Environnement)

GCOS, 1997: In situ observations for the global observing systems. Development of an integrated strategy and identification of priorities for implementation. 10-13 September 1996, Geneva, Switzerland - Edited March 1997, GCOS-28 (WMO/TD No. 793, UNEP/DEIA/MR.97.3)

Global Monitoring for Environmental Security - A Manifesto for a new European Initiative, October 1998, BNSC, CNES, DLR, EARSC, ESA, EUMETSAT, European Commission.

Graedel, T.E., Crutzen, P.J., 1993: *Atmospheric Change. An Earth System Perspective*, W.H.Freeman and Company, New York, 1993.

Gudmandsen, P., 1999: News from ESA: Granada II (October 1999), *EARSeL Newsletter N° 40*, 7-9, Dec. 1999

Gupta, J. (1998): '*International Environmental Law: An Introduction*', Course Materials for the Masters Programme in Environmental Science and Technology, 1998

Henderson-Sellers, A. 1994: Land-use change and climate. *Land Degradation & Rehabilitation 5*, 107-126.

Houghton, R.A., Unruh, J.D. and Lefebvre, P.A. 1993: Current land cover in the tropics and its potential for sequestering carbon. *Global Biogeochemical Cycles 7*, 305-320.

Houghton, R.A. 1994: The worldwide extent of land-use change. *Bioscience 44*, 305-313.

IGES/NIES (2000): Proceedings of the IGES/NIES workshop on *GHG inventories for Asia-Pacific region*, Shonan Village, 9-10 March 2000, edited by Damasa B. Magcale-Macangong.

IPCC (1994): "*Climate Change 1994: Radiative Forcing of Climate Change and An Evaluation of the IPCC IS92 Emission Scenarios*". J.T. Houghton, L.G. Meira Filho, J. Bruce, Hoesung Lee, B.A. Callander, E. Haites, N. Harris and K. Maskell (Eds). Cambridge University Press, UK. 339pp. (ISBN 0-521-55962-6).

IPCC (1996): "*Greenhouse Gas Inventory Reporting Instructions*". J Houghton, L G Meira Filho, B Lim, K Treanton, I Mamaty, Y Bonduki, D J Griggs and B ACallender (Eds.). Volume 1. IPCC/OECD/IEA. UK Meteorological Office, Bracknell.

Kasperson, J.X., Kasperson, R.E. and Turner II, B.L. 1995: *Regions at risk*. Tokyo: United Nations University Press.

Lacaux, J.P, R. Delmas, B. Cros, B. Lefeivre, and M.O. Andreae: Influence of biomass burning emissions on chemistry in the equatorial forest of Africa, in *Global Biomass Burning: Atmospheric, Climatic and Biosheric Implications*, edited by J.S.LEVINE, pp. 167-173, MIT press, Cambridge, Mass., 1991

Lacaux J.P., H.Cachier and R.Delmas, Biomass burning in Africa: an overview of its impact on atmospheric chemistry, in *Fire in the Environment: The Ecological Atmospheric and Climatic Importance of vegetation Fires*, edited by P.J.Crutzen and J.Goldhammer, pp.159-191, John Wiley, New York, 1993

Lafont, S., Chevillard, A., Kergoat, L., Dedieu, G., Maisongrande, P., 1999: Modélisation des flux de CO2 à l'aide du capteur VEGETATION: Etude de l'Europe et de la Sibérie, *Atelier de modélisation de l'atmosphère*, *Météo-France*, Toulouse, December 1999.

Lambin E.F. and Ehrlich D., 1997: Land-cover changes in sub-Saharan Africa (1982-1991): Application of a change index based on remotely-sensed surface temperature and vegetation indices at a continental scale, *Remote Sensing of Environment*, vol.61, no.2, pp.181-200.

Lambin E., 1997: Modelling and monitoring land-cover change processes in tropical regions, *Progress in Physical Geography*, vol.21, no.3, pp.375-393.

Lambin E.F., Baulies X., Bockstael N., Fischer G., Krug T., Leemans R., Moran E.F., Rindfuss R.R., Sato Y., Skole D., Turner II B.L., Vogel C.,1999: Land-use and land-cover change (LUCC): Implementation Strategy.IGBP Report 48, *IHDP Report 10*,Edited by C. Nunes and J.I. Augé.

Le Dantec V., Dufrêne E. and Saugier B., 2000: Interannual and spatial variations of maximum leaf area index in temperate deciduous stands, *Forest Ecology and Management* (in press)

Lifermann, A., 1998: POLDER: un concept original pour observer la Terre, *Lettre PIGB-PMRC France n° 8*, September 1998, 44-47.

Maisongrande P., Ruimy A., Dedieu G. and Saugier B., 1995: Monitoring seasonal and interannual variations of gross primary productivity, net primary productivity and net ecosystem productivity using a diagnostic model and remotely sensed data. *Tellus*, 47B, 178-190.

Mann et al. 1998: Global scale temperature patterns and climate forcing over the past six centuries. *Nature* 392, 779-788.

Meyer, W.B., and Turner II, B.L. 1994: *Changes in land use and land cover: A global perspective*. Cambridge, UK: Cambridge University Press.

Moore, B. 2000: Le Programme International Géosphère-Biosphère : une étude du changement global. Quelques réflexions. *Lettre PIGB-PMRC France n° 10*, February 2000.

Nemani, R.R., Running, S.W., Pielke, R.A. and Chase, T.N. 1996: Global vegetation cover changes from coarse resolution satellite data. *Journal of Geophysical Research* 101 D3, 7157-7162.

Petit et al. 1999: Climate and atmospheric history of the past 420,000 years from the Vostok ice core, Antarctica, *Nature*, 399, 429 - 436

Rasool, I., Begni,G., 1999: La responsabilite scientifique dans la recherche sur le changement global, *Bulletin de la SFPT* 156, 26-27.

Rivière, E 1999: Evaluation des puits de CO2 suivant la nouvelle méthode préconisée par le GIEC. *CITEPA*, (rapport au Ministère de l'Environnement) (in press).

Rotmans, J. and Swart, R.J. 1991: Modelling tropical deforestation and its consequences for global climate. *Ecological Modelling* 58, 217-247.

Ruimy A., Dedieu G. and Saugier B., 1994: Methodology for the estimation of terrestrial net primary production from remotely sensed data. *Journal of Geophysical Research (Atmospheres)*, 99, D3, 5263-5283.

Ruimy A., Dedieu G. and Saugier B., 1996: TURC - Terrestrial Uptake and Release of Carbon by vegetation, a diagnostic model of continental gross primary productivity and net primary productivity. *Global Biogeochemical cycles*, 10, 269-285.

Ruimy A., Jarvis P., Baldocchi D.D. and Saugier B., 1995: CO2 fluxes over plant canopies and solar radiation: a review. *Advances in Ecological Research*, 26, 1-6

Ruimy A., Kergoat L., Field C.B., Saugier B. (1996): The use of CO2 fluxes in models of the global terrestrial carbon budget. *Global Change Biology*, 2, 287-296.

Skole, D. and Tucker, C. 1993: Tropical deforestation and habitat fragmentation in the Amazon: satellite data from 1978 to 1988. *Science* 260, 1905-1910.

Sokona, Y., and Nanasta, D., 2000 : The Clean Development Mechanism : An African Delusion? *Change, Research and Policy Newsletter on Global Change from the Netherlands*, 54, 8-11.

W. Steffen, I. Noble, J. Canadell, M. Apps, D. Schulze, P. Jarvis, D.Baldocchi, P. Ciais, W. Cramer, J. Ehleringer, G. Farquhar, C. Field, A.Ghazi, R. Gifford, M. Heimann, R. Houghton, P. Kabat, C. Koerner, E.F.Lambin, S. Linder, J. Lloyd, H. Mooney, D. Murdiyarso, W. Post, C.Prentice, M. Raupach, D. Schimel, A. Shvidenko, R. Valentini (IGBP Terrestrial Carbon Working Group), 1998: The Terrestrial Carbon Cycle:Implications for the Kyoto Protocol, *Science*, vol.280, 29 May 1998, pp.1393-1394.

Tett, S.F.B. et al. 1999: Causes of twentieth-century temperature change near the Earth's surface, *Nature*, 399, 569 - 572 .

Turner II, B.L., Meyer, W.B. and Skole, D.L. 1994: Global land-use/land-cover change: towards an integrated study. *Ambio* 23, 91-95.

Turner II, B.L., Skole, D.L., Sanderson, S., Fischer, G., Fresco, L. and Leemans, R. 1995: Land-use and land-cover change: Science/research plan..*IGBP Report 35*, HDP Report 7, Stockholm: The Royal Swedish Academy of Sciences.

Zhong Xiang Zhang, 2000 : The Potential of the Market for the Kyoto Mechanisms, Change, Research and Policy *Newsletter on Global Change from the Netherlands*, 54, 5-7.

Observing our environment from Space: New solutions for a new millennium, Bégni (Ed.)
© *2002 Swets & Zeitlinger, Lisse, ISBN 90 5809 254 2*

GMES Initiative

N.D.Costa, N.Hubbard & R.Winter
Space Applications Institute, Joint Research Centre, TP261, Ispra (VA), Italy

ABSTRACT: GMES (Global Monitoring of Environment and Security) is a new European initiative to put knowledge-supporting technologies to the service of better environmental management and security concerns. It is a joint initiative between the European Commission (EC) and the European Space Agency (ESA). This presentation was aimed at describing the current status of this initiative, and the plans for the short- to medium term.

1 GMES OBJECTIVES

European determination to play a leading role in the field of environment and sustainable development is seen in the importance attached to wise management of land and resources of our common territory and our role in the governance of the global environment. Europe must also respond to new international challenges concerning crisis management, peace-keeping, humanitarian operations and development aid. To realise its aspirations the Union's decision-makers, environmental managers and the public increasingly need access to information on critical issues relating to environment and security.

Unfortunately access to the right information at the right time is a rare occurrence. Data lack coherence between topics, lack compatibility between countries, there are gaps and continuity over time is not secured. Coordinating the entire information delivery chain, relevant observations, models and information systems is required. Europe has the scientific and technical capability for this coordination, but a serious risk exists that we could become reliant on other nations to meet our information needs in key policy areas. GMES is an ambitious European initiative to address these concerns.

GMES will ensure the availability of high-quality environmental and security information on a sustained and operational basis through earth observation from space in conjunction with other information technologies. Data and information requirements arising from the integration of the environment dimension in all EU policies, the enlargement process and establishing a more integrated perspective on security issues drive the process. GMES will also provide a European contribution to international environmental monitoring efforts, yet will provide independent observations where strategically required. Building on Europe's existing capabilities and infrastructures GMES will identify and develop elements needed to establish sustainable, cost-effective operational information-gathering and distribution chains.

Thus in summary, the principal drivers behind GMES are as follows:

- Answering user requirements for data and information linked to environmental and resources management policies and specific applications at EU and Member States levels.
- The strategic necessity for Europe to play a leading role in the global stewardship of the environment and the role of information in support to such a goal.

2 HISTORY

The GMES initiative grew out of the Baveno Manifesto which was drawn up in 1998 at a meeting between the EC, ESA and the national space agencies. However the GMES stakeholders comprise a much wider group that includes user/provider agencies such as EEA, EUROSTAT, EUMETSAT, WEU-SC, the space and value-adding sectors in industry, research institutions and many national and international organisations.

In late 2000, the GMES Partnership was formed with participation of the Member States, the major Space Agencies, industry representatives and the EC. The mandate of the Partnership was to help formulate the GMES Proposal in response to the Council Request. In addition three working groups were established, viz. Environmental Conventions,

Environmental Stress and Risks and Hazards, for the purpose of collecting user requirements.

In cooperation with the Swedish Presidency, a meeting was held in Stockholm on the 21 and 22 March 2001 titled "GMES – The Users' Perspective". Half of the participants were from European environmental organisations and as end-users, they confirmed the potential of GMES and emphasized the need for support from the policy makers in the EC.

3 JOINT WORKING DOCUMENT

EC and ESA have recently issued a joint working document, which outlines the concept and action plans for GMES. This document, 'A European Approach to Global Monitoring for Environment and Security (GMES): Towards Meeting Users' Needs' (Version 2, 6 June 2001), was presented at the June meeting of the Research Council of the EC.

3.1 Priority themes

Priority areas are described in this document that have been identified according to the following criteria: (i) relevance to EU policies; (ii) possibility to produce results rapidly; (iii) complementarity with and added value to on-going activities; (iv) known interested users.

These areas have been selected in conjunction with the EEA and on the basis of the stakeholders' inputs during the consultation process. They include:

- Land cover change in Europe
- Environmental stress in Europe
- Global Vegetation Monitoring
- Global Ocean Monitoring
- Global Atmosphere Monitoring
- Support to Regional Development Aid
- Systems for Risk Management
- Systems for Crisis Management and Humanitarian Aid

In addition, a horizontal support action regarding Information Management Tools and contribution to the Development of a European Spatial Data 'Infostructure' was also selected.

3.2 Role of EO

The space segment is only one source of data for these priority themes but constitutes a critical component for building improved information systems on environment and security. The extent to which EO is utilised will vary with each application depending on its appropriateness for that particular service.

Therefore, GMES will pay particular, but not exclusive attention, to earth observation from space as far as this technology fulfils its objectives.

4 THREE IMPLEMENTATION STRANDS

GMES is viewed as an iterative process that comprises the three main strands. The actions, presented below, describe the dynamic model of GMES not only for the initial period, but for its entire implementation life.

4.1 Deliver to learn

The objective of this first strand is to deliver improved information in the selected priority themes. As such it includes activities to

- deliver specific information and information services on the basis of user driven applications and learn from these;
- deliver quality and synthesised information to users on priority topics to support Community environment and security policies;
- contribute to the identification of obstacles and solutions to the production of this information.

Through these activities, potential operational services or delivery of information within the priority themes will be identified.

4.2 Assess and structure

This second strand includes the assessment of the current information production processes and structure of the demand and supply sides.

Drawing upon the experience gained under the first strand and the awareness generated by implementing the priorities, the following objectives will be pursued :

- assessment and update of the GMES user needs;
- assessment of the obstacles to an efficient production and delivery of information for environment and security policies (e.g. the influence of data policies, the incorporation of monitoring requirements in legislation, the role of public and private funding, economic costs and social benefits);
- identification of solutions;
- establishment of a structured dialogue and collaboration between the actors and institutions involved in the process of production and delivery of information.

4.3 Develop and improve

The third strand is concerned with political, technological and scientific developments for the production and delivery of better quality information for environment and security policies.

The strengths, weaknesses, opportunities and specifications resulting from the assessment produced under the first and second strands will constitute the basis for the following actions:

- Data services: improving the quality and availability of basic data from monitoring programmes and infrastructures;
- Complementing and adjusting planned remote sensing and ground monitoring programmes and infrastructures and developing new ones to achieve coherent and complementary services;
- Implementing and developing information technologies for data transmission, access and processing and for information access;
- Research and development to produce knowledge and models on natural processes and their interactions with human activities, to transform data into information.

This will allow the development of the required infrastructure and the knowledge base in order to secure and improve a sustainable approach to the delivery of information.

5 FUTURE PLANS

A GMES conference will be held in Brussels on the 15 October under the Belgian presidency with the title 'GMES: Towards Implementation'.

Further details of planned GMES activities will be outlined during the second half of 2001 and will be presented at the December meeting of the Research Council. A detailed GMES implementation plan is foreseen at the end of 2003.

GMES is an important component of the European Strategy for Space and is in line with the orientations of the European Research Area. Preparatory work is on-going within the 5th Research Framework Programme and GMES is a priority for the 6th (2002-2006). In the ESA context, the future Earth Watch programme associated with the Earth Observation Envelope Programme should also be a major contributor.

6 FURTHER INFORMATION

The Chairman's report from the 'Users' Perspective' workshop in Stockholm and the latest version of the Joint working document on GMES are available from the GMES web site at http://gmes.jrc.it/.

Further supporting information is also available here, such as the requirements briefing documents, current status and future plans of capabilities and systems in Europe, as well as a framework for assessing GMES benefits.

1 *Land Surface Processes*

Observing our environment from Space: New solutions for a new millennium, Bégni (Ed.)
© *2002 Swets & Zeitlinger, Lisse, ISBN 90 5809 254 2*

Meteorological and Earth observation remote sensing data for mass movement preparedness

M.F.Buchroithner
Dresden University of Technology, Institute for Cartography, Dresden, Germany

ABSTRACT: One of the major problems of the efficient synergetic use of remote sensing data for natural disaster mitigation is the fusion of various meteo- and geodata sets of significantly different spatial resolution. On the other hand, different morphological types of mass movements are based on alternate concepts i.e. of generation which, again has to be initially reflected in differing methodological approaches. The unifying idea and stronghold, however, of the presented approach is the precipitation parameters which trigger the debris flows. Albeit, frequently there are no relevant precipitation climatic data available. Despite significant drawbacks in the integration of Landsat and/or SPOT data sound hazard zonation maps can be generated. For a test area in the French Alps it has been shown that remote sensing data can be used to predict potential debris flow events. High temporal frequency remote sensing data from the Meteosat series of satellites allow the identification of cloud clusters most likely to result in intense rainfall which are, in turn, likely to initiate debris flow activity. Video evidence, field observations and an empirical debris flow model linked to an instantaneous rain gauge were used to ascertain the exact times of debris flow initiation. Due to the high spatial and temporal variability of rainfall in mountainous regions and the large areas vulnerable to debris flows compared to the coverages of these observations, there are, however, restrictions on the use of these data for large regions to provide early warning of debris flow events operationally. Additionally, the possible timeliness of warnings using such observations is restricted to the relatively short interval between the onset of the triggering phenomena and the hazard event. The remote sensing techniques developed in this study, allow warning of potential debris flow events to be derived before the triggering phenomenon occurs, by attempting to recognise the evolution if intense rain-bearing clouds. In this. role the meteorological remote sensing data are not used to retrieve rainfall amounts but, instead, to derive rain cloud properties that produce debris flow triggering conditions.

1 INTRODUCTION

Worldwide, numerous mass movements occur in mountain environments. The most dangerous type of them is the debris or mud flow. Generally initiated on steep slopes, they can be described as rapid viscous flows of granular solids, water and air (Varnes, 1978). The initiation of a debris flow is commonly described to occur in two different ways. The first one considers the initiation of a debris flow as the result of a landslide that steadily transforms itself into a debris flow by dilution or liquefaction during its movement (de Graaf *et al.*, 1997). The other one attributes the initiation of a debris flow to a sudden failure of coarse debris in a high-altitude channel or gully bed (Takahashi, 1981b). The input, or trigger, for this latter mechanism stems from

either snowmelt or intense rainstorms, or rarely earthquakes.

The effects of debris flows are often catastrophic to the inhabitants of the regions in which they occur, causing serious casualties and property damage. In Armero, Columbia, in 1985, for instance, in the order of 20,000 people were killed when debris flows of up to 300 km/h buried whole valleys (Florez, 1986; Romero *et al.*, 1989). In 1996, a debris flow devastated a campsite in the Spanish Pyrenees killing more than 80 people (Garcia-Ruiz *et al.*, 1996). The casualties and economic losses due to debris flow events can be expected to increase with the projected increases in high-intensity precipitation events associated with global warming Zimmermann and Haeberli, 1992). These impacts of debris flow activity could potentially be reduced with the development of an early warning or preparedness system for communities living at hazardous locations.

Debris flows vary considerably in their size, speed and impact on the environment. Rapid debris flows are well-known episodic features in certain areas, e.g. in parts of the French Alps (Steijn van et al., 1988). The study presented here focuses on debris flows in the Bachelard Valley, south of Barcelonnette in the French Alps (Figure 1). Although not as big as those mentioned in the previous paragraph, the debris flows in this area have flow paths up to 1.5 km in length and widths up to 10 m. Their velocity has been estimated to be 7-10 km/h (Steijn van et al., 1988). In general they are initiated by short but intensive summer precipitation events. Although presenting only a minimal threat to people in this region, they provide a good opportunity to look at the provision of early warnings for mud flow hazards in general. The study area was chosen because of the relatively easily accessible mass movement source areas, and due to the relatively high frequency of debris flow events compared to other debris flow-prone regions (Buchroithner et al., 1997).

In the past, debris flow research has focused on developing initiation models and numerical simulation of the movement and deposition processes of the flow (Takahashi, 1978a; Takahashi, 1981b; Johnson and Rahn, 1970; Johnson and Rodine, 1984). In both these areas of research, prediction of debris flow events has been limited to the moment of the flow onset and dependant on the accurate description of free flowing water conditions at the time of initiation. The problems associated with this, in terms of developing an early warning system for debris flow activity for the French Alps, are threefold: First, the potential time span for the warning of an event is very short due to the relatively instantaneous nature of the onset of the triggering phenomena and the actual hazard phenomena. Second, in mountain regions precipitation and snowmelt conditions have high spatial and temporal variabilities and, hence, the accurate depiction of triggering conditions by ground-based measurements such as rain gauges is extremely difficult. Precipitation can, for instance, be extremely variable changing from zero to over 100 mm/h over the distance of a few hundred metres (de Graaf et al., 1993). This means, extrapolation of rain gauge readings is often highly error-prone. Third, very little is known about the precipitation conditions over mountainous terrain due to the lack of observations and insufficient theoretical attention

given to the weather and climate phenomena of such regions (Beniston et al., 1994).

Meteorological remote sensing data offer an opportunity to provide information on local precipitation over large areas, largely irrespective of geographical location, and details about the evolution of triggering conditions before they initiate a debris flow. Precipitation estimates derived from satellite data commonly use data of the infrared and passive microwave wavelengths. In this study infrared data measured by the Meteosat satellite series and processed according to the "Bristol approach" (Barret 1993, Barret and Cheng 1996) have been applied to provide information on potential triggering conditions for debris flow initiation. The relatively simple relationship and short-time interval between the onset of heavy rainfall, and the initiation, movement and deposition of a debris flow permits this information on the triggering conditions to be considered as an early warning of the actual debris flow event itself (Kniveton, 1995).

A key point in developing an early warning technique for rapidly occurring hazards such as debris flows, is the identification of the exact initiation time of the events. This information is required to assess, adapt and validate the technique and time of the possible early warning. Not surprisingly, given the remote location of the debris flow initiation sites, or source areas, accurate measurements of these times are rarely available. This tremendously aggravates a reverse engineering approach. In the present study this information has been obtained from a combination of video evidence, a debris flow initiation model linked to two rain gauges and field surveys. The video evidence stemmed from a video camera mounted in the debris flow source area. Unfortunately restricted to daylight hours, the camera operation was also limited to retrieving only the date of the flow event, and not the time. The exact time of the event has therefore been obtained by a debris flow initiation model using input data from automated rain gauges. Video evidence and field surveys together were used to confirm that debris flow events had occurred. Six mass movement cases have been identified in order to adapt and 'validate' this technique (Buchroithner et al., 1997).

2 PHYSIOGRAPHIC SETTING OF THE STUDY AREA

The Bachelard Valley is located 20 kilometres south of Barcelonnette in the Alpes de Haute Provence(Figure 1). The valley forms part of the

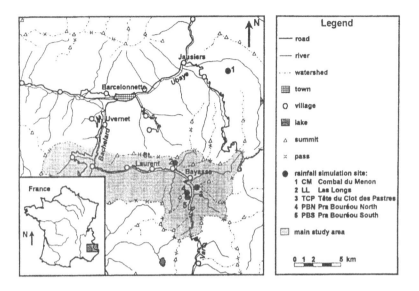

Figure 1. Location of the Bachelard Valley in the French Alps over which the DFIM has been developed by Blijenberg et al., (1996).

stream system of the river Ubaye. One of the key denudational processes along the valley walls is the initiation of debris flows. Most debris flows originate in source areas situated at altitudes between 1900 m and 2600 m. The rocks there are mainly marls and chalks of Cretaceous age. Above approximately 2000m the sandstone of the Gres d'Annot Formation is found. Vegetation in the source areas is almost entirely absent with the exception of few grasses. Human influence is minimal.

The climate in the area is characterised by both Mediterranean and oceanic influence. With respect to debris flow initiation the former is of most interest. The Mediterranean influence results in the occurrence of relatively isolated rainstorms, mainly at the end of spring and during autumn (Steijn van et al., 1988). According to field observations (de Graaf et al., 1993; Blijenberg et al., 1996) most debris flows are initiated during the heaviest of these rain storms in summer and early autumn. Rainfall data recorded in the valley show a yearly precipitation of 977 mm at an elevation of 1660 m. As a result of relief instability, rainfall varies significantly in the valley with an increase of precipitation in relation to the height.

3 DEBRIS FLOW INITIATION MODEL (DFIM)

A number of studies have linked rainfall intensity

and duration with the onset of debris flows. For the present study it was decided to use a debris flow initiation model with the variables of slope steepness and rainfall intensity (Blijenberg et al., 1996). This model has been developed after field observations of the initiation process. During these field observations it has been noted that the initiation process commenced with the formation of microscale mudflows or microslumps. These microslumps are formed under intense rainfall conditions and provide the coarse debris material in the gully bed with the necessary amounts of fine material and water to get moving. In order to investigate the complete initiation mechanism a number of rainfall simulations have been carried out in the debris flow source area by Blijenberg et al. (1996) and an empirical initiation model has been developed.

Rainfall simulation results from these observations are shown in Figure 2. From this scattergram and its best fitting borderline between actual microslumps and only superficial debris runoff it can be deduced that a critical combination of slope steepness and rainfall intensity is essential for the initiation of microslumps and subsequently of debris flows. Rainfall intensities and slope angles greater than the line drawn in the diagram are taken to indicate existing conditions for debris flow initiation. The conclusion is justified that rainfall intensities of 105 mm/h on slopes of 42 degrees are likely to initiate debris flows.

31

Measurements between 1991 - 1994 by from two automated rain gauges in the source area and applied to the above model revealed the occurrence of nine 'very likely' debris flow events. In order to verify the model derived estimations of debris flow initiation times, the readings of the rain gauges had to be matched with those of the video; both data sources recorded chronologically. The precise matching of rain gauge readings with video evidence proved difficult, due to the number of rainfall events. However, some of the cases have been successfully matched. Out of the nine events identified by the model only two have actually been witnessed by field surveys and these were labelled 'certain' debris flow events. Footage from the video camera showed that seconds after the rainfall started, superficial runoff took place. When rainfall intensities exceeded 50 mm/h for several minutes, small pebbles were detached from the valley walls. At higher intensities, micro slumps on the valley slopes were observed and in the most extreme cases a debris flow was initiated. For the present study six cases were chosen for further examination, including the two certain cases observed in the field. The reason for the omission of the three other cases was merely the availability of Meteosat data.

4 REMOTE SENSING TECHNIQUE

With the times of the estimated debris flow initiation the next step in developing an early warning approach was the application of satellite data prior to these times to identify potentially triggering rainfall conditions. For this purpose infrared data from the Meteosat satellite series have been used. They have a maximum spatial resolution of 5 km by 5 km at sub-satellite point and are sampled at the same location once every 30 minutes.

Satellite-based rainfall retrievals using infrared data rely on the indirect relationship between cloud top temperature and surface rainfall (Rasmusson and Arkin, 1992). In the past infrared-based rainfall algorithms have been successfully used to provide information on tropical and subtropical rainfall totals over daily and monthly time scales (Barrett, 1993). At higher latitudes (north of 45°), the use of infrared data has been generally restricted to summer conditions when the temperature of the Earth's surface is sufficiently different to that of the cloud tops and the land surface is free of snow cover. As stated in the introduction, the rationale for the selection of satellite data was in part because the satellite could provide information not only on precipitation conditions over a specified period but also on developing rainfall and cloud conditions likely to trigger a debris flow, thus enabling a longer early warning time until hazard event onset. An example of the use of satellite data in a similar role was shown by Barrett and Cheng (1996) with the development of the Hierarchical Operational Procedure to denote potentially high and medium rain events for flood forecasting in western Europe. In the Barrett and Cheng (1996) study, the remote sensing technique began by denoting cloud clusters using a threshold of 235 K. Pixels with temperatures higher than this were regarded as rain-free or unlikely to produce high rainfall. The next three steps in the procedure were then used to identify cloud clusters which were expanding spatially using a series of Meteosat images, and that were no cirrus clouds. In the presented study a hierarchical approach similar to that of Barrett and Cheng (1996) has been developed, to target debris flow triggering cloud conditions in the region concerned (denoted by four Meteosat pixels either side of that centred on the Bachelard Valley). An area around the Bachelard Valley was chosen to provide information on potential triggering conditions of debris flows allowing for cloud advection and in order to obtain information on general atmospheric conditions at the study site. This new technique is shown in Figure 3.

The first step in the procedure is the same as that of Barrett and Cheng (1996), identifying cloud clusters using a brightness temperature threshold of 235 K. The next step is only activated when a warning had been given from a previous image (i.e. potentially debris flow triggering conditions had been identified earlier), and simply by-past steps three to five. In step three rapidly expanding cloud clusters are identified using a normalised difference of cloud cluster areas between successive images and a threshold of only accepting cloud clusters with a normalised difference of area over 200, or new cloud clusters. In step four the heights of the cloud clusters are analysed and the coldest cloud cluster brightness temperature is compared to that of the same cloud in the temporally previous image. Likewise, in step five the technique tries to identify changes in the height of the cloud by comparing the cloud cluster temperatures as a whole to those of the same cloud cluster in a previous image. The rationale behind developing steps three to five was that in the area of interest the triggering cloud types were noted from field observations to undergo extreme rapid expansion in both size and height. Warning of potential debris flow triggering conditions is assumed on the occurrence of cloud

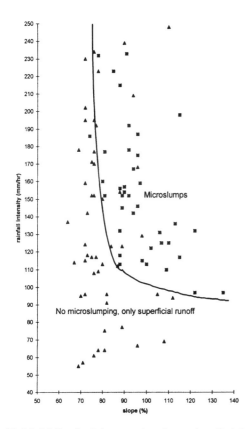

Fig.2 Rainfall simulation results from the Bachelard Valley showing the occurrence of microslumps (square symbols) as a function of slope angle and rainfall intensity from Blijenberg et al. (1996). Borderline drawing by eye.

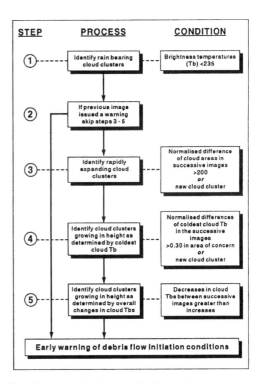

Fig. 3. Remote sensing data based early warning technique for debris flow initiation conditions.

clusters fulfilling the above requirements in the area of interest.

The tracking of cloud clusters using temporal sequences of data is a complex issue depending on the dynamics of the cloud clusters and the time interval between samples. Although Meteosat data are taken every 30 minutes, the data available for this study had a sampling rate of once every 3 hours for 1991, and once every 2 hours for the rest of the study period. In this study the comparison of cloud cluster area and height has been carried out when cloud clusters overlapped in successive images. When such a situation did not occur, it was assumed that the cloud had evolved in the period between image. In order to develop an understanding of how mapping landslide hazards could be carried out in a way other than by using a GIS, i.e. for applications when data availability is restricted, we embarked on a terrain classification using a terrain mapping method developed at ITC by van Zuidam and

Cancelado in 1979. In this method usually aerial photography is interpreted by means of a mirror-stereoscope. For our project stereo SPOT imagery has been used. Terrain classification is concerned with the arrangement and grouping of different parts of the Earth surface into a variety of categories on the basis of the similarity of the type of surface features.

For the terrain classification pursued in this project, Terrain Mapping Units (TMUs) were delineated according to geology, relief and genesis. In addition to interpreting SPOT imagery the TMU classification had to be geometrically rectified, in our case using ArcInfo, and ERDAS Imagine. The TMUs together with a supervised land use classification were then used to generate a map displaying low, medium and high degrees of mass movement hazards (Buchroithner et al., 1997).

Furthermore focus has been put on the identification and prediction of locations prone to the studied type of mass movements using a GIS data set. Digital thematic maps were prepared in order to cross these maps in a univariate analysis with mass movement distribution maps. The most notable result revealed that the presence of certain sediments is by far the

Table 1 Debris Flows Initiation and Early Warning Times (all in GMT) of the Bachelard Valley.

Certain		Very Likely		Warning Start Time	Warning End Time	Warning Length
Date	Time	Date	Time	(hours)	(hours)	(hours)
12.07.91	14:07			12:33	21:33	01:34
		09.08.91	05:03	21:33	12:33	07:30
		05.08.93	14:15	none	none	none
		15.08.93	13:38	12:33	18:33	01:05
26.06.94	15:01			02:33	18:33	12:28
		27.07.94	18:43	16:33	19:33	02:10

most important factor for the appearance and triggering of mass movements. These indications involve the distance to main streams, aspect and slope class. Expert knowledge has been exploited to assign qualitative hazard values to every relevant map class. The combination of these quantified maps led to a hazard zonation map (Buchroithner *et al.*, 1997).

5 CASE STUDIES

Debris flows, although relatively common in this particular location, are still rare events. In the four years studied, the model, rain gauges, video evidence, and field survey revealed only nine events. Of these only the initiation times of two could be defined with total confidence. The early warning times possible from the technique developed are given in Table 1 for the six case studies. The initiation times of the debris flow events are taken from the rain gauges and the model. From this table it can be seen that the remote sensing technique provided early warning in five out of the six cases. The longest warning time amounted to 12 hours 28 minutes on June 26th 1994. Unfortunately, the technique failed to provide any warning one of the cases on the August 5th 1993. This was thought to be due to the low sampling frequency of the satellite data used. The duration of the rain thought to initiate the debris flow on this day was only 19 minutes, the lowest duration of all cases. Also shown in Table 1 are the time spans over which the technique warns of triggering conditions (by step 2). The original Meteosat image and steps one, three, four and five of the remote sensing technique for the case studies on the July 12th 1991 at 12:33 GMT, and the August 27th 1993 at 02:33 GMT are shown in Figures 4 and 5 respectively. A frame indicates the target area.

In addition to the six case studies, a one-month period from 8th August to 8th September 1993 was chosen in order to assess the number of false alarms produced by the technique. This particular period was chosen because of satellite data availability and avoidance of fresh snowfall. Results from running the technique every two hours show that there were false alarms only on two days, the 13th and 27th of August. During this month rainfall occurred at the test site rain gauge, at least, once on seven occasions. Rainfall is likely to have occurred even more frequently in the region monitored as whole, including other areas susceptible to debris flows. No track has been kept of the number of cloudy days. Figure 5 shows a false alarm at 02:33 on the 27th August 1993.

6 DISCUSSION

The ability of a community to live in proximity to a natural hazard without loss of life is, in part, dependent on the earliness of warning that can be given prior to the onset of the hazardous phenomena. Preliminary results in this paper show that under certain circumstances remote sensing data can be used to provide early warning of triggering conditions for debris flow events. Of the seven cases examined in this study the triggering conditions for six were correctly identified with approximately 1 to 12 hours warning. Given the location of potentially hazardous slope conditions and source areas, this early warning information could be used to close roads and evacuate inhabitants at risk from debris flow activity. A number of methods exist for the mapping of landslide hazard zones including the creation of landslide inventory maps (Wright and Nilsen, 1974), applied geomorphological hazard mapping using aerial photo-interpretation and fieldwork (Brunsden *et al.*, 1975), slope instability

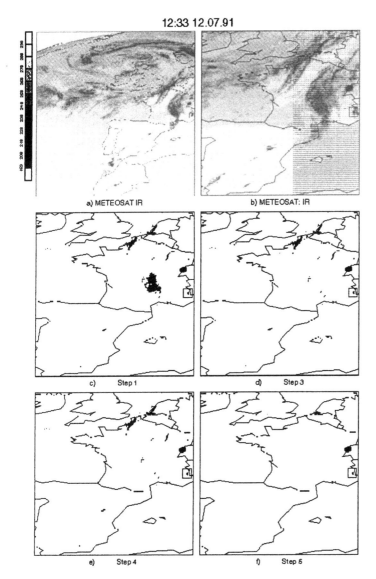

12:33 12.07.91

a) METEOSAT IR

b) METEOSAT: IR

c)   Step 1

d)   Step 3

e)   Step 4

f)   Step 5

Fig. 4. Meteosat image and early warning technique results (steps 1, 3, 4 and 5) for the 12th July 1991, 12:33 GMT. The box highlights the Bachelard Valley.

hazard mapping using terrain classifications (Carrara *et al.*, 1977) and slope instability hazard mapping by correlation of slope inventory maps with parameter maps (Zimmermann and Haeberli, 1992). The choice of the mapping methodology is often dependent on the availability of data and the required map scale, varying from synoptic or regional scale (1:100 000 to 1:250 000), to large scale (1:5 000 to 1:1 5000).

The landslides examined in this study have flow paths up to 1.5 km and widths up to 10 m. They are rapidly moving, reaching speeds of 7-10 km/h. The

Meteosat data, by contrast, have a spatial resolution of 5 - 6 km in east-west direction and 8 - 10 km in north-south direction, and a temporal resolution of 30 minutes. The data used in this study have a temporal resolution of two and three hours. The mismatch of satellite and mass movement spatial and temporal resolutions has in the past meant that remote sensing data, and in particular earth observation data, have been restricted to the monitoring of massive landslides or the zonation of regions vulnerable to landsliding (Astaras *et al.*, 1997; de Graaf *et al.*, 1997). In the present study

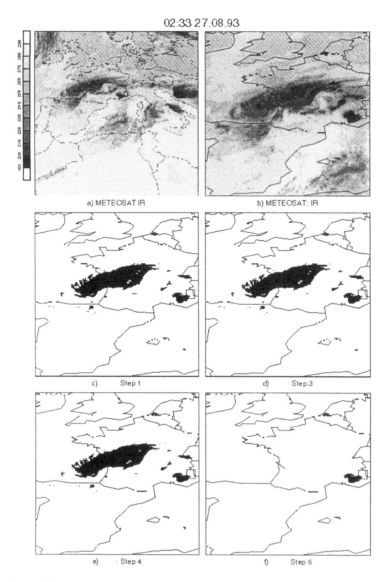

02:33 27.08.93

a) METEOSAT IR

b) METEOSAT: IR

c) Step 1

d) Step 3

e) : Step 4

f) Step 5

Fig. 5. Meteosat image and early warning technique results (steps 1, 3, 4 and 5) for the 27th August 1993, 02:33 GMT. The box highlights the Bachelard Valley.

however, it is the triggering conditions of debris flows that are monitored. They have smaller time and space scales compared to the actual debris flows and are relatively compatible with the scales of the remote sensing data used. The overlaying of a landslide hazard zonation at a large scale with rainfall information at a considerably lower scale derived from remote sensing data using a Geo-Information System would allow early warnings being targeted to small regions and, if used operationally, potentially reduce needless evacuations and road closures.

The use of cloud parameters to determine debris flow triggering conditions rather than rainfall estimates - although reducing the errors associated with retrieving rainfall estimates from satellite measurements - incurs errors associated with correctly identifying the clouds that will develop heavy rainfall, over the region concerned. An indication of this error is shown by the failure of the technique to provide early warning in one of the case studies and by the false alarms in the false alarm trial run. An increase in the accuracy of this technique would be expected based on a higher temporal frequency of the Meteosat data, especially in that case where the technique failed to identify a triggering situation and the duration of the rain event was only 19 minutes. Further investigations might look at varying the size of the time window providing cloud information for the early warning of triggering conditions. It should also be noted that the use of this technique is restricted to summer time. The appearance of fresh snow might produce false alarms. Fortunately field observations have shown that the debris flow events are generally initiated during summer.

Flash flood and debris flow warnings are of great importance for mountainous areas. In this regards the forecasting of extreme rain intensities may help, especially if overland flow or gully flow is prevalent. Further work, which undoubtedly is necessary, will also allow for improved validation, either with a mesoscale climate model to generate "synthetic storms" or using real storms. For the prediction of eminent, i.e. less than 4 hours, warnings we suggest the integration of climate model forecasts with the Meteosat warning indicators.

7 ACKNOWLEDGEMENTS

The author would like to thank Dr. H.M. Blijenberg of University of Utrecht, The Netherlands, for the rain gauge data used in this study.

8 REFERENCES

Asch van, T.W.J. and Steijn van, H., Temporal patterns of mass movements in the French Alps, *Catena*, 18, 515-527 (1991).

Astaras, T., Lambrinos, N., Soulakellis, N., Kalathas, A., and Oikonomidis, D., Delineation of landslides using GIS and digital image processing techniques on multi-temporal Landsat 5TM images: A case study from the Pindus mountain Greece, In *Remote Sensing: Integrated Appplications for Risk Assessment and Disaster Prevention for the Mediterranean* (eds. Spiteri, A., Proceedings of the 16th EARSel Symposium, Malta 20-23 May 1996) (1997).

Barrett, E.C., Satellite Rainfall Monitoring for Agrometeorology: Operational Problems, Practices and Prospects, *EARSeL Advances in Remote Sensing*, 2, 66-72 (1993).

Barrett, E.C., and Cheng, M.C., The identification and evaluation of moderate to heavy precipitation Precipitation Areas using IR and SSM/I satellite imagery over the Mediterranean Region for the STORM project, *Remote Sensing Reviews* 14, 119-150 (1996).

Beniston, M., Rebetez, M., Giorgi, F., and Marinnci, M.R., An analysis of regional climate change in Switzerland, *Theor. and Appl. Clim*, 49, 135-159 (1994).

Blijenberg, H.M., de Graaf, P.J., Hendriks, M.R., De Ruiter, J.F. and Tetering van, A.A.A., Investigation of infiltration characteristics and debris flow initiation conditions in debris flow source areas using a rainfall simulator, *Hydrologic Processes*, 10, 1527-1543 (1996).

Brunsden, D., Doornkamp, J.C., Fookes, P.G., Jones, D.K.C., and Kelly, J.M.H., Large scale geomorphological mapping and highway engineering design, *Quart. J. Engng. Geol.*, 8, 227-253 (1975).

Buchroithner, M.F., de Graaf, P., Granica, K. and Kniveton, D. R., Final Report about the project Landslides and Mudflows. Development of Operational Remote Sensing Tools for Mapping, Monitoring and Mitigation of Mass Movements, EU Contract ERBCHRXCT940 630, submitted to DG XII of the European Commission. Institute for Cartography, Dresden University of Technology, Germany, 130 pp. (1997).

Carrara, A., Pugliese, C.E. and Merenda, L., Computer based data bank and statistical analysis of slope instability phenomena, *Z. Geomorph. N.F.* 21.2, 187-222 (1977).

Florez, A., Geomorfolgia del area de Manizales-Chinchina, Cordillera Central de Columbia, *Analisis Geograficos*, Nr.9, Instituto Pirenaico de Ecologia, Zaragoza, Spain, 54p. (1986).

Garcia-Ruiz, J.M. White, S.M., Marti, C., Valero, B., Paz-Errea, M. and Gomez-Villar, A., La catastrofe del barranco de Aras (Biescas, Pirineo, Aragones) y su contexto espacio-temporal, Instituto Pirenaico de Ecologia, Zaragoza, Spain, 54p. (1996).

de Graaf, P.J., De Ruiter, J.F. and Tetering van, A.A.A., De initiatie van punistromen. Een onderzoek in het Bachelard dal, Department of Physical Geography, University of Utrect, 119 p. (1993).

Johnson, A.M. and Rahn P.H., Mobilization of debris flows, Zeitschrift fuer Geomorphologie, Supplement Band 9, 168-186 (1970).

Johnson, A.M., and Rodine, J.R., Debris flows, In *Slope Instability* (eds. Brunsden, D., and Prior, D.B.), Wiley, London, 257-361 (1984).

Kniveton, D. R., Landslides and Mudflows. Development of Operational Remote Sensing Tools for Mapping, Monitoring and Mitigation of Mass Movements: The Prediction and Monitoring of Precipitation using Satellite Data, Training Manual for EU HCM Contract No. ERBCHRXCT940630, RSU, University of Bristol, 128pp. (1995).

Kniveton, D. R., de Graff, P.J., Granica, K. and Hardy, R. J., The development of a remote sensing based technique to predict debris flow triggering conditions in the French Alps, *International Journal of Remote Sensing*, Vol. 21, No. 3, 419- 434 (2000).

Rasmusson, E.M. and Arkin, P.A., Observing Tropical Rainfall from Space: A Review, In *The Global Role of Tropical Rainfall* (eds. Theon, J.S., Matsumo, T., Sakata, T., Fugono, N.), Deepak, Hampton, VA, 105-188 (1992).

Romero, J.A., Florez, A. and Sanchez, H.A., Inventurio inicial de riesgos naturales, *Analisis a Geograficos*, Nr. 16, Instituto Geographico Augustin Codazzi, Bogota, Columbia, 56p. (1989).

Steijn van, H., De Ruig, J., and Hoozemans, F., Morphological and mechanical aspects of debris flows in parts of the French Alps, *Zeitschrift fuer Geomorphologie*, 32, 143-161 (1988).

Takahashi, T., Mechanical characteristics of debris flows, *Journal of the Hydraulics Division*, 104-HY8, American Society of Civil Engineers, 381-396 (1978a).

Takahashi, T., Debris flows, In *Annual review of fluid mechanics* **13** (eds. Dyke van, M., Wehausen, J.V., and Lumley, J.L.), 57-77 (1981b).

Varnes, D.J., Slope movement types and processes, In *Landslides, Analysis and control*. Special Report 176 (eds. Schuster, R.L. and Krizek, R.K.), Transportation Research Board of the National Academy of Sciences, Washington, 11-33 (1978).

Westen van, C.J., Medium scale landslide hazard analysis using a PC based GIS; a case study from Chinchina, Columbia, Proceedings Symp. 'Sensores remotos y sistemas de informacion geografica para el estudio de riesgos naturales', IGAC, Bogota, 16 p. (1992).

Wright, R.H., and Nilsen, T.H., Isopleth map of landslide deposits, Southern San Francisco Bay Region, California, *US Geol. Surv. Misc. Field Studies* Map MF550 (1974).

Yin, K.L. and Yan, T.Z., Statistical prediction model for slope instability of metamorphosed rocks, *Proceedings of the fifth International Symposium on Landslides*, Lausanne, Switzerland, Vol. 2, pp 1269 – 1272 (1988).

Zimmerman, M. and Haeberli, W., Climatic change and debris flow activity in mountain areas - A case study in the Swiss Alps, *Catena Supplement* 22, 59-72 (1992).

Observing our environment from Space: New solutions for a new millennium, Bégni (Ed.)
© 2002 Swets & Zeitlinger, Lisse, ISBN 90 5809 254 2

Snow avalanche risk assessment supported by remote sensing

U.Schmitt & M.Schardt
Joanneum Research, Institute of Digital Image Processing, Graz, Austria

J.Ninaus
Karl-Franzens University, Institute of Geography, Graz, Austria

.

ABSTRACT: To identify potential snow avalanche hazard zones it is necessary to apply the decisive criteria, such as critical degree of crown closure, critical size of gaps in the forest, etc., to extensive areas. The aim of the presented project is to compile standardised information for large regions. High resolution remote sensing data permits the compilation of vegetation parameters from which surface roughness can be indirectly deduced. These parameters include forestry-related variables such as the distribution of different tree species, tree age, crown closure and gaps as well as alpine pastures, scree and rocks. The digital terrain model further makes available topographical parameters, such as slope, aspect and elevation. The derived information will be compiled in an avalanche risk model which at the end will lead to an avalanche risk map.

1 INTRODUCTION

Time after time the inhabitants of alpine regions have to cope with natural disasters such as landslides, rock falls, and snow avalanches. In addition, intensive public and private construction activities during the last decades led to increased pressure on the valleys of the Alps. In order to control this development, efforts were made to prohibit uncontrolled building activities in torrent and avalanche hazard zones and hazard zone maps were prepared. Empirical data from many years of observation are available for most of the sites situated on endangered zones. Close examination and monitoring of avalanche tracks led to the identification of regular patterns which may trigger snow avalanches. Based on this experience the Austrian Institute for Avalanche and Torrent Research is developing avalanche risk models. The parameters included in the research data comprise snow, topographical parameters and type, quality and roughness of vegetation cover and composition of tree stands.

To identify potential hazard zones it is necessary to apply the criteria which have been derived from the measurement results, to larger areas. In order to adapt these criteria to regional variation, information is needed which covers all the variables which cause avalanches over extensive areas. Furthermore, the different information levels should be available in digital form and integrated in a GIS for automated processing.

So far, however, the kind of information that would fulfil the outlined requirements has not been available. While adequate data has been compiled for some areas, it is generally too heterogeneous to permit the integration of data into other databases. Moreover, the data only covers selected sites and is thus not easily applied to the analyses of larger regions. The aim of the presented project is therefore to compile standardised information for larger areas. High resolution remote sensing data permits the compilation of vegetation parameters from which surface roughness can be indirectly deduced. These parameters include forestry-related variables such as the distribution of different tree species, tree age, crown closure and gaps as well as alpine pastures, scree and rocks. The digital elevation model (DEM) further makes available topographical parameters, such as slope, aspect and elevation.

In the present study two types of avalanches were considered: the slab avalanche and the snow gliding avalanche. Since the input parameters for the regionalisation of the avalanche models, due to different release mechanisms, are varying, these types were treated separately. The investigation comprised the validation of the input parameters, the development of a rule system, that defines the dependency of snow avalanche risk on the critical vegetation and topographic parameters, the integration of the model into an Alpine Monitoring System, the production of hazard risk maps, and the verification through comparison with avalanche registers. Special attention was paid to the restrictions of remote sensing in high alpine regions.

The paper will present the concept of the study, the results derived from classification of Landsat

TM, SPOT and IRS images, and the efforts for realisation of the avalanche risk maps. The presented results have been conducted in the context of the CEO-project "Inventory of Alpine-Relevant Parameters for an Alpine Monitoring System Using Remote Sensing Data" financed by DG XII.

2 BACKGROUND

Many investigations have been carried out on the conditions and mechanisms of snow avalanche release (Ammer et al. 1985, BUWAL 1990, Fink & Drimmel 1986, Frey 1977, In der Gand 1981, In der Gand & Kronfellner-Kraus 1984, Laatsch 1977, Meyer-Grass & Schneebeli 1992, Salm 1978, Zenke 1985). Diverse parameters have been investigated under different conditions and in different areas. Comparison and combination of the results and developed models turned out to be very difficult mainly due to two factors: On one hand the findings in most cases are restricted to the respective investigation area and cannot easily be transported to other alpine regions with different topographic and climatic conditions. On the other hand, part of the findings is contradictory, and it again would need avalanche experts to assess which rules are the most reliable to be applied in the present case. However, the type of parameters implemented in the diverse avalanche risk models is similar, and the respective risk limits differ from each other. This is the reason why the study demonstrates exemplarily how parameters relevant for avalanche risk estimation can be derived by remote sensing techniques and implemented in any region based avalanche risk model.

3 CONDITIONS FOR AVALANCHE RELEASE

The conditions and release mechanisms of different types of avalanches have been investigated in detail at the Institute for Avalanche and Torrent Research (Höller 1996, 1997, 1998). Recently, special attention is put on the conditions within and near mountain forests, as avalanche release in these zones increases with the decreasing forest canopy density (Kleemayer 1993). In the present study two types of avalanches are considered. Since the input parameters for the regionalisation of the avalanche models, due to different release mechanisms, are varying, these types will be described and differentiated in the following. The parameters were made available by the Institute for Avalanche and Torrent Research.

3.1 *The Typical Slab Avalanche*

The release mechanism of slab avalanches can be described as a critical stability of the snow pack which is defined by the relation between stress and strength. The slab avalanche formation is influenced by:
☐ Topography (aspect, inclination, type of forest stand, canopy density, ...)
☐ Meteorology (temperature, precipitation, wind, ..)
☐ Snow pack (temperature gradient, weak layers, ..)
☐ The critical parameters with regard to slab avalanches are
☐ A *critical diameter of openings* ($> 50m^2$; about 7m in diameter) enables the formation of surface hoar as a necessary condition for snow slab release:
☐ A *critical canopy density* also enables the formation of hoar as a necessary condition for snow slab release. It depends on the forest type with:
   < 70-80% for pure larch stands
   < 40-70% for mixed larch - stone pine stands
   < 40%   for spruce or stone pine stands
☐ *Inclination* >30%
☐ *Aspect* probably in the North sector (WNW to ENE)

3.2 *The Snow Gliding Avalanche*

The snow gliding avalanche is a subtype of the slab avalanche. The release mechanism can be described with high gliding rates (snow gliding) due to increasing snow temperature and increasing water content. The snow gliding is influenced by the roughness of ground surface and the wet snow content near the snow-ground interface.

In open terrain snow gliding must be *expected* on
☐ abandoned pastures and meadows,
☐ long bladed grass mats,
☐ homogenous slopes.
Snow gliding is *possible* (but not as high as mentioned before) on
☐ slopes with small steps (from cows),
☐ small bushes (rhododendron, vacc.) with a height not exceeding 30 to 50cm.
Snow gliding is *very small* on
☐ dwarf pine stands,
☐ bushes > 50 cm,
☐ rocks > 30 cm.
Within forest snow gliding must be *expected* in larch stands with a very low canopy density and openings > 10m x 10m in diameter.
Additional Parameters are the
☐ Inclination >30-35%
☐ Aspect: south facing slopes.

4 DATA SOURCES

Following data sets were employed to fulfil the users requirements for avalanche risk assessment:
☐ Satellite data

- Landsat 5 TM
- SPOT 3 panchromatic
- IRS-1D panchromatic, winter scene
☐ Other data
  - CIR aerial photos (37)
  - BW aerial orthophotos (9)
  - DEM with 10m resolution
  - Digital topographic maps

A multi-sensoral and multi-temporal approach was chosen to receive the necessary parameters from the satellite data. Unfortunately, no high resolution satellite data of the new sensor generation was available for the first phase of the investigation. This is due to the fact that the chance to receive cloud free images over an Alpine area is limited. Additionally, suitable data has to be acquired during the summer period with respect to the higher sun elevation and, thus, less areas being affected by shadows.

5 PRE-PROCESSING

In alpine terrain high-quality pre-processing of the remote sensing images is essential to enable correct data analysis and overlay. The geometry of the images has to meet extremely strict requirements, firstly, because the data obtained with different sensor systems are classified multi-sensorally and secondly, because the satellite images and the classifications subsequently derived from them have to be integrated with additional digital data in a GIS. Because of the strong relief effects parametric geocoding methods were used in order to obtain the demanded sub-pixel accuracy (Raggam et al. 1999).

Topography does not only affect the geometric properties of an image but as well has a significant impact on the illumination and the reflection of the scanned area (Kenneweg et al. 1996, Schardt 1990). This effect is caused by the local variations of view and illumination angles due to mountainous terrain. Therefore, identical land cover might be represented by totally different intensity values depending on its orientation and on the position of the sun at the time of data acquisition. An optimal topographic correction levels all topographically induced variations of the illumination so that two objects characterised by same reflectance properties show the same digital number in the satellite image. In this investigation the topographic correction was successfully carried out by the use of a parametric approach namely the Minnaert and the C-correction (Meyer et al. 1993, Colby 1991).

6 CLASSIFICATION AND ACCURACY ASSESSMENT

A large number of investigations demonstrated that simple classification algorithms, such as maximum-likelihood and threshold-level procedures, represent a highly efficient, solid and operational method for the compilation of vegetation and land use classes, even in high relief terrain (Schardt 1990, Schardt & Schmitt 1996, Hill & Mehl 1995). In many investigations carried out in alpine test sites it has also been shown that by the integration of ancillary information the classification result can be significantly improved (Gallaun et al. 1999, Paracchini & Folving 1994). This is due to the fact that the occurrence of specific vegetation types is linked to the presence of suitable natural conditions. This interdependencies are especially true for the alpine regions which are characterised by major differences in altitude levels and geological/pedological conditions and consequently, different natural vegetation types. In this investigation ancillary information was integrated in order to improve the performance of the classification.

A detailed ground truth survey was performed, serving as basis for the classification of the satellite data. Reference data was collected in colour infrared aerial photographs as well as by field survey. This information was used for training of the classifier as well as for verification of the classification results, using independent reference data sets for each purpose.

The classification was based on the Landsat TM image, the IRS-1D image and on the following additional data sources:
☐ An image product derived by fusion of the Landsat TM (good spectral resolution) and the panchromatic SPOT (high spatial resolution) data. An adaptive fusion algorithm provided by Steinnocher (1998, 1999) was applied, leading to a sharpened multi-spectral image while preserving the spectral characteristics of the original Landsat TM bands.
☐ Texture features derived from the panchromatic SPOT image.
☐ Elevation, slope, and illumination derived from the DEM.
☐ Auxiliary information (e.g. a digital water layer).

A detailed signature analysis preceded the classification. The classification itself was performed in separate work steps for forest and non-forest areas after classification of the forest border.

6.1 *Classification of the forest border*

The separation of forest and non-forest areas was based on the image fusion result of the Landsat TM scene and the SPOT pan image. In this image thresholds were defined in the (fused) TM bands 2 and 3 as well as in the normalised difference vegetation index (NDVI) and combined with an AND-operation.

This approach benefits from the advantages of high spectral resolution (Landsat TM) and high spatial resolution (SPOT pan) of both data sets. In the fused image, in general, larger areas are created which are interpreted uniformly whereas the borders of these areas are defined by the high resolution of the SPOT pan image. This has the advantage that single forest pixels will not occur, which corresponds with the definition of forest. Especially in forests with a low canopy density the aggregation caused by image fusion has a positive effect on the forest demarcation. Figure 1 shows the vectorised forest border (green polygons) derived from the satellite classification superimposed to a geo-coded aerial photo illustrating a subset of the investigation area at the upper forest border line. From this result it can be stated that satellite remote sensing is an useful tool for separating forest and non-forested areas. Figure 1 demonstrates that the differentiation is even possible in open forests occurring very often at the alpine forest borderline. Nevertheless, very open "forests" with a crown closure of 10% cannot be recognised as forest areas due to the dominant reflection of the ground vegetation. A forest definition based on a threshold of 20%, however, is appropriate for remote sensing based inventory methods (Schardt et al. 1995).

The accuracy assessment of the forest / non-forest separation was performed by comparing the areas classified as forest with 1054 sample points. This sample points (in a regular grid of 100m by 100m steps) were visually categorised into forest / non-forest / forest-border by the use of CIR orthophotographs. Table 1 shows the error matrix for the forest / non-forest – points.

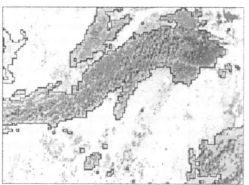

Figure 1. Forest borderline derived from classification.

6.2 Classification of forest parameters

The classification of different forest parameters was based on the categories defined by the ABIS (Alpine Monitoring and Information System) working group of the Alpine Convention as well as by the Austrian Institute for Avalanche and Torrent Research. The possibility of classifying the following parameters was investigated:

☐ Forest type (coniferous, mixed and broad-leaved forest)
☐ Natural age classes (clearing/culture, thinning/pole, timber)
☐ Crown closure (<10%, 10-40%, 41-60%, >60%), including gaps and clearings
☐ Tree species composition (esp. share of larch trees, differentiation between high growing and small trees).

The classification was based on the maximum likelihood method. In order to increase the classification accuracy a rule based approach was chosen considering digital terrain data.

Plate 1 shows two forest stands typical for the upper forest border of the Dachstein test site and how they are represented in the aerial photo and the satellite image. The classification results of the identical area demonstrate that the larch stand to a large part was classified correctly. Only part of it got mixed up with other coniferous forest types or - especially at its borders - with open forest. Also canopy closure and age category were correctly recognised as strongly mixed within this very inhomogeneous stand. The dwarf mountain pine stand covers only a narrow stripe along a debris fan. It is strongly eclipsed by the bright carbonate stones. As far as it was classified as forest, it was correctly recognised as coniferous stand, but only to a small part as dwarf mountain pine species.

For the verification of the forest classification an independent set of verification areas was selected. An area based (not pixel based) verification was performed by aggregation of the classified pixels within each verification area. The correspondence values (Table 2) prove that most of the important forest parameters can be assessed by satellite remote sensing with satisfactory accuracy. Difficulties occurred only with forest areas characterised by crown closure values of less than 60%. One of the reasons for this is that the potential differences in reflection caused by trees are superimposed by the reflection of the ground vegetation, which strongly influences the signature in open stands. Another reason is to be seen in the difficulties in finding a sufficient number of homogeneous training and verification areas representing open stands.

From Figure 1 it becomes clear that also typically low growing trees such as dwarf mountain pine and greenalder are classified as forest. This on hand corresponds with the forest definition within the overall project. On the other hand, for the avalanche risk application it would be necessary to separate between potentially high growing tree species and those small trees. However, dwarf mountain pine and greenalder cannot not be separated sufficiently from high growing coniferous or broad-leaved forest respectively with the available satellite data sets, as their spectral characteristics in the wavelength available from Landsat TM data are very similar to that

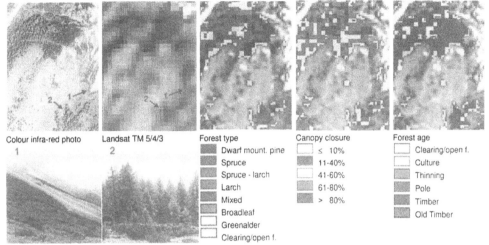

Colour infra-red photo Landsat TM 5/4/3 Forest type Canopy closure Forest age
1 2

Forest type
■ Dwarf mount. pine
■ Spruce
■ Spruce - larch
■ Larch
■ Mixed
■ Broadleaf
□ Greenalder
□ Clearing/open f.

Canopy closure
□ ≤ 10%
■ 11-40%
□ 41-60%
■ 61-80%
■ > 80%

Forest age
□ Clearing/open f.
□ Culture
■ Thinning
■ Pole
■ Timber
■ Old Timber

Plate 1. Dwarf mountain pine (1) and larch (2) in a selected area at the south face of the Dachstein massif and forest classification results superimposed to panchromatic SPOT image. (Colour plate, see p. 413)

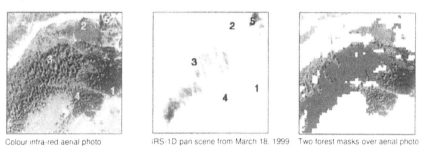

Colour infra-red aerial photo   IRS-1D pan scene from March 18. 1999   Two forest masks over aerial photo

Plate 2. Forest border based on summer Landsat TM and SPOT data (all coloured areas in right image) compared to forest border derived from combined winter IRS panchromatic and summer SPOT data (green in right image): 1 = dwarf mountain pine, 2 = greenalder, 3 = Larch stand with bushes, 4 = larch with very low crown closure + dwarf mountain pine, 5 = rock. (Colour plate, see p. 413)

Plate 3. Distribution and crown closure of forest on slopes steeper than 30%, superimposed to SPOT image: Crown closures 11-30% = red, 31-60% = yellow, >60% = green. (Colour plate, see p. 413)

of the high growing species. This is the reason why satellite data from a winter period has been introduced into the investigation.

Table 1. Quality assessment of the forest border classification.

Forest border Classified	Reference				Mean class accuracy
	Forest	Non-forest	Σpoints	Users accuracy	
Forest	547	44	591	0.93	0.97
Non-forest	2	403	405	1.00	0.95
Σpoints	549	447	996		
Producers accuracy	1.00	0.90			0.95 overall
Kappa coefficient: 0.91					

Table 2. Forest classification accuracy.

	Mean accuracy	Kappa coefficient
Forest type	92%	0.84
Crown closure	86%	0.64
Natural age class	80%	0.60

A nearly cloud free panchromatic IRS-1D image (ground resolution approx. 6m) from late winter (March 18) could be acquired for a small part of the investigated area for test purpose. During acquisition time of this image, the dwarf mountain pine and greenalder stands were still covered with snow, whereas the high growing trees were already snow free. This has been proven by field trips before and after acquisition time of the image. By the definition of simple thresholds, a so-called winter forest mask could be created which includes all high growing forest areas but excludes the snow covered small tree species such as dwarf mountain pine and greenalder, but also cultures.

Together with the previous forest border classification of the Landsat TM data it was possible to separate dwarf mountain pine and greenalder very precisely from the high growing trees. A separation from the cultures, which were also covered by snow in winter, is possible by its different spectral characteristics in the summer Landsat TM image.

Plate 2 shows an area at the upper forest border line of the investigated are. As demonstrated in the CIR aerial photo, this region is characterised by occurrence of small growing tree species such as dwarf mountain pine and greenalder, by often very low crown closures of high growing forest stands, and by mixture of high growing and small trees. The panchromatic IRS winter image of the same area shows dark grey values only for high growing forest stands with sufficient canopy density and, additionally, for some steep rocky areas. In the right image the forest mask derived from thresholds in the summer satellite data alone and the one derived from a combination of the thresholds from summer and winter satellite images are compared. From visual verification it is obvious, that with this method the typical alpine

small growing tree species can be separated from the high growing trees with high accuracy. The only limitation lays in the crown closure that can be detected within larch and broad-leaved stands, as in these stands due to lack of needles/leaves the reflection of the trunks and branches is masked by the high reflection of the snow beneath.

6.3 Classification of non-forested areas

The classification of the additional "non forest" vegetation classes was carried out with the same method as used for the classification of the forest parameters. However, more auxiliary information (elevation, slope, GIS water layer, texture features) was employed to this classification. The area based verification of the classification result shows an overall accuracy of 87% and a Kappa coefficient of 0.84.

As the verification is based only on relatively large reference areas the values are only true for regions which are characterised by large and homogeneous vegetation patterns. Small and heterogeneous patterns of the vegetation classes cannot be classified because of the limited spatial resolution of the used Landsat TM satellite data. Simulations based on infrared aerial photos showed that the availability of data with a geometric resolution of at least 5m is requested to represent the heterogeneous spatial distribution of the vegetation categories in the Austrian test site. These parameters are essential to deduce surface roughness outside of forests.

6.4 Derivatives of the DEM

Additional parameters requested by the avalanche models depend on the topography: slope, aspect, and elevation. These parameters could be derived from the digital elevation model (25m resolution) with sufficient accuracy. But another parameter relevant especially for the risk estimation of snow gliding avalanches is the homogeneity of the slope. To get this information, high resolution elevation models, such as the ones derived from laser scanner data, are necessary. However, these at present are too expensive to be applied to extended areas.

7 RESULTS

The classification results for the forest and the non-forest areas were combined by using the classified forest border line as boundary. Due to the high mountain relief the satellite images contain areas, which are totally shaded and contain no information at all. Therefore, a shadow mask was calculated based on the DEM and the sun azimuth and sun elevation for the exact acquisition time of the Landsat TM image. The fully shadowed areas were masked out in the final classification results.

The classification results then are integrated together with the topographic information in a GIS, where the rules established for avalanche risk estimation are implemented and can be retrieved for each new/updated land cover information. This at the end can serve as basis for decisions for example where to take urgent forestry measures.

A verification of the results with the Austrian avalanche register was planned. However, this register only comprises large and damaging avalanches which in most cases are released above the alpine forest border. Therefore, at the moment there is no possibility for verification of the model.

Plate 3 shows an example for a potential query to the information system. The crown cover of the forest is presented only on slopes with more than 30%, because these are the critical slopes for avalanche release. Especially at the upper forest border many areas show a crown cover smaller than 60%, which (depending on the forest type) is a critical value with respect to the protective function of the forest against avalanches.

8 CONCLUSION AND OUTLOOK

The results of the investigation showed that satellite remote sensing can furnish an important contribution for the construction of an Alpine information system as well as for the estimation of snow avalanche risk. Especially for the extensive assessment of the status and the distribution of alpine vegetation, satellite remote sensing is the only alternative, considering the extraordinary high costs that are connected to field studies or assessments based on the interpretation of aerial photography. Within the class of "forest" the most important parameters like forest border, canopy density, forest type, important tree species and age of the forest (natural age classes) can be investigated with satisfying precision. Areas outside the forest can be classified due to their spectral variation if they occur on large areas. Smaller areas, which are covered by a mixture of vegetation categories, such as rhododendron, dwarf mountain pine or smaller moor areas that do not exceed 1 ha, cannot be classified definitely because of the mixed pixel problem. Simulations have proved that a geometric resolution of about 5 to 10m in the infrared spectral range is necessary to assess the typical small-area distribution pattern of vegetation outside of forests. The future sensor systems will provide data in this resolution range.

Within the ALPMON-project, the results of the classification will be evaluated for various questions with regard to their usability. Some investigations or methodical developments respectively couldn't yet be investigated sufficiently in the ALPMON-project because of the data situation and the project subject.

They are still to be conducted with regard to the following methodical aspects:

☐ Improvement of the procedures for the topographic normalisation and the atmospheric correction of the satellite data in alpine terrain.

☐ Development of a procedure for the precise classification of the percentage of larch within a forest stand.

☐ Integration of multi-seasonal data sets in order to profit from the different phenological development of various vegetation types (larch, broadleaved).

☐ Investigation on the possible improvement of classification results by means of high resolution satellite image data probably available in the future, especially for non-forest vegetation.

☐ Investigation on the possible improvement of canopy closure estimation by means of satellite images acquired in the winter period.

☐ Development of comprehensive avalanche models which cover most alpine conditions.

9 ACKNOWLEDGEMENT

The authors thank P. Höller (Institute for Avalanche and Torrent Research) for providing the information on mechanisms and relevant parameters for snow avalanche release.

10 REFERENCES

Ammer, U., Mößmer, E.M., Schirmer, R. 1985. Vitalität und Schutzbefähigung von Bergwaldbeständen im Hinblick auf das Waldsterben. Forstwissenschaftliches Centralblatt 104 (1985), pp. 122-137.

BUWAL (ed.) 1990. Richtlinien für den Lawinenverbau im Anbruchgebiet. Bundesamt für Umwelt, Wald und Landschaft (BUWAL), Bern.

Colby J.D. 1991. Topographic normalisation in rugged terrain. Photogrammetric Engineering and Remote Sensing, Vol. 57, No. 5, pp. 531-537.

Fink, M., Drimmel, J. (ed.) 1986. Raumordnung und Naturgefahren. Schriftenreihe/ Österreichische Raumordnungskonferenz, 50.

Frey, W. 1977. Wechselseitige Beziehungen zwischen Schnee und Pflanze – Eine Zusammenstellung anhand von Literatur. Mitteilungen des Eidgenössichen Institutes für Schnee- und Lawinenforschung, n°34.

Gallaun, H.; Schardt, M., Häusler, T. 1999. Pilot Study on Monitoring of European Forests. Tagungsband der Jahrestagung des Arbeitskreises „Interpretation von Fernerkundungsdaten der DGPF, 28.-29.4.1999, Halle

In der Gand, H.R. 1981. Stand der Kenntnisse über Schnee und Lawinen in Beziehung zum Wald in Europa. Proc. XVII IUFRO World Congress 1981, Kyoto, Japan, Dev. 1, pp. 319-337.

In der Gand, H.R., Kronfellner-Kraus, G. 1984. Wind, Schnee und Wasser und gesunder Wald. Impacts de l'Homme sur la Forêt. Symposium IUFRO, Strasbourg, 17-22 Sept. 1984, Paris 1985 (Les Colloques de l'INRA, n°30), pp. 385-402.

Hill, J. & Mehl, W. 1995. Improved forest mapping by combining corrections of atmospheric and topographic effects in Landsat TM-imagery. Sensors and Environment Applications of remote sensing, Askne, 1995 Balkema, Rotterdam, ISBN 90 5410 5240.

Höller, P. 1996. Snow cover investigations and avalanche formation in mountain forests. Proceedings of the International Conference *Avalanches and Related Subjects*, Kirovsk, Russia, Sept. 2-6, 1996.

Höller, P. 1997. Das Schneegleiten auf verschieden bewirtschafteten Flächen nahe der Waldgrenze. Centralblatt für das gesamte Forstwesen, Heft 2/3, 1997.

Höller, P. 1998. Tentative investigations on surface hoar in mountain forests. Annals for glaciology 26/1998, pp. 31-34.

Kenneweg, H., Häusler, T., Schardt, M., Sagischewski, H. 1996. Large Area Operational Experiment for Forest Damage Monitoring in Europe Using Satellite Remote Sensing. Proc. XVIII ISPRS Congress, Vienna, 9-19 July 1996.

Kleemayer, K. 1993. Berechnung von Waldlawinenkarten mit GIS. Österreichische Forstzeitung 7/1993, pp. 29-30.

Laatsch, W. 1977. Zur Struktur und Bewirtschaftung der Wälder im bayerischen Alpenraum. Die Entstehung von Lawinenbahnen im Hochlagenwald. Forstwissenschaftliches Centralblatt 96 (1977), pp. 89-93.

Meyer, P., Itten, K.I., Kellenberger, T., Sandmeier, S. & Sandmeier, R. 1993. Radiometric correction of topographically induced effects on Landsat TM data in an alpine environment. ISPRS Journal of Photogrammetry and Remote Sensing, Vol. 48, No. 4, pp. 17-28.

Meyer-Grass, M., Schneebeli, M. 1992. Les avalanches en forêt et leur dépendance aux condition de la station, du peuplement et de la neige. Symposium international INTER-PRAEVENT 1992-Bern, n°2, pp 443-455.

Paracchini, M. L., Folving, S. 1994. Land use classification and regional planning in Val Malenco (Italian Alps): a study on the integration of remotely sensed data and digital terrain models for thematic mapping. Mountain Environments & Geographic Information Systems. Edited by: Martin F. Price and D. Ian Heywood, 1994.

Rachoy, W. 1994. Zukunftsorientierter Schutz vor Wildbach-, Lawinen- und Erosionsgefahren. Österreichische Forstzeitung 3/1994, pp. 13-16.

Raggam, H., Schardt, M., Gallaun, H. 1999. Geometric fusion of multisensoral and multitemporal remote sensing data. International Archives of Photogrammetry and Remote Sensing, Vol. 32, Part 7-4-3 W6, Valladolid, Spain, 3-4 June, 1999.

Salm, B. 1978. Snow forces on forest plants. Proc. Iufro Seminar Mountain Forests and Avalanches, Davos, 1978, pp. 157-181.

Schardt, M. 1990. Verwendbarkeit von Thematic Mapper-Daten zur Klassifizierung von Baumarten und natürlichen Altersklassen. DLR-Forschungsbericht (DLR-FB 90-44), Oberpfaffenhofen.

Schardt, M., Schmitt, U. 1996. Klassifikation des Waldzustandes für das Bundesland Kärnten mittels Satellitenbilddaten. VGI, Österreichische Zeitschrift für Vermessung & Geoinformation, Heft 1/96.

Schardt, M.; Kenneweg, H., Sagischewski, H. 1995. Upgrading of an Integrated Forest Information System by Use of Remote Sensing. Proceedings of the IUFRO XX World Congress, 6-12 August 1995, Tampere, Finland.

Steinnocher K. 1998. Objects, edges and texture in feature based image fusion. Proc. Expert Meeting on satellite data fusion techniques for forest and land use assessment. ISPRS, Comm. VII WG IV, Freiburg, Germany, 8.-9.12.1997, pp. 77-81.

Steinnocher, K. 1999. Adaptive fusion of multisource raster data applying filter techniques. International Archives of Photogrammetry and Remote Sensing, Vol. 32, Part 7-4-3 W6, Valladolid, Spain, 3-4 June, 1999.

Zenke, B. 1985. Der Einfluß abnehmender Bestandesvitalität auf Reichweite und Häufigkeit von Lawinen. Forstwissenschaftliches Centralblatt 104 (1985), pp. 137-145.

The use of satellite documents for studying geomorphological risks in the Romanian Carpathians

V.Loghin & C. Antohe
Valahia University, Targoviste, Romania

ABSTRACT: The qualities of the analysed satellite materials permitted us to solve the following problems: delimitation of the hydrographic basins of different orders, because the actual evolution of the relief and the impact of the geomorphologic processes with the natural and human elements of the environment (geomorphologic balance and imbalances, highly risky geomorphologic processes, degradation) are taking place according to the hierarchy of the hydrographic basins; standing out the valley and interstreams as major sculptural landforms and morphological elements of the valleys (slopes, river terraces, river bed, river meadow) in comparison with actual morphodynamics and also geomorphic processes with different degrees of risk become differentiated; identification and standing out of the zones corresponding to the different categories of land use -coniferous and deciduous forests, natural alpine and mountainous grasslands, cultures - because the geomorphologic processes (especially their intensity) are differentiated according to this component of the environment; the development a GIS for the Romanian Carpathians which can be the basis of some management projects of geomorphologic risks at different Carpathian subunit levels.

1. INTRODUCTION

In Romania, the study of natural and technological risks with the help of spatial remote sensing is at its beginning. The information provided by Earth Observation satellites is used in a small measure in the meteorological and hydrological monitoring of the territory, as well as of the sea coast processes and of the Danube Delta. These processes have a quick evolution which bring significant changes even at small time scales on the -Black Sea and the Danube Delta shores and finally on the control of the Black Sea pollution on the coast.

In this paper we try to show whether the information offered by satellite maps and images can be used for qualitative and quantitative pointing out and valuation of the surfaces under the incidence of the actual geomorphologic processes with a high degree of risk from the Romanian Carpathians.

Our paper is a part of an extensive international project, named Documentation of Mountain Disasters (DOMODIS), started by The International Association of Geomorphologists. This project aims at the normalization of methods and instruments for the documentation about mountain disasters, about the improvement and insertion of organizational structures of geomorphologic hazards management from the world's mountainous zones and especially from the highly populated and exploited ones,

modified by some major natural or technological risks.

To carry out this study we used the following methodology:

- bibliographic and cartographic documentation;

- the interpretation of some satellite documents regarding the Romanian Carpathians: satellite image map of Bucharest (1: 1.500.000), different space images from the Satellite Remote Sensing Forest Atlas of Europe (Beckel L., 1995) and Satellitenbildatlas Europa (Beckel L., Zwittkovits F ., 1997);

- the delimitation of perimeters illustrating critical environments (protected areas - natural reserves and parks) for which the exploitation of satellite information has been completed with the existing thematic maps and with practical researches, in order to estimate and sort as precisely as possible, the environmental changes produced by highly risky geomorphologic processes;

- the drawing up of geomorphologic risk maps for the perimeters selected for case studies.

2 CONSIDERATIONS CONCERNING THE CHOICE OF THIS THEME

- The Carpathian Mountains in Romania cover 66.000 km^2 which represents 27,8% of the whole area of the country;
- The Romanian Carpathians are quite populated because they have a moderate altitude (average altitude: 840 m), a great number of inter-mountain depressions (336 depressions, about 23% from the mountainous zone), some figures being significant: about 2500 rural settlements and 64 towns, with a population of about 6 million inhabitants; most of the zones have a denseness of 25-50 inhabitants per km^2 but also some zones have a denseness of over 100 inhabitants per km^2 (especially in the inter-mountain depressions where there are large towns such as Brasov, Petrosani, Resita, Hunedoara which polarize a great number of inhabitants);
- The Carpathians have many natural resources such as rich and various mineral deposits (especially non-ferrous), hydroelectric resources, forest resources (60% of Romania's forests), pastoral resources (20% from the whole natural grazing land) (Velcea V .& Savu A., 1982);
- Though the Carpathian environment is characterized by the harmony of the man-nature relations, there are many degraded areas or liable to be degraded by the action of some natural processes (including geomorphological processes) and improper human intervention.

3 THE OBJECTIVES OF THE PAPER

- The revaluation of satellite information in studying the risky geomorphologic processes of a large mountainous area - the Romanian Carpathians: distribution, genesis and evolution, impact;
- The correlation of satellite document data and those from the usual maps and practical observations;
- The development of information stratum obtained by remote sensing, in order to form a Carpathian GIS.
- The integration of this project in a regional context, illustrating the links with some international programmes such as: IDNDR (International Decade for Natural Disasters Reduction), IGBP (International Geosphere - Biosphere Programme) and DOMODIS (Documentation of Mountain Disasters), a programme initiated by the International Association of Geomorphologists.

4 THE RESULTS

The qualities of the analysed satellite materials, as far as their objectivity, truthfulness, clarity and exactness were concerned, allowed us to solve the following research items:

A. Identification and delimitation of the zones corresponding to the different categories of land use -coniferous and deciduous forests, natural alpine and mountainous grasslands, cultures. It is known that the geomorphological processes (especially their intensity) are differentiated according to this component of the environment.

Of a great importance is the delimitation of the forests, element of equilibrium in the environment, which give it a condition of stability, described by the slow modelling of the relief, without the risk of geomorphic destructive processes taking place, at least in theory. That is why, we had in view the setting out of the Carpathians' upper limit of the forested areas (1750-1800 m in the Southern Carpathians, 1600-1700 m in the Eastern Carpathians), because it separates two different morphoclimatic levels: the alpine level, characteristic of the highest mountains with favourable natural potential for the action of gelifraction and nivation, gravitational processes, spring torrential processes, and the mountain level, characteristic of moderately high and small mountains, generally well covered by forests (spruce fir, fir, beech) where the modelling of the relief and the degradation take place by the fluvial - torrential erosion and mass movement (landfalls and landslips) (figs. 1,2).

Comparing the space documents with the old maps, we could partially see a negative phenomenon in the Carpathian landscape evolution and resources potential: the retiring of the upper limit of the forest in some mountains, by natural processes (rolls, rock streams, avalanches) and human activities (deforestations).

B. Delimitation of the hydro graphical basin of different order, because the actual evolution of the relief and the impact of the geomorphological processes, with the natural and human elements of the environment (geomorphological balances and unbalances, highly risky geomorphologic processes, degradation) are taking place according to the hierarchy of the hydrographical basins; thus, the outlying basins from the alpine level and those from the mountain level where deforestations took place, were the most affected by the torrent erosion and mass movement; the former case is specific to Southern Carpathians where there is an alpine level and the latter is found in the Eastern Carpathians where flysch is extended (figs.3,4,5,6).

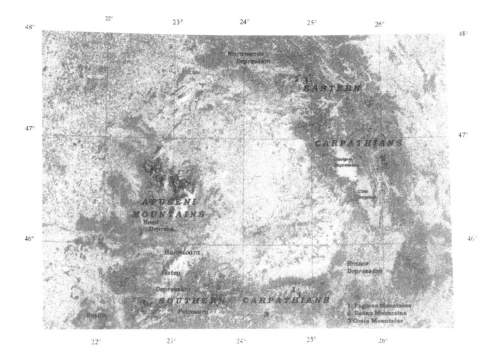

Figure1. The satellite maps of Romanian Carpathians (Satellite image map Bucharest, GSIME – 03E, 1:1.500.000). We can observe the major morphology and the land cover (alpine grasslands – beige; coniferous and beech forests, mountain grasslands – green; crops in depressions – yellow, mauve)

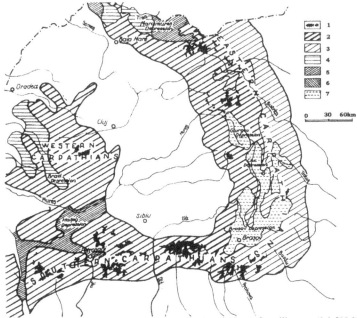

Figure 2. The geomorphologic risks map of Romanian Carpathians realized on the base of satellite map (1:1.500.000) and of the land studies. (*Geografia Romaniei, 1983*).
1. Avalanches, deflation, rill erosion; 2. Fluvio-torrential processes, rock falls, topless; 3. Fluvio-torrential processes, landslides. mudflows; 4. Sheet erosion with mass movements; 6. Mass movements with gully and sheet erosion; 7. Colluvial and aluvial accumulation.

49

Figure 3. LANDSAT TM image on the Rodna Mountains (Eastern Carphatians). It is clear the glacial and fluvial morphology, the drainage basins and the land cover.

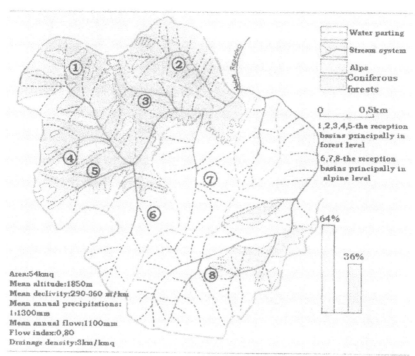

Figure 4. Some reception basins principally developed in alpine level of the Rodna Mountamis. Torrential drainage is responsible for the intense erosion and for the flood generated by the rivers from the Maramures Depression (Viseu, Iza), Tisa's tributaries. (Colour plate, see p. 414)

Figure 5. LANDSAT TM image on the Cozia Mountains (Southern Carpathians).
It is observable a deforested secondary hydrographic basin.

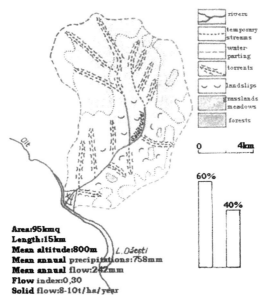

Figure 6. A deforested secondary hydrographic basin – Salatrucel. The geornorphologic consequences: mass movements (landslips, land fall) torrential erosion, alluvial transfer in collector river
(Olt), colmatage of the retention lakes. (Colour plate, see p. 414)

Figure 7 The Fagaras Mountains – twice increased fragment with the Satellite map 1:1.500.000

Figure 8 The Fagaras Mountains – aerial photo (8.000 m altitude). They have the largest alpine level within the Romanian Carpathians.

Figure 9. The Fagaras Mountains – terrestrial photo. On the background of the quaternary glacial relief it actions the actual periglacial processes: gelifraction, debris falls, avalanches, debris avalanches, solifluxions etc. These processes degrade the natural elements of the alpine landscape, the forest of the upper limit and present a high risk for the travel infrastructure.

C. The valleys and interstreams standing out as major sculptural landforms and morphological elements of the valleys (slopes, river terraces, river bed, river meadow) in comparison with which the actual morphodynamics and also geomorphic processes with different degrees of risk become differentiated. The slopes from the alpine level of the Southern Carpathians and Rodna Mountains (Eastern Carpathians), which are around the glacial circus or along the glacial valleys, or the reception basins of the drainage have a high risk of degradation because of gelifraction and nivation action which generates the detachment of blocks, rock streams, avalanches, solifluxions and torrential processes. All of them affect the tourist infrastructure, grasslands and forests at their upper limit (figs. 7,8,9). Great slanting and deforested mountainous slopes formed on flysch (grit stones, marls, clays, conglomerates) are affected by landslips which destroy roads, railroads, houses, agricultural land, etc.

Along the major valleys (Bistrita, Trotus, Buzau, Prahova, Olt, Jiu, Mures, Somes) with well developed water meadows and terraces (in basins and depressions) on which settlements and ways of communications lie, the geomorphologic risks are represented by the undermining of banks, river beds terraces and slopes basis, by flood and alluvial deposits in depressions (Brasov, Petrosani, Hateg, Brad), by a greater mobility of the river beds at confluences, because of the liquid and solid contribution of the tributaries (in these zones the risks are of floods, divagations, of the development of the dejection cones and of over- water and under-water banks in the main sewer bed).

5 CONCLUSIONS

The information gathered by analysing these satellite documents gives us the opportunity to develop a GIS for the Romanian Carpathians (by mountains, depressions, valleys, the whole mountain chain) which can be the basis of management of some geomorphological risks at different Carpathian subunits level. But it is necessary to carry out a more detailed temporal and spatial analysis, differently loaded according to priorities, giving preference to the most threatened zones and objectives. This is what we propose for the next stage, when we shall use satellite maps of a larger scale: 1:100.000, 1:50.000, 1:25.000.

REFERENCES

Beckel L. 1995. *Satellite Remote Sensing Forest Atlas of Europe.* Justus Perthes Verlag Gotha.

Beckel L. & F. Zwittkovits F. 1997. *Satellitenbild – atlas Europea*, RV Verlag, München.

Velcea V. & A. Savu. 1982. *Geografia Carpatilor si a Subcarpatilor Ramânesti.* Eds. Didactica si Pedagogica. Bucuresti: 5-12.

** *Satellite image map of Bucharest*, GSIME-03E (76 x 63 cm) 1:500.000. Geospace, Salzburg.

** *Geografia României, I, Geografia fizica.* Edit. Academiei Romăne, 1983.

2 *Ocean, Ice and Coastal Zones*

Observing our environment from Space: New solutions for a new millennium, Bégni (Ed.)
© *2002 Swets & Zeitlinger, Lisse, ISBN 90 5809 254 2*

On the use of radar imagery for coastal sea bed changes and its potential in identifying submerged hazards

I.Hennings
GEOMAR, Forschungszentrum für marine Geowissenschaften der Christian-Albrechts-Universität zu Kiel, Germany

ABSTRACT: Detailed information of the sea bed and submerged hazards in shallow water regions are important for shipping, fishery, coastal protection and management, offshore activities, environmental monitoring of ecosystems and for marine vessel traffic services (VTS). Offshore tidal current ridges or sand banks can change their positions substantially and the migration of very large sand waves is observed in shelf sea areas. Radar images of the water surface show bathymetric features as well as other oceanographic and atmospherical phenomena. Even surface manifestations of submerged hazards such as wrecks can be identified on radar imagery of shallow water areas. Radar signatures which correspond with locations of wrecks on sea charts have been analysed in relation to available hydrometeorological conditions. Information on position of these isolated dangers or wrecks is important for the savety of ship navigation. Preliminary predictions of the visibility of wreck signatures on radar images have been obtained.

1 INTRODUCTION

One important aim of national governments is to develop new management concepts for coastal seas at all relevant time and space scales. Integrated Coastal Zone Management (ICZM) is the integration of physical, ecological and socio-economical knowledge. All different functions and interests that exist in an area have to be considered. Satellite- and airborne remote sensing techniques have the potential to meet major objectives of ICZM. As recommended in 1992 by the Intergovernmental Panel on Climate Change (IPCC) and confirmed by Agenda 21 of the United Nations Conference on Environment and Development (UNCED) Earth Summit, it is the aim that coastal nations with vulnerable coastal zones develop and implement coastal zone management programmes before the year 2000. The exploitation of the coastal zone depends on the basic knowledge of their environmental parameters. Hydrographic research including the surveying of the sea bed in coastal waters is of vital societal interest because all nations are be liable according to Articles 192 and 208 of the new International Law of the Sea to protect the coastal environment. Consequently, basic knowledge of the marine environment has to be made available to identify and reduce marine pollution and accidents.

Recent research results in the field of earth observation can contribute in a concrete way to provide concepts and tools for ICZM. That this is an important fact was shown, for example, by the grounding of the lumber carrier "Pallas" off the German coast in the North Sea in the fall of 1998.

Monitoring the coastal sea, however, is not only a question of funding but just as much an operational problem. No country with several hundreds or thousands of kilometers of shoreline could possibly manage to monitor its coastal region from ships. For example, the Indonesian archipelago consists of more than 17000 islands with a total coast length of over 60000 km. Most of the countries claim the so called Exclusive Economic Zone (EEZ), an extension offshore of 200 seamiles from the coast line. Therefore, advanced space- and airborne remote sensing techniques and their data products are proposed for the management of the coastal zone. In the fifth framework programme (1999-2002) of the European Community (EC), the focus of research, technological development and demonstration funding has changed from discipline-orientated research to integrated solution under the so called "key actions". Earth observation is seen as a generic technology to contribute to the anticipated deliverables described in the contents of key actions. Basic research work in the past has shown the operational and commercial potential to combine optical and microwave data of the electromagnetic spectrum. Fusion of optical and microwave data is one of the major challenge of earth remote sensing to improve their application especially in the coastal zone. The synergetic use of infrared, optical and microwave remote sensing observations was recently investigated by Miles et al. (1998). However, this paper deals only with the use of radar imagery for coastal sea bed changes and its potential in identifying submerged hazards.

2 THE USE OF RADAR FOR BATHYMETRY AND SUBMERGED HAZARDS

Detailed information of the sea bed in shallow water regions is important for shipping, fishery, coastal protection, offshore activities, environmental monitoring of ecosystems and for marine vessel traffic services (VTS). Most of the deep ocean bottom topography has to be considered as practically unknown. Even the coastal waters of many countries are often inadequately surveyed and need to be surveyed again according to modern standards. One of the major challenges of morphodynamic modelling is the consideration with the different scale interactions of tidal current ridges or linear sand banks, sand waves, and (mega) ripples (De Vriend 1997). Typical measurement configurations for detecting the sea floor by remote sensing methods and an overview of general scales (spacings) of bedforms and ocean floor topography as a function of water depth were recently presented by Hennings (1998). Imaging radars can be used for the detection of tidal current ridges, sand waves, and other remarkable morphological changes of the sea floor in water depths of the order of less than 50 m.

Tidal current ridges or linear sand banks are the largest bedforms found on tidally-dominated continental shelves. They have maximum dimensions of 55 km in length, 6 km in width, and up to 40 m in height above the surrounding sea bed. It was shown that such large bedforms can change their positions

substantially (Hennings 1988). Long-term (> 10 years) morphological changes of the South Falls and Sandettie Bank in the Southern Bight of the North Sea are shown in Figure 1. Two sea charts have been compared which were published in 1920 and 1981. The time interval of publication of these two charts was 61 years. It can be noticed from Fig. 1 that the South Falls Bank rotated in a clockwise sense during that period. At the latitude of 51^0 27' N the 20 m depth contour migrated 1110 m eastward. During the time interval of 61 years a spatial change of 18.2 m per year was calculated. Figure 2 shows a section of the Seasat Synthetic Aperture Radar (SAR) L-band image of the southern North Sea from orbit 762 (August 19, 1978, 06.46 UTC) with the radar signature of South Falls and surface manifestations of submarine large sand waves. The radar signature of a wreck indicated by A is also visible (see next section). Several ship wakes can also be identified on the radar scene.

The results of morphological changes obtained above confirm mechanisms proposed previously for migration of sand banks including clockwise circulation of water and sand. Such clockwise residual flow, sub-parallel to the axis of the sand bank, could result from the anticlockwise orientation of the sand bank with respect to the tidal flow. The effects of bottom friction and Coriolis force would reinforce each other, producing clockwise vorticity and residual circulation (Zimmerman 1981). The migration of very large sand waves in the Strait of Dover was described by Harden Jones and Mitson (1982). All these bedforms are located within important ship routes in the northeastern entrance of the Strait of Dover. Merchant ships, especially very large carriers (VLCs) with a draught of up to 22 m navigate the so called Eurochannel from the English Channel through the Dover Strait to Rotterdam in The Netherlands. Hydrodynamic forces due to bottom topography in restricted waters known as bank effects have influences on keeping the course of a ship. Asymmetrical forces due to variations in the sea bed result in different momentums on the ship, which can have strong effects on the manoeuvring capacity. Today, 25 % of serious ship accidents are caused by touching the ground or by grounding itself. Up-to-date water depths are important for correcting sea charts or the Electronic Navigational Chart (ENC) data of modern Electronic Chart Display and Information Systems (ECDIS). This is important for the safety of navigation. The user of ECDIS can select the safety contours and safety depths and other image contents in dependence of the ship draught and other parameters of the vessel. The visualization of draught and keel clearance of the ship as well as depth contours for distinction of safety and dangerous depth areas are shown in Figure 3 (Hecht et al. 1999).

Synoptic radar signatures of the sea bed can be a contribution for hydrographic agencies to identify immediately changes of the sea floor especially after heavy storm periods. A detailed survey can then be carried out using precise shipborne echo sounders, multi-beam echo sounders and side-scan sonar systems.

In summary, the goals of using remote sensing techniques for monitoring the sea bed are, in order of priority (Kasischke et al. 1983):

1. To detect uncharted or mispositioned submerged features which are potentially hazardous to surface navigation;
2. To define the boundary and location of these hazardous features in either an absolute or relative

sense; and
3. To extract accurate, detailed, and complete water depth information.

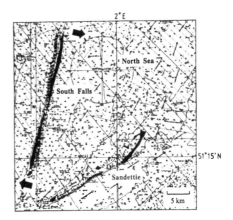

Figure 1. Morphological changes of the South Falls and Sandettie Banks in the Southern Bight of the North Sea during a time period of 61 years determined from two sea charts which were published in 1920 and 1981.

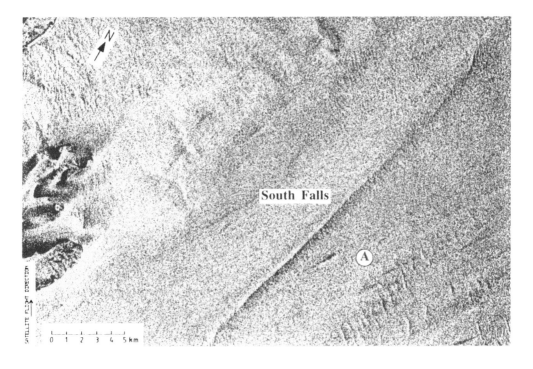

Figure 2. Section of the digitally processed Seasat SAR image of the Southern Bight of the North Sea from orbit 762 (August 19, 1978, 06.46 UTC) with frame center at $51^019'26''$ N, $1^052'51''$ E. The radar signature of South Falls, a linear sand bank, and the position of the surface manifestation of a submarine wreck indicated by A are significant features of the image. The land area in the upper left-hand corner is the English coast near Ramsgate.

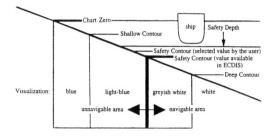

Figure 3. Visualization of draught and keel clearance of the ship as well as depth contours for distinction of safety and dangerous depth areas (modified after Hecht et al. (1999)).

3 RADAR SIGNATURES OF SEA BOTTOM TOPOGRAPHY

During the field experiment in April 1996 of the "Coastal Sediment Transport Assessment using SAR imagery" (C-STAR) project of the European Commission (EC) the Experimental Synthetic Aperture Radar (E-SAR) system of the Deutsches Zentrum für Luft- und Raumfahrt (DLR), Oberpfaffenhofen, Germany, was operated on board

a Dornier DO 228-200 aircraft. One of the main objectives was to investigate the applicability of

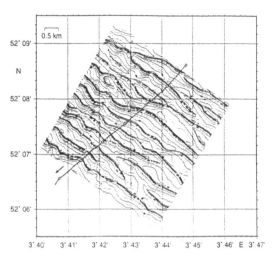

Figure 4. Bathymetric map of the C-STAR study area covered by large asymmetric sand waves off the Dutch coast in the Southern Bight of the North Sea. The analysed track of the drift buoy shown in Fig. 5 is also indicated.

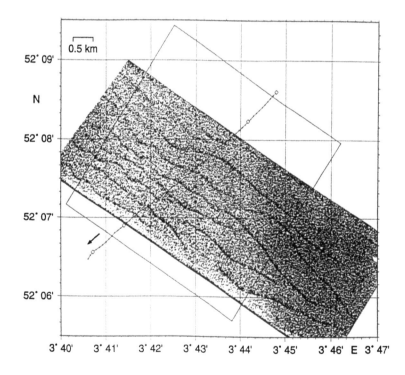

Figure 5. P-band E-SAR image (HH polarization) of the C-STAR study area taken on 16 April 1996, 08.30 UTC. The frame of the bathymetric chart shown in Fig. 4 and the analysed drift path of the ASIB system shown in Fig. 6 are also indicated.

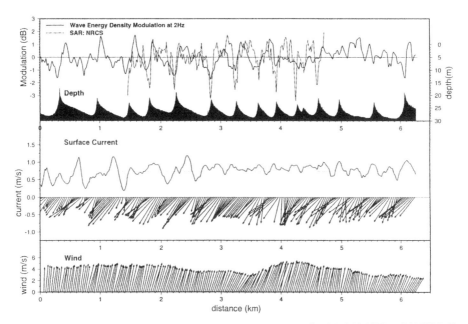

Figure 6. In situ measurements along drift buoy track (see Figs. 4 and 5) on 16 April 1996, 6.20 UTC to 8.30 UTC. Top: wave energy density modulation at 2 Hz, NRCS modulation of E-SAR image, water depth (from echo sounder). Middle: Surface current (speed and vector). Bottom: wind.

imaging radars for monitoring sea bottom topography in coastal waters in order to study its dynamical behaviour. The sea floor of the C-STAR study area is covered by large asymmetric sand waves. It is located 28 km northwest of Hoek van Holland, The Netherlands, in the southern North Sea. Figure 4 shows the bathymetric map with a mean water depth of 27 m covered by large sand waves of about 5 m height and a mean spacing of 500 m. The analysed track of the drift buoy shown in Figure 6 is also indicated in Fig. 4.

The ebb tide experiment on 16 April 1996 took place from 6:20 to 8:30 UTC with a drift distance of 6.26 km. The E-SAR system of DLR was operated in narrow swath mode with off nadir angle range between 25^0 and 55^0 at an altitude of 3000 m resulting in a scene size of about 4 km width (ground range) with an effective spatial resolution of 6 m. From a collection of 21 scenes a P-band (0.45 GHz, HH polarization) SAR image shown in Figure 5 was selected acquired on April 16, 1996 during southwestward ebb tidal current. The SAR image was further processed by TNO, Physics and Electronics Laboratory, The Hague, The Netherlands, using elaborate smoothing, and a normalization with a large scale, slowly varying

background level. The dark small streaks of reduced backscatter intensity in the SAR image (see Fig. 5) clearly correspond to the sand wave pattern shown in Fig. 4. The initial geocoding of the image (with the ship-position as reference) had to be corrected (for reasons not obvious to us): in order to make the dark streaks to coincide with the reduced wave energy density modulation measured by the ASIB system, the image was shifted 220 m towards northeast. The airborne SAR image was acquired in 2.5 minutes at about 8:30 UTC, when the buoy was at the end of the drift track.

The in situ measurements of the drift buoy are shown in Figure 6. The wave energy density is normalized with the mean value over the complete track, which is assumed to represent the equilibrium wave energy density for the present wind condition (mean wind speed of 4.3 ms^{-1}). The wave energy density was chosen at 2 Hz, as it corresponds to P-band Bragg waves with a wave length of 0.44 m, which mainly contribute to the radar backscattering. The NRCS profile of the SAR image is averaged in sand wave direction over 50 m at both sides of the buoy track.

As indicated from the buoy measurements the reduced wave energy density regions correspond to the steep slope at the upstream (northeast) side relative to the sand wave crests, where strong current divergence due to decreasing water depth is expected. The correlation of wave energy density with bottom topography is obvious in most parts along the track. The reduced backscatter regions in the NRCS profiles of the SAR image are even more pronounced: the NRCS bands are smaller (30-60 m) and the modulation depth (typically 3-4 dB, minimum to maximum) is larger than for wave energy density (typically 2 dB). Despite some noise in both profiles and the lower spatial resolution of the wave energy density, both profiles compare well with respect to the position of minima and maxima and the overall shape. The surface current in most cases changes direction when crossing the wave crests, whereas the current acceleration in the steep slope regions of the sand waves is often masked by the general current variability.

4 RADAR SIGNATURES OF SUBMARINE WRECKS

Under certain environmental conditions radar signatures of the sea surface show also the locations of submerged hazards such as wrecks (McLeish et al. 1981, Hennings & Stolte 1997). Side-scan sonar records have revealed the presence of many obstacle marks that have been generated by scour and deposition near wrecks which disturb and interrupt the tidal current flow (Caston 1979). These features have been named wreck marks. The longitudinal wreck marks are valuable as accurate pointers to the direction of dominant tidal current flow, particularly in regions where other indicators are lacking. Obstacle marks may attain 100 m in length in a unidirectional marine current, such as the comet marks described by Werner and Newton (1975). Comet marks are defined as obstacle-induced, long erosional strips occurring on current-affected sea bottoms (Werner et al. 1980). The comet marks are mainly observed on side-scan sonar records due to the acoustic backscattering contrast between the erosional zone and the surrounding sea bed.

When ship routes are planned it is necessary to investigate all wrecks in the vicinity of the route and to guarentee sufficient water depth above the dangerous wrecks. Position variations of ship wrecks and the genesis of scour hollows according to bottom currents are discussed in Nieder (1964). Van

Riet et al. (1985) described a wreck extending some 15 m above the sea bed which was undetected for 35 years.

The locations of the analysed wrecks A-E described in this paper are well known and registered on sea charts. Figure 7 gives an overview of the wreck positions which have been investigated. The positions, meteorological and oceanographic data and other information of the wrecks for the radar image acquisition times are summarized in Table 1. The positions of the wrecks, the water depth, and the minimum water depth above the wreck were taken from sea charts. Because other information was lacking, the wreck diameter d_w has been estimated according to the expression

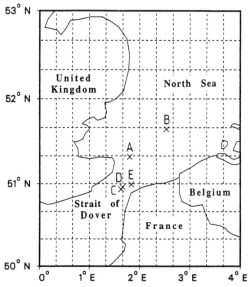

Figure 7. Overview of analysed submarine wreck positions in the Southern Bight of the North Sea and Strait of Dover.

Table 1. Positions, meteorological and oceanographic data and other information of the analysed wrecks for the radar image acquisition times. The wind speed was obtained from weather charts.

wreck	A	B	C	D	E
position	$51^0 19.3'N$	$51^0 39.0'N$	$50^0 56.0'N$	$50^0 57.6'N$	$50^0 59.5'N$
	$1^0 48.1'E$	$2^0 31.9'E$	$1^0 37.8'E$	$1^0 39.8'E$	$1^0 50.0'E$
water depth minimum	37 m	36 m	20 m	22 m	15 m
water depth above wreck	19.5 m	6.8 m	3.1 m	13.0 m	7.2 m
wreck diameter	17.5 m	29.0 m	16.9 m	9.0 m	7.8 m
water depth above wreck center	28.25 m	21.40 m	11.55 m	17.5 m	11.1 m
tidal current direction	214^0	224^0	52^0	52^0	68^0
tidal current velocity	0.7 m/s	0.7 m/s	1.0 m/s	1.0 m/s	1.1 m/s
wind direction from	140^0	270^0	326^0	326^0	326^0
wind speed	2 m/s	8 m/s	10 m/s	10 m/s	10 m/s

$$d_w = h_0 - h_{min} \qquad (1)$$

where h_0 is the water depth and h_{min} is the minimum water depth above the wreck. The water depth above the wreck center h_c can then be expressed by

$$h_c = \left(\frac{d_w}{2}\right) + h_{min} \qquad (2)$$

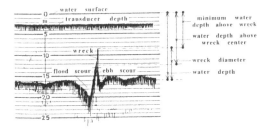

Figure 8. Example of a section of an echogram showing a wreck in a water depth of 16 m with flood and ebb scours. The water depth, wreck diameter, minimum water depth above wreck, and water depth above wreck center are indicated (modified after Nieder (1964)).

An example of a section of an echogram showing a wreck in a water depth of 16 m with flood and ebb scours is presented in Figure 8. The water depth, wreck diameter, minimum water depth above wreck, and water depth above wreck center are also indicated.

A curved wake of a submerged wreck imaged on a radar scene was identified by McLeish et al. (1981). This wake extends more than 4 km along the changing direction of tidal flow. An example of a radar signature of a submarine wreck is shown in Fig. 2 indicated by A. Other wrecks were analysed in the paper by Hennings and Stolte (1997).

All these radar signatures of wrecks are imaged like comet marks on side-scan sonar records. Radar signatures of wrecks can also be compared with a turbulent ship wake of reduced radar backscatter (dark imaged feature) with one edge of enhanced radar backscatter (bright imaged feature). The white edge is observed on that side of the wake from which the wind is blowing (luff side of the wind direction). This is an area of enhanced water surface roughness. The short water waves are modulated when they propagate into the turbulent wake region which is induced by the ship's hull and propeller. Skoelv et al. (1988) simulated such a feature using a wave-vortex interaction model. It is assumed that similar surface manifestations are also produced by submerged hazards such as wrecks.

5 DISCUSSION AND CONCLUSIONS

The development of quasi-operational tools for bathymetric radar image applications is of interest for a variety of end users especially for ICZM agencies. One aim is the application and validation of inverse modelling of the sea bottom topography to assess water depths from radar images. A research group in The Netherlands developed an operational system which has been implemented in a so-called Bathymetry Assessment System (BAS) which combines echosounding and SAR observations. A pre-operational version of this system was currently being implemented at the Rijkswaterstaat Survey Department in The Netherlands as a first step towards the routine monitoring of Dutch coastal waters.

Although surface expressions of the sea bed were discovered in radar imagery of the ocean 30 years ago there was still a need for experimental quantification to explain the coherent imaging mechanisms. This was realized within the C-STAR project where basic data of relevant hydrodynamic processes have been collected in the southern North Sea utilizing a SAR and a special buoy system which drifted across large sand waves. SAR images and in situ wave energy density measurements were acquired quasi simultaneously. This offered, for the first time, the possibility to compare the modulations of NRCS and spectral wave energy density: both profiles compare fairly well, taking into account the lower spatial resolution of the wave energy density. Although the narrow streaks of reduced sea surface roughness are undersampled, the profile of wave energy density shows the position of these streaks to coincide with the steep slope of the sand waves close to the crest. For the SAR image, although no satisfactory independent georeference was found, the comparison also helped to confirm the position of the SAR image relative to the topography with some confidence. The conclusion is, that SAR imagery and the in situ measurements provided a qualitatively consistent view of the sea suface manifestations of sea bottom topography. Quantitatively, the NRCS and wave energy density modulation depths agreed within a factor of 2. For future applications a precise georeference of the SAR image and in situ measurements of high spatial resolution would be desirable. On the other hand other investigations under weak currents and strong stratification condition showed that it is currently not possible to accurately map sand wave fields using imaging radars. Therefore, it is also necessary to investigate the influence of the sea bed topography on the surface wave field spectrum under stratified flow conditions.

Radar signatures which correspond with locations of submarine wrecks on sea charts have been analysed in relation to the hydrodynamic conditions. The following preliminary predictions of the visibility of radar signatures of submarine wrecks have been obtained:

1. Radar signatures of submarine wrecks are indicators of the current direction in tidal channels.

2. The white region of an imaged radar wreck signature correlates with the windward side.

3. Superimposed on the flow pattern of two (helicoidal) eddies behind a wreck is the turbulent region created by the shape of the wreck. Together with the effects of wind strength and direction typical radar signatures of wrecks look like comet marks on side-scan sonar records.

4. Many wrecks are not imaged on radar scenes. It is concluded that the radar imaging mechanism of wrecks depends on the size and shape of the wreck, surrounding water depth, water depth above the wreck, and the present wind and current conditions.

ACKNOWLEDGMENTS

All members of the C-STAR project are gratefully acknowledged for their excellent cooperation and assistance. We thank ESA as well as DLR for providing the Seasat SAR image. This work has been supported by the European Community (EC) as a part of the Marine Science and Technology (MAST) programme under contract MAS3-CT95-0035.

REFERENCES

Caston, G.F. 1979. Wreck marks: indicators of net sand transport. Marine Geology 33: 193-204.

De Vriend, H.J. 1997. Evolution of marine morphodynamic modelling: Time for 3-D? German Journal of Hydrography 49: 331-341.

Harden Jones, F.R. & Mitson, R.B. 1982. The movement of noisy sandwaves in the Strait of Dover. Conseil International pour l'Exploration de la Mer 40: 53-61.

Hecht, H., Berking, B., Büttgenbach, G. & Jonas, M. 1999. Die Elektronische Seekarte: Grundlagen, Möglichkeiten und Grenzen eines neuen Navigationssystems. Heidelberg: H. Wichmann Verlag.

Hennings, I. 1988. Abbildung von submariner Bodentopographie auf Luft- und Satellitenbildern im Mikrowellenbereich und im sichtbaren Bereich des elektromagnetischen Spektrums. GKSS Publication 88/E/41 GKSS-Forschungszentrum Geesthacht. Geesthacht: GKSS.

Hennings, I. 1998. An historical overview of radar imagery of sea bottom topography. International Journal of Remote Sensing 19: 1447-1454.

Hennings, I. & Stolte, S. 1997. Radar signatures of wreck marks. Proceedings of the Undersea Defence Technology (UDT) Conference and Exhibition, Hamburg, 24-26 June 1997: 84-88. Swanley: Nexus.

Kasischke, E.S., Shuchman, R.A., Lyzenga D.A. & Meadows, G.A. 1983. Detection of bottom features on Seasat Synthetic Aperture imagery. Photogrammetric Engineering and Remote Sensing 49: 1341-1353.

McLeish, W., Swift, D.J.P., Long, R.B., Ross, D. & Merrill, G. 1981. Ocean surface patterns above sea-floor bedforms as recorded by radar, southern bight of North Sea. Marine Geology 43: M1-M8.

Miles, M., Johannessen, O.M., Samuel, P., Hamre, T., Jensen, V., Sandven, S., Robinson I.S. & Scoon, A. 1998. Synergetic use of infrared, optical and microwave remote sensing observations. Final Report ESA/ESTEC Contract No. 11968/96/NL/CN. Noordwijk: ESA Publications Division.

Nieder, F. 1964. Aus der Praxis der Wracksuche. Der Seewart 25: 69-79.

Skoelv, A., Wahl, T. & Eriksen, S. 1988. Simulation of SAR imaging of ship wakes. Proceedings of IGARSS'88 Symposium, Edinburgh, 13-16 September 1988: 1525-1528. Piscataway: IEEE.

Van Riet, J.A., Kaspers, J. & Buis, B. 1985. Safety standards for a 22-metre deep-draught route. The Journal of Navigation 36: 91-113.

Werner, F. & Newton, R.S. 1975. The patterns of large-scale bed forms in the Langeland Belt (Baltic Sea). Marine Geology 19: 29-59.

Werner, F., Unsöld, G., Koopmann, B. & Stefanon, A. 1980. Field observations and flume experiments on the nature of comet marks. Sedimentary Geology 26: 233-262.

Zimmerman, J.T.F. 1981. Dynamics, diffusion and geomorphological significance of tidal residual eddies. Nature 290: 549-555.

Observing our environment from Space: New solutions for a new millennium, Bégni (Ed.)
© *2002 Swets & Zeitlinger, Lisse, ISBN 90 5809 254 2*

Some methodological enhancements to INSAR surveying of polar ice caps

A.I.Sharov & K.Gutjahr
Institute of Digital Image Processing, Joanneum Research, Wastiangasse 6, Graz, Austria

ABSTRACT: An original approach to geometric processing and thematic interpretation of ERS-1/2-INSAR data based on the calculation of interferometric phase-gradients and the generation of glacier slope maps has been offered and tested. A new GINSAR algorithm is described and the results of accuracy analysis are discussed. A conjectural hypothesis explaining some modelling errors as being due to the coherent microwave backscattering and interferential effects at the stratified glacier surface was devised. A transferential approach based on the analysis of the fast sea-ice motion forced by the glacier flow was designed as an alternative to a differential approach and has been proven a very promising technique using single SAR-interferograms for glacier change detection and ice-motion estimation. Our experiments provided significant technological simplifications to main INSAR operations, yet without compromising on accuracy and allowed, for example, several new floating ice shelves to be discovered and the frontal velocities of 36 large tide-water glaciers in the High Russian Arctic to be accurately determined for the first time in the history of their exploration.

1 INTRODUCTION

The origins of the present research date back to 1996, when we commenced studying the background of satellite radar interferometry (INSAR) and its practical applications in alpine and polar regions at the TU Graz. The authors have now been working together on methodological developments and technological improvements to INSAR surveying of polar ice caps in the Eurasian Arctic for three years. Concepts and stratagems resulting from this work served as a basis for some new research aimed at satellite monitoring and reliable topographic modelling of the glacier surface, e.g. the AMETHYST (INCO Copernicus-2 Programme), COMBINE (INTAS Programme) and OMEGA (EU EESD Programme) projects that have been positively evaluated and are now in progress. Current project results are being delivered to corresponding Commissions, but still remain largely unknown to the reading public. Some original results on the use of single SAR interferograms for studying glacier dynamics and coastal change detection in the Eurasian Arctic were preliminarily pointed out in (Sharov et al. 2000) without, however, providing thorough theoretical considerations and essential explanations.

The present paper reports on the continuation of our experiments related with the interferometric analysis of satellite SAR-data and considers some metric relations and methodological aspects, which have not been duly treated in our previous publications. Major attention is paid to the following specific items:

- improvements to INSAR modelling of the glacier surface using repeat-pass interferograms;
- accuracy analysis and preliminary interpretation of microwave backscattering effects at the stratified glacier surface in SAR-images;
- glacier change detection and ice-motion estimation using differential, transferential and gradient approaches;
- practical recommendations to the appropriate selection of SAR data for interferometric analysis, glaciological interpretation and glacier change documentation.

Seven ERS-1/2-INSAR pairs taken over several large insular ice caps in the Franz Josef Land (FJL) and Severnaya Zemlya (SZ) archipelagos under favourable environmental conditions were selected for the present methodological experiments.

The INSAR surveying, in general, and INSAR modelling of polar ice caps, in particular, is mostly focussed on the generation of digital elevation models (DEM) of glacial terrain (Joughin et al. 1996). There are, however, several principal limitations to glacier DEM generation using SAR-interferograms, which are due to

- the lack of reliable ground control,
- limited ground resolution of INSAR data,
- the influence of glacier motion,
- frequently rough surface of glacier exteriors,
- the penetration of radar signals into dry snow,
- tidal effects and difficulties in providing tide-coordinated INSAR data,
- processing errors, mostly at the stage of interferometric phase unwrapping and geocoding.

In our experience, the reliable phase unwrapping frequently becomes impossible because of complex glacier topography and significant phase noise at glacier fronts, walls and tops. The INSAR DEM of any glacier is reputed as a complex and expensive product that requires up to 10 process steps and quite a large number of computations. The general quality of such a product depends to a great extent on the environment and, at least in our practice, could not be ensured in advance, i.e. before SAR data processing.

In contrast to the interferometric DEM generation, the production of glacier slope maps (SMs) from INSAR data can be performed in a more straightforward and a much less complicated manner. Indeed, simple differentiation of the original interferometric phase $\varphi(x, y)$ results in the gradient picture called *topogram*, which can be related to terrestrial topography on a pixel-by-pixel basis unambiguously and without phase unwrapping (Brandstätter & Sharov 2000).

The magnitude of the phase gradient is defined as

$$G[\varphi(x,y)] = \sqrt{\left(\frac{\partial\varphi}{\partial x}\right)^2 + \left(\frac{\partial\varphi}{\partial y}\right)^2}, \quad (1)$$

where x and y denote the azimuth and range direction, respectively. The direction of gradient is given by the relation

$$A[\varphi(x,y)] = \begin{cases} \tan^{-1}\left[\left(\frac{\partial\varphi}{\partial x}\right)\Big/\left(\frac{\partial\varphi}{\partial y}\right)\right] + \pi\cdot n, \ \partial\varphi/\partial y \neq 0 \\ \pm\pi/2, \ if \ \partial\varphi/\partial y = 0 \end{cases}, \quad (2)$$

where $n = 0$, if the partial derivative of the interferometric phase in the range direction $\partial\varphi/\partial y$ is positive, i.e. the terrestrial slope or/and motion is directed toward the sensor, and n = 1, if $\partial\varphi/\partial y$ is negative, i.e. the terrestrial slope or/and motion is directed away from the sensor.

In digital image processing, the partial derivatives in (1, 2) are well approximated by differences, e.g. as

$$G[\varphi(x,y)] \cong \sqrt{[\varphi(x,y) - \varphi(x+1,y)]^2 + [\varphi(x,y) - \varphi(x,y+1)]^2}$$

or, by using absolute values, as follows (Gonzales & Wintz 1987)

$$G[\varphi(x,y)] \cong |\varphi(x,y) - \varphi(x+1,y)| + |\varphi(x,y) - \varphi(x+1,y)|. \quad (3)$$

In our practical work with multi-look ERS-1/2-SAR interferograms we came to the conclusion that better results can be achieved, if partial derivatives are taken along and perpendicular to the main direction of fringes in the original interferogram. Moreover, the image shift values Δx and Δy, which are set to 1 pixel in (3), can vary from one to several pixels depending on the dominant width of interferometric fringes, which in turn depends on the length of the interferometric baseline.

More general equations for the partial phase-gradients can therefore be written as follows

$$\frac{\partial\varphi}{\partial x} \approx \Delta\varphi_x(x,y) = \frac{\varphi(x+\Delta x, y) - \varphi(x,y)}{\Delta x} \quad \text{and}$$

$$\frac{\partial\varphi}{\partial y} \approx \Delta\varphi_y(x,y) = \frac{\varphi(x, y+\Delta y) - \varphi(x,y)}{\Delta y}, \quad (4)$$

or even as

$$\Delta\varphi_i(x,y) = \frac{\varphi(x+\Delta x, y+\Delta y) - \varphi(x,y)}{\sqrt{\Delta x^2 + \Delta y^2}} \quad \text{and}$$

$$\Delta\varphi_j(x,y) = \frac{\varphi(x-\Delta y, y+\Delta x) - \varphi(x,y)}{\sqrt{\Delta x^2 + \Delta y^2}}, \quad (5)$$

where the index i generally denotes an arbitrary direction $\vec{i} = [\Delta x \ \ \Delta y]^T$ (in our case it is perpendicular to the main fringe direction) and the index j implies the orthogonal direction $\vec{j} = [-\Delta y \ \ \Delta x]^T$ (along fringes).

Increasing increment values Δx and Δy deteriorates the ground resolution of the topogram, but improves its contrast and decreases the phase noise. In practice, the increment values Δx and Δy are manipulated separately in range and azimuth direction within the interval of $\Delta x, \Delta y = 0...6$ pixels. The topogram can be directly converted to the glacier slope map by using the next algorithm (Eqs. 6-8). The terrestrial slope value ε and the relative height increments Δh_x and Δh_y in the specified directions $\vec{x}$ and $\vec{y}$ are related by the simple formula

$$\cos\varepsilon = \frac{1}{\sqrt{\Delta h_x^2 + \Delta h_y^2 + 1}} . \qquad (6)$$

Considering the discrete character of the interferometric phase, the height increments Δh_x and Δh_y can be defined from the following equations

$$\Delta h_x \cong \frac{\lambda}{4\pi B} \cdot \frac{r \cdot \sin\theta}{\cos(\zeta - \theta)} \cdot \Delta\varphi_x \qquad \text{and}$$

$$\Delta h_y \cong \frac{\lambda}{4\pi B} \cdot \frac{r \cdot \sin\theta}{\cos(\zeta - \theta)} \cdot \Delta\varphi_y , \qquad (7)$$

where λ = 5.66 cm - is the wavelength of SAR signal, $B(x)$ - is the length of the interferometric baseline, $r(y)$ - is the slant range of the master image, $\theta(y)$ - is the incidence angle and $\zeta(x)$ - is the angle between the baseline and the horizontal plane.

The slope value in azimuth and range direction can be defined separately from the next equations

$$\tan\varepsilon_x = \frac{\Delta h_x}{\Delta x} \quad \text{and} \quad \tan\varepsilon_x = \frac{\Delta h_y}{\Delta y} . \qquad (8)$$

Thus, by inserting values of partial phase-gradients $\Delta\varphi_x$ and $\Delta\varphi_y$ from (4) or (5) into equation (7) we obtain the partial height increments and the absolute slope value on a pixel-by pixel basis. The plus or minus sign of the slope ε corresponds to the sign of the height increment Δh_y, which in turn depends on the sign of the partial phase-gradient $\Delta\varphi_y$.

The technique offered, called *gradient-INSAR* (GINSAR), excludes the areal error propagation and the resultant slope values are quite tolerant of local phase errors. Such an approach therefore provides rapid access to results, yet without compromising on accuracy even under significant phase noise. The stage of phase filtering, which is usually included in all known phase unwrapping algorithms, in our case, becomes optional and is used mostly for cosmetic reasons. Good results were obtained with a spectral filter described in (Goldstein & Werner 1998) that makes use of phase filtering in the Fourier domain. The response of this filter appears as follows

$$f(u,v) = f_o(u,v) \cdot |f(u,v)|^\alpha , \qquad (9)$$

where $f_o(u,v) = F[\varphi(x,y)]$ - is the Fast Fourier transform of the interferometric phase function and α = [0,1]. In the case of GINSAR modelling of polar ice caps we usually applied α = 0.9.

Some examples illustrating the performance of the GINSAR technique are given in Figure 1. There are several fragments from different ERS-1/2-INSAR products showing a huge Vostock-1 Ice

Dome situated on La Ronciere Island, FJL with a maximum height of 431 m a.s.l. (Fig.1, b, c, d). Both, the topogram and the glacier SM in Figures 1, c and d provide valuable information on the possible character of glacier flow and sun radiation, though they cannot be directly used for determining glacier elevations. Nevertheless, any glacier SM can be transformed into a glacier DEM by integrating height increments along the path, which has to be defined as it is done, e.g., in interferometric phase unwrapping using the branch-cut method.

Simplified flowcharts specifying principal stages of both procedures for INSAR DEM and GINSAR SM generation are given in Table 1 for the purpose of comparison. It is clear that the procedure of GINSAR SM generation (B) involves less complicated operations than that of producing the INSAR DEM (A). Both procedures have been implemented in the new RSG 3.6 software package distributed by JOANNEUM RESEARCH.

Table 1. Flowcharts of INSAR DEM
and GINSAR SM generation

A. Principal stages of the INSAR DEM generation	
1	Co-registration of SAR images and calculation of the interferogram
2	Flat terrain phase correction and phase filtering
3	Interferometric phase unwrapping
4	Model adjustment using ground control points
5	DEM generation and geocoding. Quality control

B. Principal stages of the GINSAR SM generation	
1	Co-registration of SAR images and calculation of the interferogram
2	Phase filtering and topogram calculation
3	SM generation and quality control

3 MODELLING ERRORS & ACCURACY CONTROL

The methodological accuracy of the GINSAR technique was verified by comparing GINSAR SMs with the SM generated from the traditional INSAR DEM, the latter being obtained from the same original interferogram after phase unwrapping and geocoding. At the moment of writing this paper, we had no accurate results from other accuracy tests using SMs generated from optical data and available maps. The mean difference between the GINSAR SM and the INSAR SM over the ice dome was estimated at 0.01° and the root mean square difference was given as 0.3° (Fig. 1, e).

Careful comparison between two GINSAR SMs obtained from original interferograms of 3/4 September 1995 ($B_\perp$ = - 49 m) and 8/9 October 1995 ($B_\perp$ = 129 m), however, has shown significant differences in the form of bright periodical circular structures (Fig. 1, f). Within these structures, the mean slope difference reaches up to 11° and the rms. slope difference amounts to up to 14°, which could not be explained by the glacier flow or technical errors. One possible explanation for such structures can be given on the basis of physical interactions of the C-radar signal with the stratified glacier surface. The effect of penetration of the C-radar signal into dry snow to the depth of up to several tens of meters impairs the vertical accuracy of the INSAR DEM of the glacier surface (Guneriussen et al. 1999). This influence is reduced to some degree in the INSAR SM.

Coherent backscattering of microwaves from different layers of the snowpack can, however, result in interferential effects at the glacier surface, which in turn, might essentially influence the intensity of the return signal. Depending on the incidence angle, and the relief, depth and density of the snow cover, one can expect constructive or destructive interference (multiple-ray interference) of scattered signals and high or low return echo from the glacier surface. Thus a quite homogeneous glacier surface may appear alternately light and dark even under similar environmental conditions (Fig.2, a, b).

The result can be observed in a single SAR-image, which is why we termed this effect *monointerference*. Occasionally different amplitudes of two SAR-images constituting the INSAR-pair reduce the contrast of the interferometric picture and can lead to the local decorrelation and significant phase errors. In this context, it is worth noting the similarity between the image structures in Figures 1, f and 2, b. Apart from modelling errors, the penetration of radar signals into snow provides valuable thematic information on the vertical structure of the snowpack and underlying relief.

Typical fragments from amplitude (IN)SAR images showing microwave backscattering effects in the snowpack are given in Figure 2. An example of the glacier surface representation in the ERS-1-SAR PRI-image is shown in Figure 2, a. Figure 2, b shows a unique circular structure with a radius of about 5 km that was discovered in the ERS-1/2-SAR interferogram taken over the flat snow-covered area of Pioneer Island, SZ on 22/23 March, 1996. Another interesting structure of ice divides forming a star system on Kropotkina Ice Dome, FJL, was detected in the ERS-1-SAR image of Sept. 4, 1991.

An exact explanation of the origin of these structures has yet to be found, but we think that their occurrence in SAR images is related with the effect known in optics as "total internal reflection". The penetration of microwave from the air into the medium with a larger refractive index (snow) and further enhanced backscattering from the underlying interface with the denser medium (ice) can improve the contrast in the representation of the underlying

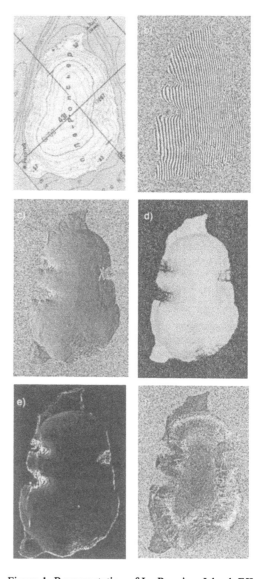

Figure 1. Representation of La Ronciere Island, FJL in Russian topographic map (a), original interferogram (b), topogram (c), GINSAR SM (d), difference GINSAR SM - INSAR SM (e), difference GINSAR SM_1 – GINSAR SM_2 (f).

surface and reveal its fine details. It should be noted that this hypothesis does not rule out an interferometric point of view on the origin of periodical structures in Figures 1, f and 2, a, b, c.

We think that both effects (monointerference and enhanced internal backscattering) do coexist and both interpretations are valid. A simple graphic scheme explaining both effects is given in Figure 2,d.

3 GLACIER CHANGE DETECTION AND ICE-MOTION ESTIMATION USING INSAR DATA

The INSAR technique making use of satellite SAR-images with a typical ground resolution of several tens of meters has the one big advantage that it allows quite small glacier changes / motions in the centimetre range to be detected and measured. A full separation between the impacts of glacier topography and glacier motion on the interferometric phase is necessary in order to attain such a high performance.

This key procedure is traditionally based on differencing between the original interferogram and the reference interferogram of the same glacier, which does not contain the phase term related to the ice motion. Apart from the necessity of obtaining and processing several suitable INSAR pairs, which is not easy in the first place, there are, however, some limitations to this approach called DINSAR.

Since we are discussing glaciers in permanent motion, it is practically impossible to find out the real interferometric model of such glaciers without

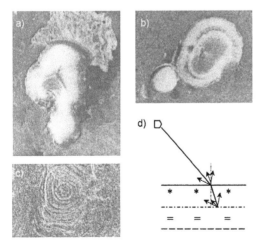

Figure 2. Graphic scheme (d) and examples of backscattering effects on Kropotkina Ice Dome, FJL (a), Vostock-4 Ice Cap, Eva-Liv Island, FJL (b) and Pioneer Island, SZ (c);

motion fringes. The reference interferogram without motion effects can be synthesised from accurate glacier DEMs or contemporary topographic maps, but they are usually unavailable in the High Arctic. An original interferogram without topographic phase can be obtained, if the length of the interferometric baseline is, by lucky chance, equal to zero. In this case, however, the glacier motion can be determined directly and there is no need for the DINSAR approach at all.

We did not obtain ERS-1/2-SAR interferograms with zero-baselines for our study areas and, therefore, decided to use the GINSAR approach for detecting small changes resulting from ice motion. This idea is based on the fact that the INSAR scheme is much more sensitive to displacements than to the surface topography. Due to the relatively flat topography of ice caps and rather high velocities of glacial flow, the rate of motion fringes in the interferential picture is usually much higher than that of topographic fringes. These differences are well emphasised in the topogram because of the differentiation process.

Figure 3 shows, for example, interferometric pictures of the western coastline of Armitage Peninsula (Fig. 3, a, mid) and the northern margin of the Dzegudze Ice Dome on Prince George Land, FJL, with the large outlet glacier No.5 flowing into Geographer's Bay (Fig. 3, a, bottom left). In the topogram (Fig. 3, c), all areas of fast ice and outlet glaciers undergoing motion within the 1-day interval between SAR surveys are reproduced by dark- or bright-grey values depending on the motion direction, and can be distinguished from the steady land area and immovable glacier parts given in grey. Thus, the sea ice attached to the ice shore is well recognisable and the coastline can be reliably delineated. Cracks in the sea ice are also well detectable (Fig. 3, c).

Moreover, it can be seen that the character of the glacier motion in its marginal part is quite different to that of the fast ice attached to the glacier front (marked with an arrow in Fig. 3, c). The dates of 31.12.1996 / 01.01.1997 correspond to the moon phase of ¾, and it can be supposed that there were no strong tidal effects in FJL at that time. Thus, the radiometric differences in the topogram across the glacier front can be explained by different inertial properties and correspondingly different reactions of the glacier and the fast ice to tidal effects. An inferred longitudinal profile of the outlet glacier No.5 is given in Figure 3, d.

Joint analysis of an original interferogram, fringe image (Fig. 3, b) and topogram (Fig. 3, c) revealed that a quite large (up to 15 km²) seaward marginal part of the outlet glacier No.5 undergoes both, horizontal motion due to the glacier flow and local vertical movements caused by tidal effects. This provides a good verification for the assumption made in (Dowdeswell et al. 1994) and allows the conclusion to be drawn of this glacier as being the largest floating ice shelf in FJL. Furthermore, it has been discovered that two smaller outlet glaciers Nos. 18 and 89 also undergo tidal movements. Other outlet glaciers in Prince George Land did not show essential vertical motions in winter 1996 - 1997. There is also some evidence of a floating marginal part of outlet glacier No.6 in Salisbury Island, FJL.

Nonetheless, the analysis of brightness characteristics of the topogram allows only the qualitative estimation of ice motion to be performed. For the accurate measurement of frontal glacier velocities in single ERS-1/2-SAR interferograms we devised an original approach called *transferential* (Sharov et al. 2000). This approach is based on the interferometric analysis of the fast-ice motion away from the shore forced by the glacier flow. According to A.F.Glazovskiy and V.S.Koryakin (pers. comm. 2001) M.M.Ermolaev (AARI, St. Petersburg) already tried to evaluate glacier velocities on Novaya Zemlya by terrestrial measurements of fast-ice motions in the 1930s.

In our case, the effect of the fast-ice motion forced by glacial flow is manifested in the form of "outflows" that can be observed in winter interferograms at fronts of active tide-water glaciers (Fig. 3, b and 4, c). The orientation of "outflows" mostly coincides with the SAR-range direction. In the tide-coordinated INSAR data without significant tidal effects, the frontal glacier velocity along the SAR-range direction can be directly determined by calculating the number of interferometric fringes k within the outflow as follows

$$V_{hr} = \frac{\lambda \cdot k}{2t \cdot \sin \theta}, \qquad (10)$$

where t is the temporal baseline of the interferogram.

For example, Figure 4 represents four outlet glaciers Nos. 1, 2, 3 and 4 on La Ronciere Island in the winter interferogram of 17/18 December 1995 ($B_\perp$ = - 41 m, Fig. 4, c), when the FJL straits are completely covered with fast ice, and in the summer interferogram of 3/4 September 1995 ($B_\perp$ = - 49 m, Fig.4, a), when coastal waters are mostly released from their ice cover, and the coastline is well detectable in the coherence image (Fig. 4, b). Frontal velocities of these outlet glaciers were measured in the winter fringe image as 11.6, 7.6, 44.0 and 18.5 cm/day, respectively.

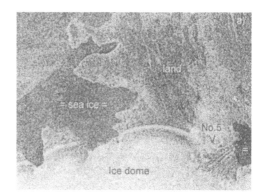

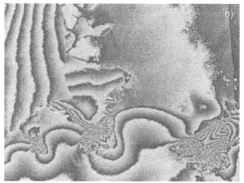

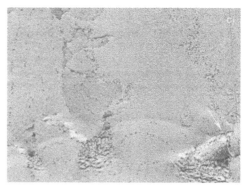

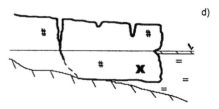

Figure 3. Interferometric products generated from SAR data taken over Prince George Land in 31.12.1996 / 01.01.1997 ($B_\perp$ = 89 m): amplitude image (a), fringe image (b), topogram (c); inferred longitudinal profile of the outlet glacier No.5 (d).

It should be noted, however, that the velocity of glacier No.2 can be underestimated because of the unfavourable glacier orientation with respect to the SAR-range direction (R). A very large crevasse, about 300 m wide, was discovered in the westernmost part of the ice dome by joint interpretation of the topogram and coherence image.

Many outlet glaciers in FJL and SZ are oriented along the SAR-range direction and daily velocities of 36 glacier snouts were determined as given in the Table 1. An application of the transferential approach to the interferogram of glacier No.5 (Fig. 3, b) gives the horizontal velocity of glacier flow as 30 cm/day while the amplitude of tidal motion is given as 15 cm. The velocities of several glaciers marked in bold in Table 2 were measured several times using different interferograms. For example, daily velocities of Impetuous and Milky glaciers were measured with interferograms of 17/18 December 1995 (1 day interval) and 22/25 February 1994 (3 days). The velocity of glacier No.6 in SZ was measured in interferograms of 22/23 March and 18/19 May 1996.

Quite good correspondence between the glacier velocities determined in different tandem interferograms can be noticed, notwithstanding some differences, which can be explained by natural changes in the glacier flow and somewhat different orientation of satellite orbits. Differences between velocities from 1-day and 3-day interferograms reflect the well-known effect of decreasing mean velocities with increasing periods of observation.

The main drawback to the interferometric method of change detection in single SAR interferograms is that it is not suited for the detection and measurement of large glacier changes. Nevertheless, simple differencing, for example, between multitemporal INSAR products such as DEMs or SMs, coherence or amplitude images is the simplest way to detect large glacier changes and is being used operationally by different research groups. Good results in detecting moderate changes in the meter range were obtained by differencing partial phase-gradients $\Delta\varphi_x$ and $\Delta\varphi_y$ from equations (4, 5).

5 ON THE SELECTION OF INSAR DATA

Generally speaking, the quality of ERS-1/2-SAR interferograms cannot be evaluated in advance, i.e. before ordering data, and care must be taken in order to ensure adequate properties of the SAR imagery to be purchased from archives. In our experience, all SAR-images obtained on dates with unreliable and unfavourable weather conditions characterised with low atmospheric pressure, high winds, high temperature, heavy clouds and precipitation were ultimately excluded from consideration. Weather conditions during INSAR surveys were evaluated by using meteorological maps and archival optical images simultaneously taken over the same area by NOAA satellites.

Table 2. Daily velocities of several outlet glaciers in FJL and SZ

ISLAND, LAND	GLACIER NAME OR NO.	VELOCITY, cm/day	DATE
Champ, FJL	No.5 and No.6	32.6 and 21.7	17/18 December 1995
Hall, FJL	Sonklar, No.7, No.17	30.2 , 47.4, 18.9	17/18 December 1995
Payer, FJL	No.2, No.3, No.4, No.7	4.8, 1.4, 33.4, 12.9	17/18 December 1995
La Ronciere, FJL	No.1, No.2, No.3, No.4	11.6, 7.6, 44.0, 18.5	17/18 December 1995
Prince George Land, FJL	No.5, No.18, No.23	29.8, 32.5, 46.9	31/01 December 1996
Salisbury, FJL	Eastern, No.13	18.1 and 34.8	17/18 December 1995
Wiener Neustadt, FJL	No.3 and No.5	5.1 and 21.7	18/19 October 1995
Wilczek Land, FJL	**Impetuous, Milky,** Karo n/s, No.2, No. 9 **No.5, Renown**	**27.8 - 45.1, 15.5-27.8,** 47.5/85.0, 9.4, 36.2 **10.4 - 16.6, 27.9 – 31.4**	Feb. 1994 – Dec. 1995 17/18 December 1995 Feb. 1994 - Oct. 1995
Komsomolets, SZ	No.8, No.13, No.15, **No.16**	43.5, 26.8, 3.6, **29.0 – 30.5**	22/23 March 1996 Mar. 1996 – May 1996
October Revolution, SZ	No.12, No.15, No.22, No.48	8.3, 33.5, 16.7, 13.3	22/23 March 1996

Since cloudless weather in the High Arctic approaches the improbable, we chose dates with "favourable" clouds of forms Ac, Sc, Cu, Cc, Ci, and avoided periods with unfavourable clouds such as Ns, heavy St and especially frontal multilayered cloudiness of forms As/Ns and Cs/As/Ns. This approach to SAR-data selection ensured that nearly 90% of our 27 interferograms showed quite good quality. The best interferograms of the FJL and SZ archipelagos were produced from SAR-images taken in October and in the winter period from December till March.

There is a certain controversy between the demand for good weather conditions (high atmospherical pressure) and the cartographic/ hydrographic requirement for high water levels during surveys because the mean high water line is charted to represent the coastline in topographic maps. According to the universally accepted practice of Arctic surveying, the sea level recorded at the time of survey is usually used as a datum plane. Nevertheless, current sea-level changes must be taken into account and additional tide-gauge data or, at least, data on the moon phase are necessary for the preliminary evaluation of tidal effects. In order to distinguish between the impacts by vertical and horizontal ice motion and to interpret (G)INSAR data reliably, some reference objects with the motion character known *a priori*, e.g. steady objects, areas of fast ice attached to the coastline with prevailing vertical (swaying) motions, outlet glaciers and fast-ice areas in profound bays oriented in the SAR-azimuth direction so that the impact by horizontal motion can be neglected etc., must be identified.

CONCLUSIONS

We will continue our experiments in the Arctic using INSAR data. Some additional, more effective means, e.g. a new technique of *"running fringes"*, are presently being developed by our group for glacier change detection and distinction between vertical and horizontal motions in single SAR-interferograms. Apart from methodological enhancements, our studies provide an essential contribution to the AMETIST *metainformation system* representing actual and potential changes of ice shores in the Eurasian Arctic.

REFERENCES

Brandstätter G. & Sharov A.I. 2000. Metric relations in INSAR topographic modelling. In J.L.Casanova

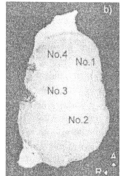

Figure 4. Vostock-1 Ice Dome, La Ronciere Island, FJL on fringe image (a) and coherence image (b), fringe image (c) and topogram (d) generated from SAR data taken in Sept. 3/4 and Dec. 17/18, 1995.

(ed), *Remote Sensing in the 21ˢᵗ Century*: 523-532. Rotterdam: Balkema.

Dowdeswell J. et al. 1994. Evidence for floating ice shelves in Franz Josef Land, Russian High Arctic. *J. Arctic and Alpine Research* 26(1): 86-92.

Goldstein R.M. & Werner C.L. 1998. Radar interferogram filtering for geophysical applications; *Geoph. Res. Letters* 25(21): 4035-4038.

Gonzales R. & Wintz P. 1987. *Digital Image Processing*. Addison – Wesley.

Guneriussen T. et al. 1999. Snow water equivalent of dry snow derived from InSAR – theory and results from ERS tandem SAR data. Presented at the Fringe'99 Conference, Liege, Belgium. http://www.esa.int/fringe99/

Joughin I., Winebrenner D., Fahnestock M. 1996. Measurement of ice-sheet topography using satellite-radar interferometry. *J.Glac.* 42(140): 11-21.

Sharov, A., Raggam, H., Schardt M. 2000. Satellite hydrographic monitoring along the Russian Arctic Coast. *IAPRS* XXXIII(B4): 947-955.

Observing our environment from Space: New solutions for a new millennium, Bégni (Ed.)
© 2002 Swets & Zeitlinger, Lisse, ISBN 90 5809 254 2

Remote sensing of chlorophyll and temperature in coastal upwelling system

J.M.Torres Palenzuela, M.M. Sacau Cuadrado & T.Losada Doval
Universidad de Vigo, Lagoas-Marcosende, Spain

ABSTRACT: The northwest coast of Iberian Peninsula (north of Portugal and Galicia, region of Spain) has a high productivity which is a logical consequence of its western oceanic bound together with the seasonal northward winds, generating strong upwelling of rich nutrient cold waters during the period from May to September. The Atlantic coast of Iberian Peninsula offers an excellent opportunity to study upwelling filaments and their importance in lateral exchanges of nutrient and biota. Simultaneous acquisition of *SeaWIFS* (Sea Viewing Wide Field Sensor) and *AVHRR* (Advanced Very High Resolution Radiometer) NOAA imagery of the north-west of the Iberian Margin were obtained from June to September 1998. Satellite-derived chlorophyll-a from *SeaWiFS* data, which have been provided by the Remote Sensing Group at Plymouth Marine Laboratory and sea surface temperature *(SSI)* from *A VHRR* data were compared in order to find a correlation between sea surface temperature, chlorophyll-a and upwelling indexes which were calculated from geostrophic winds.

1 INTRODUCTION

The north-western coast of the Iberian Peninsula and, in particular, the Atlantic coast, constitutes an important seafood producing area, not only in terms of quality, but also in terms of variety, and abundance of some species.

The morphology of this coastline is highly varied, the most characteristic coastal feature being the *Rias*. The *Rias* are a kind of extended bay that reach into the coast in approximate north-eastern direction.

The high productivity is a logical consequence of its western oceanic bound together with the seasonal northward winds, generating strong upwelling of rich nutrient cold waters during the period from May to October. In opposition to these favourable conditions, during a period in late September, winds change to a southward direction resulting in a down-welling phenomenon which causes upwelling relaxation and advection of warmer surface water towards the coast.

In summer, strong but variable upwelling and associated offshore advective filaments are notable features (Haynes *et al*.., 1993).The eddies and filaments may export large quantities of organic matter produced on the shelf. Filaments tend to recur every year at the same sites (Haynes *et al*.,1993). The filaments extend up to 200 krn offshore. Typically, they first develop in June and occur principally in late July to extend 200-250 Km offshore in September, subsequently disappearing towards October, which ends the upwelling season, some may remain until December. Filaments have potential importance as exporters of coastal waters and its contents to the ocean, carrying a seasonally intense cross-slope flux. This causes a high productivity of phytoplankton, the source of a very productive mussel culture industry. The in-water chlorophyll-*a* concentration is calculated using *Sea Wifs* sensor data.

The cooler upwelled waters (two or three degrees lower than the original surface water) appear as a characteristic band of low temperatures close to the coast. These coastal events can be detected by satellite sensors using thermal infrared bands *(NOAA-AVHRR* data).

2 UPWELLING

Upwelling is the most important physical event on the contribution of basic nutrients for developing of marine life in this area. Wind acts over the sea surface, in the same direction in which it blows. Due to Earth's rotation, the water layers are acted upon by the *Coriolis* force which follows the normal direction of the movement and causes the water currents to deviate 45° *cum sole* from the wind direction at the surface and the angle of deviation increases with increasing depth. The current vectors form a spiral pattern known as *Ekman* spiral (Bakun, A., 1973). In these areas, the increase in nutrients is due to the upwelling of waters from depths of 50 to 200 metres to the surface. The upwelled water is cooler and saltier than the original surface water, and typically has much greater concentration of nutrients such as nitrates, phosphates and silicates that are key to sustaining biological production. It is for this reason that marine ecosystems in the ocean's eastern boundary currents are highly productive, and capable of maintaining large standing crops of plankton and massive fish stocks

3 UPWELLING INDEX

The coastal upwelling indices are calculated based upon *Ekman's theory*. Assuming homogeneity, uniform wind and steady state conditions, the mass transport of the surface water due to wind stress is 90° to the right of the wind direction in the Northern Hemisphere. *Ekman mass transport* is defined as the wind stress divided by the *Coriolis* parameter (a function of the earth's rotation and latitude).

Ekman transports are resolved into components parallel and normal to the local coastline orientation. The magnitude of the offshore component is considered to be an index of the amount of water upwelled from the base of the *Ekman* layer. Positive values are, in general, the result of equatorward wind stress. Negative values imply downwelling, the onshore advection of surface waters accompanied by a downward displacement of water.

Wind speed, V_g, can be obtained equating the magnitudes of *PGF* (pressure gradient force) and *CF* (*Coriolis Force*):

$$V_g = \frac{1}{\rho f} \frac{\Delta p}{\Delta n} \qquad (1)$$

where V_g = the strength of the wind; Δp = pressure difference between two points separated by Δn; ρ = air density (typical value on surface pressure 1.2 kg m^{-3}); and f = *Coriolis* parameter that depends on latitude. The sea-surface stress was computed for each daily wind according to the classical square-law formula:

$$\tau = C_d \rho_a V_g v_g \qquad (2)$$

where C_d = dimensionless "drag coefficient" (Hidy 1972.); ρ_a = air density (about 1.2 kg m^{-3} at mid-latitudes); V_g = wind speed; and v_g = north-south component of wind speed.

The upwelling index is a rough estimate of the water flow upwelled per kilometre of coast. It can be obtained according to the next equation:

$$I_w = \frac{\rho_a C_d V_g}{f \rho_w} v_g \qquad (3)$$

The mass transport has come to be called the *Ekman Transport*, where ρ_a = density of air.

4 NOAA-AVHRR SST DATA

The thermal infrared bands (channels 3, 4 & 5) are used for estimation of sea surface temperatures.

The radiance for each pixel is calculated per channel using the formula:

$$L = S_i C + I_i \qquad (4)$$

Where L = radiance (mW m^{-2} sr^{-1} cm); C = input data value (ranging from 0 to 1023); S_i and I_i are the derived men calibration values.

4.1 Conversion from radiance to brightness temperatures.

Radiance is converted into brightness temperature using the inverse of Planck's radiation equation:

$$T = \frac{c_2 \eta}{\ln\left(1 + \dfrac{c_1 \eta^3}{L}\right)} \qquad (5)$$

where T = brightness temperature (°K); L = radiance from equation (6); η = wavenumber for the AVHRR channel (cm^{-1}); $c_1 = 1.1910659 \times 10^{-5}$ (mW m^{-2} sr^{-1} cm^{-4}); and $c_2 = 1.438833$ (cm °K).

4.2 Conversion from brightness temperatures to sea surface temperatures.

The existing operational algorithms for providing *SST* from *AVHRR* data are based on infrared multichannel methods, employing "empirical" formulas (McClain et al.1986) whose coefficients, a_0, a_1 and a_2, are obtained by linear regressions between pairs of simultaneously measured "in situ" temperatures and satellite radiance.

In absence of clouds and observations at nadir, sea surface temperature may be written as:

$$T_s = a_0 T_i + a_1 (T_i - T_j) + a_2 \qquad (6)$$

where T_i and T_j are brightness temperatures determined from the radiance values in two different infrared window channels i and j.

5 CHLOROPHYLL DATA

Sea-viewing Wide Field-of-view Sensor (*SeaWiFS*), was designed to interpret the bio-optical properties of the surface layers of the ocean by measuring reflectance for specific wavebands in the visible part of the spectrum. A chlorophyll algorithm was assembled from various sources during the *SeaWiFS* Bio-optical Algorithm Mini-Workshop (*SeaBAM*) (McClain, C.R.,1997). The in-water chlorophyll-a concentration is calculated using only a ratio of 490 nm. to 555 nm. (remote sensing reflectance), i.e., Chlorophyll algorithm:

$$Chl = -0.0929 + 10^{(0.2974 - 2.2429\,X + 0.8358\,X2 - 0.0077\,X3)} \qquad (7)$$

with:

$$X = \log_{10}\left(\frac{R_{rs}(490)}{R_{rs}(555)}\right). \qquad (8)$$

and Remote Sensing Reflectance:

$$(Rrs) = nLw / F0 \qquad (9)$$

where *F0* is the extraterrestrial downwelling solar irradiance. Recent comparisons between pairs of *in situ* optical and pigment concentration resulted in a third order polynomial relating chlorophyll-a and the ratio of *SeaWiFS* bands 3 and 5 with very small apparent error or bias (O'Reilly et al.,1998). The use of ratios is advantageous for two major reasons. Firstly, it normalises the data by removing any systematic variation that occurs across all the wavebands being used. This includes the effects of variable solar elevation, although this is not strictly wavelength independent. Secondly, the use of ratios allows the elimination of the scattering from the algorithm. Scattering is a good approximation wavelength independent, and hence, a ratio of the reflectance includes the ratio of two similar quantities, which approximates to unity and the reflectance ratio can subsequently be approximated by the ratio of the absorption.

6 RESULTS

Daily geostrophic winds and upwelling indexes were calculated from 15th June until 15th September (see Fig.-1). It shows clearly that the highest positive indices are reached during the summer months as a result of equator-ward wind stress. There are two big peaks (July & August) separated by a short period of low values

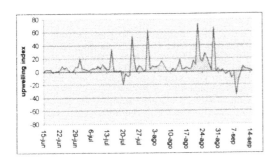

Figure 1. Coastal upwelling index by day.

We have also made use of temperature data from *NOAA-AVHRR* and chlorophyll indexes from *SeaWifs* sensor of following dates: *23rd June; 3rd August and 15th September*.

Images were georeferenced and masked. *SST* and chlorophyll correlation was studied in the upwelling area during the days said previously.

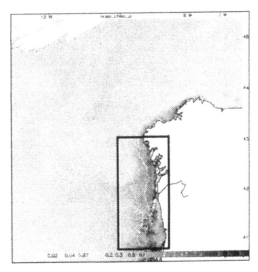

Figure 2. The SeaWifs image shows the upwelling area where the SST and chlorophyll correlation was studied (black box).

23rd June:

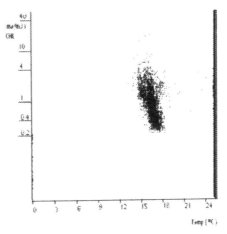

Figure 3. Correlation between the Chl. Index and *SST* data for the 23rd June.

Figure 3 corresponds to *23rd June* (winds in the previous week were weak). It shows a close correlation between temperature and chlorophyll data with values of 15-18°C and 0.2-4 mg/m^3 respectively.

3rd August:

Previous week was characterised by strong north component winds leading again to an increase in the upwelling indexes and a temperature decrease which reach values less than 14 °C.

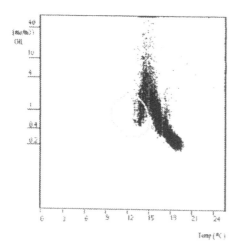

Figure 4. Correlation between the *Chl.* Index and *SST* data for the *3rd August*.

The Minho river is a natural boundary between Portugal and Galicia region (Spain). It shows up like a line in the ground mask of the *SST* or *Chlorophyll* images. The oceanic waters, close the coast in the southern part of the river Minho's mouth, are cold and poor in chlorophyll pigments (see the circle of the Figure 4).

15th September:

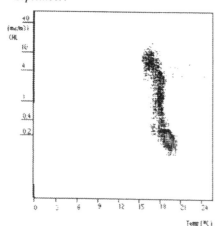

Figure 5. Correlation between the Chl. Index and *SST* data for the *15th September*.

A strong downwelling phenomenon has been detected during the first ten days of September. The previous four days to *15th September*, the upwelling index was positive again. The correlation graph shows a general warming which reaches temperatures values of 20°C (Figure 5).

7 CONCLUSIONS

From the above results we may conclude that:

We were able to establish a relation between upwelling processes and wind patterns using geostrophic winds and upwelling indexes.

The *SST* data from *NOAA-AVHRR* images show a close relation to the upwelling indices obtained from 15th June until 15th September.

The chlorophyll index follows a clear correlation to the *SST* in upwelling water owing to the fact that upwelled water is cooler than the original surface water. It has much greater concentration of nutrients such as nitrates, phosphates and silicates than surrounding waters. This leads to an important biological production.

From the knowledge of geostrophic winds and upwelling indices, we established a relation between upwelling processes and chlorophyll index obtained from *SeaWifs* data.

Similar oscillation patterns between chlorophyll and upwelling indexes were found.

Finally, the temperature with maximum chlorophyll index was between 14.5-15.5°C. This temperature rank corresponds to atlantic water close the coast and bounded to the west by the continental platform.

8 ACKNOWLEDGMENTS

We are grateful to Tim Smyth and Peter Miller from Remote Sensing Group of Plymouth Marine Laboratory for providing us the SeaWifs images used in this work.

9 REFERENCES

Bakun, A., 1973. Coastal upwelling indices, west coast of North America 1946-71. *NOAA Tech. Rep.* NMFS SSFR-671. U.S. Department of Commerce. pp. 103.

Barton, I. J., Zavody, A. M., O'Brien, D. M., Cutten, D. R., Saunders, R. W., and Llewellyn-Jones, D. T., 1989. Theoretical algorithms for satellite-derived sea surface temperatures. *Journal of Geophysical Research*, 94, 3365-3375.

Fiúza, A. F. G., 1992. The measurement of sea surface temperature from satellites. *In Space Oceanography*, edited by A. P. Cracknell (Singapore: World Scientific), pp. 197-279.

Haynes, R., E.D. Barton and Pilling, I., 1993. Development, Persistence, and Variability of Upwelling Filaments off the Atlantic coast of the Iberian Peninsula. *Journal of Geophysical Research*, vol. 98, NO. C12, pp. 22,681-22,692

Hidy, G., M., 1972. A view of recent air –sea interaction research . *Bull Am met Soc* ,53:1083-1102.

McClain, C. R., Chao, S., Atkinson, P. L., Blanton, J. O., and Frederico C., 1986. Win-Driven Upwelling in the Vicinity of Cape Finisterre, Spain. *Journal of Geophysical Research*, 91, 8470-8486.

McClain, C.R., 1997. SeaWiFS Bio-optical Mini-workshop (SeaBAM) Overview.

O'Reilly, J. E., Maritorena, S., Mitchell, B. G., Siegel, D. A., Carder, K. L., Garver, S. A., Kahru, M., and McClain, C. ,1998. Ocean color chlorophyll algorithms for SeaWiFS. . *Journal of Geophysical Research*, 103: 24,937-24,953.

Njoku, E. G., Barnett, T. P., Laurs, R. M., and Vastano, A. C., 1985. Advances in satellite sea surface temperature measurement and oceanographic applications. *Journal of Geophysical Research*, 90, 11,573-11,586.

Tenore K.L., Boyer, L.F., and Cal R.M., 1982. Coastal upwelling in the Rías Bajas, NW Spain: contrasting the benthic regimes of the Rías de Arosa and Muros. *Journal. Marine Research*, 40, (3): 701-772.

Wooster, W., Bakun, A., and McLain D. 1976. The seasonal upwelling cycle along the eastern boundary of the North Atlantic. *Journal Marine Research*, 34 (2): 131-141.

Observing our environment from Space: New solutions for a new millennium, Bégni (Ed.)

Coastal zone hazards in India: Study based on remote sensing and GIS techniques

R.Krishnamoorthy, G.S.Bharathi & P.Periakali
Department of Applied Geology, University of Madras, Chennai, India

S.Ramachandran
Institute for Ocean Management, Anna University, Chennai, India

ABSTRACT: Remote sensing and GIS tools were used to study the coastal zone hazards in India. The impact of anthropogenic and biophysical forces on coastal hazards were studied in detail. The high-resolution optical remote sensing data such as IRS, SPOT and Landsat found to be more useful in deriving both qualitative and quantitative information on coastal hazards with limited ground truth. Multidate satellite data and the GIS tools are more essential in coastal hazards assessment. This study concludes that the coastal pollution is a severe environmental hazard in many parts of India's coastal zone followed by shoreline erosion/accretion, seawater intrusion and land use change due to urban expansion and industrial development. The outcome of this study will form useful information for the preparation of integrated coastal zone management (ICZM) plan for the important "hot spots" in coastal India.

1 INTRODUCTION

India has a coastline of more than 7500 km in length including its island territories. The coastal areas in India face wide range of problems induced by human and natural forces. Both biophysical and human impacts have altered considerably the coastline and also the coastal resources. The major biophysical causes are cyclones, changes in land use and land cover in the adjacent watershed, changes in the flow of freshwater, man-made coastal activities like port development, growth of coastal villages, transport network, etc. A lot of land-based activities are destroying the coastal resources and also maximise the coastal pollution. Coastal pollution is attributable to a number of factors. The worst offender is the untreated industrial effluent. It is estimated that about 8000 industries are discharging waste in to the sea all over coastal India. The most serious problem in the east coast is shrimp farming. The problems due to aquaculture along the coasts include damage to sensitive areas such as mangrove forests, marshes and salination of ground water. Pollution of coastal waters can have great impact on the production of fish since it affects the coastal nursery grounds and other valuable wetland habitats. The mainland coast of India is remarkably unintended and generally emergent. The east and west coasts are markedly different in their geomorphology. The west coast is generally exposed to heavy surf and rocky shores and headlands. The east coast is generally shelving with beaches, lagoons, deltas and marshes. It is also situated relatively low with huge alluvial plains and deltas and endowed with coastal ecosystems like mangroves, coral reefs, sea grasses, salt marshes, etc. (Ramachandran, 1999). The International Ocean Institute has conducted a regional survey that shows that the population pressure has been considered as the most important problem (Rajagopalan, 1996). Two megacities (Mumbai and Kolkata) with a population of 16 million each are located on the coast. Various developmental activities and population pressure have witnessed the degradation of coastal resources like mangroves, coral reefs, seagrasses, etc. In the state of Tamil Nadu, between the years 1986 and 1993, 0.36 sqkm area of mangroves in Pichavaram was lost and nearly 2500 sqkm area of mangroves lost in entire India between the years 1986 and 1994 (Krishnamoorthy, 1995).

2 STUDY OBJECTIVES

In this study the remote sensing data mainly the optical sensors such as SPOT, IRS LISS-I, LISS-II and LISS-III sensors were analysed to study the coastal environmental hazards in selected parts of India (Figure 1). The advantages in using remote sensing data and GIS tools were emphasised based on the experience gained from various site specific studies and the review of literature with particular reference to India's coastal zone.

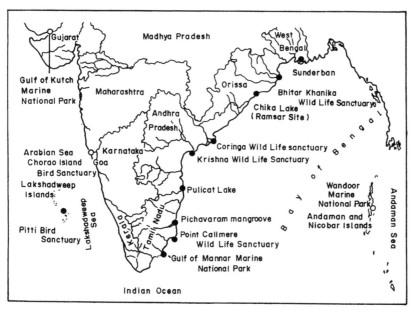

Figure 1. Study area location map showing "hot spots " along coastal India

3 COASTAL ZONE HAZARDS

3.1 *Shoreline changes*

The factors which cause coastal erosion in India are wave breaking, reduction in sediment input to coasts, tectonic upheavals and sea level rise. Coastal erosion is very severe along the west coast of India rather than the east coast. Large-scale erosion was observed along the southern part of west coast especially in Kerala coast, where the coast has been receding at the rate of nearly 5 m per year. The east coast is an emerging coast and only selected parts are undergoing erosion, which is mainly due to coastal developmental activities like construction of port, harbour, etc. Nation wide mapping on shoreline change has been carried out by the Space Applications Centre jointly with the regional centres based on multidate (preferably low tide) imagery on 1:250,000 scale with reasonable positional accuracy. Large-scale changes in the coastline especially along the river and estuarine mouths were observed in site specific studies carried out using high-resolution sensor data of IRS LISS-III. Most of the river and estuarine mouths were partly or almost closed during summer season, which is mainly due to the formation of sand spit. Also most harbours have the problem of sand deposition due to littoral drift. The major coastal mangrove wetlands along the east coast are facing severe consequences due to sand

spit formation and closing of its estuarine mouth. It is very essential to include/consider the coastal geomorphology in every aspect of coastal management plan preparation. Coastal geomorphology is directly influenced by various factors like coastal hydrodynamics, river mechanics, regional tectonics, sea level rise, coastal hydrology, man-made structures i.e. ports, harbours, etc. Local biotic and climatic factors also influence the geomorphology of the coast. IPCC (1992) has reported that a one-meter rise in sea level would inundate about 1,700 sqkm area of coastal zone in India in Orissa and West Bengal, of which 1,500 sqkm is highly productive agricultural land. The coast along Tamil Nadu is most vulnerable to cyclones and strong storm surges, which cause inundation of low-lying coastal areas resulting in damages to crops and property. Cyclonic erosion and accretion are experienced along the entire coast almost every alternate year. Major steps have been put forward by the Erosion Control Directorate of Central Water Commission to prevent the hazards due to coastal erosion and accretion along the entire Indian coast. The road transport and adjoining urban areas were highly affected along the Ennore coast. Based on the analysis of shorelines of 1970 and 1992 it was estimated that about 1.7 sqkm area of coastal zone was eroded in this particular coastal zone. Figure 2 shows a sinking temple during 1995, which is now completely submerged in the sea.

Figure 2. A temple sinking in the Bay of Bengal.

Using 1915 and 1968 topographical maps and 1986 and 1990 satellite imagery, coastal erosion and accretion sites were demarcated using a GIS. There four sites vigorously undergoing coastal erosion in Tamil Nadu are (i) Ennore, (ii) Mahabalipuram, (iii) Rameswaram and (iv) the areas north of Kanyakumari in Tamil Nadu. The coast of Mahabalipuram is UNESCO heritage site, where as the Coleroon and Rameswaram areas have mangroves and coral reefs, respectively. The closure of Coleroon river mouth has witnessed the natural felling of mangroves in Pichavaram since it affects the influx of seawater coming inside the mangrove wetlands. Both erosion and accretion are taking place in the Gulf of Mannar. Many parts of Rameswaram are undergoing severe erosion and accretion. The other islands in Gulf of Mannar are facing the problem of severe accretion. Most of the materials from Rameswaram main island are settling over the coral reef areas around neighboring islands.

Coastal erosion is a calamity because it casuses (i) loss of invaluable land areas and lives (ii) habitation displacement/relocation (iii) loss of beaches (iv) loss of transport network, infrastuctures and other installations. Especially in North Chennai, the shoreline is encroaching rapidly on to the highly populated metropolitan area for a length of about 10 km. One of the important features of coastal accretion is the Marina beach in Chennai coast. It is the most natural and second largest beach in the world, which has a length of about 6 km. The construction of a pier by the British East India Company and construction of an artificial Madras harbour obstructed the littoral drift of Bay of Bengal, which resulted in the erosion of seacoast on the northern side of harbour and accretion of sand on the southern side. The Chennai fishing harbour was constructed in 1975 and the Chennai Port has extended the break-

water and installed a harbour. With the construction of Madras (Chennai) Port in 1875 to 1905, severe modifications of the shoreline occurred. Thereafter, the shoreline has been changing very aggressively, especially to the north of Royapuram. Besides the Ennore Express Highway, about 15 fishing villages adjacent to the shoreline are facing the severe problem of shoreline erosion. In this study, it is estimated that about 77 sqm area of land is sacrificed to the sea every year due to shoreline erosion. Earlier studies carried out for this area also confirm that there is a net loss of around 749 m land eroded in a 10 year period. On the contrary, the southern part of Chennai harbour is showing an accreting (progradation) trend. The area of Marina beach is increasing 40 sq.m every year. This has also resulted in the closure of Adyar and Cooum river mouths. Similarly, the Ennore creek mouth is also getting closed frequently after the construction of Ennore Satellite Port to its north. The closure of Ennore creek mouth requires regular dredging. In order to meet the growing demand of power in India, the Central Electricity Authority has planned to set up a number of coal based thermal power stations in the coastal regions of south India. Under this plan, the Tamil Nadu Electricity Board (TNEB) has proposed to set up a new thermal station at Ennore. The facilities available at Chennai Port to handle the coal for power stations are insufficient. In order to meet the required coal to the thermal power stations, the port facilities have been developed at Paradip (Orissa) as loading port and Ennore (North Chennai) as unloading port. The modern port at Ennore has been operational since January 2001 for handling coal for the exclusive use of TNEB. The satellite port at Ennore is now adding a new dimension to the already existing problems. The northern part of north breakwater is now facing the problem of shoreline erosion and will ultimately affect the Pulicat lake and could even merge this lake with the sea, displacing the entire coastal dunes and beaches. Based on the severity of shoreline erosion and accretion problems, the North Chennai coast has been divided in to four zones as given below.

(i)   zone of severe accretion (Adyar to Cooum)
(ii)   zone of severe erosion (North of Royapuram to south of Ennore creek)
(iii) zone of accretion (Ennore creek mouth)
(iv)   zone of erosion (North of Ennore Satellite Port to Pulicat).

3.2 Coastal pollution

The satellite data by synoptic view provides regional insight on the geomorphological settings of the area. The evolutionary trends of rivers, their migration behavior and nature of sediment discharge in to the coastal waters are well brought out using remote sensing data. The sediment movement and deposition along the coast depends on the concentration of sediments discharged by the rivers and wave dynamics near the coast. Some details are explained in the following paragraphs.

The distribution and transport of suspended sediments in any coastal marine environment is highly complex. Many studies have shown that suspended sediments concentration (SSC) can be detected and mapped using remote sensing data both qualitatively and quantitatively. Remote sensing can provide timely and repeated information concerning suspended sediments and pollution in coastal waters. Since early 1970s, a large number of researchers have used satellite and airborne sensors to estimate and map water quality parameters. With data from Landsat MSS, Nimbus CZCS and NOAA AVHRR it has been possible to study coastal water quality. Regression models and digital chromaticity techniques were developed to study the salinity, chlorophyll-a, turbidity and total suspended solids using Landsat MSS and IRS LISS-II data (Chauhan et al., 1996). The distribution and transport of suspended sediment pattern, distribution and spread has important applications in modifications of harbour channels, dredging operations, beach erosion, water quality, aquaculture site selection, etc. Based on the analysis of Landsat TM and IRS LISS-I data, it was observed that high plumes of suspended sediments are coming from the nearshore wetlands of Cauvery delta and finally moving towards the Jaffna coast (Figure 3). The discharges from salt industries contribute a thick plume of sediments entering in to coastal waters. The density sliced image of LISS-I provides the regional view on qualitative information on SSC in coastal waters. Density slicing of band 4/2 output was given colour code to show the variations in SSC. This qualitative study was undertaken both during monsoon and nonmonsoon seasons for comparison and also to note the variations in concentration, distribution pattern, etc. Heavy discharge of suspended sediments was noted in Pulicat and Vedaranyam nearshore and offshore waters during both monsoon and nonmonsoon. Considerable amount of suspended load from Palk Bay enters in to the Gulf of Mannar and moves further towards Tuticorin harbour and get reduced in its spread beyond this area probably due to the obstruction by the harbour structure. Qualitative mapping has identified that the distribution and transport of SSC converges in to two specific sites along Tamil Nadu coast i.e. Chennai and Tuticorin harbours where the harbour structure obstructs the flow of suspended sediments; also the deposition of sediments finally leads to siltation in harbours. These two sites were chosen for synchronous SSC sample collection for quantitative study at the time of IRS satellite overpass. Digital

chromaticity technique was used to prepare quantative map (Chauhan et al., 1996). The above study shows that the different classes of high to low SSC ranging from less than 5 mg/L in offshore areas to 21 mg/L in near shore of Tuticorin were also delineated. Both principal components analysis (PCA) and convolution filtering enhance the suspended sediments distribution but only the density slicing technique could achieve qualitative variations in concentration. Since several parts of Tamil Nadu coast receive high discharges from coastal wetlands, disposal from industries, etc. there is high concentration of suspended sediments in the nearshore waters. The best spectral wavelength for suspended sediment sensitivity is ~560 nm. IRS band 2 provides best sensitivity and band 4 provides maximum land/water contrast hence the ratio of bands 4/2 provides appreciable amount of information on suspended sediments.

Due to the low spatial resolution (72 m) of LISS-I sensor, it was not possible to carry out qualitative studies without applying digital enhancement techniques. However, IRS LISS-III sensor of 23.5 m resolution has been used to demarcate the spread of fly ash discharges from Ennore thermal plant without applying any digital enhancement. It is also estimated that the fly ash slurry spreads over linearly up to a distance of about 5.5 km inside the Bay of Bengal. The fly ash is continuously deposited in the sea as also the discharge of effluents of certain industries along this coastal belt. Additionally, both east and west coasts receive untreated sewage from the outfalls and untreated/treated industrial effluents.

Sea surface temperature (SST) and turbidity are the other important water quality parameters, which can also be measured by satellite sensors reliably and accurately. The ability to measure "ocean colour," which is translatable in to surface pigment concentrations (chlorophyll) contributed by photosynthetic phytoplankton, has enabled wide-scale mapping of algal blooms i.e. "red tides". Toxic red tides have long been recognised as problems for human health and the fishing industry. Red tides are natural phenomena; they can be stimulated by human activities such as pollution or habitat alteration. There are not many earlier records regarding the observation of red tides in Indian coastal waters. However, the effect of fly ash discharge from thermal power plants has been reported earlier. In this study, the oil slick and sewage disposal in Chennai harbour waters have been identified using SPOT PAN data by applying PC analysis as shown in figure 4. The basic characteristics of enhancement of maximum information from specific band (blue to green region of EMR) by applying PC analysis highly enhance the oil slick signature with tonal variations. However, the LISS-

III sensor data was found to be highly suitable for SSC qualitative studies in coastal waters (Figure 5). Also few enhancement techniques identified as most suitable for water quality studies are given in Table 1.

Table 1. Suitable digital analysis for water quality studies.

Digital Analysis	Water quality parameters
Local optimisation	Chlorophyll-a concentration
Ratio of IRS bands 4/2 and density slicing	Suspended sediments concentration
Resampling and supervised classification	Salinity variations
PC analysis of SPOT	Coastal pollution especially oil slick, fly ash discharge, etc.

3.3 Industrialisation and urban expansion

Both industrialisation and population dynamics are very important driving forces for the land use changes along coastal India. The land use change-detection study shows that there is a tremendous increase in built-up areas along the Chennai coast (Parviz, 1997). Industrialisation has also led to over exploitation of ground water in North Chennai. For example, a potential well field is located around Minjur town; about 7 million gallons per day of groundwater is being pumped. Detailed investigations in this area have revealed that the freshwater–seawater interface, which was originally at 2 km from the sea coast, has moved to about 9 km during the course of 20 years (Raju, 2000). Development of Ennore Satellite Port has also led to the acquisition of a substantial extent of land, affecting land use patterns in its vicinity. A good example of this is the acquisition of salt pans and other vacant lands for developmental purposes. It is expected that port development will bring a substantial change in the land use of the nearby area due to the influx of large-scale commercial operations, which will require development of large storage and transportation facilities (IOM, 2000). The over exploitation of coastal aquifers has already led to the problem of seawater incursion; it has already been observed that the seawater has intruded up to a maximum of 13 km near Kattur in North Chennai.

3.4 Impact of cyclone

The Bay of Bengal coast is more prone to cyclones than the Arabian sea coast. Available data indicates that cyclones are 5 times more frequent in the Bay of Bengal than in the Arabian Sea. Annually, around three cyclones under depressions affect the coast. The tidal waves, heavy rainfall and winds that accompany the cyclones intensify coastal erosion. The coast is more affected during the SW monsoon from June to September. The erosion is reported to be very severe during these months. In the NE monsoon cyclonic winds ranging up to 120 km/hour blow in these region accompanied by heavy rains and waves. During these periods the waves of around 4 m occur and break very near the coast resulting in heavy shoreline erosion.

3.5 Coastal ecosystem degradation

The coastal mangroves are in the process of degradation due to various anthropogenic and biophysical causes. India has a mangrove area of about 4000 sqkm and major mangrove areas occur along the east coast and in the Andaman and Nicobar Islands. All major areas of mangrove development are associated with sources of terrigenous sediment, relatively high freshwater input and protected or rapidly growing shorelines. Relatively small patches of mangroves occur in India where the tidal amplitude is less with reduced supply of freshwater and sediments. Multi-date satellite data have been used to analyse the changes in area of mangrove vegetation cover apart from identifying the degradation areas and assessing its underlying causes. LISS-III sensor data is found to be more useful to demarcate the degradation areas and also the regeneration areas in mangrove wetlands since this sensor has additional band in the short wave infrared region. The other IRS sensors such as LISS-I, LISS-II and PAN are not equivalent to LISS-III in mangrove vegetation studies because of its spatial resolution and spectral bandwidth. The LISS-I sensor data of IRS 1A and 1B are found to be useful for demarcating entire mangrove wetlands including associated mudflats, waterways, etc. due to its coarse resolution (72 m). Digital analysis especially band ratioing of LISS-III bands 3 and 4, is found to be more useful to demarcate mangrove zonations. The advancement of agriculture and aquaculture activities around the mangrove areas could be assessed using this sensor data as shown in figure 6. Mangroves in the southern part of Mahanadhi delta in east coast of India are facing very severe threat due to conversion for agriculture.

The advancement of agriculture activities and the converted mangrove areas clearly appeared on the satellite image. The impact of human activity is more on mangroves in the Andaman and Nicobar islands. Both the discharge of sediments from agricultural lands adjacent to mangroves and growth of coastal villages are found to be major threats for mangroves.

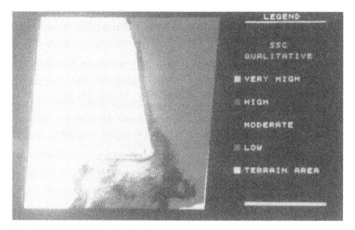

Figure 3. IRS LISS-I density sliced image showing the spread of SSC from Vedaranyam

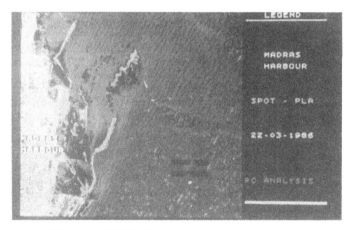

Figure 4. PC analysis of SPOT PAN showing pollution in Chennai harbour waters

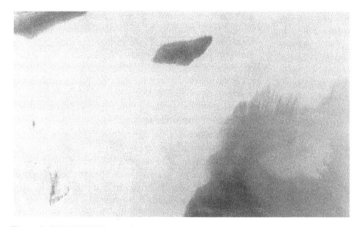

Figure 5. IRS LISS-III image showing plume of sediments coming in to Bay of Bengal

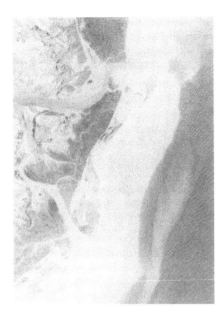

Figure 6. IRS LISS-III shows mangroves and other land use and land cover categories in Godavari delta.

4 DISCUSSION

The health of coastal-terrestrial and marine ecosystems is essential to the maintenance of marine biodiversity and the ability of oceans to provision humanity. The majority of the world's known marine species reside within near shore zones or depend on coastal habitats for part of their life cycles. Most of the world's marine fish catch – an important source of animal protein consumed by humans – is taken from coastal areas. The coastal zone of India is now posed with various problems, mainly from biophysical factors and the human activities. The major biophysical factor is coastal geomorphology. Population pressure and its related causes such as encroachments, conversion of coastal areas for various activities is having high influence on coastal hazards.

Considering the intensity of environmental hazards along Indian coast, all the activities are in one way or other, leading to pollution in the coastal zone and also in coastal waters. Shoreline erosion and accretion also contribute considerably to suspended sediments load in coastal waters. Based on the results obtained from this study, coastal pollution is the major environmental hazard, followed by the natural calamity due to shoreline erosion/accretion. The prioritisation of coastal environmental hazards, based on the review of literature, results obtained from this study and field visits has been given in figure 7.

An increasing body of evidence suggests that in the coming decades global warming due to the greenhouse effect will lead to a substantial rise in sea level. Estimates for the next century range from 0.5 to 2 meters. The expected rise in sea level will raise the water table along the coast and result in increased salination of coastal aquifers. The coastline is changing very dynamically which is having more influence on the coastal ecosystems. This study concludes that the various driving forces for coastal zone hazards could be studied with reasonable accuracy using satellite data particularly optical sensor data. The recent application of satellite technology to the synoptic detection and monitoring of global ocean features has permitted the tracking of large and small-scale surface properties of the coastal ocean waters with exacting detail. Considerable amount of studies carried out to measure "ocean colour", have enabled the mapping of primary productivity in ocean waters and the other physical features (upwelling, squirts, jets, fronts, intrusions, rings, etc.). Earlier, various investigators have used Nimbus-7 CZCS data to map phytoplankton pigment and sea surface temperature (SST). These two ocean parameters are very useful to demarcate Potential Fishery Zone (PFZ).

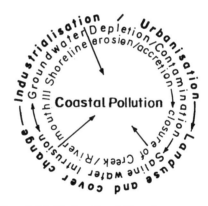

Figure 7. Prioritisation of coastal hazards

In the absence of CZCS, the efforts have largely been focused towards using the SST derived from NOAA-AVHRR data. The Department of Ocean Development (DOD) is funding the national programme on dissemination of PFZ information for the benefit of local fishermen all along the Indian coast through the Department of Space since early 1993. The OCM of IRS-P4 is also being used for extracting physical parameters such as phytoplankton, yellow substance and suspended sediments in ocean waters. The near shore coastal features, especially the distribution and transport of suspended sediments and pollutants, are highly complex. Suspended sediments, which float in coastal waters, are an important parameter that modifies the biological productivity. Increase in suspended sediments concentration will adversely affect photosynthesis and productivity of coastal waters. Also, the sediments load coming from non-point sources will bring nutrients and enrich the biological growth in coastal waters. Therefore, the studies on concentration, spread, and distribution pattern of SSC in coastal waters are more important to understand the coastal water quality. Littoral currents influence the distribution and movement of suspended sediments. In the east coast of Bay of Bengal, the littoral current moves in the clockwise direction for nine months in a year (except NE monsoon period). Remote sensing can provide timely and repeated information concerning suspended sediments and pollution in coastal waters. Since early 1970s, a large number of researchers have used satellite and airborne sensors to estimate and map water quality parameters. Regression models and digital chromaticity techniques were developed to study the salinity, chlorophyll-a, turbidity and SSC using Landsat MSS, IRS LISS-II data (Chauhan et al., 1996). Additionally, the distribution and transport of suspended sediments has important applications in modifications of harbour channels, dredging operations, beach erosion, water quality, etc. The present study reveals that IRS LISS-III digital data is more suitable to derive the above information qualitatively.

5 CONCLUSIONS

Under various national programmes, the entire coastal zone of India has been mapped to estimate its resources, degradation areas, etc. using the multisensor data of IRS, Landsat and SPOT series. Nation wide mapping on shoreline change have been carried out by Space Applications Centre jointly with the regional centres based on multidate (preferably low tide) imagery on 1:250,000 scale with reasonable positional accuracy. The outcome of these national programmes has been found to be more useful to the coastal managers/planners and decision-makers in the country. However, the traditional single sector resource management studies for the ecologically and environmentally "hot spots" along the coastal India are more warranted for ICZM. Hence, in the present study, a clear insight on various environmental hazards is provided along with assessment / prioritisation of the hazards based on their severity.

ACKNOWLEDGEMENTS

The authors are grateful to the Department of Ocean Development, Government of India. The presentation of this paper in the 21st EARSeL Symposium during 14-16 May 2001 is funded by the EARSeL secretariat and the author (R.K) is most thankful to the Organisers and also to the University of Madras for according permission to present this work.

REFERENCES

Chauhan, P., Nayak, S.R., Ramesh, R., Krishnamoorthy, R. and Ramachandran, S. 1996. Remote sensing of suspended sediments along the Tamil Nadu coastal waters, *Indian J. Remote Sensing,* Vol. 24, 105-114.

IOM. 2000. Coastal management plan for North Chennai coast, *ICZOMAT short course report,* Institute for Ocean Management, Anna Univ., Chennai, 22 pp.

IPCC. 1992. Report on global climate change and the rising challenge, *IPCC,* 12 pp.

Krishnamoorthy, R. 1995. Remote sensing of mangrove forest in Tamil Nadu coast, India. *Ph.D. thesis,* Anna University, 202 pp.

Parviz, Z. F. 1997. Digital approaches for change detection in Urban environments using remote sensing data, *Ph.D. thesis,* Anna Univ., Chennai.

Rajagopalan, R. 1996. Coastal zone management, *In: Voice for the oceans,* International Ocean Institute, 36-38.

Raju, M. B. 2000. Groundwater depletion and its impact on coastal zone, *In: Proc. Symp. Management of problems in coastal areas,* OEC, IIT, Chennai, 114 – 125 pp.

Ramachandran, S. 1999. Coastal zone management in India – problems, practice and requirements, *In: Prespectives on integrated coastal zone management,* 211-225.

Observing our environment from Space: New solutions for a new millennium, Bégni (Ed.)
© 2002 Swets & Zeitlinger, Lisse, ISBN 90 5809 254 2

Sea surface temperatures of the northwest coast of the Iberian peninsula using A VHRR data

M.Sacau Cuadrado & J.M.Torres Palenzuela
University of Vigo, Spain

ABSTRACT: The main aim of this work is the study of the physical environment of the system using A VHRR data to look at differences in water temperature and circulation patterns of the Atlantic Ocean northwest of Iberian Peninsula region, in order to identify the coastal upwelling. Daily sequences of thermal images ftom NOM-A VHRR have been studied in order to monitor the upwelling events and find the seasonal variations which are associated with changes in wind regime during 1998. Since winds are usually one of the main causes of upwelling they have been calculated during the previous days in which the SST image was obtained. Results obtained indicate that there is a close relation between upwelling events and meteorological parameters such as geostrophic winds. This relation has been used in order to make a reliable prediction system of upwelling events and toxicity blooms in the northwest coast of the Iberian Peninsula region.

1 INTRODUCTION

The Galician Rias and the northwest coast of Portugal are areas strongly affected by seasonal wind-driven upwelling from March through October (Wooster et al. 1976). The high productivity of these coasts is a consequence of their western oceanic bound together with the seasonal northerly wmds. When strong northerly winds dominate, a north to south current forms parallel to the coast. As a result, upwelling is the main factor responsible for the high primary production, from 700 to 1200 mg C.m^{-2} .day^{-1}, which means the most intensive raft culture of edible mussels in the world (Tenore *et al.*, 1982).

The meteorological conditions of the area of study are related to the position of the Azores anticyclone. When the first low pressure announces the end of summer, winds change from northerly (upwelling-favourable) to southerly (downwelling-favourable). This change in winds causes upwelling relaxation and advection of warmer surface water towards the coast, that coincides with dinoflagellate blooms. (Fraga 1993).

The infra-red sensors measure the radiation emitted from the sea surface in the wavebands 3.5-3.9μm, 10.3-11.3μm, and 11.5-12.5μm. Data obtained from NOAA-AVHRR can be used to detect upwelling areas. Due to the fact that winds are usually one of the main causes of coastal upwelling they have been calculated during the days prior to obtaining the SST image.

2 GEOSTROPHIC WINDS

Winds act on the upper layer of the water column. When this layer (called the surface Ekman layer) is affected by longshore winds with the coast on their left, it is moved offshore due to the Coriolis effect. Therefore, upwelling is caused by surface wind stress in combination with the effect of the Earth's rotation. The strength of upwelling depends on wind characteristics such as speed, duration, fetch, and direction. Since all of these fluctuate throughout the year, upwelling varies with the seasons.

In order to get an idea about the wind patterns we have used sea level pressure charts from which geostrophic wind vectors have been estimated during 1998.

The strength of the wind is determined by the pressure gradient, the reciprocal of the air density and the Coriolis parameter that depends on latitude:

$$V_g = \frac{1}{\rho f} \frac{\Delta p}{\Delta n} \qquad (1)$$

where V_g = the strength of the wind; Δp = pressure difference between two points separated by Δn, ρ = air density (typical value on surface pressure 1.2 kg m^{-3}); and f = Coriolis parameter that depends on latitude.

Vector geostrophic wind velocity can be expressed with two components:

- East- West component:

$$u_g = -\frac{1}{\rho f}\frac{\Delta p}{\Delta y} \qquad (2)$$

- North-South component:

$$v_g = \frac{1}{\rho f}\cdot\frac{\Delta p}{\Delta x} \qquad (3)$$

where Δx and Δy = north-south and east-west distance respectively.

The procedure used in this work was to take the geostrophic wind derived from the pressure chart, multiply that value by 0.7 (to approximate frictional effects), and rotate cyclonically the vector by 15° to estimate the surface wind velocity.

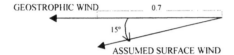

GEOSTROPHIC WIND   0.7

15°

ASSUMED SURFACE WIND

Figure 1. Diagram showing the transformation of the geostrophic wind vector to form an estimate of the wind near the sea surface.

3 EKMAN TRANSPORT

The Ekman model of coastal upwelling intensity, i.e. offshore near-surfat transport, from a knowledge of only the local wind stress Tau (τ).

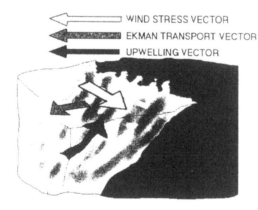

WIND STRESS VECTOR

EKMAN TRANSPORT VECTOR

UPWELLING VECTOR

Figure 2. Coastal upwelling.

Under Ekman's assumptions of steady state motion, uniform wind, and infinite homogeneous ocean, the mass transport per unit width of ocean surface is directed 90 degrees to the right (Northern Hemisphere) of the direction toward which the wind is blowing and is relted to the magnitude of the wind stress by the expression (Wooster *et al.* 1976):

$$M = \frac{\tau}{\rho f} \qquad (4)$$

where M = mass transport resulting from a wind stress, τ; ρ = water density ($\sim$1025 kg m^{-3}); and f = Coriolis parameter, $9.73 \cdot 10^{-5}$ s^{-1} at Latitude 42°.

This mass transport has come to be called the E*kman Transport.*

The sea-surface stress was computed for each daily wind according to the classical square-law formula:

$$\tau = C_d \rho_a V_g v_g \qquad (5)$$

where C_d = dimensionless "drag coefficient". It varies from about 0.001 to 0.0025 depending on the air-sea temperature difference, the water roughness, and on the wind speed itself. A median value of 0.0013 was used for the calculations (Hidy 1972); ρ_a = air density (about 1.2 kg m^{-3} at mid-latitudes); V_g = wind speed; and v_g = north-south component of wind speed.

Thus, Ekman transport was calculated according to the equation (4). The units are m^3 sg^{-1} per 100m of coastline.

Ekman transports are resolved into components that are parallel and normal to the local coastline orientation. The magnitude of the offshore component is considered to be an index of the amount of water upwelled from the base of the Ekman layer. Positive values are, in general, the result of equaterward wind stress. Negative values imply downwelling, the onshore advection of surface waters accompanied by a downward displacement of water.

4 SATELLITE IMAGES

4.1 NOAA-AVHRR SST images

Sea surface temperature (SST) is one of the most widely observed ocean parameters. It is important for quantitative studies of the structure of the ocean and the thermal boundary between the atmosphere and the ocean.

Thus, SST measurement is very important in environmental studies (Njoku *et al.* 1985, Barton *et al.* 1989). It is an indicator of the heat content of the upper ocean and is a tracer of surface current velocities (Fiúza 1992).

The best source of data for calibrating SST is the Advanced Very High Resolution Radiometer (AVHRR) flying on the operational polar-orbiting satellites of the U.S. National and Atmospheric Administration (NOAA).

This scanning radiometer uses 6 detectors that collect different bands of radiation wavelengths as shown in Table 1. It has been in operation since 1978 and has become a world-wide standard source of SST information.

Table 1. AVHRR sensor characteristics.

Channel Number	Resolution at Nadir (km)	Wavelength (μm)	Typical Use
1	1.1	0.58-0.68	Daytime cloud and surface mapping
2	1.1	0.725-1.00	Land-water boundaries
3 A	1.1	1.58-1.64	Snow and ice detection
3 B	1.1	3.55-3.93	Night cloud mapping, SST
4	1.1	10.3-11.3	Night cloud mapping, SST
5	1.1	11.5-12.5	SST

The processing steps involved in this calculation are:
1 Cloud detection and elimination.
2 Geometrical rectification
3 In-flight calibration of AVHRR thermal bands (3,4 & 5) and conversion of satellite received radiances into brightness temperature.
4 Atmospheric corrections and conversion of water-leaving radiance into sea surface temperature.

4.2 SST algorithms

All AVHRR algorithms share the general form described in next equation:

$$SST = aT_i + g(T_i - Tj) + c \qquad (6)$$

where T_i and T_j are brightness temperature measurements in channels i and j, and a and c are constants. The g term is defined as:

$$g = \frac{(1 - t_i)}{(t_i - t_j)} \qquad (7)$$

where t is the transmittance through the atmosphere from the surface to the satellite.

McClain et al. (1985) developed algorithms for SST retrieval based on linear differences in brightness temperatures among AVHRR channels.

Since the atmospheric contributions are different at night from the day time, it is necessary to use different algorithms for night and day.

In general there are three classes of SST algorithms:

– The "SPLIT-WINDOW" algorithm uses channels 4 (11 μm) & 5 (12 μm) brightness temperatures.
– The "DUAL-WINDOW" algorithm uses channels 3 (3.7 μm) & 4 (11 μm) brightness temperatures.
– The "TRIPLE-WINDOW" algorithm uses channels 3 (3.7 μm), 4 (11 μm) & 5 (12 μm) brightness temperatures.

In all the three algorithms the input satellite brightness temperatures are in K while the output SST values are in °C. These algorithms provide SST data with a resolution of 0.1°C.

Sequences of SST images corresponding to the coast of Galicia and northwest Portugal were studied in order to detect upwelling areas during 1998. These areas are easily detected because deeper, cooler water is forced to the surface. Thus, the sea surface water temperature is well below the normal temperature. Infrared satellite data can provide information on the position and strength of the surface temperature gradient associated with the upwelling phenomenon.

5 RESULTS

The following figure displays the coastal indices obtained by day.

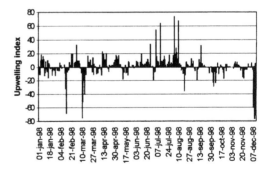

Figure 3. Coastal upwelling indices by day

It shows clearly that the highest positive indices are reached during the summer months as a result of equator-ward wind stress. There are two big peaks (July & August) separated by a short period of low values. Graphs represent the SST values along three longitudinal transects where filaments of cold upwelled water usually appear (Figure 4):

- (1) placed in front of Finisterre coast.
- (2) placed in front of Viana do Castelo coast.
- (3) placed in front of Aveiro coast

Data were represented in 2 axis graphs: X, longitude and Y SST (°C).

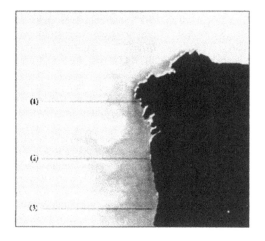

Figure 4. NOAA satellite image showing the three SST longitudinal transects (3rd August, 1998).

5.1 SST longitudinal variations.

5.1.1 8th July 1998.

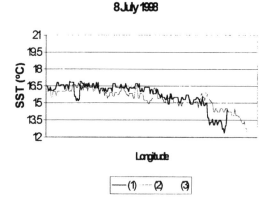

Figure 5. SST longitudinal variation (8th July 1998)

This date corresponds with the beginning of the strong upwelling period. The three longitudinal transects show that temperatures have a decreasing trend toward the shoreline. The temperature differences between the upwelling zone, close to the coast, and the open ocean are about 4.5 (transect 1), 4.4 (transect 2) and 3.3 °C (transect 3).

Transect (3) shows a gap which corresponds with an area covered by clouds.

5.1.2 3rd August 1998.

Figure 6. SST longitudinal variation (3rd August 1998)

Satellite image (Figure 4) and SST calculation show one of the strongest near-shore upwelling events that has been detected during 1998 . Figure 5 shows clearly that there is a marked diminution in temperatures toward the coast. The temperature differences between the upwelling zone and the open ocean is about 6.6, 5.5 and 6.7 °C respectively.

The satellite image depicts three long plumes of cold upwelled water which extend zonally for hundreds of kilometers offshore along the western coast. The most extensive filament corresponds to the one in front of Viana do Castelo coast (Portugal).

5.1.3 7th September 1998.

Figure 7. SST longitudinal variation (7th September 1998)

SST values are very uniform along the three longitudinal transects showing no significant variations between the open ocean and the shoreline. All temperatures obtained range between 19.5 and 21 °C. Figure 7 shows lots of gaps corresponding to

little areas where cloudiness does not allow SST calculation. As a result of the change in the wind regime there is a noticeable increase of the water temperatures along the transects.

6 CONCLUSIONS

From the above results we may conclude that:

1 From the knowledge of geostrophic winds and upwelling indices, we are able to establish a relation between upwelling processes and wind patterns.
2 SST values obtained from NOAA-AVHRR images show a close relation to the upwelling indices obtained from 1998.
3 The beginning of the upwelling period is detected on 8th July. Three upwelled filaments are detected in the satellite image.
4 The strongest near-shore upwelling event is detected on 3rd August, as is clear on the SST longitudinal transects. The temperatures gradient along these three lines is very strong.
5 In September the wind regime changes from northerly to southerly, resulting in a remarkable warming of SST.

REFERENCES

Barton, I. J., Zavody, A. M., O´Brien, D. M., Cutten, D. R., Saunders, R. W., and Llewellyn-Jones, D. T., 1989. Theoretical algorithms for satellite-derived sea surface temperatures. *Journal of Geophysical Research*, 94, 3365-3375.

Fiúza, A. F. G., 1992. The measurement of sea surface temperature from satellites. In Space Oceanography, edited by A. P. Cracknell (Singapore: World Scientific), pp. 197-279.

Fraga, S. And Bakun, A. 1993. Global climate change and harmful algal blooms: The example of *Gymnodinium catenatum* on the Galician coast in: Toxic Phytoplankton Blooms in the Sea. T.J. Smayda and Y. Shimizu [eds]. Elsevier. Amsterdam. pp: 59-65.

Hidy, G., M. 1972. A view of recent air –sea interaction research . *Bull Am met Soc* ,53:1083-1102.

McClain, C. R., Chao, S., Atkinson, P. L., Blanton, J. O., and Frederico C. 1986. Win-Driven Upwelling in the Vicinity of Cape Finisterre, Spain. *Journal of Geophysical Research*, 91, 8470-8486.

Njoku, E. G., Barnett, T. P., Laurs, R. M., and Vastano, A. C., 1985. Advances in satellite sea surface temperature measurement and oceanographic applications. *Journal of Geophysical Research*, 90, 11,573-11,586.

Tenore K.L., Boyer, L.F., and Cal R.M. 1982. Coastal upwelling in the Rías Bajas, NW Spain: contrasting the benthic regimes of the Rías de Arosa and Muros. *J. Mar. Res.* 40, (3): 701-772.

Wooster, W., Bakun, A., and McLain D. 1976. The seasonal upwelling cycle along the eastern boundary of the North Atlantic. *J. Mar. Res.* 34 (2): 131-141.

3   *Fires*

Observing our environment from Space: New solutions for a new millennium, Bégni (Ed.)
© 2002 Swets & Zeitlinger, Lisse, ISBN 90 5809 254 2

A neural network approach for monitoring spatial-temporal behaviour of multiple fire phenomena with NOAA-AVHRR images

K.R.Al-Rawi, J.L.Casanova & El M.Louakfaoui
University of Valladolid, Spain

ABSTRACT: This study focuses on multiple fire phenomena in Valencia, Spain. Monitoring took place for the whole of the fire period from 21 May to 13 July 1994. The temporal and spatial evolution of the fire have been addressed on a daily basis, by means of approaches for mapping burned areas, fire detection, and a state-of-the-art "integrated fire evolution monitoring system"

1 INTRODUCTION

Fire usually goes through three distinct stages, namely, the ignition phase, the propagation phase, and the extinction phase (Clarke *et al.* 1994). The prediction of fire danger is a desirable goal as regards the prevention of fire occurrence. Many studies have addressed this issue (Paltridge and Barber 1988, Lopez *et al.* 1991, Gonzalez-Alonso and Casanova 1997), but we still have some way to go. However, fire detection in the ignition stage or early in the propagation stage will be of great help regarding fire suppression.

This study has been conducted for monitoring the spatial-temporal behaviour of multiple fire occurrences in Valencia, Spain, from 26 May to 13 July 1994. The Burned Area Mapping System (BAMS), Fire Detection System (FDS) (Al-Rawi *et al.* 2000a), as well as the Integrated Fire Evolution Monitoring System (IFEMS) (Al-Rawi *et al.* 2000b) have been employed. These systems are based on a Supervised ART-II artificial neural network (Al-Rawi *et al.* 1999)

2 SUPERVISED ART-II NEURAL NETWORK

2.1 *Architecture of Supervised ART -II*

Supervised ART-II has been built from an input layer, category layer, and memory field (Al-Rawi *et al.* 1999). The number of nodes in the input layer is equal to the number of input features. In case of using complement coding, to avoid the category proliferation problem (Carpenter *et al.* 1997), the number of input nodes should be doubled. The category layer is divided into stacks. The number of stacks is equal to the number of classes. Each stack contains all category nodes that represent its class. The length of the memory field is equal to the number of classes. All input nodes are fully connected to all committed category nodes. The full architecture of Supervised ART-II is shown (fig. 1).

2.2 *Training phase*

For each input pattern, the score for all committed category nodes is computed:

$$T_{j_k k} = \frac{\sum_{i=1}^{2M} (A_i \wedge w_{ij_k k})}{\alpha + \sum_{i=1}^{2M} w_{ij_k k}}; \, j_k = 1 \ldots C(k); k = 1 \ldots L$$

where $w_{ij_k k}$ are the weights, which connect each committed category node j_k in each stack k with input nodes i (i=1...2M), M is the dimension of the normalised input vector A[0,1]. C(k) is the number of committed category nodes in stack number k, L is total number of classes, and α is the choice value parameter ($\alpha > 0$).

If the matching value of the category node that has the highest choice value is greater or equal to the predetermined vigilance parameter ρ :

$$\sum_{i=1}^{2M} (A_i \wedge w_{iJK}) / M \geq \rho$$

and class matching occurs, all weights of the winning node should be trained:

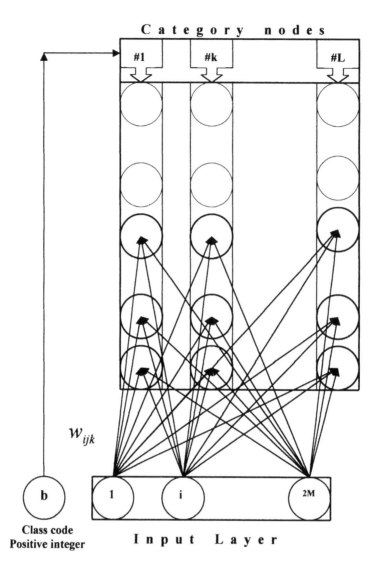

Figure 1: Architecture of Supervised ART-II. Committed category nodes are shown in dark. Uncommitted category nodes are shown in light. Weights are connected to all committed category nodes. Uncommitted category nodes are not connected.

$$w_{IJK}^{new} = \beta(A_i \wedge w_{IJK}^{old}) + (1 - \beta)w_{IJK}^{old} \; ; i = 1...2M$$

Where β [0,1] is the dynamic learning rate. Otherwise, a value of -1 is assigned to the choice value of this category node to put it out of competition. In this case if the matching value of the previously winning node is greater than the vigilance parameter then it is assigned to the vigilance parameter. This has been suggested by (Carpenter *et al.* 1997) in order to make the network able to classify rare events. If non-of the committed category node can represent the current input, a new node should be committed.

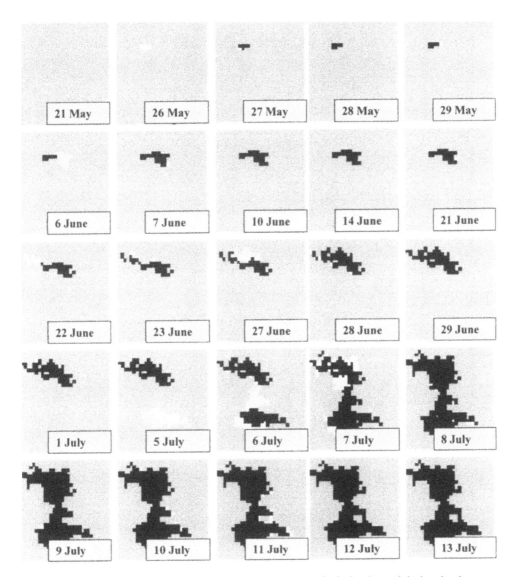

Figure 2: Fire monitoring using BAMS. White represents pixels that burned during the time between the previous image and the current image. Fire has been detected in 18 images. Black represents pixels that burned before the time of the previous image.

2.3 Testing phase

For each input pattern, the committed category node with the highest choice value is determined. The stack number of this node is the class code.

3 DATA

Images of the AVHRR sensor have been employed. The images represent the daily afternoon passing of the satellite, and covered the period from 1 May to 13 July 1994 - a period of 74 days. However, a total of 35 images have been used because other images were either contaminated by clouds or are not included in our archives. Images of all channels have been employed here except channel 5. The images of channel 1 and channel 2 have been used to make up the Normalized Difference Vegetation Index (NDVI) images with a view to monitor the fire using BAMS. The images of both channel 3 and channel 4 have been employed to monitor the fire via FDS.

Figure 3: Fire monitoring using FDS. White represents pixels that are in fire at the time of the image. Fire has been detected in 10 images.

4 METHOD

For monitoring the fire evolution using BAMS, Maximum-Value-Composite (MVC) images from the available NDVI images for a ten-day period have been constructed. The NDVI-MVC image for the period from 1 May to 21 May has been constructed. This represents the NDVI image before the fire occurred. NDVI-MVC images have been constructed for a ten-day period from all available NDVI images, starting with the image of the

corresponding day. The normalised values [0,1] for NDVI images and their complements are incorporated simultaneously into the system. The fire evolution map has been constructed using BAMS.

For monitoring fire evolution using FDS, the normalised values [0,1] for the images of channel 3 and channel 4 for the corresponding day, the MVC for both channel 3 and channel 4 for the period from 1 May to 21 May, together with their complements, have been incorporated into the network. An

98

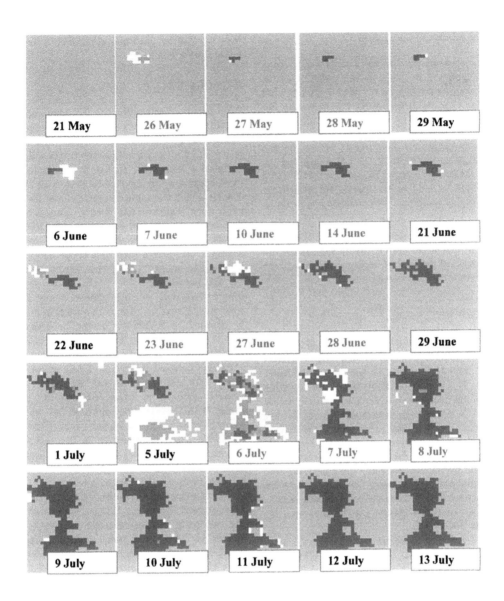

Figure 4. Fire monitoring using IFEMS. Red, black, white, and yellow represent fire front, area that burned completely, at the time of the current image, area that burned between the time of the previous image and the current image, and area that are located beneath the flames, respectively. (Colour plate, see p. 415)

examination has been made of the performance of the FDS related to the BAMS in monitoring the spatial and temporal evolution of the fire on a daily basis.

The IFEMS (Al-Rawi et al. 2000b) which represent the integration of the BAMS and FDS will be employed for fire monitoring.

5 RESULT AND DISCUSSION

The spatial and temporal evolution of the fire for the period from 21 May to 13 July has been presented. The fire was monitored for a total of 25 days during this period: 21,26,27,28,29 May, 6,7, 10, 14, 21, 22,23,27,28,29 June, 1,5,6,7, 8, 9, 10, 11, 12, and 13 July.

5.1 Fire monitoring with RAMS

The evolution of the fire on a daily basis, using BAMS, is shown in figure 2. The evolution of the fire for the first ten images (21 May to 21 June) has been clearly recognised (see the corresponding images in figure 2). The black area represents the pixels burned before the time of the last image. The white area represents those pixels burned for the period between the time of the last image and the time of the current image. The fire was first detected on 26 May. A total of five pixels are seen to be burned in this image (see the corresponding image in figure 2).

The fire has been detected using BAMS for 22 June, 5,6,7 , 8, 9, 10, and 11 July. It detected the fire in 18 images of the total 24 images that represent the duration time of the fire (see figure 2).

5.2 Fire monitoring with FDS

While BAMS detected the fire on images of 27 May, 29 May, 6 June, 7 June, and 21 June, the FDS failed to do so until 22 June (see the corresponding images in figure 3). The reason for this is that fire occurring between the time of two images can be detected by the BAMS. The system detects the trace of the fire rather than the fire itself. Therefore a fire with a short lifetime can be detected by BAMS but cannot be detected by using the FDS. The FDS can detect only those pixels, which are in a fire area at the time the satellite passes over. Using the FDS, we lose fire data when we have missing images or cloudy ones. However, the burned pixels can be recovered by using the HAMS.

The fire was detected one more time on 22 June. The fire is more visible when using the FDS rather than the HAMS, for the images of 22 and 23 June. This is because the first one detects the pixels in a fire as well as the pixels beneath the flames of the fire. That is why FDS is a poor approach as regards drawing up a fire evolution map. The FDS failed to detect fire in the following two images. The fire has been detected using FDS for 29 June, 1,5,6, 8, and 9 July. It detected the fire in only 10 images for the duration of the fire (figure 3).

5.3 Fire monitoring with IFEMS

It is clear that HAMS is more powerful than FDS for monitoring forest fires. However, the exact location of the fire front cannot be determined using the HAMS. Integrating both the HAMS and FDS is recommended in order to locate the fire front at the time the satellite passes over, see (figure4). This has

been reported by (AI-Rawi et al. 2000b) for monitoring fire evolution.

6 DISCUSSION

FDS failed to detect fire on 27 May; however, BAMS detected one pixel that burned completely between 26-27 May (white area), the fire not detected on 28 May. We believe the fire is still on but occupied a very small area or it was beneath the trees, because it was detected by BAMS on 29 May. The large area that burned completely on the image of 6 June supports this theory. This area was detected by BAMS but not detected by FDS. The reason for this is that the burned area mapping approach can detect fires that occur between the time of two images. The system detects the scar of the fire, which lasts for weeks, rather than the infrared radiation, which is emitted during fire life span only. Therefore a fire with a short lifetime can be detected by BAMS but cannot be detected by FDS. FDS can detect an area where there is a fire at the time the satellite passes over.

BAMS detected the fire on 7 July in two spots, while FDS did not. Fire not detected by both systems in images of 10 and 14 June. On 21 June image, BAMS detected two spots of fire, which are located approximately at the same area of the image of 6 June. This indicates that the fire was still on despite the fact that it was not detected in the images of 10 and 14 June. As we explained before it might be very small or it was beneath trees.

A fire detected one day later (22 June). The active fire occupied five pixels. It is located just at one of the two spots that was detected by BAMS a day before. If BAMS had been employed, such fire could have been prevented. The fire was detected by both systems on 23 June. The image of 27 June shows a quite large area that burned completely between 23 and 27 of June. BAMS detected four spots on 28 June. One spot was detected by both systems on 29 June. Two spots in active fires were detected on 1 July. However, a huge fire broke out on 5 and 6 of July.

Hot spots that appear and then disappear before the satellite passes over cannot be detected using FDS. If a hot spot is not observed by FDS this does not mean it has been put out. It might still be burning but to undetected because it might occupy a very small area or it has started in a light fuel area then has moved under dense trees.

More details can be found in (AI-Rawi *et al.* 2001).

7 CONCLUSIONS

Although the performance of BAMS for monitoring forest fires is better than FDS, using IFEMS is preferable since it differentiates between burned area, area burned between two consecutive images, area in active fire, and area beneath flames.

REFERENCES

Al-Rawi, K. R., Gonzalo, C., and Arquero, A. 1999. Supervised ART-II: A new neural network architecture, with quicker learning algorithm, for classifying multi-valued input patterns, In proceedings of the European Symposium on Artificial Neural Network ESANN; Bruges, Belgium, 289-294.

Al-Rawi, K. R., Casanova, J. L., and Calle, A. 2000a. Burned areas mapping system and fire detection system, based on neural networks and NOAA-AVHRR imagery. International Journal of Remote Sensing (in press).

Al-Rawi, K. R., Casanova, J. L., and Romo, A. 2000b. IFEMS: New approach for monitoring wildfire evolution with NOAA-AVHRR imagery. International Journal of Remote Sensing (in press).

Al-Rawi, K. R., Casanova J. L., and Louakfaoui, M., 2001. IFEMS for monitoring spatial-temporal behaviour of multiple fire phenomena, International Journal of Remote Sensing (in press).

Carpenter, G. A., Gaja, M. N., Gapa, S., and Woodcock, C. E., 1997, ART neural networks for remote sensing vegetation classification from Landsat TM and terrain data, *IEEE Transaction on Geoscience and Remote Sensing*, 35, 308-325.

Clarke, K. C., Brass, J. A., Riggan, P. J., 1994, A cellular automation Model of wildfire propagation and extinction, *Photogram- metic Engineering and Remote Sensing*, 60, 1355-1367.

Gonzalez-Alonso, F., Casanova, J. L., 1997, Application of NOAA-AVHRR images for the validation and risk assessment of natural disasters in Spain. *Remote Sensing 96*, Spiteri (Ed.), 1997, Balkema, Rotterdam, 327-333.

Lopez, S., Gonzalez, F ., Llop, R. and Cuevas, M. 1991, An evaluation of the utility of NOAA-AVHRR images for monitoring forest fire risk in Spain. *International Journal of Remote Sensing*, 12, 1841-1851.

Paltridge, G. W., and Barber, J. 1988. Monitoring grassland dryness and fire potential in Australia with NOAA-AVHRR data. *Remote Sensing Environment*, 25, 381-394.

Observing our environment from Space: New solutions for a new millennium, Bégni (Ed.)
© 2002 Swets & Zeitlinger, Lisse, ISBN 90 5809 254 2

Monitoring of active fire by the airborne imaging spectrometer MIVIS

A.Barducci, P.Marcoionni, I.Pippi, M.Poggesi
C.N.R.-I.R.O.E. "Nello Carrara", Firenze, Italy

A.Cavazzini, R.de Paulis & G.Pizzaferri
Compagnia Generale Ripreseaeree, Fontana (Parma) - Italy

ABSTRACT: We have investigated the problem of fire recognition and management assuming that the fire may produce noticeable effects in almost the entire optical spectral range (from the visible until the thermal infrared). Accordingly, we have chosen to exploit the application to the concerned problem of airborne, high resolution hyperspectral sensors that provide the user with a large amount of information about the spectral properties of the imaged target. The main problems addressed by this work are the fire-front and burned-area recognition and the extraction of information useful for the management of the burned area. The paper shows data gathered by the sensor MIVIS over the fire, the data-processing result and is completed with a brief theoretical discussion of the involved topics. Some hints are given about the diagnostic capabilities of other hyperspectral devices.

1 INTRODUCTION

The monitoring and the management of the fire and other natural hazards by means of optical remote sensing instruments is an important topic that in the past has been intensively investigated (e.g.: Muirhead, & Cracknell 1984, Matson, & Holben 1987, Kasischke & Harrell 1993, Cahoon & Stocks 1994, White & Ryan 1996, Pereira & Setzer 1996, Chuvieco 1997, Nakayama & Liew 1999, Michalek & Kasischke 2000, Kant & Badarinath 2000, Chrysoulakis & Cartalis 2000). The option to reliably reveal faint traces of fire and to monitor its growth has remarkable applications for the management of forests and the protection of the environment. The remote sensing application for fire recognition and monitoring purposes should achieve the following aims (see Salvador & Valeriano 2000, Dwyer & Gregoire 2000, and Stroppiana et al. 2000):

1. a measure of the localisation and the geographic extension of the fire-front;
2. an estimate of the fire intensity;
3. the monitoring of the burned area in order to detect traces of latent fire as well as the presence of residual vegetation not entirely burned;
4. the mapping of the burned area for the cartography and the scheduling of the restoration activity.

The recognition and the monitoring the fire locations is often approached by selecting some spectral range (e.g.: the thermal infrared) that is thought to hold useful information for the proposed end (e.g.: Robinson, 1991). Moreover the fire monitoring is often attempted by means of traditional space-borne multispectral sensors (e.g.: the Landsat TM), that have a high probability to grab the fire location but provide a little information amount and coarse spectral resolution (see Pereira & Setzer 1993, Chuvieco 1994, Kasischke & French 1995, Pereira & Setzer 1996, Justice & Scholes 1996, Fernandez & Casanova 1997, Nakayama et al. 1999, Boles & Verbyla 1999, Koutsias & Karteris 2000, Fuller & M. Fulk 2000).

The main drawback of the depicted approach is due to the very coarse resolution and to the very low data amount provided by traditional satellite sensors, which produce only rough estimates of the fire extension and intensity. Moreover the target fire is imaged periodically and monitoring of the evolution of a given fire over time is exceptionally difficult. Due to both the lesser resolution of the collected images and the absence of validation measurements, the utilised space-borne images are not corrected for atmospheric effects and the thermal infrared data as well as the visible and middle infrared data are converted to ground temperature or ground albedo by means of a simple and not effective lookup-table conversion law (e.g.: Chrysoulakis & Cartalis 2000). Alternatively, the target temperature is retrieved using the inverse Planck's law (e.g.: Boles & Verbyla 1999), but always the effects introduced from a varying land surface emissivity are wholly neglected. We point out that neglecting both atmospheric and ground emissivity effects may originates huge errors in temperature estimates (e.g.: Barducci & Pippi 1996), no matter the nature and type of data-processing applied.

We have investigated the problem of fire recognition and management assuming that the fire may produce noticeable effects (information) in almost the entire optical spectral range (from the visible until the thermal infrared), but that these fire effects may be reliably measured only when high spectral, spatial and radiometric resolution are achieved. Accordingly, we have chosen to exploit the application to the concerned problem of an airborne, high resolution hyperspectral sensor (the MIVIS) that provides the user with a large amount of information about the spectral properties of the imaged target.

The main problems addressed by this work are the fire-front and burned-area recognition and the extraction of information useful for the management of the burned area. The paper shows the image-data gathered over a fire occurred in the Northern Italy that was occasionally imaged by the sensor. The data-processing result is discussed and is completed with a brief theoretical discussion of the involved topics. Some hints are given about the diagnostic capabilities of other hyperspectral devices.

It is worth noting that the high resolution images shown along this paper represent a rare scientific event and that very often fires are observed by means of space-borne sensors only.

metrically calibrated data. As can be seen the smoke originated by the fire strongly reduces the fire-front visibility in the entire visible spectral range.

Figure 1. True-colour (Red: 11th channel, Green: 6th channel, Blue: 1st channel) image acquired by the MIVIS over the Alps (Italy) and showing a natural fire affecting a large wood. The fire-front and the burned areas may be identified even in this visible picture, however the smoke spread around heavily dims the visibility and sometimes hinders the fire. (Colour plate, see p. 416)

2 CASE-STUDY

The case-study considered is a natural fire broke out in July 1999 that spread over a forest of the Alps mountains, in the Northern part of Italy. The fire front was imaged in the late morning (10:55 local time) by the MIVIS sensor that was flown aboard of a CASA 212 in the West-East direction from a relative ground height between 1 and 2 km. No ground truth or other ancillary measurement was acquired during the overflight and the image-processing was mainly devoted to investigate the fire effects on the broad spectral range covered by this sensor. The MIVIS is a hyperspectral airborne sensor that operates from 0.44 µm up to 14 µm of wavelength, with a spectral resolution ranging from about 10 nm in the SWIR (1.5 µm of wavelength) until 20 nm in the visible and 350 nm at thermal infrared wavelengths. The MIVIS sampling provides 102 independent spectral channels (measures of monochromatic radiance), digitised with 12 bits of quantization accuracy. The broad spectral coverage, the huge volume of gathered data and the high digitalisation accuracy allowed by the MIVIS, make this sensor a valuable tool for hyperspectral investigation of the Earth surface.

Figure 1 shows a true-colour picture of the observed scene, obtained from the MIVIS radio-

3 IMAGE-PROCESSING: EARLY RESULTS

In this paper we investigate the effects produced by the fire front in the radiation spectra gathered by the MIVIS sensor in the visible, infrared and thermal infrared spectral ranges. The research has been devoted to highlight the different effects the fire produces in different spectral channels, in order to obtain useful information that might be successfully employed for monitoring and detection purposes. We have found that the fire has strong effects at about any optical wavelength, from the visible up to the thermal infrared radiation.

While smoke plumes may be easily seen at a shorter wavelength (e.g.: in the visible, see also Figure 1), the short wave infrared (in the interval 1 µm – 2.5 µm) seems be able to precisely track the fire front. This option is not easily achieved by means of thermal infrared data that due to the high temperatures involved might be saturated even in the neighbourhood of the fire-front. This phenomenon is shown in Figures 2 and 3.

Figure 2 shows the MIVIS image acquired in the 23rd spectral channel (at a wavelength of 1.275 µm) where the fire-front appears bright and well distinguished from the burned region. The thermal infrared range instead shows both the fire-front and the previously burned areas as bright regions, making their recognition complex. We analyse how the fire-front and the burned area appearance change with changing the selected spectral band and we study the relevant information brought by the different channels.

We have found that the H2O (vapour) - CO2 blended absorption feature located around 1.4 μm is a reliable indicator of the fire-front that appears bright and well localised at the related wavelengths. This property, that may be useful for monitoring and management purposes, is probably due to intense black-body emission of radiation from vaporised water and carbon dioxide molecules that have been excited by the energy release caused by the fire.

Figure 2. MIVIS image radiometrically calibrated acquired over the Alps (Italy) in the 23rd spectral channel at 1.275 μm of wavelength. The image clearly shows the fire-front that appears as a bright area while already burned areas are mapped to dark regions. This characteristic may be useful for the discrimination of these two types of pixels.

We were not able to investigate this property for the two other H2O+CO2 blends, located at 1.9 μm and 2.4 μm respectively, due to the low signal-to-noise ratio and to some coherent noise patterns obtained at these wavelengths.

Figure 4 shows one example of processing obtained from the channels 23 (at 1.275 μm) and 27 (at 1.475 μm) of the MIVIS, located respectively in the continuum (left) and on the infrared flank (right) of the absorption band originated from water vapour and carbon dioxide at 1.4 μm.

The image represents the difference between the two concerned channels normalised to the continuum intensity (radiance in the 23rd channel). In a different wording the image-signal represents the partial band-depth computed on its longer wavelength side. Let us note how this simple processing

is able to enhance in the resulting image the burned and the burning areas.

Figure 3. MIVIS image radiometrically calibrated acquired over the Alps (Italy) in the 102nd spectral channel at 12.470 μm of wavelength. The image clearly shows the fire-front and the burned areas as bright regions however, as explained in the text, their difference is subtle.

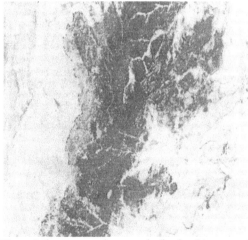

Figure 4. Image processing example (band ratio) computed from the radiometrically calibrated MIVIS data acquired over the Alps (Italy). The imaged area is a large wood affected by natural fire. The image clearly shows the fire-front and the burned areas as dark regions as explained in the text.

A similar processing is shown in Figure 5 that was obtained from a straight band ratio, between the bands 21 and 27 of the MIVIS. The 21st MIVIS channel maps the residual radiation revealed in the centre of the H2O absorption band located at 1.175

µm of wavelength, while the 27th one is on the right flank of the H2O+CO2 blend at 1.4 µm.

Figure 5. Image processing example (21st to 27th band ratio) computed from the radiometrically calibrated MIVIS data acquired over the Alps (Italy). The image clearly shows the fire-front and the burned areas as bright regions. This processing ideally enhance the effects of CO_2 absorption and emission only, as corrected from water-vapour effects.

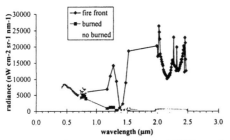

Figure 6. MIVIS spectra collected by the MIVIS over the fire-front, the burned area and the unaffected soil in the wavelength range 0.4 µm – 2.5 µm. The main difference between these spectra is due to the strong black-body emission starting from about 900 nm of wavelength. This strong emission is however modulated by the noticeable extinction produced by the above cool atmospheric (inversion) layer. (Colour plate, see p. 416)

The picture of Figure 5 therefore closely accounts for the extinction/emission balance of the CO2 only. As can be seen the processing enhance both the burned and the burning areas, which appears bright in the obtained image.

The physical meaning of the processing we have executed may also be inferred from the analysis of the radiance spectra registered by the MIVIS on the fire-front, burned areas and the unaffected soil, that are shown in Figures 6 and 7. Figure 6 shows the

typical spectra collected over these kinds of target in the visible and medium infrared spectral range; the thermal infrared spectra are shown in Figure 7. From a rough analysis of these spectra we point out the following points.

1. The difference between the already burned areas and the unaffected soils is easily recognised in the thermal infrared range, where the higher temperature of the burned region is revealed as a brighter image-signal.

2. The fire-front visibility is improved at wave-lengths in the medium IR (1.1 µm – 2.5 µm) where a strong black-body emission is observed. This strong emission, probably related to source temperature approaching 800 – 900K, is modulated by the absorption feature originated by the above atmosphere, which behaves like an inversion layer. We point out that if low-height overflights are executed a clearer black-body line emission should be observed at the wavelengths corresponding to the stronger emission lines of the main atmospheric green-house gases.

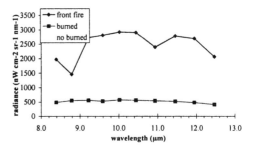

Figure 7. MIVIS spectra collected by the MIVIS over the fire-front, the burned area and the unaffected soil in the thermal infrared spectral range. The main difference between these spectra is due to the strong black-body emission produced by the fire that, however, is only partially appreciated at these wave-lengths. This emission seems to be modulated by two absorption features that could be produced by O_3. (Colour plate, see p. 416)

Another interesting way of viewing the gathered images is to give prominence only to the burning areas. This is reached by applying a tool of stretch – processing upon the image acquired in the 54th channel. The result is shown in Figure 8.

Differently from the other processed images we are able to recognize fires as very bright spots. This particular filter allows us to retrieve some spatial information we had lost with the previous calculation such as isolated small fires. Their identification is very important because they represent dangerous starting point from which new fires may be generated.

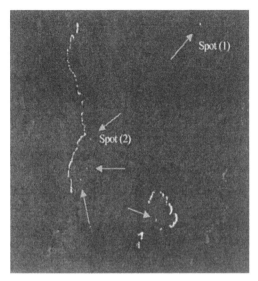

Figure 8. MIVIS image radiometrically calibrated acquired over the Alps (Italy) in the 54th channel at 2.196 μm of wavelength. Together with the fire-front we can recognize isolated small fires as indicated by green arrows.

This assessment is also verified from a spectral point of view. The spectra generated by the pixels labelled as spot (1) and spot (2) confirm our suggestion: they actually represent two fires acting like as "hotbeds" that must be controlled. The results are outlined in the following Figure 9.

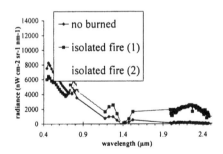

Figure 9. Spectra collected by the MIVIS over the two isolated fire spot indicated in the previous Figure8 as spot (1) and spot (2), in comparison with spectra of an unaffected soil in the visible and near infrared spectral range. (Colour plate, see p. 416)

From the above analysis we are also able to identify the profiles of the fire-fronts under the smoke spread around. The Figure 10 shows how it is relatively easy to recognize the burning areas.

The presence of smoke plumes, formed mainly by aerosol and particles as direct products of the com-

Figure 10. True-colour, (Red: 11th channel, Green: 6th channel, Blue: stretched 54th channel) image acquired over the Alps (Italy). The fire-fronts are now visible even if the smoke spread around is present.

Figure 11(a). True-colour (Red: 11th channel, Green: 6th channel, Blue: 1st channel) image acquired by the MIVIS over the Alps (Italy) and showing a natural fire affecting a large wood. (b) Zoom of the area around the isolated fire. The smoke spread around doesn't affect the fire.

bustion together with a greater concentration of absorbing gases, gives rise to an interesting effect on the acquired radiance from burning areas. The radiance emerging from the isolated fire indicated as Spot (1) in Figure 8 reaches the sensor passing through an atmospheric layer which is not heavily affected by aerosol and smoke as shown in Figure 11. So its spectral content reproduces the Planck distribution of a hot black-body at a temperature between 700K and 900K with a fine modulation due to absorption by gases such as H_2O, N_2O, CH_4, CO_2 and so forth.

The spectra presented in Figure 12 are averaged on four pixels around the isolated fire of Figure 11. As can be seen, the spectral region from 1.9 μm up to 2.5 μm represents the initial tail of the Planck distribution that is brighter than the reflected solar radiance observed in the visible range. Even if smoke plumes are present as shown in Figure 11(b) their geometrical position allows the emerging radiance to reach the sensor. The lower intensity of the solar radiance component reflected by the burning ground may be explained considering that the burning areas have a negligible albedo.

However if we measure the spectra emerging from an isolated fire that is heavily affected by plumes and smokes because it is near other fires more intense, then the situation becomes more interesting. The plumes act as an opaque layer which prevents most of black-body emitted radiance from reaching the sensor but reflects the solar radiance impinging on its upper boundaries. If we assume negligible the radiance emitting from the plumes itself, the net result is a significant reduction of the intensity in the spectral region 1.9µm – 2.5µm due to the black-body emission and a corresponding magnification of the reflected solar radiance.

From a physics point of view the interpretation of the observed effect is easily approached if we take into account the Fresnel equations which assess that at normal incidence for a monochromatic plane wave the reflectance R of a separation boundary between two media with refraction indexes n_i and n_t is:

$$R = \left(\frac{n_i - n_t}{n_i + n_t} \right)^2 \qquad (1)$$

where n_i refers to the medium in which the electromagnetic wave is travelling and n_t refers to the medium on which the wave is impinging. In addition the complex refraction index n (with m and K as the real and imaginary part respectively) is connected to the absorption coefficient $\alpha(\omega)$ according with the Beer's law in this way:

$$n(\omega) \sim m(\omega) - iK(\omega) \qquad (2)$$

$$2\alpha(\omega) = \frac{2K\omega}{c} \qquad (3)$$

where K is defined as the absorbance, ω is the frequency of the considered wave and c is the speed of light. It is necessary to underline these statements are accepted if we suppose to treat the problem of an absorption event with the classical physics.

It should be reasonable to suppose the plumes, generated from the combustion of wood and leaves, as a medium with a great refraction index due to its large absorption coefficient.

In this way from the above analysis the larger is the refraction index, the larger will be the reflectance R; so we would expect the reflected component more intense than that transmitted one.

In Figure 13 we have showed spectra collected over the isolated fire indicates as spot (2) in Figure 8 and averaged upon four pixels around the fire. We have

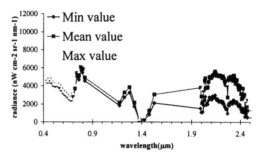

Figure 12. Spectra over the isolated fire spot indicated in the previous Figure8 as spot (1) and averaged upon four pixels around the spot fire. Statistical elaboration is also showed. (Colour plate, see p. 416)

also presented a statistical elaboration about these spectra: minima, maximal and mean values. The comparison between the spectra shown in Figure 12 and Figure 13 confirmed the qualitative theoretical analysis above described.

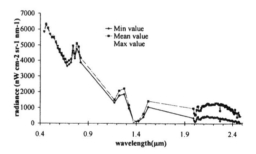

Figure 13. Spectra over the isolated fire spot indicated in the previous Figure8 as spot (2) and averaged upon four pixels around the spot fire. Statistical elaboration is also showed. The solar radiance results more intense than the radiance emitted from fire. (Colour plate, see p. 417)

These results may be predicted with a line by line simulation of radiance emerging from a fire. By means MODTRAN4 we have simulated this event choosing a proper distribution of gases and aerosol as suggested by some scientific reports (National Institute of Standards and Technology 1998). Some of these simulations are summarized in Figure 14. At low height (~50-100m) the radiance reaching the sensor contains only those photons emitted from a hot black body at an estimated temperature of about 800K. Seen that the considered area is burning we would expect that its albedo is rather low causing a small intensity in the visible part of the electromagnetic spectrum.

As the height grows the radiance coming from the ground intercepts atmospheric layers filled with absorbing gases and plumes which reduce the near in-

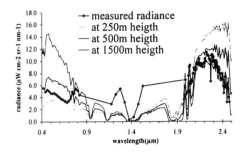

Figure 14. Spectra simulated with MODTRAN 4 of radiance emerging from a 800K fire at different heights. In the graphic it is also depicted the measured radiance from fire spot(1).(Colour plate, see p. 417)

frared radiance and reflect mainly the solar component coming from the upper layers of the atmosphere.

It's worthy to notice that as the photons emerging from fire pass through the upper atmospheric layers they "see" the same gas but with a colder temperature i.e. with the corresponding Planck distribution shifted towards the longer wavelengths. In other words the absorption coefficients of the two gases don't match perfectly and it causes the radiance in the range 2.35 - 2.45 µm to be attenuated more quickly than in the range 2.10 – 2.20 µm.

4 CONCLUSIONS

The problem of fire detection and monitoring has been re-examined in order to investigate the main physical features involved in the radiation-matter interaction for fires and burnt areas and how these features affect the observed data.

Moreover our research was aimed to assess some reliable experimental methodologies which would account for these features and improve the monitoring capabilities.

Our research has employed data gathered by a high resolution hyperspectral sensor (MIVIS) operated on board of a Casa 212 airplane.

From the analysis of these images, referred to a natural fire broke out in July 1999 in the northern part of Italy, we have outlined the effects of smoke plumes on the observed radiance and a practical tool of distinguishing front-fire from isolated fires that represent hot areas from which new fire may be generated.

We have also developed a simple theoretical model in terms of absorbing and reflecting media whose predicted results are in good agreement with the observed spectra.

5 REFERENCES

E. Chuvieco, "A Review of Remote Sensing Methods for the Study of Large Wildland Fires", Alcalá de Henares, Spain: Universidad de Alcalá, 1997.

Y. Kant and K.V.S. Badarinath, "Studies on land surface temperature over heterogeneous areas using AVHRR data", International Journal of Remote Sensing, 2000, Vol. 21, No8, 1749 – 1756.

N. Chrysoulakis and C. Cartalis, "A new approach for the detection of major fires caused by industrial accidents, using NOAA/AVHRR imagery", International Journal of Remote Sensing, 2000, Vol. 21, No8, 1743 – 1748.

J. M. Robinson, "Fire from space: global fire evaluation using infrared remote sensing", International Journal of Remote Sensing, 1991, No12, 3 – 24.

K. Muirhead, and A.P. Cracknell, "Identification of gas flares in the North Sea using satellite data", International Journal of Remote Sensing, 1984, 5, 199-212.

M. Matson, and B. Holben, "Satellite detection of tropical burning in Brazil", International Journal of Remote Sensing, 1987, 8, 509-516.

E. Dwyer and J. M. Gregoire, "Global spatial and temporal distribution of vegetation fire as determined from satellite observations", International Journal of Remote Sensing, 2000, Vol. 21, No6 & 7, 1289 – 1302.

N. Koutsias and M. Karteris, "Burned area mapping using logistic regression modeling of a single post-fire Landsat-5 Thematic Mapper image", International Journal of Remote Sensing, 2000, Vol. 21, No4, 673 – 687.

R. Salvador and J. Valeriano, "A semi-automatic methodology to detect fire scars in shrubs and evergreen forests with Landsat MSS time series", International Journal of Remote Sensing, 2000, Vol. 21, No4, 655 – 671.

A. Barducci and I. Pippi, "Temperature and Emissivity Retrieval from Remotely Sensed Images Using the 'Grey Body Emissivity' Method", IEEE Transactions on Geoscience and Remote Sensing, vol.34, n.3, pp.681-695, 1996.

E. Chuvieco, "Global fire mapping and fire danger estimation using AVHRR images", Photogrammetric Engineering and Remote Sensing, 1994, Vol. 60, 563 – 570.

A. Fernandez and J. M. Casanova, "Automatic mapping of surfaces affected by forest fires in Spain using AVHRR NDVI composite image data", Remote Sensing of Environment, 1997, Vol. 60, 153 – 162.

M. C. Pereira and A. W. Setzer, "Spectral characteristics of fire scars in Landasat-5 TM images of Amazonia", International Journal of Remote Sensing, 1993, Vol. 14, 2061 – 2078.

M. C. Pereira and A. W. Setzer, "Comparison of fire detection in savannas using AVHRR's channel 3 and TM images", International Journal of Remote Sensing, 1996, Vol. 17, 1925 – 1937.

E. S. Kasischke and H. F. Harrell, "Monitoring of wildfires in boreal forests using large area AVHRR NDVI composite image data", Remote Sensing of Environment, 1993, Vol. 45, 61 – 67.

J. L. Michalek and E. S. Kasischke, "Using Landsat TM data to estimate carbon release from burned biomass in an Alaskan spruce forest complex", International Journal of Remote Sensing, 2000, Vol. 21, No2, 323 – 338.

E. S. Kasischke and N. H. French, "Locating and estimating the areal extent of wildfires in Alaskan boreal forests using multiple season AVHRR NDVI composite data, Remote Sensing of Environment, 1995, Vol. 51, 263 – 275.

Cahoon and Stocks, "Satellite analysis of the severe 1987 forest fires in northen China and southeastern Siberia", 1994, Journal of Geophysical Research, Vol. 99, 18627 – 18638.

J. D. White and K. C. Ryan, "Remote sensing of forest fire severity and vegetation recovery", International Journal of Wildland Fire, 1996, Vol. 6, 125 – 136.

D. O. Fuller and M. Fulk, "Comparison of NOAA-AVHRR and DMSP-OLS for operational fire monitoring in Kalimantan Indonesia", International Journal of Remote Sensing, 2000, Vol. 21, No1, 181 – 187.

C. O. Justice and R. J. Scholes, "Satellite remote sensing of fires during the SAFARI campaign using NOAA advanced very high resolution radiometer data", Journal of Geophysical Research, 1996, Vol. 101, 22851 – 23863.

S. H. Boles and D. L. Verbyla, "Effect of scan angle on AVHRR fire detection accuracy in interior Alaska", International Journal of Remote Sensing, 1999, Vol. 20, No17, 3437 – 3443.

M. Nakayama and S. C. Liew, "Contextual algorithm adapted for NOAA-AVHRR fire detection in Indonesia", International Journal of Remote Sensing, 1999, Vol. 20, No17, 3415 – 3421.

United States Department of Commerce, National Institute of Standards and Technology, "Fire Detection Using Reflected Near Infrared radiation and Source Temperature Discrimination", NIST-GCR-98-747.

W. E. Mell, H. R. Baum and K. B. McGrattan, "Simulation of fires with radiative heat transfer", Proceeding for the Second International Conference on Fire Research and Engineering, August 3-8, 1997, Gaithersburg, MD 20899, USA.

Observing our environment from Space: New solutions for a new millennium, Bégni (Ed.)
© 2002 Swets & Zeitlinger, Lisse, ISBN 90 5809 254 2

Relationship between drought indicators based on remote sensing and forest fires incidence in Spain

A. Vázquez, F. González-Alonso, J.M. Cuevas
Laboratorio de Teledetección, CIFOR-INIA, Crta. A Coruña km 7, Madrid, Spain

A. Calle & J.L. Casanova
LATUV, Universidad de Valladolid, Prado de la Magdalena s/n, Valladolid, Spain

ABSTRACT: In this communication we combine different products obtained from the NOAA-AVHRR sensor in order to evaluate the relationships between the anomalies in the NDVI in peninsular Spain and the Balearic islands during the four summer months of the year 2000 and the active "hot spots" detected in the same time period and available in the World Fire Web. The analysis is carried out using the relative proportions of pixels affected and not affected by fire in the several anomaly classes defined for the whole study area and additionally for the different Autonomous Communities. Our starting hypothesis is that forest fires are registered in higher proportions in the negative anomaly classes, that is, with NDVI values lower than the mean for a reference period. The results obtained present a great agreement with this hypothesis, in such a way that the relative proportion of pixels affected by fire is greater in the negative anomaly classes.

1 INTRODUCTION

Forest fires are a recurrent disturbance that conditions, in some cases in a determinant way, the vegetation of a large number of areas (Moreno *et al.*, 1998). Fire incidence over a defined territory depends on multiple factors. Among these factors the moisture content of vegetation could determine the ignition probability and have a strong effect in the fire spread (Burgan *et al.*, 1998). Vegetation status can be assessed, at least in relative terms, by means of the deviations in the values of the Normalized Difference Vegetation Index (NDVI) for a defined time period in relation to mean values obtained from a reference period, that is, the anomalies in the NDVI (López-Soria *et al.*, 1991; González-Alonso *et al.*, 1998). NDVI is very related to several descriptors of the vegetation cover and is considered as a good indicator of the degree of development and the amount of vegetation which frequently depends on the water availability. Additionally, there are several international initiatives, such as the World Fire Web (WFW), that provide the daily location of active hot spots, mainly active forest fires, at a global level based on the information provided by the infrared and visible bands of the NOAA-AVHRR sensor (Stroppiana *et al.*, 2000).

The objective of this work is to combine the two previous information sources and evaluate the relationships between the NDVI anomalies during the summer of 2000 and the "hot-spots" detected in the same time period. NDVI anomalies were calculated from two different AVHRR data sets using unequal reference periods for each one. Additionally we have explored these relationships at a monthly basis and also individually for each of the Autonomous Communities (CCAA) of Spain. The analysis is carried out by means of the relative proportions of pixels affected by fire in the several anomalies classes defined. Our starting hypothesis was that fires should be registered in higher proportions in areas with negative anomalies, that is, with NDVI values lower than the means for the reference period.

2 MATERIAL AND METHODS

2.1 *Study area and fire data*

The study area is peninsular Spain and the Balearic islands. The forest area (with or without a tree canopy) was obtained resampling the CORINE Land Cover data (NATLAN, 2000) at a original resolution of 250 m to a pixel size of 1 km. Out of the 49 original categories we have discarded the ones considered as non forest territory. Those pixels that contain streams or water bodies according to the hydrology vector layer (1:1,000,000) provided by the Instituto Geográfico Nacional were also excluded. The final study area was close to 24 million hectares.

WFW is a system to detect and locate globally the vegetation fires developed by the Space Applications Institute (SAI) at the Joint Research Center. The NOAA-AVHRR images are acquired by a network of reception stations that process the data and provide daily, at each of the network nodes, the coordinates of the "hot spots" detected. Detection is based on a contextual algorithm using the five visible and infrared channels (Flasse & Ceccato, 1996). In this work we have used the daily data available at the Maspalomas node. It is necessary to take into account that the WFW is under development and some of the detected "hot-spot" could not be due to real forest fires (Dwyer *et al.*, 2000). Daily data of "hot-spot" located in a geographic window covering the study area were merged into monthly files. Geographic coordinates were projected to the UTM zone 30 using the European Datum. These locations were rasterized over images with a pixel size of 1 km. The procedure considered as affected by fire the pixels with one or more "hot-spots" detected.

Table 1.- Number of pixels affected by fire from June to September ("hot spots" detected by WFW) in the whole study area (All the CCAA) and in each of the CCAA and percentage of pixels affected in relation to the reference area (forest area). CCAA with less than 15 pixels affected were excluded. The CCAA codes are displayed in Figure 1.

Autonomous Communities		Pixels with fire	
Code	Name	Number	%
	All the CCAA	2.809	1,2%
1	Galicia	408	2,2%
2	Asturias	82	1,3%
3	País Vasco	15	0,4%
4	Navarra	27	0,5%
5	Castilla-León	444	1,0%
6	Aragón	76	0,4%
7	Madrid	59	1,5%
8	Castilla-La Mancha	182	0,5%
9	Extremadura	432	1,7%
10	Cataluña	79	0,5%
11	Valencia	91	0,8%
12	Baleares	23	1,5%
13	Murcia	61	1,7%
14	Andalucía	826	2,1%

2.2 NDVI data and procedures

The NDVI anomalies for the period 1995-98 are based on NOAA-AVHRR data provided by the Deutsches Fernerkundungsdatenzentrum (DLR) and retrieved using the ISIS system (Intelligent Satellite Data Information System). The data used are monthly composites of daily maximum values from June to September (1995-98 and 2000). One of the images was geometrically corrected to topographic maps (39 GCP and RMSE = 0.89) using a nearest neighbour resampling method. The rest of the images were registered to this one. NDVI anomalies were calculated for the four months period and also

for each of the four months separately. The analysis was performed for the whole study area and individually for the different CCAA.

The NDVI anomalies for the period 1993-99 are based on NOAA_AVHRR images received by the LATUV (Laboratorio de Teledetección de la Universidad de Valladolid). Initial data were composites of maximum values for 10-days periods. The methods applied to the calculation of the NDVI anomalies can be found in several publications (e.g. González-Alonso *et al.*, 1998). In the present work, the period of interest started in the last 10-days period of May and ended in the last 10-days period of September. The reference period for the LATUV data was 1993-99, larger that the one used previously with the DLR data.

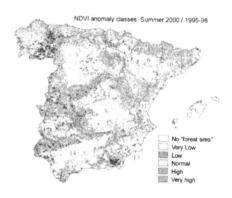

Figure 1.- Map of the NDVI anomalies for the 2000 summer in relation to the mean value for the 1995-98 (June to September) reference period based on NOAA-AVHRR data distributed by DLR. Red colors indicate negative anomalies and green one positives. Black points represent the pixels with "hot spots" based on the WFW (Colour plate, see p. 418)

NDVI anomalies are calculated as a ratio for each pixel, once excluded those with clouds, between the interest and the reference periods. The ratios obtained with both data sets (DLR and LATUV) were reclassified into five anomalies classes. In the class "Normal" are aggregated the pixels with a maximum variation of ± 2.5% between the NDVI values in the period of interest and the mean value for the reference period. Negative anomalies are classified into "Very low" (<82,5%) and "Low" (≥82,5% and <97,5%). Positive anomalies are classified into "Very high" (≥117,5%) and "High" (≥102,5% and <117,5%). These anomalies classes were masked using the forests and rivers layers and later crossed with the forest fires ("hot spots") thematic layers obtained from the WFW data.

3 RESULTS

The number of pixels affected by fire during the summer of 2000 is shown in Table 1 for all the CCAA and for each one. The highest fire incidence was registered in Galicia and Andalucía with a percentage of affected pixels greater than 2%.

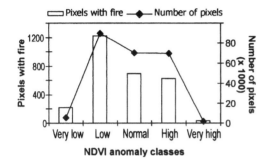

Figure 2. Number of pixels affected by fire and total number of pixels in each of the NDVI anomaly class defined for the summer 2000 vs. 1995-98 (DLR data).

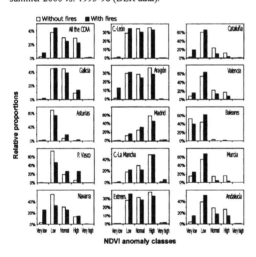

Figure 3.- Relative proportions of pixels affected and not affected by fire in each of the anomaly classes defined for the summer 2000 vs. 1995-98 (DLR data) for the all study area (upper left graph) and for the 14 CCAA listed in Table 1. The sum of proportions is 100 for both categories in each graph

3.1 *NDVI anomalies 2000 vs. 1995-98*

Using the 1995-98 reference period from June to September (Figure 1), most of the pixels have NDVI values lower than the means for the reference period. Negative anomalies, in red in the color plate, dominate in the north, east and other areas of the west of the peninsula. Figure 2 shows the distribution of pixels with fires and the total number of pixels in each of the anomaly classes as displayed in Figure 1.

The upper-left graph of Figure 3 ("All the CCAA") shows, in relative terms, that the proportion of pixels affected by fire is higher than the proportion of not affected pixels in the two negative anomaly classes, and the contrary in the rest of classes.

Looking at the patterns obtained for each CCAA (Figure 3) we can see that in most of them the relative proportion of pixels with fire is higher in the "Low" anomaly class and the contrary in the positive anomalies classes. Nevertheless, some CCAA, mainly those of the northern part of the country, do not follow this pattern. These northern CCAA are shown in the graphs located in the left column of Figure 3.

Trends in the distribution of anomaly classes during the summer (from June to September) show an increase in the area covered by negative anomalies, from 15% of pixels in June to 76% in September (not shown). Figure 4 shows the comparisons, in relative terms, between pixels affected and not affected by fire in each of the anomaly classes. For the months of June and July (Fig. 4A, B), the patterns are the same than the one commented for the four month period. In August (Fig. 4C) the proportion of pixels with fire is higher in the "Very low" class, but also in the two positive anomaly classes. During September (Fig. 4D) it is recovered the expected pattern and the proportion of pixels with fire is a little higher in the negative anomaly classes and the contrary in the rest.

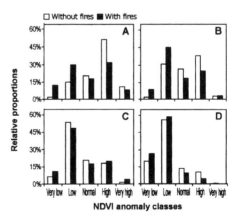

Figure 4. Relative proportions of pixels affected and non affected by fire in each of the anomaly classes defined for the 1995-98 reference period for the four summer months of 2000: (A) June; (B) July; (C) August and (D) September. The sum of proportions is 100 for both categories in each graph.

3.2 *NDVI anomalies 2000 vs. 1993-99*

The period of interest in the calculation of the NDVI anomalies is in this data set very similar to the one analyzed previously, but the reference period is longer and thus, the anomaly classes have a very different distribution (Fig. 5A). Based on these data, most of the pixels of the forest areas have NDVI values higher than the means for the reference period. The analysis of the fire incidence in relation to the anomaly classes (Fig. 5B) shows that the proportion of pixels with fire was higher in the "Low", but also in the "Normal" classes, giving different patterns to the one obtained for the 1995-98 period.

We have performed several independence test (χ^2), between the anomaly classes and pixels affected or not affected by fire, for both data sets (DLR and LATUV). The results allow to say ($P \leq 0.05$) that both variables are not statistically independent. Similar, except in Madrid and Baleares, results were obtained applying this test separately for each of the CCAA.

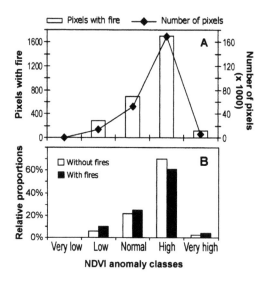

Figure 5. NDVI anomalies for the summer 2000 vs. 1993-99 (LATUV data) against the pixels affected by fires. (A) Number of pixels affected by fire and total number of pixels in each anomaly class and (B) Relative proportions of pixels affected and not affected by fire in each anomaly class. The sum of proportions is 100 for both categories.

4 DISCUSSION

Our starting hypothesis was that fires are registered in higher proportions in negative anomalies areas, that is, with NDVI values lower than the mean for the reference period. The results obtained present a great agreement with this hypothesis, in such a way that the relative proportion of pixels affected by fire is greater in the negative anomaly classes. Nevertheless, the results obtained from both analysis with

different periods of reference, point to the dependence of the anomaly classes to the reference periods. Excluding dry years like 1994 from the calculations performed for the first analysis, leads to important changes in the reference period and therefore in the results.

The fact that the expected pattern was not obtained in the northern CCAA could be related with the relative role of temperature and precipitation in the NDVI evolution. In warmer areas the role of precipitation could be more important than in cooler areas, more related to temperatures (Cihlar et al., 1991). In a regional analysis, Vázquez and Moreno (1993) found that the predictive power of precipitation variables in relation to fire incidence was very low in the cooler and humid northern areas of Spain while the predictive power of temperature variables was very high. Assuming that precipitation patterns control the NDVI evolution in Spain, this can explain why the negative anomalies were not related in the expected way with fire incidence in the northern communities.

Finally, there are several limitations in the fire detection using data from only the AVHRR sensor (Dwyer et al., 2000). AVHRR images are a instantaneous view of the active fires in a given period of 24 hours. Additionally, detected fires can be over- or infra- estimated due to confusion with hot surfaces or reflection over surfaces such as clouds or water. In the present analysis we have tried to minimize such errors using only forest areas and excluding streams and water bodies. In any case, and in spite of the previous limitations, the present validation shows the relevant role that NDVI anomalies have on the spatial distribution of fires and their value as a tool in the system of fight against forest fires.

ACKNOWLEDGEMENTS

This work was possible due to the financial support provided by the SC96_96 and SC50_00 INIA projects. We thank also the SAI and DLR organizations for the facilities in using their processed data

REFERENCES

Burgan, R.E., Klaver, R.W. & Klaver, J.M. 1998. Fuel Models and Fire Potential from Satellite and Surface Observations. *Int. J. Wildland Fire* 8(3): 159-170

Cihlar, J., St.-Laurent, L. & Dyer, J.A. 1991. Relation between the normalized difference vegetation index and ecological variables. *Remote Sens. Environ.* 35: 279-298.

Dwyer, E., Pinnock, S., Grégoire, J.M., & Pereira, J.M.C. 2000. Global spatial and temporal distribution of vegetation fire as determined from satellite observations. *Int. J. Remote Sensing* 21: 1289-1302.

Flasse, S.P. & Ceccato, P.S. 1996. A contextual algorithm for AVHRR fire detection. *Int. J. Remote Sensing*, 17(2): 419-424.

González-Alonso F., Calle, A., Casanova, J.L., Cuevas, J.M. & Vázquez, A.; (1998). Drought monitoring in Spain using NOAA-AVHRR multitemporal images. Proceedings de *ECO BP'98 ISPRS VII Symposium*, Budapest, Sept. 1998, pp. 709-719

López-Soria, S., González-Alonso, F., Llop, R. & Cuevas, J.M. 1991. An evaluation of the utility of NOAA AVHRR images for monitoring forest fire risk in Spain. *Int. J. Remote Sensing* 12(9): 1841-1851.

Moreno, J.M., Vázquez, A., & Vélez, R. 1998. Recent history of forest fires in Spain. In: *Large Forest Fires*, edited by J.M. Moreno (Leiden: Backhuys Publishers), pp. 159-185.

NATLAN, 2000. CORINE Land Cover 250 m. European Environment Agency.

Stroppiana, D., Pinnock S. & Grégoire J.M. 2000. The Global Fire Product: daily fire occurrence from April 1992 to December 1993 derived from NOAA AVHRR data. *Int. J. Remote Sensing* 21(6-7): 1279-1288.

Vázquez, A. & Moreno, J.M.. 1993. Sensitivity of fire occurrence to meteorological variables in Mediterranean and Atlantic areas of Spain. *Landscape and Urban Planning* 24: 129-142.

Observing our environment from Space: New solutions for a new millennium, Bégni (Ed.)
© *2002 Swets & Zeitlinger, Lisse, ISBN 90 5809 254 2*

Use of satellite images in researching the impacts of forest fires.
A case study: Gallipoli National Park

N.Musaoglu
Istanbul Technical University, Civil Engineering Faculty, Remote Sensing Division, Maslak/ Istanbul, Turkey

S.Kaya
Gebze Institute of Technology, Faculty of Engineering, Geodesy and Photogrammetry Engineering Department, Gebze, Kocaeli, Turkey

ABSTRACT: In this study, the forest fire that occurred in 1994 in the Gallipoli Peninsula has been studied through LANDSAT TM satellite images belonging to the dates before and after the fire. Classification algorithms have been applied to satellite images; end images have been modeled three-dimensionally together with the digital elevation model and transferred to a geographical information system. A GIS to be formed for the purpose of preventing a fire in the forests that are among the most important natural resources, measures to be taken and things to be done in the event of a fire and contributions of the satellite images to update such a system have been investigated.

1 INTRODUCTION

Rich natural resources occupy an important place in the economies of the nations. Determining the natural resources and compiling their inventories and planned use of such resources are among the important parameters in protecting the ecological balance of national economy.

Natural catastrophes such as forest fires and floods pose the biggest threat for the process of sustainable growth. As our country is located in the Mediterranean climatic belt, forest fires occurring especially in summer months appear to be a big problem for our country. Determining the aftermath effects of the forest fires, re-planning of the affected area and monitoring the operations after the fire require rather time consuming efforts for large areas. Satellite remote sensing data, acquired before and after the fire, have been used successfully to map burned areas, species affected, and severity levels of damage, as well as monitoring the vegetation regeneration status after the fire (Chuvieco and Congoltan, 1988, Özkan 1998, Koutsias et al., 2000). Remote sensing data offer regular information for fire prevention and damage assessment and provide large scale environmental monitoring and updating capabilities (Musaoğlu 1999, Honikel and Wegmüller, 2000). The field of remote sensing has added greatly to our ability to understand forest ecosystems, with the production of increasingly detailed maps and attribute sets, from the level of the stand to the landscape (Mickelson et al., 1998).

As the areas where forest fires occur are fire-prone locations, application projects are being carried out in the direction of rendering the regenerated forests as fire-resistant. For this purpose, reforestation area is split into pockets of 20-30 hectares considering the topography and dominant wind direction of the fire area. On both sides of the fire prevention roads of 10-15 m wide established on the borders of these pockets of land, green bands of 15-30 meters wide are formed and fire-resistant tree types are planted in these bands. Also the in-stream vegetation is improved, enhancing the fire resistance of the forest by ensuring a mixture of leafed and needled trees, using the natural type of the area in suitable ecological conditions of the stream valleys (AGM, 2001).

In this study, change and impact to the environment were examined for the Gallipoli peninsula, located in western Turkey with extensive forest cover, where a huge forest fire occurred in 1995. For this purpose, LANDSAT TM satellite images of the dates before and after the fire were used. Furthermore, for the purpose of examining the current position, IKONOS satellite images have also been utilized in the selected control areas. In order to study the condition of the area and the changes, various image enhancement and classification algorithms will be applied to satellite images. The result obtained from satellite images will be transferred to a database together with other data such as forest maps, soil maps, population and digital elevation model of the area in order to set up geographic information system. By means of queries to be applied to the geographical information system, land use in the area according to the formation and slope groups will be determined and suggestions will be made about future planning.

2 STUDY AREA

Forests cover around 20.6 m hectares of land in Turkey, constituting 26 % of the country's surface area. Forests of Turkey have a rich mixture in terms of tree types. Most of the forests in Turkey are located along the Black Sea, The Mediterranean Sea and The Aegean Sea.

The Dardanelles connecting the Aegean Sea with the Marmara Sea is one of the major waterways of strategically and geopolitical importance. The Gallipoli Peninsula, on the other hand, is of historic importance not only for Turkey but also for many nations. Having an important history, the Gallipoli Peninsula was declared as a National Park on 02.11.1973.

On 25.07.1994, a forest fire lasting for more than fifty hours broke out in the area covered with thick forests. Burnt down area was 4.049 hectares. According to management plan data, 674 hectares of the burnt down land were spoiled forests and 3375 hectares were productive forests. Of the burnt down forests, an area of 2251.5 hectares covered the area where the Wars of the Dardanelles had taken place, close to the British and Turkish memorial cemeteries. In order to preserve and develop the historical, cultural values as well as the forest and vegetation cover of the Historical National Park of the Gallipoli Peninsula where the naval and land battles of the Dardanelles took place in 1915, and to arrange the rules and regulations governing the administration of the area, for the purpose of making it an international site of peace, a piece of legislation, The Gallipoli Peninsula Historical National Park Act No. 4533 was passed on 20th February 2000. The area affected by the fire is part of The Gallipoli Peninsula Historical National Park which covers 33 000 hectares.

In the area with rugged topography, climate of Marmara Region, which has crossroads climatic, features between the Mediterranean and The Black Sea climates. As general characteristics, springs and autumns are rainy, winters are cold and summers are hot. One of the major factors in the size of the forest fires is the wind. Although daily average windspeed in the area is 4.5 meters per second, the actual windspeed at the time of the fire was 75 km/h per second according to the meteorological records (Hoşgör et al., 1995). Forests of the area consist of mainly Norway spruce, and other leafed types, predominantly Coccifera, as well as brushwood groups. Immediately after the fire, intensive reforestation works were carried out in the area. In this connection needled trees were planted in an area of 2025 hectares and leafed trees were planted in an area of 1421 hectares.

3 DATA USED

3.1 Satellite images

In this study LANDSAT TM satellite images dated 1992 and 1998 were used in order to analyze the situation before and after the fire. IKONOS satellite images, too, have been utilized in selecting a sample area in classification.

3.2 Maps

In order to analyze the topography of the area and to form Digital Elevation model, 11 topographic maps at 1/ 25 000 scale were used. Also the maps showing the boundaries of the fire area were used to determine the fire area.

3.3 Ground truth studies

In analyzing the area, results of the projects carried out in the fire area, personal information and photographs of the area have been utilized (Fig. 1).

4 METHOD

4.1 Rectification

In order to ensure integration of the satellite images with other data groups, all data groups have to be defined within the same co-ordinate group (Musaoğlu, 1999). Especially in the data, which would be evaluated in an information system, it is not possible to jointly evaluate the data, which are defined in different systems.

To be mostly useful to forestry users, any image products must directly integrate with GIS applications that is, they must be georeferenced (Warner et al., 1998, Quackenbush et al., 2000). In geometric conversion of the satellite data generally 1st degree Affin conversion is used and it is found to yield sufficient results (Welch and Usery 1984). Rectification

Figure 1. Burned area from Kanlisirt. (Colour plate, see p. 419)

was made to the satellite images by using 1/25000 and 1/5000 scale maps of the study area in digital media. The images were transformed into the co-ordinate system by using 1st degree Affin transformation and 0.5 pixels RMS.

4.2 *Digital elevation model (DEM)*

As the digital elevation models bear altitude information, they are used in obtaining resolutions, which can not be obtained by two-dimensional analysis. By means of the interpolation methods applied to the altitude values which have been digitized from topographic maps, ground measurements, photogrammetry or remote sensing data and them transferred to computer media; brightness values at raster data at the desired pixel size and radiometric resolution can be obtained from the altitude values. Selection of the data sources, frequency of the points selected on the land and the mathematical method used in conversion are important for the quality and accuracy of the digital elevation model. Therefore, before the digital elevation model is formed, a method has to be established according to the purpose of the study and the desired accuracy (Musaoğlu, 1999).

In this study , digital elevation model was created by digitizing at every 50 meters the contour lines from the topographic maps.

4.3 *Classification*

Different feature types in digital images constitute combinations, which contain different digital values depending on the natural spectral reflection and emission features. The aim of the classification is to group the objects carrying same spectral features. In classifying the images belonging to the area, first of all classes giving high level of accuracy in uncontrolled classification were combined with the results obtained through ground studies. Then, controlled classification was applied to the images. Results of the classification are indicated in Figure 2 and 3.

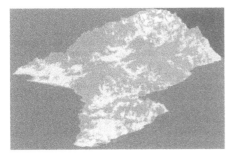

Figure 2. Classification result (1992). (Colour plate, see p. 419)

Figure 3. Classification result (1998). (Colour plate, see p. 419)

4.4 *Geographic information systems (GIS)*

Geographic information systems provide opportunities for creating more dynamic and meaningful analysis for potential users by integrating multiple of spatial information (Green et al., 1994). Interfacing of GIS technology with remote sensing will provide maximum information content and analysis capabilities to the users (Seker et al., 2000).

Fire danger is composed of ecological, human and climatic factors. Therefore, the systematic analysis of the factors including forest characteristics, meteorological status, topographic condition causing forest fire should be made. The relationships between biophysical factors and fire danger are paid more attention to (Huang et al., 2000).

In this study, the layers obtained as a result of classifications were transferred to an information system to evaluate the situation before and after the fire depending on their distribution in terms of the area and elevation (Fig. 4, 5, 6).

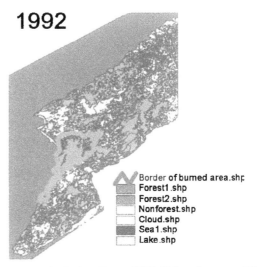

Figure 4. Distribution of layers (1992). (Colour plate, see p. 419)

119

1998

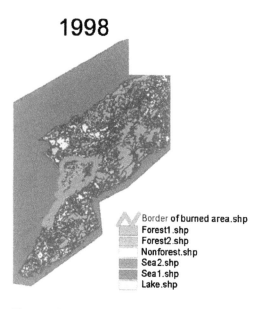

Border of burned area.shp
Forest1.shp
Forest2.shp
Nonforest.shp
Sea2.shp
Sea1.shp
Lake.shp

Figure 5. Distribution of layers (1998). (Colour plate, see p. 419)

1992

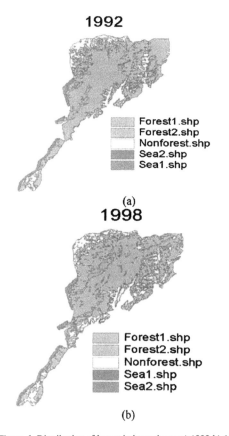

Forest1.shp
Forest2.shp
Nonforest.shp
Sea2.shp
Sea1.shp

(a)

1998

Forest1.shp
Forest2.shp
Nonforest.shp
Sea1.shp
Sea2.shp

(b)

Figure 6. Distribution of layers in burned area a) 1992 b) 1998.
(Colour plate, see p. 420)

5 CONCLUSIONS

As natural catastrophes can not be predicted, carrying out unplanned works in case of a catastrophe leads to many mishaps. A large fire occurring and lasting for a long time in such a historically important area as Gallipoli, which was also, declared a national park is a natural consequence of such an approach. Therefore, in areas with particular risk factor, all parameters that may affect the area have to be ascertained and incorporated into an information system so as to be prepared for potential catastrophes and quick solutions for the problems must be found through a planning to be made.

In making the analysis of the area through satellite images before the fire, area-related data in terms of both land use classes and tree types or density can be elicited. Monitoring and evaluating the works carried out after a fire can also be done through satellite images. Furthermore, as the forest lands cover wide areas, use of satellite images in updating and evaluating the executed works will reduce costs. Besides the results obtained from satellite images; entering parameters such as topographic features, direction of the wind and other climatic characteristics can help to locate the risk-prone areas. Joint evaluation of all these data, establishing information systems which include also the areas likely to be affected by a catastrophe and the measures to be taken as well as extending these works to cover all the risk-prone areas will surely lead to a decrease the frequency and severity of such catastrophes.

ACKNOWLEDGEMENTS

We hereby extend our thanks to Prof. Dr. Doğan Kantarcı, Instructor at Istanbul University, Faculty for his valuable and supportive contributions to this paper in terms of information of data pertaining to the area and to NIK Company who kindly provided IKONOS images.

REFERENCES

Ağaçlandırma Genel Müdürlüğü (AGM), 2001. Orman Yangınları ve Ağaçlandırma.http://www.agm.gov.tr
Chuvieco, E., and Congoltan, R.G., 1988. Mapping and Inventory of Forest Fires From Digital Processing of TM Data, Geocarto International, Vol.4, pp.41-53.
Green, K., Kempka, D., Lackey, L., 1994. Using Remote Sensing to detect and Monitor Land Cover and Land Use Change, PE /RS, Vol.60, No. 3, pp. 331-337.
Honikel M., Wegmüller U., 2000. A Method For the Forest Fire Damage Assessment with SAR and Optical Remote Sensing Data, Fusion on earth Data, Sophia Antipolis, France, 26-28 January.
Hoşgör, S., Özer, A.E., Fener, N., Gül, S, 1995. Gelibolu Yarımadası Tarihi Milli Parkı Ağaçlandırma ve Peyzaj projesi, Marmara Bölge Müdürlüğü,

Huang F., Liu X. N., Yuan, J. G., 2000. Study on forest fire danger model with remote sensing based on GIS, Chinese Geographical Science, Vol. 10, pp. 61-67.

Koutsias, N., Karteris, M., Chuvieco, E., 2000. The Use of HIS Transformation of Landsat TM Data for Burned Land Mapping, PE & RS, Vol. 66, No. 7,pp. 829-839.

Mickelson, J. G., Civco, D. L., Silander, J. A., 1998. Delineating Forest Canopy Species in the Northeastern United States Using Multi-Temporal TM Imagery, PE&RS, Vol. 64, No. 9, pp. 891-904.

Özkan, C., 1998. Uzaktan Algılama Verileriyle Orman Yangını Analizi, Yüksek Lisans Tezi, İTÜ Fen Bilimleri Enstitüsü, İstanbul.

Quackenbush, L. J., Hopkins, P. F., Kinn, G., 2000. Developing Forestry Products From High Resolution Digital Aerial Imagery, PE & RS, Vol. 66, No.11, pp. 1337-1346.

Şeker D. Z., Musaoğlu, N., Kaya Ş., 2000. Investigation of Vegetation in Turkey by Using Remote Sensing Data and GIS, *XIXth Congress of ISPRS*, 16 - 23 July, Amsterdam, Holland, Proceedings pp. 1357-1363.

Warner, T. A., J. Y. Lee, and J. B. Mc Graw, 1998. Delineation and Identification of Individual trees in the eastern deciduous forest, Proceedings of the International Forum on Automated Interpretation of High Spatial Resolution Digital Imagery for Forestry, 10-12 February 1998, Victoria, British Columbia, Canada, pp. 81-91.

Welch, R., Usery, E.L., 1984. Cartographic Accuracy of Landsat 4 MSS and TM Image Data, *IEEE Transactions on Geoscience and Remote Sensing*, Vol.GE -22, No: 3.

4   *Atmospheric Risks*

Observing our environment from Space: New solutions for a new millennium, Bégni (Ed.)
© *2002 Swets & Zeitlinger, Lisse, ISBN 90 5809 254 2*

Evaluation of "Mitch" hurricane damages in Central America using satellite imagery

P.Romeijn, F.Yakam Simen, E.Nezry & I.Supit
PRIVATEERS N.V. Private Experts in Remote Sensing, Philipsburg, Netherlands Antilles

ABSTRACT: The states of Honduras, Nicaragua, and El Salvador have been toughly hit by the "Mitch" hurricane in the first days of November 1998. The extent of damages due to this hurricane, as well as their impact upon local economy were exceptional. In this framework, a remote sensing project was immediately initiated to provide a large scale evaluation of the damages suffered by the three countries. To reach this objective, new remote sensing products called ™DYNAMIC-RADAR and ™DYNAMIC-OPTICAL products have been designed. These products are, based on using, either radar satellite images, or optical/radar satellite data fusion. The most efficient techniques have been applied to produce these products, thus enabling change detection at a very fine spatial scale (from 20 meters x 20 meters to 10 meters x 10 meters). Project schedule and operations, as well as the validation of its products are reported in this article.

1 INTRODUCTION

In the first days of November 1998, Central America was devastated by the hurricane "Mitch". Although the three states of El Salvador, Honduras and Nicaragua had already often experienced hurricanes in the past, they had never before suffered such total devastation combining wide area floodings, huge forest cover destruction, important landslides, and large scale destruction of housing and infrastructures.

To answer the public call for assistance expressed by these three countries, a joint project was put up by PRIVATEERS N.V., SpotImage and the European Space Agency (ESA) in order to carry out a wide area / high spatial resolution evaluation of the damages caused by the hurricane, using optical (presently, SPOT) and radar (presently, ERS) satellite imagery.

The aim of the project was to deliver as quickly as possible accurate high spatial resolution digital and printed maps of the damages that could be used as straightforward as possible by local photo-interpreters.

2 TECHNICAL IMPLEMENTATION

To reach this goal, the following data processing procedure was applied:
1) The ERS SAR (Synthetic Aperture Radar) data were used to detect the damages caused by the hurricane. First, calibration (Laur *et al.* 1998) of the whole ERS data set has been done to allow further radiometry-based comparison and change detection. In the change detection process, the ERS-1 SAR archive data acquired before "Mitch" was compared to the ERS-2 SAR data acquired just after "Mitch".
2) Combined speckle filtering and super-resolution techniques developed by PRIVATEERS N.V. (Nezry & Yakam Simen 1999) have been applied to the ERS SAR data, to detect and map existing targets, before and after "Mitch", with particular emphasis to housing and infrastructures, at a spatial resolution of 10 meters x 10 meters (SPOT Panchromatic resolution).
3) Change detection techniques appropriate to SAR images have been applied to detect and map the damages.
4) SPOT panchromatic archive images or speckle filtered ERS Synthetic Aperture Radar (SAR) images were used as mapping background.
5) Finally, two sets of cartographic (Universal Transverse Mercator projection) maps of before/after "Mitch" changes have been produced, on 10x10 meters (SPOT/ERS) and 20 meters x 20 meters (ERS only) grids, respectively.
6) A documentation describing the photo-interpretation keys and procedures for the end-users in Central America was issued and distributed.

Since change detection is carried out on the bi- or multitemporal set of RADAR images, the radiometric resolution of the RADAR images used must be of very high quality. Since the radiometric resolution of RADAR images is naturally corrupted by the presence of speckle, speckle filtering (a too seldom mastered technique !) is the most critical issue for the success of the whole operation.

The filter that has been used in this project is an adaptive speckle filter for multi-channel detected SAR images, recently developed by PRIVATEERS N.V.: the Distribution-Entropy Maximum A Posteriori (DE-MAP) filter (Nezry & Yakam Simen 1999).

The superiority of Bayesian speckle filters is mainly due to the introduction of A Priori scene knowledge in the filtering process. Nevertheless, in the presence of very strong texture or of mixed textures, as it is often the case in SAR images of dense tropical forest, and/or in the presence of relief (which is the case in Honduras, Nicaragua and El Salvador), it may be hazardous to make an assumption about the probability density function of the radar reflectivity of the scene. In this context, the A Priori knowledge with regard to the observed scene can hardly be an analytical first order statistical model. However, in the DE-MAP speckle filters, a Maximum Entropy constraint on texture is introduced as A Priori knowledge regarding the imaged scene.

The new DE-MAP filter is particularly efficient to reduce speckle noise, while preserving textural properties and spatial resolution, especially in strongly textured SAR images (Nezry & Yakam Simen 1999). It adapts to a much larger range of textures than the previous MAP filters (Lopes *et al.* 1993, Nezry *et al.* 1996) developed under the assumption of K-distributed SAR intensity (Lopes *et al.* 1993).

From the theoretical point of view, it is noteworthy that this filter:
- presents the very attractive properties of a control system (Nezry *et al.* 1997);
- allows to super-resolve SAR images (Nezry & Yakam Simen 1999, Nezry *et al.* 1997), which is of high particular interest for ERS (PRI: pixel size: 12.5 meters x 12.5 meters, spatial resolution 22 meters x 25 meters) and SPOT panchromatic (pixel size: 10 meters x 10 meters) data fusion.

3 PROJECT SCHEDULE

The SPOT and ERS satellite images were made available by SpotImage and ESA/Eurimage respectively in the course of November and December 1998. Altogether, these data allowed to study more than 120.000 km² in the most devastated areas of Honduras, Nicaragua and El Salvador. PRIVATEERS N.V. was in charge of project implementation and production of the value added products.

The first ERS SAR archive images have been received by PRIVATEERS N.V. on November 17th, 1998. Before the end of December 1998, all the ERS (19 archive images, and 7 recent images acquired in November 1998) and SPOT (11 archive images, and 2 recent images acquired in November 1998) images had been produced and delivered to PRIVATEERS N.V. by SpotImage and ESA. The whole ERS / SPOT coverage used in this project is shown in Figure 1.

All ERS-based value-added cartographic products and part of the SPOT/ERS-based value-added cartographic products have been delivered by PRIVATEERS N.V. already during the month of December 1998. All the cartographic high spatial resolution value-added products had been delivered in the first days of 1999.

This efficiency in managing and operating the project enabled rapid delivery of these results to the United Nations for validation planned in January/February 1999 in El Salvador. Validation was also carried out in Nicaragua by SpotImage in January/February 1999 and by PRIVATEERS N.V. in Honduras in March 1999.

4 DETECTION OF THE DAMAGES USING SAR DATA ONLY

For every ERS frame, a set of products called ™"DYNAMIC-RADAR" is produced to enable the detection and the photo-interpretation of the changes and damages caused by the "Mitch" hurricane. The ™DYNAMIC-RADAR products are cartographic products (Universal Transverse Mercator projection, georeference WGS 1984). The sampling rate adopted in the present project is 20 meters x 20 meters.

The ™DYNAMIC-RADAR products are defined as follows:
1. A multi-date composite black and white image called "CARTO". This image is formed by averaging the calibrated and speckle filtered ERS archive data acquired before the "Mitch" hurricane.
2. A black and white image file illustrating the change in radar backscatter, and called "RATIO". This image is produced by comparison between the calibrated and speckle filtered ERS image acquired after the "Mitch" hurricane and the CARTO product.
3. A 3 channels color image file called ™DYNAMIC-RADAR.

A ™DYNAMIC-RADAR product combines:
- the CARTO image (black and white), which is the background radar map.
- the positive changes in radar backscattering (encoded in red) build using the RATIO image.
- the negative changes in radar backscattering (encoded in blue) build using the RATIO image.

The ™DYNAMIC-RADAR product results from the color composition of the CARTO image and of the two masks of temporal changes, in green, red and blue, respectively.

Figure 1: The whole ERS and SPOT coverage (from the Atlantic Ocean coast in the North, to the Pacific Ocean coast in the South) used in the "Mitch" project: cartographic (UTM projected) mosaic of the [TM]DYNAMIC RADAR and [TM]DYNAMIC-OPTICAL products (original in color).

Regarding the photo-interpretation keys, in the [TM]DYNAMIC-RADAR product, red tones denote in most cases, either the areas where the vegetation has suffered heavy damages, or destroyed infrastructures. Blue tones denote in most cases flooded areas, or water saturated areas (see Figure 2).

5 DETECTION OF THE DAMAGES USING SAR AND OPTICAL DATA

For every SPOT panchromatic frame (SPOT-P or SPOT-M), a set of products called [TM]"DYNAMIC-OPTICAL" is produced to enable the detection and the photo-interpretation of the changes and damages caused by the "Mitch" hurricane. The [TM]DYNAMIC-

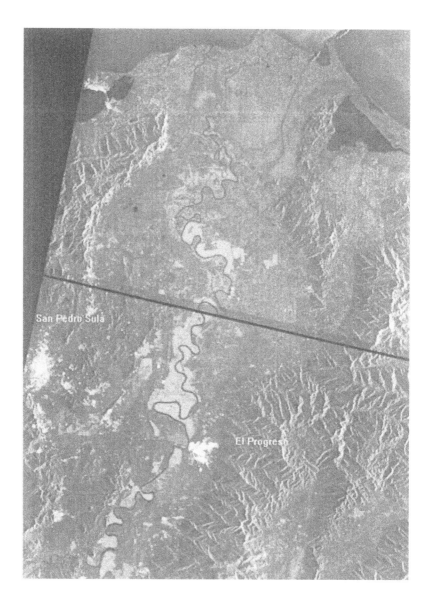

Figure 2: ™DYNAMIC-RADAR product: Provinces of Cortès, Yoro, Atlantida in Honduras.
The main banana plantations area in Honduras, near San Pedro Sula, the second city in Honduras. Situation on November 5, 1998.
The area represented is 56 x 77 kilometers, in UTM cartographic projection.
In blue, the flooded areas. In red, the damages to vegetation (original in color).

OPTICAL products are cartographic products (Universal Transverse Mercator projection, georeference WGS 1984). Their sampling rate is 10 meters x 10 meters.

The ™DYNAMIC-OPTICAL products are designed as follows:
1. A black and white image called "CARTO". This image is a panchromatic image acquired before

(archive data) or after (recent acquisitions) the "Mitch" hurricane.
2. A black and white image file illustrating the change in radar backscatter, and called "RATIO" (same as for the ™DYNAMIC-RADAR products). This image is produced by comparison between the calibrated and speckle filtered SAR image acquired after the "Mitch" hurricane and the

Figure 3: ᵀᴹDYNAMIC-SPOT product: Pacific coast (Honduras/Nicaragua)
The area represented is 51 x 52 kilometers, in UTM cartographic projection.
In Blue: Floodings of Rios Choluteca and Rio Negro, Gulf of Fonseca, Honduras.
In Red: damages to pisciculture and agriculture.
(original in color)

calibrated and speckle filtered SAR image acquired before the "Mitch" hurricane.
3. a 3 channels image file called ᵀᴹDYNAMIC-OPTICAL.

Thus, the ᵀᴹDYNAMIC-OPTICAL product combines:
- the panchromatic image (black and white), which is the background map.
- a mask of the positive changes in radar backscattering (encoded in red) build using the RATIO image.
- a mask of the negative changes in radar backscattering (encoded in blue) build using the RATIO image.

The ᵀᴹDYNAMIC-OPTICAL product results from the addition of the panchromatic (CARTO) optical image and of the two masks of temporal changes, in green, red and blue, respectively (see Figure 3).

It is therefore the result of a common-sense data fusion process integrating SAR change detection and the panchromatic data to facilitate intuitive but correct photo-interpretation of the final product by non-remote sensing specialists.

Regarding the photo-interpretation keys in ᵀᴹDYNAMIC-OPTICAL products, they are similar to those of the ᵀᴹDYNAMIC RADAR products.

6 VALIDATION OF RESULTS

6.1 *Nicaragua*

The mission of the SpotImage team in Nicaragua in January/February 1999 was the first opportunity to confront the damage identification and evaluation products to ground reality.

After SpotImage team members taking part to this validation campaign reported, these products were reflecting correctly the situation, and were suitable to assess the material damages caused by the "Mitch" hurricane. In particular, the damages caused by the deadly mudflow descending from the Casitas volcano had been correctly detected, as well as the damages in and around the city of Chinandega.

Since then, it seems that the ™DYNAMIC-RADAR and ™DYNAMIC-OPTICAL products concerning Nicaragua were intensively used there.

6.2 *El Salvador*

As mentioned earlier, the results of this project have been transmitted to the commissioned United Nations body. Original intention was to validate and use these results in El Salvador. However, no feedback has been received.

6.3 *Honduras*

During a mission in Honduras in March 1999, PRI-VATEERS N.V. team members went to compare the damage identification and evaluation products to ground reality in the area around Tegucigalpa, and along the road from Tegucigalpa to Juticalpa.
- In Tegucigalpa, missing houses could be identified, and destroyed houses could be found relying on the corresponding ™DYNAMIC-RADAR product. The outstanding performances of the new PRIVA-TEERS N.V. processing techniques preserving the spatial resolution of the SAR could be appreciated there.
- Along the road Tegucigalpa-Juticalpa, destroyed bridges could be accurately identified on both the ™DYNAMIC-RADAR and ™DYNAMIC-OPTICAL products. The evaluation team relied on these products to plan in advance overcoming these difficulties when traveling in this area.
- Huge areas (mainly north of Tegucigalpa, easily identifiable in red on Figure 1) where the forest cover has been completely wiped out by the hurricane have also been identified as change areas detected (in red tones) by the DYNAMIC products. In particular, the changes observed on the slopes, originally thought to be artifacts, reflect with a good accuracy the disappearance of forest cover on the top and on the slopes of hills and mountains.

A striking example of these changes has been observed on the hills south of Tegucigalpa, where both the forest cover and the housing were completely destroyed, as detected in the ™DYNAMIC-RADAR product.

7 TRANSFER OF INFORMATION TO LOCAL END-USERS

Whenever possible, and as much as possible, the cartographic products were communicated by the project partners to the concerned institutions.

Transmission of the results in Nicaragua has been done quickly, already in February 1999, due to good prior knowledge of the country.

In Honduras, results have been transmitted after a rather long delay. Due to our limited knowledge of the country, the difficulty was to identify the bodies and institutions who could use remote sensing prod-

ucts. Nevertheless, since the fall of 1999, Honduran end users (mainly national development agencies) are using them for the reconstruction, as well as to update the cartography of the country.

The ™DYNAMIC-RADAR and ™DYNAMIC-OPTICAL cartographic products have also been used to plan reconstruction missions send to Honduras by humanitarian NGO's from 1999 to 2001.

8 CONCLUSION

The "Mitch" project proved that, in order to produce useful results very shortly after a natural disaster such as a huge hurricane, the building of a professionally serious team, good project partners coordination and the use of appropriate and well mastered techniques lead to rapid project implementation and execution, even if the concerned area is very wide, and the requirements in terms of spatial resolution are very strict (70,000 km^2 were covered at the spatial resolution of 20 meters x 20 meters, and an additional 55,000 km^2 were covered at the spatial resolution of 10 meters x 10 meters).

Regarding the products themselves, the ™DYNAMIC-RADAR and the ™DYNAMIC-OPTICAL products proved to be reliable photo-interpretation tools to locate and identify the changes and the damages caused by the hurricane: flooding (San Pedro Sula, Gulf of Fonseca, etc.), destruction of housing (Tegucigalpa, Chinandega, Leon, etc.) and infrastructures (bridges), forest damages and brutal deforestation (especially in Honduras).

However, in future projects of the same nature, improvements must be made with regard to the dissemination of results and the procurement of appropriate technical support to the concerned end-users. This applies in particular to some developing countries where remote sensing is still a little known technology.

The above conclusions were confirmed during the recent operation in support of disaster relief workers following the El-Salvador earthquake of January 13, 2001 (Nezry *et al.* 2001b, BBC 2001).

Finally, since the "Mitch" hurricane, similar techniques have been successfully used to assess the damages caused by:
- the giant floods of the Yang-Tse-Kiang river in China (July/August 1999),
- the Izmit earthquake in Turkey (August 17, 1999, see Sarti *et al.* 2000),
- the "Lenny" hurricane (Guadeloupe Island, November 1999, see Fellah *et al.* 2000),
- heavy rains in Venezuela (December 1999 - January 2000, see Nezry *et al.* 2001a),
- the earthquake of January 13, 2001 in El Salvador (Nezry *et al.* 2001b).

9 REFERENCES

BBC, 2001: "Aid from Space". *BBC News | SCI/TECH | Aid from space, by BBC News Online's Ivan Noble, http://news.bbc.co.uk/hi/english/sci/tech/newsid_1172000/ 1172346.stm, 16 February 2001.*

Fellah K., Nezry E., Bally P., Bequignon J., Herrmann A., Bestault C. & de Fraipont P. 2000. Rapid mapping system for hurricane damage assessment. *Proceedings of the ERS-ENVISAT Symposium, ESA SP-461 (CD-ROM), paper nr.339, Gothenburg (Sweden), 16-20 October 2000.*

Laur H., Bally P., Meadows P., Sanchez P., Schaettler B., Lopinto E. & Esteban D. 1998. ERS SAR calibration: derivation of the backscattering coefficient in ESA ERS SAR PRI products. *ESA document nr.ES-TN-RS-PM-HL09, Issue 2, Rev.5b, 47 p.*

Lopes A., Nezry E., Touzi R. & Laur H. 1993. Structure detection and statistical adaptive speckle filtering in SAR images. *International Journal of Remote Sensing, 14(9) :1735-1758.*

Nezry E., Romeijn P., Sarti F., Inglada J., Zagolski F. & Supit I. 2001b. Breaking new grounds for remote sensing in support of disaster relief efforts: detecting and pinpointing earthquake damages in near real-time (El-Salvador, January 2001). *To be presented at the 8th SPIE-EUROPTO Symposium on Remote Sensing, Toulouse (France), 17-21 September 2001.*

Nezry E., Sarti F., Yakam Simon F., Romeijn P. & Supit I. 2001a. The devastation of Venezuela by heavy rains in December 1999: Disaster monitoring and evaluation of damages using ERS SAR and SPOT images. *Proceedings of the 21th EARSeL Symposium, Marne La Vallée (France), 14-16 May 2001. This issue.*

Nezry E. & Yakam Simen F. 1999. Five new distribution-entropy MAP speckle filters for polarimetric SAR data, and for Single or Multi-Channel Detected and Complex SAR Images. *Proceedings of SPIE, 3869 :100-105.*

Nezry E., Yakam Simen F., Zagolski F. & Supit I. 1997. Control systems principles applied to speckle filtering and geophysical information extraction in multi-channel SAR images. *Proceedings of SPIE, 3217 :48-57.*

Nezry E., Zagolski F., Lopes A. & Yakam Simen F. 1996. Bayesian filtering of multi-channel SAR images for detection of thin details and SAR data fusion. *Proceedings of SPIE, 2958 :130-139.*

Sarti F., Nezry E. & Adragna F. 2000. Complementarity of correlation and interferometry for the analysis of the effects of the Izmit earthquake with radar data. *Proceedings of SCI'2000, Orlando (Florida, USA), 6 p., 23-26 July 2000.*

Observing our environment from Space: New solutions for a new millennium, Bégni (Ed.)
© 2002 Swets & Zeitlinger, Lisse, ISBN 90 5809 254 2

Influence of the three-dimensional effects on the simulation of landscapes in thermal infrared

T. Poglio
System Architecture Division, Alcatel Space, Cannes-la-Bocca, France; Groupe Télédétection & Modélisation, Ecole des Mines de Paris, Sophia Antipolis, France

E. Savaria
System Architecture Division, Alcatel Space, Cannes-la-Bocca, France

L. Wald
Groupe Télédétection & Modélisation, Ecole des Mines de Paris, Sophia Antipolis, France

ABSTRACT: The paper deals with the modelling of landscapes for the simulation of very high spatial resolution images in the thermal infrared range, from 3 to 14 µm. It focuses on the influence of the 3-D effects on the simulation. The major relevant physical processes are described. Examples are made, comparing simulations obtained with 2-D and 3-D representation of the landscape. They help in classifying the relative influence of each process. The necessity to take into account a 3-D landscape representation for the simulation of very high spatial resolution images in the infrared range is also demonstrated.

1 INTRODUCTION

There is a large demand of very high spatial resolution imagery in the infrared range in many and various fields, like meteorology, farming or military information. Such imagery with a spatial resolution of a meter or so is not yet available but new spaceborne systems are under development. Critical points are the assessment of the capacity of such systems and users training to the use of such imagery. The simulation is a crucial tool in this respect. It helps to reproduce the characteristics of the observing system. The output is a simulated image such as it would be delivered by the system. An essential point of the simulation of the observing system is an accurate knowledge of the input scene parameters, which are provided by a simulator of landscapes.

Variable meteorological conditions, different places, different landscapes, different times and different spectral bands should be simulated. Instead of gathering large sets of real images, a landscape synthesis method has been selected in order to meet better these requirements. The simulator of landscapes requires as input the knowledge of objects in the scene (spatial distribution and physical parameters) and simulation conditions. For each object, the simulator predicts the heat exchanges between objects, the temporal evolution of heat balance, the spectral emission and the spectral reflection of all incident fluxes. In thermal infrared, the flux coming from an object is partly emitted by the object because of its own temperature, and partly due to the reflection of incident rays on the surface of this object. Depending on the surface material and the

spectral band, emission or reflection process dominates the signal.

For remotely sensed images, even if the relief indirectly influences the image whatever the resolution is, this paper focuses on very high spatial resolution where the objects generating the relief can be described. Impacts of this 3-D description on the synthesised image should be studied.

Jaloustre-Audouin (1998) and Jaloustre-Audouin *et al.* (1997) have developed a simulator of any type of landscape in 2-D. It models very efficiently the physical behavior of the objects taken separately. Image simulators with 3-D landscape representations as input exist, but only for specific application like thermal behavior of vehicles (Johnson *et al.*, 1998) or radiative budget modeling for vegetated areas (Guillevic, 1999).

Physical processes playing a part in the signal coming from the scene are described in the first section. The following section presents the importance of physical phenomena in a 3-D representation. Then, a quantification of these phenomena is proposed with help of examples, making comparisons between simulations obtained with 2 and 3-D representations. A classification of impact is proposed for the main relevant phenomena. Some other phenomena may have a noticeable impact but depend too much on the spectral range, the meteorological or geographical conditions to enter this general classification. Their impacts are discussed in regard with the condition of the simulation, especially with the spectral range. It is conclude that the image sampling rate is linked to the necessity of an adapted landscape representation for the very high spatial resolution image simulation in the infrared range.

2 THE PHYSICAL PROCESS

In thermal infrared, the flux coming from an object in a given spectral range is both due to its own temperature and to spectral reflection of incident fluxes in this range. Depending on the spectral band and the surface material, the emission or reflection process dominates the signal. So, both processes have to be computed carefully.

2.1 Emitted flux

The emitted flux or irradiance (in $W.m^{-2}$) in a given spectral range, from λ_1 to λ_2 is given by:

$$L^e_{\lambda_1\lambda_2}(\theta,\varphi) = \pi \int_{\lambda_1}^{\lambda_2} \varepsilon_s(\lambda,\theta,\varphi) L^{bb}(\lambda,T_s) d\lambda \qquad (1)$$

$$L^{bb}(\lambda,T_s) = \frac{2 \cdot hc^2}{\lambda^5 \cdot \left[\exp\left(\dfrac{hc}{\lambda k T_s} \right) - 1 \right]} \qquad (2)$$

where L^{bb} is the flux emitted by a blackbody; T_s the surface temperature; and (θ,φ) the angles of the viewing direction. h, c and k are respectively the Planck constant; the light velocity and the Boltzmann constant. The quantity ε_s is the spectral emissivity of the object.

Under the thermodynamical conditions usually encountered in landscapes, the heat equation may be written as:

$$\frac{\partial T_s}{\partial t} = \kappa \cdot \Delta T_s \qquad (3)$$

where κ is the thermal diffusivity, *i.e.* the ratio between the thermal conductivity and the product of the material density by the specific heat. This equation may be solved using e.g., the finite difference method, knowing:
- the temperature at the previous moment,
- the deep temperature of the object,
- the flux balance at the surface of the object.

The two temperatures result from inertia phenomena, whereas flux balance is an instantaneous phenomenon. Large temperature variations are due to large changes in flux balance. The flux balance at the surface of an object is given by the difference between radiative and convective fluxes. This difference corresponds to the conductive flux within the depth of the object, which gives rise to variations in surface temperature.

Table 1 presents typically values of these different fluxes. Radiative fluxes are very dependent on hourly conditions; solar irradiance can go from 0 at night from 1000 $W.m^{-2}$ at midday. Sudden change in solar irradiance very much impacts flux balance value, whereas longwave irradiance does not. Ra-

diative losses depend on the 4th power of surface temperature; there are very large for high surface temperature (50 °C) and smaller for weak temperature (0 °C). Convective fluxes mainly depend on gradient temperature between air and ground, and wind velocity. In addition, heat latent flux is very dependent on the ground moisture and the difference between relative humidity of the atmosphere and the ground.

Table 1: typical possible values for different fluxes ($W.m^{-2}$) at 45 °N.

Radiative fluxes:	Day	Night
Solar flux (for normal radiance direction):	900	-
- diffuse component	140	-
- direct component	760	-
Longwave flux	410	210
Radiative losses	700	280
Convective fluxes:		
Sensible heat	600	-50
Latent flux	800	-100

2.2 Reflected flux

All incident fluxes coming from other objects in the scene have to be considered as potential sources. An object i reflects a part of the received radiations H_i of various origins:
- solar radiation,
- atmospheric emission,
- reflected and emitted radiations from the surrounding objects.

The flux RR_i reflected by the object i is given by:

$$RR_{i\,\lambda_1\lambda_2} = \pi \int_{\lambda_1}^{\lambda_2} \rho_i(\lambda,\theta,\varphi) H_i(\lambda,\theta,\varphi) d\lambda \qquad (4)$$

For computational reasons, H_i is written as a sum of contributions ordered by the number of consecutive reflections. Assuming Lambertian reflection and isotropic emission:

$$H_i = B_i + \sum_{V_i} \rho_j F'_{ij} \left\{ B_j + ... \left\{ B_p + \sum_{V_p} \rho_q F'_{pq} B_q \right\} ... \right\} \qquad (5)$$

$$F_{ij} = \xi_{ij} \cdot F'_{ij} \qquad (6)$$

where B_i represents the different sources for the object i: solar radiation, atmospheric emission, and emission of the surrounding objects; ρ_i is the fraction of flux scattered or reflected; V_i is the environment of the object i; F_{ij} and ξ_{ij} are respectively the form factor and the transmission coefficient between the objects i and j.

It follows that the computation of the reflected flux received by an object requests the knowledge of the distribution and the orientation of the objects in the surrounding and their interactions.

3 2-D LANDSCAPE REPRESENTATION VERSUS 3-D

A 2-D representation of a very high spatial resolution scene disregards:
- solar shades; due to buildings, houses, trees…,
- wind disturbance around buildings,
- influence between objects: multiple reflections, heat conduction, obstructions of the horizon…,
- all the 2-D effects (especially humidity assessment) affected by those described as previous points.

Solar shade effect is certainly the natural effect giving the most important differences between a 2-D and a 3-D representation. In a 2-D representation, each object receives from the Sun the same solar global flux. The global flux is the sum of the direct and the diffuse components. The direct component expresses the flux coming directly from the Sun attenuated by the atmospheric path. The diffuse component corresponds to the part of radiation reflected, diffracted or diffused by molecules constituting the atmosphere and aerosols. The ratio of these two components is varying all the day long, mainly with the Sun elevation angle. This is illustrated in Figure 1, which displays the global irradiance and its components received on a horizontal plane without horizon obstruction. The simulation is made using the ESRA model (Rigollier et al., 2000).

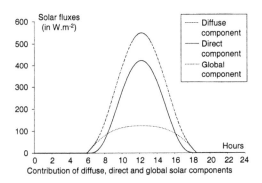

Figure 1: typical values of the different contributions of solar components for a horizontal object; the 21st of March, 45 °N.

Because the solar radiation is not isotropic, orientation between object and the Sun is also playing a role in the difference of solar fluxes balance between objects; 2-D representation cannot reproduce this reality. In addition, the diffuse radiation received by an object depends on the fraction of the sky dome, which is viewed from this object.

A shaded object will only receive a part of the diffuse component depending on the viewed fraction of the sky dome. Since its flux balance is generally negative in such a situation, its temperature will decrease, and its own emission too. Furthermore, because of the shadow, it will not reflect as much flux as if it was sunny. These two effects add to change considerably the global flux emitted by this object. Note that if this effect is very important it is nevertheless local, and does not affect more objects than the shaded ones.

In a 2-D description, the wind velocity is the same for each object. Reality is different, because each 3-D object disturbs the airflow. Solid objects, like houses or buildings, prevent wind from blowing with the same direction and velocity everywhere. This affects the entire scene; it is dependent on the distribution and the orientation of the objects in the scene. Since it is a convective effect, it only affects temperature of the object, and emitted flux as a consequence.

In a 2-D description there is no radiative interaction between objects. This is not true in 3-D. In the infrared range, each object acts both as a source and as a reflecting object. The assessment of the flux balance or the spectral emission coming from an object is highly dependent of its environment. A radiosity method (Watt, 2000; Sillion et al., 1994) can be used to solve such a problem. In order to decrease the computational efforts, consideration of the physical processes help in reducing the environment V_i (eq. 5) of an object to a few objects in its nearest neighborhood. Among these processing is the decrease of the influence of an object onto another one as their distance increase. The low reflectance value also contributes to this decrease; for most objects it ranges from 0,05 to 0,3 (ASTER, 2000).

When two objects in different thermodynamical states are in contact, heat exchange occurs until equilibrium establishes. Heat exchange occurs from the warmest object to the coldest. A temperature gradient exists through the boundary, related to a gradient in emitted radiance. Heat conduction tends to reduce these gradients. In most cases, due to different boundary condition on each object, thermal equilibrium is not reached.

In a 3-D representation, two cases are observed where heat conduction plays a noticeable role:
- a temperature gradient exists between two materials. Such a gradient may arise from shading effects,
- the scene geometry permits local heat accumulation or local heat loss.

Considering a building with a north-oriented facade and a sun-drenched roof, heat conduction will process from the roof to the ground trough the facade,

135

due to temperature gradient. Considering a house with a south-oriented corner, *i.e.* a south-east oriented facade and a south-west oriented one, it will produce a heat accumulation in the corner all the day long and its temperature will locally increase.

Although differences of several degrees in temperature exist between situations simulated taking into account heat conduction or not, it only affects a small area. Expressed in terms of distance, even if flux balance is very different on both objects, an important temperature difference only exists on a few tens of centimeters. Considering or not heat conduction as an important phenomenon for infrared landscape simulation highly depends on the sampling rate expected for the final image. Available images in the infrared range taken at very high spatial resolution exhibit smooth temperature transitions. Hence simulation should reproduce such observations and heat conduction should be considered locally.

4 ILLUSTRATING THESE PHENOMENA WITH EXAMPLES

An example is given in order to illustrate our discussion. It is a grass ground, without rain during the simulation, on the 21st of March, at 45 °N, with relative moisture of 50 % for ground and 60 % for air. The wind is blowing at 1 m.s^{-1}, air temperature is oscillating between 7 and 18 °C. The albedo is 0,15 and the average emissivity integrated on the total spectral range is 0,84.

In accordance with the previous works and analyses made by Jaloustre-Audouin (1998), we used the following models. For the global radiative fluxes, models used were ESRA for solar radiation (Rigollier *et al.*, 1999), Swinbank for longwave radiation (*In* Jaloustre-Audouin, 1998), the blackbody function and average material emissivity for emission. A sensible heat model (Louis, 1979), and a latent heat model (Noilhan & Planton, 1989) were used for the prediction of convective fluxes. Spectral irradiance was assessed by the use of MODTRAN for solar radiation (Kneizys *et al.*, 1996) and the Berger's model (1988) was used for atmospheric radiation assessment.

Impacts of physical phenomena were studied in two spectral bands: band II from 3 to 5 μm, and band III from 8 to 12 μm. These two bands correspond to atmospheric windows, where the signal is the less affected by the atmosphere transmission. The spectral reflectance is given by the ASTER database (ASTER, 2000). Using the Kirchoff's law we assumed the spectral complementarity between emissivity and reflectance:

$$\varepsilon(\lambda,\theta,\varphi) = 1 - \rho(\lambda,\theta,\varphi) \qquad (7)$$

Although thermal equilibrium never occurs in the simulations for remote sensing applications (Salis-

bury, 1994), Kirchoff's law can be used to link spectral reflectance and emissivity without making noticeable errors (Korb, 1999).

In most cases, the physical effects combine in a complex manner and cannot be separated. For illustration, we selected cases for which this separation is possible.

4.1 *Solar shade effect*

Two situations corresponding to a 3-D representation of the landscape were simulated. The first one is the grass ground shaded from 9 to 12 h local solar time (LST) and the second one the grass ground shaded from 13 to 16 h LST. These simulations are compared to the simulation for a 2-D representation without shades. According to the ESRA model, the average diffuse solar component is respectively 25 % and 20 % of global irradiance for these two periods. These three simulations are shown in Figure 2 for band II and Figure 3 for band III.

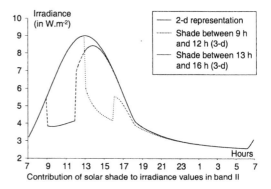

Figure 2: solar shade effects in band II. In full line, daily surface temperature evolution considering a 2-D representation. In dotted lines, the temperature evolution in two cases: a shading effect between 9 and 12 h and a shading one between 13 and 16 h LST.

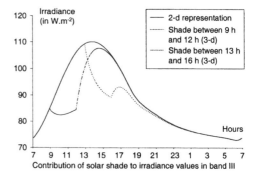

Figure 3: as Figure 2 but for band III.

For both bands, even if the irradiance values differ, the 2-D landscape representation approximation overestimates the irradiance. It not only occurs during the shaded time, but also in the following moments due to temperature inertia phenomenon. This overestimation may be very important, especially if the shaded areas cover several pixels in the high resolution simulated image.

The following tables (Tables 2 and 3) present the average values obtained for 2-D and 3-D landscape representation.

Table 2: average temperature and irradiance values from 9 to 12 h LST in 3-D and 2-D landscape representation cases.

	Average temperature	Average irradiance in band II	Average irradiance in band III
Ground shaded (3-D representation)	12,6 °C	4,0 W.m^{-2}	83 W.m^{-2}
Ground without shade (2-D representation)	21,6 °C	7,2 W.m^{-2}	97 W.m^{-2}
Relative Increase	-	+ 80 %	+ 17 %

Table 3: average temperature and irradiance values from 13 to 16 h LST in 3-D and 2-D landscape representation cases.

	Average temperature	Average irradiance in band II	Average irradiance in band III
Ground shaded (3-D representation)	19,6 °C	4,8 W.m^{-2}	95 W.m^{-2}
Ground without shade (2-D representation)	28,2 °C	8,3 W.m^{-2}	108 W.m^{-2}
Relative Increase	-	+ 73 %	+15 %

In band II, spectral reflection of incident solar flux constitutes the main part of the signal. Thus, as shade affects both temperature and spectral reflections, it is in this spectral range that approximating 3-D by a 2-D representation causes the largest errors. The error is always larger than 70 % in this spectral band. In band III, relative errors are smaller than in band II, but absolute values of differences are not. Anyway, these gaps are very important and shades must be taken into account in high spatial resolution image simulation in this spectral range.

4.2 *Wind disturbance*

Wind velocity and direction are affected by the pattern of all objects building the scene. Wind velocity will decrease if the object is protected from wind-blow whereas it will probably increase in a street between buildings. A simulation was done with the same conditions than previously described. The wind velocity is set to 1, 3 and 8 m.s^{-1}. Irradiance values in band III are presented on Figure 4. The larger the wind velocity, the more the surface temperature behavior comparable with air temperature. In case of inaccurate estimation of the wind velocity, Figure 4, shows that the largest errors in predicting

irradiance are reached for the highest surface temperature values. It can be explained by the fact that irradiance follow the blackbody law, and an increase of a degree at a high temperature (30 °C) has more impact on the irradiance than an increase of a degree for a small temperature (5 °C). The maximal difference between temperature and irradiance are observed at 14 h LST. These differences are presented in the Table 4.

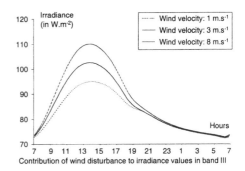

Figure 4: wind disturbance effects on irradiance values in band III.

Table 4: maximal temperature and irradiance values for different wind velocity.

Wind velocity (m.s^{-1})	Maximal temperature (°C)	Maximal irradiance in band II (W.m^{-2})	Maximal irradiance in band III (W.m^{-2})
1	29.1	9.0	110.1
3	24.6	8.2	102.6
8	19.8	7.5	95.1

Irradiance variations are not proportional to wind velocity. Particularly, irradiance is very sensible to wind velocity variations for small velocity. In band III, the difference is larger than 7 W.m^{-2} between a situation simulated with a wind velocity of 1 m.s^{-1} and 3 m.s^{-1}. It can be explained by the impact of the wind velocity on other phenomena. Not only wind velocity variations change sensible heat flux, but also latent heat flux and ground moisture. Without any precipitation during the simulation, ground will dry quicker under a strong wind than it will do under a light one. An accurate prediction of wind velocity on each object, *i.e.* an accurate prediction of velocity and direction of wind in the scene is necessary.

4.3 *Multiple reflections*

The aim here is to assess influence of object on each other. In this simulation, the landscape (Figure 5) is composed of three buildings enclosing a place. Walls of each building (objects n°1, 2 and 3) are made of construction concrete, and insulating material (polystyrene). The ground between buildings is in asphalt above earth. The simulation takes place on the 12th of October. Other simulation conditions are the same than in the previous example. Dimensions indicated on the Figure 5 are such as: $c = 1.5d = 2a = 3b$. Emission and reflection of incident flux are presented in Table 5. The nth order reflection for an object is computed by considering the n-1th reflections of its environment, *i.e.* all other objects, the atmosphere and the Sun.

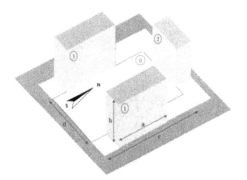

Figure 5: the scene considered in this example; the object 0 (framed) is the ground in asphalt, the objects 1, 2 and 3 are buildings. The interest for the simulation is the West-oriented facade of the building 1.

For each object, emission and reflected fluxes are computed. The interest here is the increase of global flux coming from an object, due to the computation of multiple reflections.

Table 5: emitted and reflected fluxes computed at order 1, 2 and 3 for the building 1. Contribution of current order compared with the sum of the previous ones (in percentage).

	Midnight		Midday	
Temperature	14,9 °C		21,6 °C	
Spectral flux (W.m⁻²)	Band II	Band III	Band II	Band III
Emitted flux	3,34	93,09	4,32	104,52
Reflected flux; order 1	3,68 +9,2 %	96,43 +3,5 %	5,56 +22,3 %	109,25 +4,3 %
Reflected flux; order 2	3,69 +0,3 %	96,49 +0,1 %	5,62 +1,6 %	109,33 +0,1 %
Reflected flux; order 3	3,69 +0 %	96,49 +0 %	5,62 +0 %	109,34 +0 %

During the day, although it is illuminated by the Sun with nearly parallel incidence (due to azimuth angle at this date), reflected flux in band II represents 23 % of the global flux coming from the building n° 1. Compared to global flux, 77 % of the flux coming from the building is due to emission, 22 % is due to reflection computed at the first order, and 1 % is due to reflection computed at the second order. More generally, the first order for reflection is the most important whatever time and spectral bands are. Contribution of a given order in reflected flux decreases with the increase of the order of diffusion. In band III, computing reflected flux at the first order of diffusion might be enough. This is due to the

very small influence of solar incident irradiance in this spectral range. In band II, the second order represents only about 1 % of the global flux. Nevertheless, it represents approximately 1/50th of irradiance given in Table 2 or 3. It is a significant value with respect to the encoding properties of the envisioned observing system.

Multiple reflections are nevertheless very dependent on the spectral reflectance. In the example above, for the concrete construction, average reflectances are about 11 % in band II and 6 % in band III. Other materials may have larger reflectance values. For such cases, multiple reflections become noticeable; computing reflections at the first order of diffusion during the night and the second order during the day is necessary.

Another effect due to multiple reflection will appear in the interpretation of the images. During the night, the asphalt ground and the building number 2 are the warmest in the scene. There are also neighbors to buildings 1 and 3. The contributions of objects 0 and 2 to the flux emitted by the object 1 and 3 will reduce the differences between the buildings 2 and 1 (or 2 and 3), which would have been observed without the multiple reflections. This renders the accurate assessment of the actual temperature of each building more difficult.

5 CLASSIFICATION OF THE IMPACT OF PHYSICAL PHENOMENA ON THE SIMULATION

The approximation made by using a 2-D instead of a 3-D representation leads to errors that depend upon the considered physical phenomenon and its local relevance. For example, the wind effects will be found everywhere while the heat conduction will be noticeable for a few objects in a limited distance to their boundary. In another example, not only buildings can shade the ground, but they can also decrease the wind velocity. If these two effects add, simulations considering or not 3-D representation will differ drastically. On the contrary, combined influence of phenomena like multiple reflections, obstruction of the horizon or heat conduction can be null. Nevertheless, a classification of the impact of the physical phenomena can be performed by considering the average influence in a typical case of interest, such as an urban area. This classification is given in Table 6.

Table 6: classification (order of importance) of the impacts of the main physical phenomena influencing the simulation.

	Percentage
Solar shade effect	80 %
Wind disturbance around 3-D objects	20 %
Multiple reflections	
Horizon obstructions	} ≅ 2 %
Heat Conduction	

The percentages are only indicative. As a whole, the shade affects are predominant, followed by the wind disturbances. Temperature of a given material expresses recent past of flux balance at the surface of the object including shade effects. Not only these effects are important at a given time, but also in a near future because of inertia phenomenon. So, the computation of shades has to be done all along the simulation, during the heat equation integration process. Though it only affects temperature and emitted flux, an accurate wind velocity modeling is also necessary, especially if the simulation has to be done during the night, when solar flux contribution is null. Phenomena directly attached at the environment of an object are more difficult to classify, because far too dependant on the simulation conditions. Without any particular knowledge of the simulation conditions they have to be taken into account. But, two criteria can be applied. If image simulation have to be done for thermal wavelengths (about 10 µm), computation of multiple reflections is usually not necessary. If the final image sampling rate is about a few meters the simulation of heat conduction between objects will not carry out a lot additional information and is not necessary.

6 CONCLUSION

In thermal infrared image simulation, as the spatial sampling rate increases, realistic representation of the landscape is only possible with a 3-D representation. In this respect, objects that build the landscape interact each other. Realistic image simulation will have to reproduce faithfully these physical interactions between objects. In the infrared range, the main physical phenomena affecting specifically the signal emitted by a 3-D scene are shade effect, wind disturbance linked to landscape relief, and depending on the simulation conditions, multiple reflections or heat conduction. Consequently, results presented here will be used as a starting point in the specification, the design and the development of a simulator of landscapes in the thermal infrared range with a very high spatial resolution.

REFERENCES

ASTER, 2000. ASTER spectral library Ver 1.2, CD-ROM, Jet Propulsion Laboratory, NASA, October 2000, http://spectib.jpl.nasa.gov/archive/jhu.html.

Berger X., 1988. A simple model for computing the spectral radiance of clear skies. *Solar Energy*, **40**, n°4, 321-333.

Guillevic P., 1999. *Modélisation des bilans radiatif et énergétique des couverts végétaux*. Thèse de Doctorat, Université P. Sabatier, Toulouse, France, 181 pp.

Jaloustre-Audouin K., 1998. *SPIRou : Synthèse de Paysage en InfraRouge par modélisation physique des échanges à la surface*. Thèse de Doctorat, Université de Nice-Sophia Antipolis, Nice, France, 169 pp.

Jaloustre-Audouin K., Savaria E., Wald L., 1997. Simulated images of outdoor scenes in infrared spectral band, *AeroSense'97*, SPIE, Orlando, USA.

Johnson K., Curran A., Less D., Levanen D., Marttila E., Gonda T., Jones J., 1998. MuSES: A new heat and signature management design tool for virtual prototyping, *Proceedings of the 9th Annual Ground Target Modelling & Validation Conference*, Houghton, MI.

Korb A.R., Salisbury J.W., D'Aria D.M., 1999. Thermal-infrared remote sensing and Kirchoff's law. 2. Fields measurements. *Journal of Geophysical Research*, **104**, 15,339-15,350.

Kneizys F.X., Shettle E.P., Gallery W.O., Chetwind J.H., Abreu L.W., Selby J.E., Clough S.A., Fenn R.W., 1988. *Atmospheric transmittance/radiance: computer code LOWTRAN 7* (AFGL-TR-88-0177), Hanscom AFB Massachusetts, Air Force Geophysics Laboratories, USA.

Louis J.-F., 1979. A parametric model of vertical eddy fluxes in the atmosphere. *Boundary Layer Meteorology*, 17, 187-202.

Noilhan J., Planton S., 1989. A simple parametrization of land surface processes for meteorological models. *Monthly Weather Reviews*, **117**, 536-549.

Rigollier C., Bauer O., Wald L., 2000. On the clear sky model of the ESRA - European Solar Radiation Atlas - with respect to the Heliosat method. *Solar Energy*, **68**, n°1, 33-48.

Salisbury J.W., Wald A., D'Aria D.M., 1994. Thermal-infrared remote sensing and Kirchoff's law. 1. Laboratory measurements. *Journal of Geophysical Research*, **99**, 11,897-11,911.

Sillion F.X., Puech C., 1994. *Radiosity & Global Illumination*. Morgan Kaufmann Publishers, Inc., ISBN 1-558-60277-1, San Francisco, CA, U.S.A., 251 p.

Watt A., 2000. *3-D Computer Graphics*, Third Edition, Addison-Wesley Publishing Compagny Inc, ISBN 0-201-39855-9.

Observing our environment from Space: New solutions for a new millennium, Bégni (Ed.)
© *2002 Swets & Zeitlinger, Lisse, ISBN 90 5809 254 2*

Evaluating offshore wind energy resource by spaceborne radar sensors: Toward a multi-source approach

N.Fichaux & T.Ranchin
Groupe Télédétection & Modélisation, Ecole des Mines de Paris, Sophia Antipolis cedex, France

ABSTRACT: In the framework of the current development of offshore wind energy exploitation, an accurate evaluation of the wind potential is crucial for sitting windmills. Nowadays, the resource is evaluated from discrete measurements, which must be extrapolated in order to provide a global wind resource map. But in this case, local conditions and variations of wind are not expressed.
This paper deals with a method that enables to obtain spatially accurate wind maps from ERS2 SAR (Synthetic Aperture Radar) images using advanced signal processing techniques. Concerning the temporal accuracy, the poor repetitiveness of data from ERS SAR sensor has been considered. Future prospects are presented, that concerns the design of a data fusion algorithm. This work is done in the framework of our study, which aims at designing a method to obtain spatially and temporally accurate wind statistics over a given area of interest.

1 INTRODUCTION

Using renewable energies is a new priority for Europe and especially for France since the Kyoto protocol. Consequently, France has developed a program that aims at increasing the part of renewable sources in the production of energy. The objective is an amount of 21% of energy produced from renewable sources in 2010. In this framework, wind energy has an important role to play as the wind potential is really important in France (51 TWh/year, representing 14% of the energy produced in 1997)[1] and wind energy involves well known and relatively cheap technologies.

Moreover, offshore technologies are bound to be developed because of their advantages: the wind potential is as important in offshore as in land (21600 TWh/year in offshore and 29600 TWh/year in land)[2], the lower turbulence of wind reduces the fatigue of the materials, and the nuisances in terms of noise and landscape are lower too.

In a given area of interest, offshore wind turbine parks are currently located in the 5 to 10 Km zones from coast at 10 to 20 meters depth. In order to provide energy at attractive costs, it is necessary to locate the windiest places, and then to locally evaluate the wind resource, that is not done at present.

Considering the surface covered by wind farms, the accuracy of the wind grid needed is about 1 Km or less. But nowadays, the resource assessment is performed by direct measurements of the wind phenomenon by the mean of onshore meteorological stations, buoys... Discrete and scarce measurements are then obtained which must be extrapolated, using models such as WASP©, in order to provide a global wind resource grid with an accuracy of about 50 to 100 Km. Thus, local condition and variations of wind are not expressed. Moreover, the uncertainties of the models used are great over coastal areas.

In this paper, the spatial and temporal variability of wind fields over coastal areas will be shown. The help of remotely sensed data to assess the wind potential will then be presented. Then, an algorithm is described, which has been designed to provide spatially accurate wind maps over coastal areas from ERS2 SAR (Synthetic Aperture Radar) images. And finally, considering the poor repetitiveness of data from ERS2 SAR (*i.e.* 35 days), future prospects are introduced concerning a multi-source approach to improve the temporal resolution of data. The last point is crucial to enable us to obtain temporally and spatially accurate wind statistics over the area of interest.

---

[1] see http://www.espace-eolien.fr
[2] see http://www.espace-eolien.fr

2 AREA OF INTEREST

The area studied is the French 'Golfe du Lion'. This area is situated in South France, more precisely in 'Languedoc-Roussillon' (Figure 1). The area of interest is the 5 to 10 Km zone situated along the coastline and represents a surface of about 100 Km² (ADEME 1999, unpubl.).

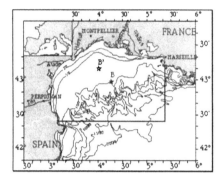

Figure 1: Area of interest from the FETCH database - B & B' represent the position of the buoy before and after 1998/03/25.

3 DATA USED

The FETCH experiment[3] was carried out during March and April 1998 and took place in the French Golfe du Lion. This experiment aimed at studying the exchanges between ocean and atmosphere and improving the wind detection techniques from remotely sensed data. Several data sources were available during the FETCH experiment: a boat, 3 buoys and 2 planes each carrying numerous sensors. The database has been completed with the images acquired by the ERS satellites.

The data used in this paper are from two sources: the SAR PRI (precision image) images (resolution 12.5 m) have been acquired during the FETCH experiment and the wind data were provided by four inland meteorological stations (METEOFRANCE). These stations located in the French Golfe du Lion are named: Grande Motte (GM), Leucate (Leu), Sete, Sainte-Marie (Ste Marie).

4 LOCAL VARIATIONS OF WIND COMPONENTS

In this section, the spatio-temporal variability of wind components over the area of interest will be studied.

Figure 2 has been obtained from data provided by the four meteorological stations. Mean wind speed

and variance were computed for data acquired every hour during the whole year 1998.

It must be specified that Leucate has a specific location compared to the other stations because of its higher position more far from coast. But it must be considered that this is typically the kind of problem one can encounter when evaluating the wind potential.

As can be seen, even if the mean speed is similar, the variance of the wind speed is important and is different from one station to another. So, it can be said that the temporal variations of wind components are varying over a given area of interest. This phenomenon can easily be explained by the difference of topology from a station to another. This remark induces that local coastal effects are present and have to be taken in account when statistically evaluating the wind potential.

Concerning the present management of the spatial variability, it must be considered that the wind maps are presently obtained from discrete measurements that are extrapolated over the whole area of interest. Table 1 shows the correlation coefficient between the four stations. The coefficients are weak, inducing that there is no real connection between the wind condition for the four locations. So, it seems difficult to determine the wind components for a point situated between a station and another, as its wind conditions will be independent of the wind conditions of the measurement stations.

Considering these points, it can be concluded that to obtain accurate wind maps over the area of interest, discrete scarce measurements are not sufficient. So, a high resolution analysis of the area at a global scale has to be done.

Remote sensing technologies provide a global vision of the area of interest and are helpful to resolve that kind of problem (Johannessen & Korsbakken 1998). In that framework, a method of wind component extraction based on remote sensing technologies will now be presented.

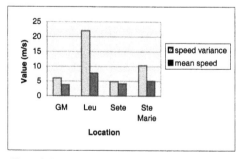

Figure 2: Mean wind speed and its Variance for four locations in the French Golfe du Lion year 1998

---

[3] see http://dataserv.cetp.ipsl.fr/FETCH

142

Table 1: Wind speed and direction intercorrelation coefficient

Wind speed correlation	GM	Leu	Sete	Ste Marie
GM	1	0,3774522	0,4611465	0,6532521
Leu	0,3774522	1	0,5925337	0,4680313
Sete	0,4611465	0,5925337	1	0,4933199
Ste Marie	0,6532521	0,4680313	0,4933199	1

Wind direction correlation	GM	Leu	Sete	Ste Marie
GM	1	0,3187045	0,5498439	0,4933999
Leu	0,3187045	1	0,4662512	0,3479979
Sete	0,5498439	0,4662512	1	0,5489907
Ste Marie	0,4933999	0,3479979	0,5489907	1

5 BACKGROUND

In this section is briefly summed up the physical background of our studies and the main concepts which will used throughout our discussion.

Wind phenomenon creates sea waves with a given wavelength (1 cm to 1 Km). By observing the waves constituting the sea surface at these wavelengths it is possible to extract the wind components (speed and direction). The sea is an interface, there, where the wind can indirectly be measured from its impact on the sea surface.

SAR sensors, such as those carried by ERS1 & 2, are 'any time, any weather' sensors. For these sensors, the main effect contributing to backscattering is Bragg scattering. In the Bragg hypothesis, the wavelength of wind waves interacting with the radar signal is proportional to the wavelength of the radar echo. Only considering the Bragg effect, as the ERS SAR sensors are C-Band sensors (frequency: 5.3 GHz), the only waves that contributes to the imaging process have a 5 to 10 cm wavelength, which is inside the wind wave wavelength domain. The lowest wind able to generate such a phenomenon is a 3 to 4 m/s wind. So, it will only be possible to detect wind at higher speed with such a sensor.

Furthermore, some other phenomena step in the SAR imaging processing (Kerbaol, 1997) such as the tilt effect. This effect is a geometric effect: the Bragg's backscatters are seen at various angles because of longer waves. Then, the backscattering coefficients are modulated, and wind waves appear on the radar image. It will be shown that this phenomenon enables us to extract wind direction from SAR images.

In order to extract both wind speed and direction from radar images, two main C-band models exist that have been calibrated for scatterometer data (Stoffelen & Anderson 1993) and shown to be efficient when used on SAR data (Furevik & Korsbakken 2000; Horstmann et al. 1998; Korsbakken & Furevik 1998; Vandermark et al. 1998; Horstmann et al. 1997; Rosenthal et al. 1995; Quilfen et al. 1994). They are named the CMOD4 algorithm (Stoffelen & Anderson 1997), which has been developed by ESA, and the CMOD_ifr2 algorithm developed by the Ifremer organisation. They provide a backscattering coefficient for a given wind direction, incidence angle and wind speed. The wind speed range is defined between 4 and 30 m/s and these models have a precision of about 2 m/s for wind speed and 15° for wind direction. Moreover it has been shown that these models agree within a precision range of 1.6 m/s. The ESA's CMOD4 model has been shown to be more accurate than CMOD_ifr2 for low wind speed and less accurate for high wind speed. But Donnelly et al. (1999) recalibrated the CMOD4 model for high wind speed, thus creating a CMOD4HW (High Wind) model. In this study, the CMOD4 model is used.

6 COMBINED WIND SPEED AND DIRECTION EXTRACTION ALGORITHM

The CMOD models provide a backscattering coefficient from given wind speed, wind direction and incidence angle. So, to obtain wind speed and direction, the model has to be inverted. As the incidence angle and backscattering coefficient are known at each point of the SAR image, one of the parameter (speed or direction) must be known to obtain the other one (Mastenbroek, 1998). Furevik & Korsbakken (2000) indicated that, if the wind direction is well known, it is possible to extract wind speed from SAR images with 500 meter accuracy. Wackerman et al. (1996) developed a wind direction extraction method and obtained a map of wind direction at 16 Km resolution. His approach is based on the tilt effect: Bragg's scatters are modulated by wind waves, thus low frequency features appear on the SAR image. As these features are wind dependent, their direction provides a good estimation of wind direction (Section 5). The feature's direction extraction can be done by using signal processing techniques that will be detailed now.

A property of the two-dimensional FFT (Fast Fourier Transform), when applied on a two-dimensional signal (an image) is its ability to provide the orientation of the spatial frequencies in the signal. Wackerman et al. (1996) applied a local FFT on the image to extract the wind features' direction. The image had to be averaged to get rid of the speckle. This explains the low resolution of the wind direction grid obtained. The approach the main points will be detailed now (Figure 3) is nearly the same but adapted to high resolution wind fields extraction: a FFT has been applied locally on the image after a Wavelet Transform (WT) has been processed.

6.1 Extracting wind direction

In this section, is presented a method for extracting the wind direction from SAR images. This method is based on the application of local FFT on an image

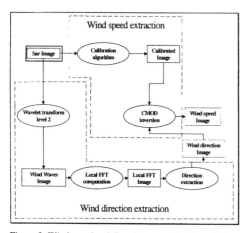

Figure 3: Wind speed and direction extraction from SAR image

processed by WT. The algorithm used to compute the WT is now described.

6.1.1 *The Wavelet Transform algorithm*

Mallat (1989) first introduced the concept of multi-résolution algorithm (MRA) for multiscale representation. Figure 4 is a very convenient description of pyramidal algorithms. From the original image (bottom of the pyramid), MRA allows the computation of successive approximations with coarser and coarser spatial resolutions. Climbing the pyramid, the different floors represent successive approximations. To describe difference of information between two successive approximations, Mallat has mathematically associated MRA with the WT. In this case, WT describes the difference of information between two consecutive approximations. Hence, when applying a MRA using WT, one can describe, hierarchically, the information content of a remotely sensed image (see *e.g.* Ranchin 1997).

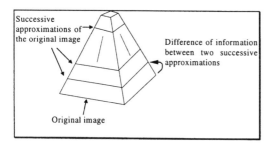

Figure 4: Representation of the successive approximations of an image by the means of a multirésolution algorithm.

The algorithm used for this purpose is the one proposed by Dutilleux (1987). From the original image, the "à trous" algorithm allows the computation of successive approximations by smoothing. The

wavelet coefficients (results of the application of a wavelet transform on a signal or image) are computed by a pixel-to-pixel difference between two successive approximations.

The SAR image is processed this way. The second level of approximation will now be processed. This image of the wavelet coefficients characterises the features with a spatial scale of 25 to 50 meters. Note that one of the properties of the WT is that all the scales are present at any approximation level, although more or less represented. So, it can be said that any wave frequency is represented in the image of the wavelet coefficient but the spectrum is "centred" preferentially on the wind waves wavelength.

6.1.2 *Applying FFT on the processed image*

In order to extract the wind direction from the SAR image, a sliding window FFT is applied on the wind wave image obtained by wavelet transform: the FFT is processed on successive vignettes of the wavelet coefficients image. As the wind waves' wavelength is the principal wavelength represented in the image, the maximum of intensity relies to the wind waves. So, the direction indicated by the maximums of the spectra is the wind waves' propagation direction. An example is shown below (Figure 5): the wind direction is clearly visible.

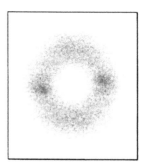

Figure 5: Local image spectrum obtained by FFT

This principle has been applied on an ERS2 SAR PRI from the FETCH database. That image has been acquired the 1998/03/19 at 10:29. The orbit number is 15221 and frame is 2745. This orbit was descending, this explains that the East is on the left of the image.

The wind direction has been measured near this area at the same moment by four sensors aboard the boat. The wind direction measured was about 300° from North (positive on the East).

A 1024x1024 pixels extract of the entire image has been done. The wind vectors have been computed over 128x128 pixels vignettes representing, as the resolution is 12,5m, a wind direction map with an accuracy of 1600 m. Note that the algorithm has been designed to be able to provide maps at higher accuracy.

Figure 6 represents the results obtained: the local wind directions obtained (white segments) have to be compared with the direction obtained by the boat's sensors (arrow). It can be seen that the direction computed are coherent all together and moreover relies well to the wind direction measures within the accepted range of 15°. Local variations of wind direction can then be seen.

Some errors are present, explained by the presence of phenomena with long wavelength. These artefacts involve further developments to be done: for example, a multi-resolution analysis could permit to improve the coherence of the wind directions obtained.

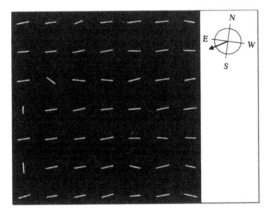

Figure 6: Wind direction vector from SAR image

As the local wind direction vectors have been obtained, these data can be provided to the CMOD4 model.

6.2 *Inverting the CMOD4 model*

The CMOD4 model provide a backscattering coefficient from given wind speed, wind direction and incidence angle. As this is a model with one output and three inputs, it can not be directly inverted.

As shown Figure 7 the backscattering coefficient increases continually with the wind speed at a given wind direction and incidence angle. Thus, provided the wind direction and backscattering coefficient at a given point and the local incidence angle are known, the local wind speed can be extracted.

6.3 *Expected results*

As show in paragraph 6.1, accurate wind direction maps can be obtained, and considering the knowl-

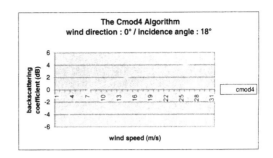

Figure 7: backscattering coefficient as a function of wind speed computed by the CMOD4 model at given wind direction and incidence angle

edge that has been collected from the bibliography, accurate wind speed maps over areas of interest will soon be obtained.

The originality of this algorithm remains in two points: firstly, wind maps will be generated at a resolution less than one kilometre and secondly this algorithm will extract wind speed and direction from only one data source, as shown Figure 3.

7 THE NEED FOR A DATA FUSION ALGORITHM

The objective of our studies should be considered now, that is to say to develop the ability to obtain statistics of the wind potential over a given area. But considering the low repetitiveness of data provided by the ERS SAR *i.e.* 35 days, other data sources should be considered.

Bentamy et al. (1995) compared the wind field obtained from ERS1 scatterometer, altimeter and Special Sensor Microwave Imager (SSM/I). It was shown that the wind field obtained from these three sensors show an agreement to better 2 m/s although the spatial resolution of these sensors was far different: 50 Km for scatterometer, 7 Km for altimeter and 25 Km for SSM/I. This example shows that the sensors available for wind field retrieval can be of different accuracy. Other sorts of data sources available could be meteorological models, which provide low resolution wind field estimation every three hours, or meteorological stations, which provide hourly wind estimates.

Regarding the numerous data source available, it seem reasonable to think that it is possible to design an algorithm that will enable us to generate high resolution wind maps by combining high and low resolution data sources. The principle would be to extract knowledge on local variability of the wind components over the area of interest from high resolution data and to apply this knowledge to low resolution data, thus obtaining daily high resolution wind maps from low resolution data sources.

8 CONCLUSION

In this paper the aim of our studies have been exposed, it is to say designing a method enabling to obtain temporally and spatially accurate wind statistics over a given area of interest. In this framework, the importance of remotely sensed data for wind field retrieval has been shown.

Then, concerning the spatial resolution, a method that will enable the extraction of high resolution wind fields from SAR data has been exposed. That method takes use of advanced signal processing techniques and wind fields maps at a spatial resolution better than 1 Km² will be soon available. The originality of this method remains in the fact that wind speed and direction are extracted from the same image.

Finally, the problem of the temporal resolution has been exposed. As the SAR data repetitiveness is insufficient, it has been explained why a multi-source approach has to be designed.

As a conclusion, it can be said that, although the method has not been entirely implemented yet, the capability to obtain high resolution wind fields from SAR images has been demonstrated. Moreover a method of data fusion for wind field retrieval will be developed for increasing the temporal resolution.

9 ACKNOWLEDGMENT

This work is supported by the French ADEME (Agence pour le Developpement Et la Maitrise de l'Energie).
We would like to acknowledge Daniele Hauser (CETP) for the access to data of the FETCH experiment.

REFERENCES

ADEME, 1999. *Etude de l'eolien offshore en languedoc-roussillon.*

Bentamy, A. Queffeulou, P. Chapron, B. Katsaros, K. 1995. Wind fields from scatterometer, altimeter & Special Sensor Microwave/Imager. *IGARSS'94 International Geoscience and Remote Sensing Symposium.* XXX(2): 1116-1118.

Donnelly, W.J. Carswell, J.R. McIntosh, R.E. Chang, P.S. Wilkerson, J. Marks, F. Black, P.G. 1999. Revised ocean backscatter models at C & Ku band under high wind conditions. *Journal of Geophysical Research* 104(C5): 11485-11497.

Furevik, B. & Korsbakken, E. 2000. Comparison of derived wind speed from SAR and scatterometer during the ERS tandem phase. *IEEE transaction on geoscience and remote sensing* 38(2): 1113-1121.

Horstmann, J. Koch, W. Lehner, S. Rosenthal, W. 1998. Ocean wind fields and their variability derived from SAR. *ESA special earth observation quarterly* 59: 8-12.

Horstmann, J. Lehner, S. Koch, W. Rosenthal, W. 1997. Wind fields from ERS SAR compared with a mesoscale atmospheric model near the coast. *Space at service of our environment; Proc. intern. symp., Florence, 17-21 March 1997:* 1205-1209.

Johannessen, O.M. & Korsbakken, E. 1998. Determination of Wind energy from SAR images for sitting windmills locations. *ESA special earth observation quarterly* 59: 2-4.

Kerbaol, V. 1997. *Analyse spectrale et statistique vent-vague des images radar à ouverture synthétique (ROS) —Application aux données des satellites ERS1/2.* Thèse de doctorat. Rennes I, France, pp. 185.

Korsbakken, E. & Furevik, B. 1998. Wind field retrieval compared with scatterometer wind field during the ERS tandem phase. *ESA special earth observation quarterly* 59: 23-26.

Mastenbroek, K. 1998. High resolution wind fields from ERS SAR. *ESA special earth observation quarterly* 59: 20-22.

Quilfen, Y. Bentamy, A. Queffeulou, P. Chapron, B. 1994. Calibration / Validation of ERS_1 wind scatterometer precision products. *IGARSS'94 International Geoscience and Remote Sensing Symposium.* XXX(2): 945-947.

Rosenthal, W. Lehner, S. Horstmann, J. Koch, W. 1995. Wind measurements using ERS1 SAR. *Proceedings of the second ERS applications symposium, London, 6-8 December 1995.*

Stoffelen, Ad. & Anderson, D. 1997. Scatterometer data interpretation: estimation and validation of the transfer function CMOD4. *Journal of Geophysical Research* 102(C3): 5767-5780.

Stoffelen, Ad. & Anderson, D. 1993. Characterisation of ERS1 scatterometer measurements and wind retrieval. *Space at the service of our environment; Proc. intern. symp., Hamburg, 11-14 October 1993:* 997-1001.

Vandermark, D. Vachon, P.W. Chapron, B. 1998. Assessment of ERS1 SAR wind speed estimates using an airborne altimeter. *ESA special earth observation quarterly* 59: 5-8.

Wackerman, C.C. Rufenach, C.L. Schuman, R.A. Johannessen, J.A. Davidson, K.L. 1996. Wind vector retrieval using ERS1 Synthetic Aperture Radar Imagery. *IEEE transaction on geoscience and remote sensing* 34(6): 1343-1352.

Observing our environment from Space: New solutions for a new millennium, Bégni (Ed.)
© 2002 Swets & Zeitlinger, Lisse, ISBN 90 5809 254 2

Satellite data for air pollution mapping over a city – virtual stations

A.Ung, L.Wald & T.Ranchin
Groupe Télédétection & Modélisation, Ecole des Mines de Paris, Sophia Antipolis, France

C.Weber & J.Hirsch
Laboratoire Image et Ville, Université Louis Pasteur, Strasbourg, France

G.Perron & J.Kleinpeter
ASPA (Association pour la Surveillance et l'étude de la Pollution atmosphérique en Alsace), Schiltigheim, France

ABSTRACT: Atmospheric pollution becomes a critical factor of anticipated deaths, which concerns ambient air quality as well as air quality in houses and places of work. For the well being of human and for the population information, we need to evaluate the actual exposure of persons to ambient pollution. One way to perform it is to evaluate the space time budget of air pollution exposition. Hence an information on the spatial distribution of pollutant concentrations is required. Several tools exist; most of them provide maps of pollutant concentrations but over a regional scale with a grid of 1 km, which is insufficient. Measuring stations, scarcely distributed in the city, provide a complete surveillance but their costs limit the knowledge of pollutant concentration to specific points of the city. To overcome this problem, several notions are defined: "identity card of a measuring station", "pseudostation" and "virtual station". Based on a multi-sources approach, this paper presents a methodology using remotely sensed data for the mapping of pollutant concentrations over a city.

1 INTRODUCTION

Nowadays most large cities in Europe have acquired a surveillance network for air quality. A network is composed of static measuring stations, which allow a continuous surveillance of air pollution on station location. Pollution data are collected in near real time and used to compute an atmospheric pollution indicator (ATMO). This indicator aims at informing local authorities, as well as the population, of the atmospheric air quality. In answer to a high rating of the ATMO indicator, public authorities are able to take restrictive measures with car traffic and with some air polluting companies.

Car traffic is increasing continuously. Atmospheric pollution hampers human breathing and impacts on life quality. Furthermore it becomes a critical factor of anticipated deaths, which concerns ambient air quality as well as air quality in houses and places of work. Studies suggest that current levels of particulate pollution in urban air associated not only with short-term, but also long-term increases in cardio respiratory morbidity and mortality (Nevalainen 1998). However the real exposure of persons to ambient pollution cannot be estimated with the present network. The costs of a measuring station and of its maintenance limit the knowledge to specific points of the town.

To evaluate the exposition to atmospheric pollution, an accurate knowledge of the spatial distribution of pollutants over the city is required. Several tools exist to perform it. Most of them provide maps of pollutant concentrations but over a regional scale with a grid of 1 km, which is insufficient. The first section offers a state-of-the-art showing the limitation of present tools. Actually, no accurate knowledge of the spatial distribution of atmospheric pollutants, over a city, is currently available. To overcome this problem, a methodology using remotely sensed data and other data sources is presented. It leads to define the notion of "*virtual station*". Then an application of this methodology to the city of Nantes is shown to reader. The last section deals with on a discussion of the results obtained and the future works.

2 STATE-OF-THE-ART

The aim of air quality monitoring is to get an estimate of pollutant concentrations in time and space. Several models exist and provide a map of pollutant concentrations by modelling its transport, deposition and/or transformation processes in the atmosphere. Those models are classified with regard to the scales of atmospheric processes. Spatial scales may range from the very local scale (e.g. street level, direct surroundings of a chimney) to the global scale (up to 100 km); time scale may range from minutes (estimation of peak concentrations) up to days (estimation of trends). Those models can be distinguished on the treatment of the transport equations (Eulerian, Lagrangian models) and on the complexity of various processes (chemistry, wet and dry deposition). Further descriptions of those models are given in a report of the European Topic Centre on Air Quality (Moussiopoulos 1996).

To evaluate the population exposition to air pollution, a mesoscale (urban scale) model is required. Existing models differ with regard to the structure of the computational domain, the utilised parameterisations, the method of initialisation, the imposed boundary conditions and the applied numerical techniques. Such models require at input considerable meteorological, geographical information, and emission data. Hence they require numerical simulations, often in conflict with the limited data processing resources, and not enough accurate yet. Those model evaluations are impossible without appropriate experimental data at proper locations in Europe.

The other ways to reconstruct the signal are models by means of interpolation and extrapolation of measurements. The problem is to reconstruct the spatial distribution of the pollutant considered within a geographical area, given limited values. Some scientists use the kriging method (Frangi *et al.* 1996, Carletti *et al.* 2000); some others recommend the use of the thin plates interpolation method (Ionescu *et al.* 1996). The quality of the mapping can be judged by comparing predictions and appropriate measurements. But the validity and accuracy of such approach depend on the number of measurements. It results that no accurate knowledge of the spatial distribution of atmospheric pollutants, over a city, is currently accessible.

Air pollution in cities is a complex phenomenon involving local topography, local wind flows and microclimates. For its better description and understanding, the study of its space-time variability should include different data sources related to urban morphological and environmental features. In this context, satellite images are certainly a valuable help in getting urban polluting features. Satellite imagery improves the monitoring of cities in a wide range of applications, e.g., the detection of urban changing, mapping roads and streets (Blanc 1999, Couloigner 1998), mapping urban demarcation (Weber 1995), mapping of physic parameters such as albedo and heat fluxes (Parlow 1998), and also mapping urban air pollution (Basly 2000). Several studies have shown the possible relationships between satellite data and air pollution (Finzi & Lechi 1991, Sifakis 1992, Poli *et al.* 1994, Brivio *et al.* 1995, Sifakis *et al.* 1998, Retalis *et al.* 1999, Wald & Baleynaud 1999, Basly 2000). According to those studies, a methodology for the mapping is developed.

3 THE METHODOLOGY

The proposed methodology combines point measurements and different data sources such as geographical database BDTOPO®IGN© and remotely sensed data provided by the Landsat Thematic Mapper (TM) sensor. It allows a definition of virtual stations, which provide additional measurements that are used for densifying the input parameters of interpolating and extrapolating methods. For a better understanding, it could be separated into 3 steps.

3.1 Step 1: Characterization of measuring stations and of pseudostations

The urban area is divided into cells by a regular grid. The cell size defines the spatial resolution of the mapping. A cell size less than 100 meters is required for a city mapping. The methodology makes use of measuring stations and others data to locate places of the city, which have the same environmental, morphological and polluting features than a real measuring station. Such places are defined as *"pseudostations"*. Their location could be performs using the notion of *"identity card"* of a cell. Actually cells are characterised by a list of parameters (its identity card) and pseudostations are cells having a same identity card as a real station.

Using remotely sensed data, those cells are materialised by pixels of the satellite image. The spatial resolution of the mapping of pollution and the spatial resolution of the satellite image are the same: 30m for the Landsat TM. It follows that digital numbers outputs from the satellite are parameters of the identity of a cell.

3.2 Step 2: Estimation of pollution - virtual stations

Pseudostations are places in the city where static measuring stations could be installed to increase the surveillance network. To make use of those pseudostations for the mapping of air pollution, a relation between satellite data and ground measurements must be derived. This relation aims at predicting pollutant concentrations in those pseudostations. We define *"virtual stations"* as pseudostations for which concentrations of pollutants can be predicted. So virtual stations are operational stations that companies in charge of measuring air quality in the city will have, in the case of an unlimited budget. They form a subset of the set of pseudostations.

3.3 Step 3: mapping by interpolation and extrapolation using virtual stations

The mapping is performed using an interpolating and/or extrapolating model: thin plates, polynomial, linear or Hsieh-Clough-Tocher. The accuracy is controlled with the help of correlation, error and mean square error between the estimated and the observed data.

4 APPLICATION AND DISCUSSION

The study area is the city of Nantes. The objectives of this first application to the city are twofold. The application of the methodology to the city of Nantes illustrates the benefits of Earth observation data for the knowledge of the atmospheric pollutant distribution. For lack of morphological indicator, only satellite data and ground measurements are used to define the identity card of a cell and to estimate its pollution. Therefore the set of pseudostations identify with the set of virtual stations. It will be shown that mapping pollutant concentration is possible using remotely sensed data and virtual stations. The results demonstrate the benefits of virtual stations.

4.1 *The study area and data used*

The city of Nantes is located in Western France at 54 km from the Atlantic Ocean. Geographical coordinates are 47.23° latitude North and 1.55° longitude West. That location explains the oceanic climate, which is mildly. Except for an oil refinery 20 km to the west, Nantes has no polluting industry. Atmospheric pollution is mostly due to motor vehicles. The local organization in charge of the air quality network in the city and vicinity is Air Pays de la Loire.

The studied day (22 May 1992) is a sunny day with clear sky and light wind from NorthEast. Pollution level was below critical levels. A Landsat satellite image was acquired on that day at 1017 UT. A portion of this image is represented in Figure 1. This size is 15.36*15.36 km, with a pixel size of 30*30 m. The channel TM4 (at 0.8 μm) reveals the patterns of the streets and roads in the city (in dark tones).

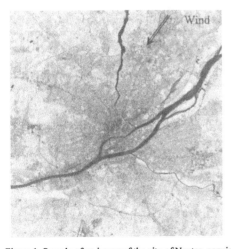

Figure 1. Sample of an image of the city of Nantes, acquired by satellite Landsat (TM4, at 0.8 μm), for May 22, 1992. The network of streets is seen fairly well in dark tones, as well as the Loire River and the airport in the south. ©Eurimage, reproduced with permission.

4.2 *Application to the city of Nantes*

Digital numbers output from satellite images are used to define the identity card of a pixel. In that case, pseudostations are pixels having same digital numbers as a measuring station in several combinations of satellite bands. For the day under concern, 12 measuring stations were operating. A strong correlation was found with some pollutants. To evaluate the accuracy of the method we only use 6 points out of 12. We apply the method and compare outputs to the remaining 6 stations.

Poli *et al.* (1994) studied the relationship between a map of apparent temperature of Rome (Italy) and the total particulate matter suspended in the air in the winter season. The particulate matter is assumed to be a significant tracer of the atmospheric pollution, as well as a good indicator of the air quality. Wald and Baleynaud (1999) focussed their studies on the correlation between black particulate concentration and digital numbers output from the infrared thermal band TM6. According to their studies, a linear regression is applied on pseudostations, as being a first approximation of the relation between satellite data and ground-based measurements. That defines virtual stations.

4.3 *Results*

Figure 2 presents an example of map of concentration in black particulates obtained with the thin plate method applied to 6 measurements.

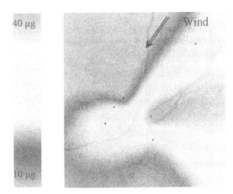

Figure 2. Extract of a map of concentration in black particulates over the city of Nantes for May 22, 1992 obtained from thin plates interpolation of the 6 measurements – in black dots.

In Figure 2, we can see that a veil of black particulate smoke is dispersing in the opposite direction of the wind. That certainly does not reflect the reality of polluting situation. This image could be compared to the image of Figure 3.

Figure 3 is a map of concentrations in black particulates obtained with the proposed methodology. The combination of satellite band, used to define the pseudostations, is the combination [1,2,3,4]. 43 pseudostations are located. Virtual stations (including real stations) were derived by a linear regression. Virtual stations are homogeneously located over the city of Nantes. In figure 3, virtual stations provide a better spatialisation of the studied phenomenon.

The resulting image, in figure 3, is obtained with the same thin plate method applied to the 43 virtual stations. Compared to figure 2, this image gives a more realistic view of the distribution of pollutants over the city. Actually zones of high and low levels of air pollution appear in the map. The veil of black particulate smoke is dispersing in the same direction than the wind direction. The authors are aware of the limitations of such a map and stress that there is currently not validated map of air pollution to compare to.

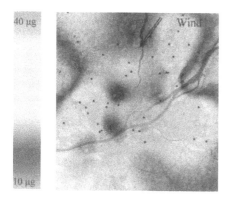

Figure 3. Extract of a map of black particulates over Nantes for May 22, 1992 obtained from thin plates interpolation of the 43 virtual stations – in black dots.

The discrepancies between the estimated concentrations and real measuring concentrations in black particulates are evaluated using the 6 remaining stations (table 1).

Table 1. Correlation, bias and RMSE (Root Mean Square Error) in percentage between estimated and real black particulate concentration over the city of Nantes for May 22, 1992.

	correlation	error %	RMSE %
without virtual stations	0,47	-76	89
with virtual stations	0,7	-32	54

The table 1 points out that virtual stations allow a better accuracy of the mapping of black particulates over the city of Nantes. Positive correlation is found between estimated and observed data for the 6 remaining stations and the use of virtual stations increase this correlation. In addition, the use of virtual stations

4.4 Limitations

The results show a great influence of the choice of the combinations of satellite bands of the generalized vector on the number and on the location of the virtual stations (table 2). The greater the number of spectral bands, the lower the number of virtual stations. In addition, the validation results (correlation, error and RMSE) depend on initial set of measuring stations.

Table 2. Numbers of virtual stations (including real station) located in the city of Nantes.

combination	n_virtual stations
[1,2,3]	385
[1,2,3,4]	43
[1,2,3,5]	30
[1,2,3,7]	48

Pseudostations are places of the city having same morphological, environmental and polluting features as a real fixed station. Their location should not be variable. It results that further studies are required for a better description of stations, through the notion of their identity card.

Previous works have shown that a fairly accurate linear relationship exists between satellite outputs and concentrations in black particulates. Further studies are needed to prove the significance of a relationship between satellite data and the other pollutants.

5 CONCLUSION AND PERSPECTIVE

This first application of the methodology for the mapping of the concentration in black particulates illustrates the potentialities of Earth observation data for the mapping of pollutant concentration. It shows that the mapping of concentration of black particulates is possible using the thermal band TM6 image of the Landsat satellite. The use of remotely sensed data for the mapping of pollutants over a city brings a better spatialisation of the phenomena under study. The notion of virtual stations improves the mapping. This encourages us to continue our investigation.

Further studies are needed to fully understand the links between the pollutant measurements and satellite data. Next investigation will focus on a better analyze of measuring stations and its neighborhood to improve the definition of an identity card of a station, of pseudostations and of virtual stations. The research will include a geographical data source: the BDTOPO®IGN© (figure 4) containing physical and morphological properties of the site (building heights, positions of roads...).

Figure 4. 3D dimension view on a location in the city of Strasbourg (France). An example of BDTOPO®IGN©, 1989.

This data source will allow
- a georeferencing of satellite images for a better localisation of stations.
- a better definition of the identity card of a stations

Morphological indicators for "distance to the road", "rate of building" or "wind exposition" will be defined.

ACKNOWLEDGEMENTS

The authors are indebted to Eurimage which has provided the Landsat TM data. The authors also thank Air Pays de la Loire for providing the pollution data.

REFERENCES

Basly L., 2000. Télédétection pour la qualité de l'air en milieu urbain. Thèse de doctorat, sciences de technologies de l'information et de la communication, Université de Nice Sophia Antipolis, France. 182 p.

Blanc P., 1999. Développement de méthodes pour la détection de changement. Thèse de doctorat, informatique temps réel – automatique – robotique, Ecole des Mines de Paris, France.

Brivio P.A., Genovese G., Massari S., Mileo N., Saura G. et Zilioli E., 1995. Atmospheric pollution and satellite remotely sensed surface temperature in metropolitan areas. *In: Proc. EARSeL symposium: Advances in Remote Sensing: pollution monitoring and geographical information systems*, Paris: EARSeL, pp. 40-46.

Carletti R., Picci M. and Romano D., 2000. Kriging and bilinear methods for estimating spatial pattern of atmospheric pollutants. *Environmental Monitoring and Assesment*, 63, 341-359.

Couloigner I., 1998. Reconnaissance de formes dans des images de télédétection du milieu urbain. *Thèse de doctorat, propagation – télécommunication et télédétection, Université de Sophia Antipolis. 129p.*

Finzi G. and Lechi G. M., 1991. Landsat Images of urban air pollution in stable meteorological conditions. *Il Nuovo Cimento*, 14C,433-443.

Frangi J.P., Jacquemoud S., Puybonnieux-Texier V. and Lazard H., 1996. Suivi spatio-temporel de la concentration en NO_x en île de France. *C.R. Acad. Sci. Paris 323*, série II a, pp. 373-379.

Ionescu A., Mayer E. and Colda I., 1996. Méthodes mathématiques pour estimer le champ de concentration d'une polluant gazeux à partir des valeurs mesurées aux points dispersés. *Pollution Atmosphérique*, janvier-mars 1966, pp. 78-89.

Sifakis N., 1992. Potentialités de l'imagerie à haute résolution spatiale pour le suivi de la répartition de pollutions atmosphériques dans la basse troposphère. Etude du cas d'Athènes'. *Thèse de doctorat, chimie de la pollution atmosphérique & physique de l'environnement, université de Paris VII*, 287 p.

Poli U., Pignatoro F. Rocchi V. and Bracco L., 1994. Study of the heat island over the city of Rome from Landsat-TM satellite in relation with urban air pollution. *In: R. Vaughan ed., Proc. 13 th EARSeL Symposium, Remote sensing – From research to operational applications in the new Europe, Dundee, Scotland, UK, 28 June – 1st July 1993*, Springer Hungarica, pp. 413-422.

Moussiopoulos N., Berge E., Bohler T., de Leeuw F., Gronskei K.E., Mylona S., Tombrou M., 1996. Ambient air quality, pollutant dispersion and transport models. *Report of the European Topic Centre on Air Quality.* 94 p.

Nevalainen J. and Pekkanen J., 1998. The effect of particulate air pollution on life expectancy. *The Science of the Total Environment 217*, pp. 137-141.

Parlow E., 1998. Net radiation of urban areas. *In: Gudmansen P. Ed., Future trends in remote sensing, Rotterdam: Balkema Publishers*, pp. 221-226.

Retalis A., Cartalis C. and Athanassiou E., 1999. Assesment of the distribution of aerosols in the area of Athens with the use of Landsat Thematic Mapper data. International Journal of Remote Sensing, 20, 5, pp. 939-945.

Sifakis N., Soulakellis N.A. and Paronis D.K., 1998. Quantitative mapping of air pollution density using earth observations: a new processing method and application to an urban area. *International Journal of Remote Sensing, 19, 17*, pp. 3289-3300.

Wald L. and Baleynaud J.M., 1999. Observing air quality over the city of Nantes by means of Landsat thermal infrared data. *International Journal of Remote Sensing, 20, 5*, pp. 947-959.

Observing our environment from Space: New solutions for a new millennium, Bégni (Ed.)
© 2002 Swets & Zeitlinger, Lisse, ISBN 90 5809 254 2

The devastation of Venezuela by heavy rains in December 1999: Disaster monitoring and evaluation of damages using ERS InSAR tandem data and SPOT images

E.Nezry, F.Yakam Simen & I.Supit
PRIVATEERS N.V. Private Experts in Remote Sensing, Philipsburg, Netherlands Antilles

F.Sarti
CNES (Centre National d'Etudes Spatiales, QTIS/SR), Toulouse, France

P.Romeijn
TREEMAIL International Foresty Advisors, Heelsum, Netherlands

ABSTRACT: After the devastation caused by two weeks of torrential rain in Venezuela in the second half of December 1999, it is estimated that up to 30,000 people have died in floods and landslides; a further 200,000 people have been left homeless as whole towns along the Caribbean coast have been washed away (Centeno 1999). Concerned about the dramatic extent of this human tragedy, PRIVATEERS N.V. has immediately offered its support to Venezuela. Supported by the French Space Agency (CNES) and the European Space Agency (ESA) who provided the sets of SPOT and ERS SAR (Synthetic Aperture Radar) images acquired before and after the disaster, this project aims to the provision of cartographic maps designed to assess the situation and to help with the rehabilitation of the country. These maps are designed to be easily exploitable by local photo-interpreters.

1 INTRODUCTION

The satellite data made available for this project are summarized in the table below:

Table 1: Satellite imagery made available for the project.

Satellite	Product	Date	Coverage
SPOT-1	P	18-01-1999	60 x 78 km
SPOT-4	M	25-12-1999	60 x 122 km
ERS-1	RAW	29-11-1998	112 x 107 km
ERS-2	RAW	30-11-1998	112 x 107 km
ERS-1	RAW	23-01-2000	112 x 107 km
ERS-2	RAW	24-01-2000	112 x 107 km

All four ERS SAR images where acquired in repeat-pass conditions. Therefore, they cover the same area. These ERS data form two interferometric couples, acquired before and after the disaster.

The SPOT data are single-channel images. The SPOT-P image acquired before the disaster is a panchromatic image (0.50-0.90 μ), whereas the SPOT-M image after the disaster is acquired in the red part of the optical spectrum (0.61-0.69 μ).

Regarding the data acquisitions during and after the heavy rains, the month elapsed between SPOT and ERS acquisitions will provide a dynamic view of the event, despite of the different nature of the data.

The experience earned in past similar occasions such as "Mitch" hurricane in November 1998 (Nezry *et al.* 2001a), China floods in the summer of 1999,

Izmit earthquake in August 1999 (Sarti *et al.* 2000) and "Lenny" hurricane in November 1999 (Fellah *et al.* 2000), has been extensively exploited in this project. Moreover, new techniques to detect changes due to a natural disaster using homogeneous (same sensor) or heterogeneous (different sensors) data sets have been further developed.

2 SITUATION ON DECEMBER 25, 1999 (SPOT DATA ONLY)

To detect flood extension, violent sedimentation phenomena, and the main flood streams, change detection is carried out using the SPOT-P archive and the SPOT-M present images. To this aim, the techniques used in (Nezry *et al.* 2001a, Fellah *et al.* 2000) are insufficient, and a split-window technique is applied to enable a comparison of the radiometries in SPOT-P and SPOT-M data.

Some inconveniences with the data hampered the complete detection of important changes. First, the presence of a stripe of clouds along the coastal cordillera in the SPOT-P archive image hindered the detection of sedimentation phenomena between the coast and the city of Caracas. Second, the fact that the radiometry of the SPOT-M image was saturated within urban areas hindered the detection of changes in such areas; it also causes some serious confusion between built-up areas and violent sedimentation events at the vicinity of urban areas.

Nevertheless, a good detection of the flooded areas, of the main flood streams and violent sedimentation events is achieved in most part of the common coverage of the archive and recent SPOT data.

To obtain a more readable map, the detected changes are overlaid in color over the SPOT-M image acquired on December 25, 1999.

This final SPOT change detection map is produced at a spatial resolution and a pixel size of 10 meters x 10 meters. It is georeferenced in Universal Transverse Mercator (UTM) - WGS 1984 cartographic projection. A subset of this map is shown in Figure 1.

3 SITUATION ON JANUARY 24, 2000 (ERS SAR DATA ONLY)

Based on the calibrated radar reflectivity, the ERS SAR data are used to detect the damages. In the change detection process, the ERS SAR archive data are compared to the ERS SAR data acquired just after the rains.

A combination of speckle filtering and super-resolution techniques (Nezry et al. 1996) is applied to the ERS SAR data, to detect and map existing targets at a spatial resolution and a pixel size of 15 meters x 15 meters, before and after the rains, with particular emphasis to housing and infrastructures.

To achieve this objective, the radiometric resolution of the SAR images used must be of very high quality. Since the radiometric resolution of SAR images is naturally corrupted by the presence of speckle, speckle filtering is the most critical issue for the success of the whole operation.

The filter that has been used in this project is an adaptive speckle filter for multi-channel detected SAR images, recently developed by PRIVATEERS N.V.: the Distribution-Entropy Maximum A Posteriori (DE-MAP) filter (Nezry & Yakam Simen 1999).

The superiority of Bayesian speckle filters is mainly due to the introduction of A Priori scene knowledge in the filtering process. Nevertheless, in the presence of very strong texture or of mixed textures, as it is often the case in SAR images of dense tropical forest, and/or in the presence of relief (which is the case in coastal Venezuela), it may be hazardous to make an assumption about the probability density function of the radar reflectivity of the scene. In this context, the A Priori knowledge with regard to the observed scene can hardly be an analytical first order statistical model. However, in the DE-MAP speckle filters, a Maximum Entropy constraint on texture is introduced as A Priori knowledge regarding the imaged scene.

The new DE-MAP filter is particularly efficient to reduce speckle noise, while preserving textural properties and spatial resolution, especially in strongly textured SAR images (Nezry & Yakam Simen 1999).

It adapts to a much larger range of textures than the previous multi-channel MAP filters (Nezry et al. 1996, Nezry et al. 1997) developed under the assumption of K-distributed SAR intensity (Lopes et al. 1993).

From the theoretical point of view, it is noteworthy that:
- this filter presents the very attractive properties of a control system (Nezry et al. 1997),
- it allows to super-resolve SAR images (Nezry & Yakam Simen 1999, Nezry et al. 1997), which is of high particular interest for ERS and SPOT-Panchromatic/SPOT-M (pixel size: 10 meters x 10 meters) data fusion.

Detection of the changes due to the heavy rains is carried out in a similar way than in Nezry et al. 2001 and in Fellah et al. 2000. A set of ™"DYNAMIC-RADAR" products (Nezry et al. 2001a) is produced to enable the detection and the photo-interpretation of the changes and damages. Nevertheless, the ™DYNAMIC-RADAR products are presently improved: a combination of more change indices is used, and a specific detection of changes occurred to point targets (built-up areas) is performed on the original complex SAR data.

Finally, the spatial sampling rate adopted for these improved ™"DYNAMIC-RADAR" products is 15 meters x 15 meters. They are georeferenced in UTM-WGS 1984 cartographic projection. A subset of this map, featuring an area around the city of Maracay (on the shores of the lago de Valencia), is shown in Figure 2.

Regarding the photo-interpretation keys, in these improved ™"DYNAMIC-RADAR" product, red tones denote the areas where the vegetation has suffered heavy damages, or sedimentation areas. Dark blue tones denote in most cases flooded areas, or water saturated areas. Light blue areas denote the location of heavily damages built-up areas. Grey tones, from white to black, denote undamaged areas (see Figure 2).

4 GLOBAL ASSESSMENT: FUSION OF SPOT AND ERS SAR DATA

In the present project, fusion of SPOT and ERS change detection products makes sense if we intend to get a synoptic view of the situation over the timeframe of one month, i.e. from the date of acquisition of the last SPOT image to the date of acquisition of the last ERS SAR image.

To this end, the changes detected on December 25, 1999 using SPOT data and the changes detected on January 24, 1999 using ERS SAR images are combined in a single map.

For a better representation, the SPOT-P archive image is used as mapping background. The final product is the result of a common-sense data fusion process integrating ERS SAR change detection and SPOT-P archive data, to facilitate photo-interpretation of the final product: above all, it allows a better understanding of the total extent of the floods and of the damages suffered in December 1999 and January 2000.

The SPOT/ERS change detection map is produced at a spatial resolution and a pixel size of 10 meters x 10 meters. It is georeferenced in UTM-WGS 1984 projection. A subset of this map, featuring the coastal area north of Caracas, is shown in Figure 3.

5 ARE THE ERS INSAR DATA USEFUL ?

With the intent to detect additional changes caused by the rains, interferometric ERS tandem pairs acquired before and after the disaster have been used to produce coherence maps.

These coherence maps have been filtered using a specific statistically adaptive filter developed by PRIVATEERS N.V. (Nezry 1996), to improve their radiometric quality.

Although the coherence maps are unexploitable in most of the covered area, mainly due to the presence of strong relief (from sea level to 2430 meters), a comparison of these coherence maps reveals dramatic changes over flat terrain, as shown in the Figure 4, featuring the Maiquetia international airport.

The strong increase of the ERS coherence observable on the runways can be attributed to the presence of sediment deposits left by the erosion consequent to the rains.

Nevertheless, from a global point of view, coherence maps have proven here their limitations for the detection of even important changes in the presence of strong relief, as it is the case in most of the area covered by our ERS images.

6 CONCLUSION

After the success of several previous similar projects, the present project proved that, in order to produce useful results very shortly after a natural disaster, the use of appropriate and well mastered techniques leads to rapid project implementation and execution, even if the concerned area is very wide, and the requirements in terms of spatial resolution are very strict.

In this project, 12,000 km^2 were covered at the resolution of 15 meters x 15 meters, and an additional 3,000 km^2 were covered at the resolution of 10 meters x 10 meters, in the most toughly affected part of Venezuela - coastal area, Caracas, etc.

Regarding the products themselves, they provide the local photo-interpreters and/or end-users with a clear overview of the situation at the time of the satellite data acquisition, during or just after the crisis, at high spatial resolution. In this overview the main changes of interest, such as damages to housing and infrastructures, and damages to agriculture and forest cover are put into evidence.

Moreover, when data (even heterogeneous data) are available at different times during and/or just after the crisis, it is possible to provide a clear synoptic overview of the whole event by fusion of the information acquired over time, by the same or by different spaceborne sensors.

For instance, on the example shown in Figure 3, one can immediately evaluate the huge damages suffered by the coastal area of Venezuela, and the devastation caused to the coastal cities by the violent flood streams and sedimentation.

Previously, techniques similar to those presented in this paper have been successfully used to assess the effects of the Mitch hurricane (Honduras, El Salvador, Nicaragua, November 1998, see Nezry *et al.* 2001a), the damages of the giant floods of the Yang-Tse-Kiang river in China (July/August 1999), the damages caused by the Izmit earthquake (Turkey, August 1999, see Sarti *et al.* 2000), the damages due to the "Lenny" hurricane (Guadeloupe Island, November 1999, see Fellah *et al.* 2000).

More recently, similar techniques proved successful to evaluate in near real time the damages caused by the earthquake of January 2001 in El Salvador (Nezry *et al.* 2001b).

In the course of these projects, the techniques used have been continuously improved by PRIVATEERS N.V., with the aim to obtain a better detection of changes and damages, as well as to get a better interpretation of the detected changes in terms of nature of the damages. This technical progress will be pursued in the future.

These techniques are now fully operational (Nezry *et al.* 2001b). In a world increasingly affected by natural disasters, they must, and will, be more intensively exploited to speed-up post-crisis reaction and establish rehabilitation projects.

However, in future projects of the same nature, improvements must be made with regard to the decision-taking processes at the level of the political bodies on which the major data providers depend, in order to shorten their delay of reaction, and with regard to the data delivery delays.

Improvements must also be made with regard to the dissemination of results and the procurement of appropriate technical support to the concerned end-users. This applies in particular to most of the developing countries (but not only) where remote sensing is still a little known, or poorly mastered technology.

The above conclusions were confirmed during the recent operation in support of disaster relief workers following the El-Salvador earthquake of January 13, 2001 (Nezry *et al.* 2001b, BBC 2001).

Figure 1: Change detection using SPOT panchromatic images: situation on December 25, 1999 (Original in color).

Figure 2: Change detection using ERS SAR images: situation on January 24. 2000. Eroded areas. sedimentation areas, and damages to the vegetation cover are in red (Original in color).

Figure 3: Change detection using SPOT panchromatic and ERS SAR images: situation on January 24. 2000.
Still intact built-up areas are in white. Damaged built-up areas are in light blue. Areas that have been flooded and coastal sedimentation plumes are in dark blue. Flood streams and violent sedimentation areas are in violet. Eroded areas and damages to the vegetation cover are in red (Original in color).

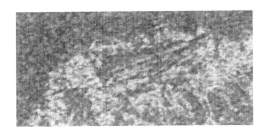

Figure 4: Comparison of the ERS-1/ERS-2 tandem coherence images of Maiquetia international airport, acquired before (29/30 November 1998 in red), and after (23/24 January 2000 in blue) the heavy rains (Original in color).

7 ACKNOWLEDGMENTS

We acknowledge herewith the Centre National d'Etudes Spatiales (CNES, the French Space Agency), the European Space Agency (ESA), and SpotImage for the provision of the ERS SAR data and for the SPOT data which have been used in this project.

8 REFERENCES

BBC, 2001: Aid from Space, *BBC News | SCI/TECH | Aid from space, by BBC News Online's Ivan Noble, http://news.bbc.co.uk/hi/english/sci/tech/newsid_1172000/ 1172346.stm, 16 February 2001.*

J.C. Centeno, J.C. 1999. Venezuela viste de luto, *Open Letter circulated over the Internet, 7 p., 24 December 1999.*

Fellah K., Nezry E., Bally P., Bequignon J., Herrmann A., Bestault C. & de Fraipont P. 2000. Rapid mapping system for hurricane damage assessment. *ERS-ENVISAT Symposium, Gothenburg (Sweden), ESA SP-461 (CD-ROM), 16-20 October 2000, Paper nr.339.*

Lopes A., Nezry E., Touzi R. & Laur H. 1993. Structure detection and statistical adaptive speckle filtering in SAR images. *International Journal of Remote Sensing, 14 (9) :1735-1758.*

Nezry E. 1996. New polarimetric - interferometric filters using prior knowledge of phase difference and degree of coherence first order statistics. *10 p., Unpublished.*

Nezry E. & Yakam Simen F. 1999. A family of distribution-entropy MAP speckle filters for polarimetric SAR data, and for single or multi-channel detected and complex SAR images. *Proceedings of the CEOS SAR Workshop, ESA SP-450 :219-223, Toulouse (France), 26-29 October 1999.*

Nezry E., Yakam Simen F., Romeijn P., Supit I. & Bally P. 2001a. Evaluation of "Mitch" hurricane damages in Central America using satellite imagery. *21th EARSeL Symposium, Marne La Vallée (France), 14-16 May 2001. This issue.*

Nezry E., Yakam Simen F., Zagolski F. & Supit I. 1997. Control systems principles applied to speckle filtering and geophysical information extraction in multi-channel SAR images. *Proceedings of SPIE, 3217 :48-57.*

Nezry E., Zagolski F., Lopes A. & Yakam Simen F. 1996. Bayesian filtering of multi-channel SAR images for detection of thin details and SAR data fusion. *Proceedings of SPIE, 2958:130-139.*

Sarti F., Nezry E. & Adragna F. 2000. Complementarity of correlation and interferometry for the analysis of the effects of the Izmit earthquake with radar data. *Proceedings of the Multi-Conference On Systemics, Cybernetics and Informatics (SCI2000), 6 p., Orlando (Florida, USA), 23-26 July 2000.*

Nezry E., Romeijn P., Sarti F., Inglada J., Zagolski F. & Supit I. 2001b. Breaking new grounds for remote sensing in support of disaster relief efforts: detecting and pinpointing earthquake damages in near real-time (El-Salvador, January 2001). *To be presented at the 8th SPIE-EUROPTO Symposium on Remote Sensing, Toulouse (France), 17-21 September 2001.*

Observing our environment from Space: New solutions for a new millennium, Bégni (Ed.)
© *2002 Swets & Zeitlinger, Lisse, ISBN 90 5809 254 2*

Spatio-temporal pattern formation in rainfall from remote sensing observation

L.N.Vasiliev
Institute of Geography, Moscow, Russia

ABSTRACT: Self-organised criticality (SOC) has been proposed as a mechanism for generating complexity and power laws. One characteristic of many natural systems is power law statistics. That is the likelihood of an event occurring decreases as some power of the event's size. Our focus here is on rainfall in the context of sand pile model. Rainfall is probably one of the direct examples of SOC phenomenon in the environment. The methodology fully treats the non-linearity of the problem. The characteristic features of SOC in rainfall include: 1) dynamic scaling 2) a power law dependence on spatio-temporal patterns. The multiscaling hypothesis may be expressed as a relatively compact statement: there exist two numbers termed the size and time scaling powers such that precipitation obeys the functional equation. Rainfall forms geometric space-time bodies which are topologically equivalent (homeomorphic) to the balls.

1 INTRODUCTION

In several recent years remote-sensing spaceborne procedures have yielded considerable achievements in the timely acquisition of data about global precipitation. Measurements from several space systems in orbit, combined together, make it possible to get regular daily data about global space-time precipitation and, due to this, to describe this process at a global level (Huffman 2000). New concepts in the understanding of rainfall behavior formed when the idea of self-organized criticality (SOC) had been developed (Bak et al. 1987). They are be implemented in the computer model which characterizes the behavior of a non-linear many-body system in a state of interrupted equilibrium. The sand pile model provides the most attractive and fairly well studied version of SOC. Interrupted equilibrium means that periods of rest interchange with avalanche-type perturbations which do not have fixed regular periodicity. The advantage of this model lies in the fact that universal relations arise in it, which characterize avalanche dynamics behavior. The dimension of avalanches, their duration and intervals between events obey the power law. The exponents may vary from system to system, but the scaling relations may be more universal. SOC signatures have been observed in precipitation in the 49-year rain-gauge measurements (Vasiliev 2000). These results, however, refer only to the time domain. At the center of our understanding of space-time pattern formation in precipitation is dynamic scaling. Interest to SOC in

the environment, in particular in precipitation, is explained by attempt to pose deep questions about our ability to predict natural phenomena. Though the SOC theory is not yet completed, its idea reflects a fairly simple, robust principle of interrupted equilibrium.

Let us imagine a rainfall process in space and time coordinates X, Y, t. A time series of precipitation images may be ordered in a 3D array which elements are determined by estimated precipitation $P(X, Y, t)$. The value P that refers to a fixed point X, Y, is equivalent to rain gauge measurements. Such time series are characterized by multifractality and power law in precipitation, its duration and intervals without rainfall. Space-time concept of precipitation leads to the formation of 3-dimensional self-affine fractal set. Scaling in a horizontal and a vertical cross-section therein should be different (Mandelbrot 1986). This implies that the precipitation field in the XY –cross-section, with Δt fixed, is self-similar, but it is self-affine in terms of variables X (or Y) and t. Scaling exponents for horizontal and vertical cross-sections will be different. The same is true about the exponents of global precipitation measured in segments over the spherical surface of the Earth.

The step that follows next is associated with a powerful and vivid picture of space-time precipitation bodies topologically equivalent (homeomorphic) to balls. This can be imagined as watering from a sprinkler when the latter is being moved along a vegetable patch. The length of an affine ball

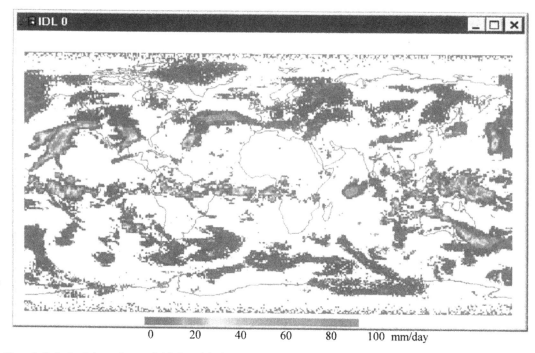

0   20   40   60   80   100 mm/day

Figure 1. Daily 1 x 1 degree image of global precipitation on 21 May 1997. (Colour plate, see p. 421)

thus formed is proportional to time, whereas its mass is proportional to the volume of the ball. What follows presents this approach to precipitation in terms of spaceborne and radar measurements.

2 INPUT DATA

The 15 minute instantaneous radar reflectivity data provided by WSI is created from the US National Weather Service 5 cm and 10 cm radars. The radar reflectivity data is in image form with a size of 1837 rows by 3661 columns and has a resolution of 2 km by 2 km. Rainfall rates were derived from the Reflectivity-Precipitation relationship (Woodley et al 1975). The One-Degree Daily (1DD) precipitation data set is the first approach to estimating global precipitation at the 1x1-degree scale strictly from observational data (Huffman 2000). An example of a 1DD image is shown in Figure 1.

3 DYNAMIC SCALING

Presentation of space-time precipitation pattern as a self-affine fractal is associated with at least two scaling exponents a_L, a_t, as shown in Figure 2. The grid dimension of the self-similar field of precipitation measured in the Δt interval is determined by a_L, whereas the dimension along the time axis, by a_t.

The grid dimension sets up the dependence between the size of the cube within which precipitation was recorded and the number of such events $N(L, \Delta t)$. Here an event means precipitation within a cube with the size $L \times L \times \Delta t$. For positive λ_1, λ_2 that determine the coefficients of cell-size L increase in the horizontal cross-section and along the time axis t, the dynamic scaling is described by:

$$N(L, \Delta t) = \lambda_1^{-a_L} \lambda_2^{-a_t} N(\frac{L}{\lambda_1}, \frac{\Delta t}{\lambda_2}), \qquad (1)$$

where $N(L, \Delta t)$, $N(\frac{L}{\lambda_1}, \frac{\Delta t}{\lambda_2})$ - are the numbers of cubes with sides $(L, \Delta t)$, $(\frac{L}{\lambda_1}, \frac{\Delta t}{\lambda_2})$, that cover the self-affine set of precipitation. If $\lambda_1 = \lambda_2 = \lambda$ we have $a = a_t + a_L$. However, it is only a partial case, and a *cannot* be regarded as a fractal dimension of a self-affine set.

Equation 1 determines dynamic scaling in precipitation by transformation from one to another resolution in space and time λ_1, λ_2. The exponents a_t, a_L are not independent. For a self-similar field of precipitation, measured within the interval $\Delta t = 15$ minutes, $a_L = 1.85$. The a_t value depends on the size L that is associated with a resolution in space. The reflectivity images have a resolution of 2 km by

160

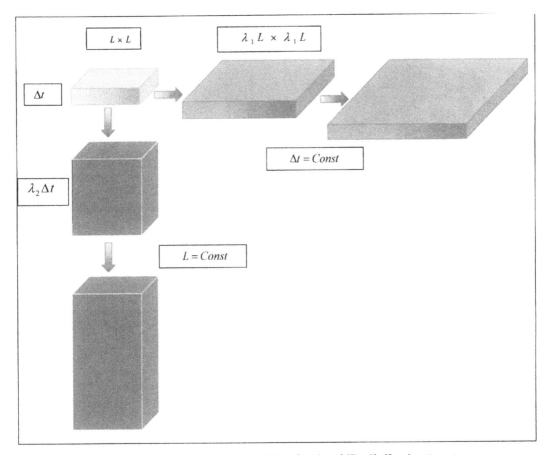

Figure 2. Schematic illustration of dynamic scaling in precipitation. Covering of 3D self-affine fractal set of space-time precipitation with cubes of sizes $L \times L \times \Delta t$. (Colour plate, see p. 421)

2 km. Global precipitation is measured on the segments in the 1-degree solid angle whose area is larger than 10000 km. Covering of 3D space-time related to the rain-gauge measurements. The crossover L-size where the exponent $a_t = 1$ is above 100 km. Hence, for one-degree daily precipitation $a_t = 1$. It means that the number of events is proportional to λ_2 and depends on time.

4 VIRTUAL BALLS OF PRECIPITATION

Rainfall is presented in terms of avalanche dynamics using time series of global images which record precipitation in fixed time intervals equal to 24 hours. Each image referred to time t shows precipitation P as concentrated in separate isolated regions that cover the area $A_t(P)$. If two successive images $A_t(P)$ and $A_{t+1}(P)$ overlap, that is $A_t(P) \bigcap A_{t+1}(P) \neq \varnothing$, then within this time equal to

two time intervals the precipitation is presented as a volume $V(P) = \sum_t A_t(P)$. In space-time such groups form bodies topologically equivalent (homeomorphic) to balls. To experimentally confirm the signatures of self-organized criticality in the rainfall images of one-degree daily global precipitation over 1997 to 1999 made within the Global Precipitation Climatology Project (GPCP) and US Composite Reflectivity Data were used.

3D spatio-temporal presentation of the global precipitation is shown in Fig. 3 where the z-axis is time. The space-time distribution is subdivided into two regions (at treatment V and P values) corresponding to air mass showers and those produced by organized systems such as fronts, squall lines and low centers. Power law distribution of the ball volumes and their masses is shown on Figs 4 and 5. The distribution is given by $N(V) \sim P^{-\tau}$ and $N(P) \sim P^{-\gamma}$. On the territory of the USA $\tau = 2.042$

161

Figure 3. Daily 1x1 degree 3D spatio-temporal pattern of precipitation balls on a global scale during 1997.(Colour plate, see p. 422)

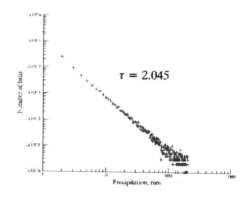

Figure 4. Power law dependece on precipitation balls: An air mass shower regime.

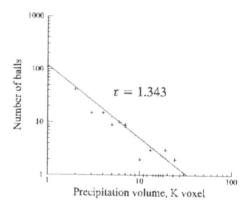

Figure 5. Power law dependece on precipitation balls: Rainfall produced by organized systems such as fronts and low ceners.

and $\gamma = 2.045$ for air mass showers. For precipitation produced by fronts and low centers $\tau = \gamma = 1.343$.

5 CRITICAL POINT IN PRECIPITATION

Scaling derived from Equation 1 makes it possible to get an idea about the inner structure of experimental data concerning the rainfall dynamics. However, it does not give the critical-point exponent. To clear out the origin of the τ exponent in the distribution of virtual balls scaling, numerical simulation and spatial-temporal pattern of precipitation should be considered together. Here we put to use percolation and renormalization transformation. The percolation theory is used for interpreting various physical phenomena reaching critical points. This is a phase-transition model that was formulated only in comparatively recent time (Stanley et al. 1982).

The simplest model of percolation can be formulated as follows. Given is a periodic cubic lattice embedded in 3-dimensional space with the probability p for each site $a \times a \times \Delta t$ of lattice to be occupied (meaning precipitation) and the probability $1 - p$ to be empty. For small p most of the occupied sites are surrounded by vacant neighbor sites. However as p increases, many of the neighbor sites become occupied and the sites form clusters of volumes V and length $\xi(p)$. As p increases, $\xi(p)$ increases monotonically, and at a critical value of p_c it diverges $\xi(p) = |p - p_c|^{-\nu}$ with the exponent $\nu = 0.9$ for $d = 3$. For $p > p_c$ there appears, in addition to the finite clusters, a cluster that is infinite in extent. The p_c value depends on the space dimension d and on the type of a lattice; as to ν, however, it depends only on d.

Here we transfer the percolation model into a virtual space to explain the mechanism obeying which power law is generated in precipitation balls. In the model of percolation the volumes V of clusters have the power law distribution $N(V) \sim V^{-\tau}$ and the power exponent $\tau = 2.17$ for $d = 3$. However, we stress once again that precipitation forms a self-affine fractal set and the exponent τ should not coincide. This is the consequence of a self-affine fractal which could not have one dimension. Comparison of τ values for the percolation model and for precipitation shows that an atmospheric process may be interpreted in terms of percolation.

To find a critical point we should use renormalization transformation (Fisher 1998). The renormalization group theory explains the dynamic scaling in precipitation and permits calculations of critical values. For virtual balls it is a way to answer the ques-

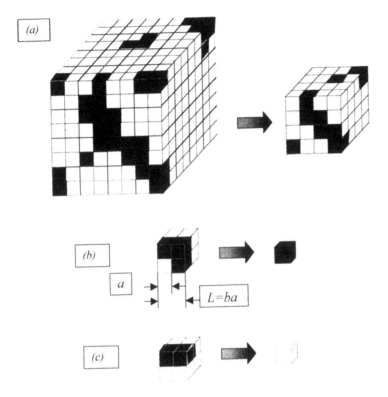

Figure 6. The construction of renormalization transformation for time-space cell decomposition for precipitation. 3D cubic lattice of spatio-temporal pattern of precipitation divided up into blocks or cells of dimensions $(L = ba) \times (L = ba) \times (L = b\Delta t)$.
(a) - Rescaling a cubic lattice by forming cells out of groups of sites. *(b)* - Cell-to-cube/site construction with $b = 2$: occupied.
(c) - Cell-to-cube/site construction with $b = 2$: empty.

tion concerning the existence of a ball of infinite length and the condition for it to appear. Just as each cube (site) in the 3D lattice is described by parameter p, its probability of being occupied, so each 3D cell of edge b, $(L = ba) \times (L = ba) \times (L = ba)$, each containing b^3 sites is described by a parameter p', which we may regard as being the 3D occupation probability. The relationship between p and p' obeys the rule under which the cell formed is considered occupied or vacant, that is, whether or not precipitation is there.

The construction of renormalization transformation for time-space decomposition for precipitation is shown in Figure 6. After partitioning a cubic lattice into cells of edge b, these cells play the role of renormalized cubes. If the cubes are occupied with probability p, then the cells may be defined as occupied with probability p', where $p' = R(p)$. It is simply the probability of there being a path that spans the cell from top to bottom. If such a spanning path exists, then, according to this choice of weight function the cell is occupied. After the first step of decomposition it repeats over and over again. The essential step in the renormalization transformation is

the construction of a functional relation between the original parameter p and the renormalized parameter p'. The function $p' = R(p)$ is termed a renormalization transformation. Its actual choice varies from one problem to the other. In the case of precipitation the cell is occupied if vertical connection of two occupied sites exists. For 3D cubic lattice

$$R(p) = p^8 + 8p^7(1-p) + 28p^6(1-p)^2 + 56p^5(1-p)^3$$
$$+ 54p^4(1-p)^4 + 24p^3(1-p)^5 + 4p^2(1-p)^6$$

(3)

It follows from (Eq. 3) that $R(p = 0.282) = 0.282$, that is, for space-time probability of precipitation $p = 0.282$ in $(L \times L \times \Delta t)$- size cubes $p' = p$. This value is termed the fixed point $p*$ of the renormalization which determines the percolation threshold (Fig. 6). All that implies that there exists a critical precipitation concentration p_c below which at $p < p_c$ only self-affine finite-volume balls of precipitation exist and above which at $p > p_c$ an infinite ball is present as well as finite-volume ball distributed by a power law. The correlation length $\xi(p)$ has a certain geophysical interpretation. It

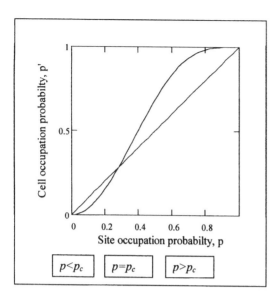

Figure 7. Position-space renormalization transformation. In the renormalization group approach to percolation in 3D spatio-temporal pattern of precipitation naturally divides into three regimes $p < p_c$, $p = p_c$ and $p > p_c$.

agrees with treating precipitation as infinite-length balls at $\xi(p)$ values reaching the percolation threshold.

For one-degree daily (1DD) precipitation on the spherical segments the probability of events is $p > p_c$, that is space-time precipitation exceeded the critical value. It means that a combination of air mass showers and precipitation produced by organized systems such as fronts, squall lines and low centers, leads to the formation of infinite-length balls. The theoretical conclusion is fully confirmed by the one-degree daily precipitation data. If the process is considered at the regional level, the boundary condition may lead to the $p < p_c$ case. Then precipitation forms only isolated balls with the power law distribution.

Dynamic scaling in precipitation non-trivially affects the p-value, a probability of precipitation in the cubic lattice. Transformation from low to high resolution in space and time changes the site $L \times L \times \Delta t$ to $L/\lambda_1 \times L/\lambda_1 \times \Delta t/\lambda_2$. In accordance with (Eq. 1) the probability of precipitation on a new cube lattice p_h with a higher resolution is related to p as

$$p_h = \lambda_1^{-2+a_L} \lambda_2^{-1+a_t} p \qquad (4)$$

For global precipitation 4-times higher resolution (6 hours, 0.25 degrees) changes the probability $p_h = 0.81p$, with which the system remains in the state above critical with $p > p_c$. At the regional level the higher resolution may characterize the system in the state below critical with $p < p_c$.

Comparison of the avalanche dynamics in precipitation with the SOC model makes obvious their similarities and differences. One of the main reasons why they differ is boundary conditions in the model which generates avalanche dynamics at the local level. If we are to consider rainfalls over the separate region then signatures of interrupted equilibrium and the power law distribution of precipitation agree with the sand pile model of SOC. However, on the global level when boundary conditions are observed, the SOC model should meet new requirements brought about by the equilibrium violation at several places.

The another comment concerns the interrupted equilibrium in terms of SOC. It may concern only separate parts of the system characterizing the local behavior. On the global level the system cannot be at rest, since precipitation arise and continue. Thus, although precipitation does demonstrate certain common SOC features, it is still necessary that new models be developed, those adapted to specific natural phenomena and to their scale level, as SOC models of forest fire and evolution are adapted.

6 CONCLUSION

New images of global precipitation relying upon space technology give new approaches to the process. Hence within the spatio-temporal pattern precipitation forms a self-affine fractal structure. It is characterized by dynamic scaling–as well as power law dependence of water masses manifesting themselves as virtual balls for air mass showers and precipitation which were produced by organized systems such as fronts, squall lines and low centers. Dynamic scaling, percolation theory and renormalization group approach have given a new insight into the properties of global space-time precipitation. Common features of avalanche dynamics include virtual balls, power law and critical point. An important consequence of scaling is the development of transformation for space-time precipitation from low to high resolution along the path toward increasingly realistic model of self-organized criticality in precipitation.

ACKNOWLEDGMENTS

Data provided by the Global Hydrology Resource Center (GHRC) at the Global Hydrology and Climate Center, Huntsville, Alabama. The 1DD data were provided by the NASA/Goddard Space Flight Center's Laboratory for Atmospheres, which develops and computes the 1DD as a contribution to the GEWEX Global Precipitation Climatology Project.

REFERENCES

Bak, P., C. Tang & K. Wiesenfeld 1987. Self-organized criticality: an explanation of 1/f noise. *Phys. Rev. Lett.* 59: 381-384.

Fisher, M. E. 1998. Renormalization group theory: Its basis and formulation in statistical physics. *Rev. Mod. Phys.* 71, (2): S358-S366.

Huffman, G. J., R. F. Adler et al. 2000. Global Precipitation at one-degree daily resolution from multi-satellite observations. *Journal of Hydrometeorology.* In print.

Mandelbrot, B. 1986. Self-affine fractal sets. In L. Pietronero & E. Tosatti (eds.), *Fractals in physics*: 3-28. Amsterdam: North-Holland.

Stanley, H. E., P. Reynolds, S. Redner, and F. Family. 1982. Position-space renormalization group for models of linear polymers, branched polymers, and gels. In T. W. Burkhardt & J. M. J. van Leeuwen (eds.). *Real-Space Renormalization.* 169-206. Springer-Verlag, Berlin.

Vasiliev, L. 2000. Self-organized criticality in the environment and remotely sensed data. In J. L. Casanova (ed.). *Remote Sensing in the 21st Century: Economic and Environmental Applications.* 255-259. Balkema, Rotterdam.

Woodley, W. L., A. R. Olsen et al. 1975. Comparison of gage and radar methods of convective rain measurement. *Journal of Applied Meteorology.* 909-928.

5 *Sensors, Systems and Methods*

Observing our environment from Space: New solutions for a new millennium, Bégni (Ed.)
© 2002 Swets & Zeitlinger, Lisse, ISBN 90 5809 254 2

A numerical analysis of the accuracy of laboratory and field goniometer measurements

N. Widen
Finnish Geodetic Institute, Dep. of Remote sensing and Photogrammetry

R. Kuittinen
Finnish Geodetic Institute

ABSTRACT: This paper presents a numerical method for estimating the accuracy of field and laboratory measurements of bidirectional reflectance. After a detailed analysis of the measuring instrument and data acquisition process, the variables which determine the accuracy of the output are identified, and different techniques for laboratory and field spectrometry are compared. The results of the analysis show how both the measuring instrument itself, and the measuring technique in field experiments determine the accuracy of experimental data.

1 INTRODUCTION

Spectral reflectance and albedo of land cover and land surface are important parameters for environmental monitoring and global change research. Heat exchange between the Earths surface and the atmosphere depends on the cover fraction and land cover type, due to different spectral absorption and scattering properties. Changes in the land surface properties and land coverage (i.e. deforestation, desertification, erosion, forest fires, snow melting, deglaciation) occur continuously, and are frequently either directly or indirectly caused by human activities. High quality local and global scale land surface monitoring is thus essential to produce accurate and up-to-date input parameters for ecologic, hydrologic and atmospheric models used to develop the environment in a sustainable way.

Ground data is needed for thematic mapping based on hyper- or multispectral air- or spaceborne images to provide spectral signatures for different land cover types. The interpretation of the images and classification is also hampered by angular dependence of spectral reflectance, and during the last decade, great effort has been made by the scientific community to quantitatively characterise the bidirectional reflectance functions of different land cover types. There is, however, very little information available about the accuracy of the results, presented in the form of databases, or as sets of parameters retrieved by model inversion.

2 BIDIRECTIONAL REFLECTANCE FACTOR

BRF is defined as the ratio of radiance L_r reflected from the target surface into direction (θ, ϕ) to the radiance L_{ref} reflected into the same direction from a reference surface with known reflectance characteristics R, in identical illumination conditions. Thus, the BRF is derived from

$$R(\theta_s,\theta_v,\varphi,\lambda)=\frac{L(\theta_s,\theta_v,\varphi,\lambda)}{L_{ref}(\theta_s,\theta_v,\varphi,\lambda)}R. \quad (1.)$$

The reference surface is usually a diffuse reflectance standard. Reference panels of different types can be used, such as Goretex, $BaSO_4$ or Spectralon. BRF characteristics of the reference panel have to be characterized in laboratory measurements if they are not provided by the manufaturer with the calibration certificate, to account for any anisotropic scattering propertes of the reflectance standard.

Bidirectional reflectance is measured using a goniometer or other mechanical device which allows viewing the same surface from many directions, providing an accurate reference frame. The viewing instrument is a single or multiband optical detector, and BRF is measured either over the full spectrum of visible and near infrared light, or for a single band in this range.

3 A NUMERICAL ANALYSIS OF EXPERIMENTS

3.1 Introduction

A measurements system is in general characterized by its static and dynamic properties (Bentley, J., 1988). Static properties describe the output of the measurement system at constant or slowly varying input, and the way the system responds to sudden changes in the input is called dynamic properties. The most commonly used static properties of a sensing element are its sensitivity, resolution, and linearity. Sensitivity is defined as the rate of change in output vs. input, and resolution is the largest input which causes no measurable response. Repeatability of a measurement refers to the ability of the measurement system to give the same output at repeated measurements with the same input, in steady-state conditions.

The input is a linear combination of the required input variable, and modifying or interfering environmental effects. A modifying effect changes the linear sensitivity of the system, and an interfering effect causes a constant bias. The impact of random or systematic errors due to environmental effects can be quantified by specifying the accuracy of the measurement system. Absolute accuracy is defined as the difference between measured value and true value. Common techniques for reducing the error caused by environmental effects are isolation, i.e. elimination of the environmental variables, or by introducing an opposing environmental input to cancel an environmental input.

3.2 Identifying variables which influence measuring accuracy

The measurement system used for measuring BRF of a surface is a goniometer, a spectrometer connected to a portable computer, and a reflectance standard. The surface is illuminated from the direction (θ_s, φ_s) by a collimated halogen lamp (laboratory), or by the sun. The spectrometer is mounted on the goniometer to point towards the center of the construction, and can be moved on the surface of a half hemisphere in two independent directions specified by angles (θ_v, φ_v). Angular movement of the detector is either automated or manually controlled. Viewing and illumination directions can be determined with a certain precision denoted by $(\Delta\theta_v, \Delta\varphi_v)$ and $(\Delta\theta_s, \Delta\varphi_s)$, which depends on the construction of the goniometer. Usually the direction $\varphi = 0$ is defined by the direction of illumination (principal plane), and only one coordinate φ_r is used. The sys-

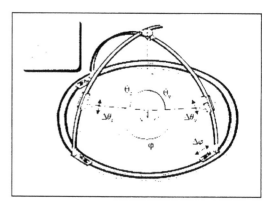

Picture 1 Overview of a goniometer and the angles used in calculations. (Picture from the Joint Research Center, Ispra, Italy.)

tem dependent measuring accuracy, which is equivalent to measuring accuracy of an isolated laboratory experiment, is limited by positioning precisions of detector and lightsource, and mechanical stability of the construction, assuming steady-state illumination. In field measurements environmental effects are present, the most significant environmental variables are diffuse light scattered by the atmosphere, and drifting of the sun during data acquisition. In BRF field measurements it is important to suppress environmental effects either in the data processing algorithms after measurement, or using error reduction techniques during experiments.

3.3 Estimating measuring accuracy numerically and error reduction

In the following sections the data acquisition process in field and laboratory experiments is analysed numerically, and quantitative estimates for measuring accuracy are calculated for three different types of experiments.

BRF is a function of three angles and wavelength, hereafter denoted by $\rho(\theta_s, \theta_v, \varphi)$, where wavelength dependence is suppressed. Differentiation is a common method for estimating measuring accuracy, i.e. (2.)

$$d\rho = \frac{\partial\rho}{\partial\theta_s}d\theta_s + \frac{\partial\rho}{\partial\theta_s}\frac{\partial\theta_s}{\partial t}dt + \frac{\partial\rho}{\partial\varphi}d\varphi + \frac{\partial\rho}{\partial\varphi}\frac{\partial\varphi}{\partial t}dt + \dots$$
$$\dots + \frac{\partial\rho}{\partial\theta_v}d\theta_v.$$

This expression accounts for the environmental variables (excluding atmospheric effects), and for the measuring instrument itself. Partial derivatives in time account for drifting of the sun during a BRF field measurement with complete angular coverage.

A numerical root-mean-square estimate for the absolute error $\Delta\rho$ is calculated using (3.),

$$\Delta\rho=\sqrt{\left(\frac{\partial\rho}{\partial\theta_s}\right)^2\Delta\theta_s^2+\left(\frac{\partial\rho}{\partial\theta_s}\frac{\partial\theta_s}{\partial t}\right)^2\Delta t^2+\left(\frac{\partial\rho}{\partial\varphi}\right)^2\Delta\varphi^2+...\\...+\left(\frac{\partial\rho\partial\varphi}{\partial\varphi\partial t}\right)^2\Delta t^2+\left(\frac{\partial\rho}{\partial\theta_v}\right)^2\Delta\theta_v^2}$$

where $\Delta\theta_s$, $\Delta\theta_v$, and $\Delta\varphi$ are precisions of source and detector positions, and Δt is time from beginning of the experiment. Absolute and relative errors for a BRF measurement with full angular coverage are defined as (4.)

$$\frac{1}{N}\sum_{i=1}^{N}\Delta\rho_i \text{ , and } \frac{1}{N}\sum_{i=1}^{N}\frac{\Delta\rho_i}{\rho_i}.$$

N is the total number of measurements.

3.3.1 Laboratory measurements

The laboratory measurement is an isolated experiment, i.e. the environmental variables are negligible. In this case, when there is no diffuse illumination, and the lightsource is stationary;

$$\frac{\partial\theta_s}{\partial t}=\frac{\partial\varphi}{\partial t}=0.$$

Assuming steady-state illumination, we have calculate $\Delta\rho$ using the model and equation (3.), with parameters

$\Delta\theta_v=0.1°,1.0°,5.0°,10.0°$
$\Delta\varphi=0.1°,1.0°,5.0°,10.0°$
$\Delta\theta_s=1.0°,3.0°,5.0°,10.0°$
$\theta_s=30°,50°,60°$

Partial derivatives with respect to θ and φ are calculated numerically using empirical or semiempirical models presented by many authors (Rahman et. al, 1993, Roujean et. al, 1992, Wanner et. al., 1995, etc.). In this study the model of Rahman is used, with input parameters corresponding to a vegetation

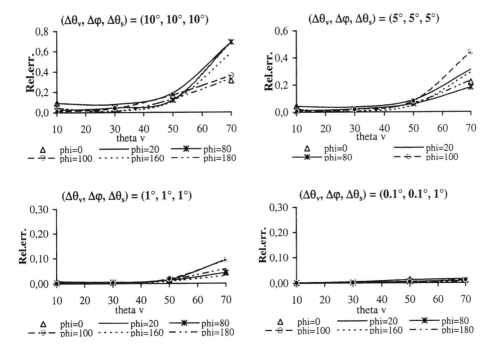

Figure 1 Calculated relative errors of laboratory experiment with different instrument parameters. Direction of illumination 30° from zenith.

target and above mentioned angles of illumination. The hot-spot area is left out from the calculations.

A few examples of the results are shown in Figure 1. The figure indicate that low-precision instruments can be used for measurements at small viewing angles, but a high-precision instrument is required for measuring BRF with a complete angular coverage. In this example the accuracy tends to decrease remarkably at viewing angles of 50° or more, but as the relative error depends on the input parameters to the BRF model (the shape of the function), it is not a general trend.

3.3.2 Field measurements

The accuracy of field measurements is estimated with the same method as laboratory measurements. Now all the terms in equation (3.) are significant. The most important error reduction technique in field experiments concerns atmospherical effects. Diffuse light is possible to estimate experimentally by means of blocking the direct solar beam manually, and measuring target and reference (dark measurement). Dark measurements are then subtracted from the illuminated target and reference data, and BRF is calculated:

$$R = \frac{L - L^{dark}}{L_{ref} - L_{ref}^{dark}} \ . \tag{5.}$$

The following estimate for measuring accuracy of field experiments does not account for atmospheric effects.

The second important error reduction concerns drifting of the sun during data acquisition. The solar elevation as function time is approximately a parabola, and θ_s is continuously changing during experiments. If the direction $\varphi=0$ is defined by the position of the sun at the first measurement, there will be a time-dependent error in φ also during data acquisition.

If the measured surface is approximately isotropic, the azimuthal drift of the sun can be compensated for by performing the measurements in series of constant viewing angle. Each time the viewing angle is changed, the origin of the azimuthal angle is redefined at the current sun position. This is equivalent to introducing an opposing environmental input to cancel $\partial\varphi/\partial t$. The best time for measurements is around noon, when the solar zenith angle changes slowly. Fast data acquisition in field experiments is essential, in order to minimize environmental efects, and a fast manually operated goniometer is more appropriate than slow automated systems for accurate measurements.

Field experiment I

Assume a field measurement in clear-sky conditions with fixed principal plane. Data acquisition starts from nadir, and a total of N observations are measured. If the interval between successive datapoints is 20° in azimuthal direction, 18 angles are measured for each zenith angle.

At solar noon, we have assumed that

$$\frac{\partial\theta_s}{\partial t} \approx 0, \ \frac{\partial\varphi}{\partial t} \approx 15°/h$$

In the morning or afternoon,

$$\frac{\partial\theta_s}{\partial t} \approx 7°/h, \ \frac{\partial\varphi}{\partial t} \approx 15°/h \quad \text{approximately.}$$

The input parameters for error estimation are

$\Delta\theta_v = 0.1°, 1.0°, 5.0°, 10.0°$
$\Delta\varphi = 0.1°, 1.0°, 5.0°, 10.0°$
$\Delta\theta_s = 1.0°, 3.0°, 5.0°, 10.0°$
$\theta_s = 30°, 50°, 60°$
$\delta t \approx 20s, 60s, 120s$
$\Delta t = n \times \delta t, n = 0, 1, 2, \ldots$

Field experiment II

Field experiment I is performed with redefined principal plane (or origin $\varphi = 0$) for each viewing zenith angle. $\partial\theta_s/\partial t$ and $\partial\varphi/\partial t$ are as previously defined, but $\Delta t=0$ each time viewing zenith is changed.

Figure 2 illustrates how measuring accuracy of field experiments successively decreases with increasing measuring time (δt = 20s, 60s, 120s). The last figure shows how the opposing environmental input of Field experiment II restores the accuracy of a slow field measurement (δt = 120s) to laboratory level. It is evident that both the choice of measuring instrument and the measuring technique are important factors for a succesful field experiment. If field data is used for parameter retrieval by running radiative transfer models in reverse mode, the accuracy of output parameters can be significantly improved by using high-accuracy input data.

4 CONCLUSIONS

This paper presents a numerical model for estimating the measuring accuracy of goniometer measurements. Both the measuring system itself and the most important environmental effects are analysed. The model has been tested for three different types

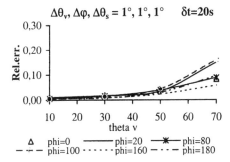

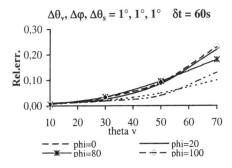

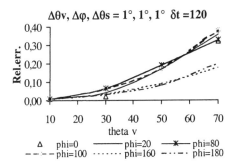

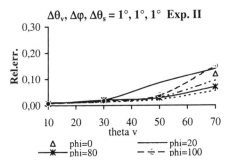

of experiments with a large range of input parameters.

The calculations reveal the importance of a high-precision measuring instrument when BRF measurements with complete angular coverage, including slant viewing angles, are performed. Manually operated goniometers are an advantage in field experiments, where the shortest possible measuring time is required to improve measuring accuracy. Fast field experiments (measuring time 30 min) are comparable to laboratory experiments, and slow field experiments (measuring time 2 hrs) should only be performed at high solar angles using a high-precision instrument. The method can be further developed by including estimates for detector accuracy and environmental effects which where suppressed in this study.

REFERENCES

Bentley, J., 1988. Principles of measurements systems, 2nd ed. *Longman Group UK Limited.*

Rahman, H., Pinty, B., and Verstraete, M., 1993. Coupled Surface-Atmosphere Reflectance (CSAR) Model 2. Semiempirical Surface Model Usable With NOAA Advanced Very High Resolution Radiometer Data. *J. Geophys. Res.* 98(11), pp.20791 - 20801.

Roujean, J., Leroy, M., and Deshamps, P., 1992. A bidirectional reflectance model of the earth's surface for the correction of remote sensing data. *J. Geophys. Res. 97, 20455-20468.*

Wanner, W., Li, X., and Strahler, A., 1995. On the derivation of kernels for kernel-driven models of bidirectional reflectance. *J. Geophys. Res.*

Figure 2 An illustration of how the measuring accuracy decreases with increasing measuring time in field experiments. The last figure shows how the method of field experiment II ($\delta t = 120s$) improves the accuracy.

Observing our environment from Space: New solutions for a new millennium, Bégni (Ed.)
© *2002 Swets & Zeitlinger, Lisse, ISBN 90 5809 254 2*

Accuracy of high-pulse-rate laser scanners for digital target models

E. Ahokas, H. Kaartinen, L. Matikainen & J. Hyyppä
Finnish Geodetic Institute, Department of Remote Sensing and Photogrammetry, Masala, Finland

H. Hyyppä
Helsinki University of Technology, Espoo, Finland

ABSTRACT: This paper evaluates the feasibility of high-pulse-rate laser scanners in urban and rural environment. A TopoSys-1 measurement campaign was organized in summer 2000 in southern Finland in several locations. This paper deals with the planimetric and vertical accuracy of a laser-derived digital surface model (DSM) and how much the terrain type affects the elevation accuracy. Also the ability to detect street lamps is reviewed. Reference points were measured with a RTK GPS system. The results are based mainly on the flying height of 400 m and partly on 800 m. A planimetric error of 26 cm and a vertical error of 23 cm were detected.

1 INTRODUCTION

Three-dimensional infrastructure models are needed in a vast range of applications of civil engineering and modern information society. Civil engineering sector needs detailed three-dimensional information in planning, visualization and maintenance of the construction process. Collection of such information with conventional mapping means is time consuming, costly and slow. If it is not very costly, then it is typically not very accurate. Traditional surveying methods include tachymetric mapping and stereoscopic photogrammetry.

Advances in laser scanning technology have challenged the conventional measurements in urban and rural environments. The principle of the airborne laser scanning is as follows, Figure 1. A laser beam is moved across the flight direction e.g. by a rotating or oscillating mirror. The flying height and the scan angle define how wide an area is covered on the ground. When an airplane is moving forward during scanning, the ground will be observed in the along-track direction. The laser transmits pulses to the ground and measures the travelling time of these energy pulses returning to the laser receiver unit. The distance to the ground is determined. Differential GPS and inertial measurement systems are needed to determine the position and orientation of the laser scanner. Ground coordinates can then be calculated for the laser-recorded points.

Typical subjects within laser scanning deal with digital elevation or terrain models in various conditions (forest and urban areas), such as works by

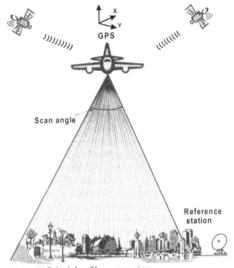

Figure 1. Principle of laser scanning.

Kraus & Pfeifer (1998), and Pyysalo (2000), forest inventory (Hyyppä & Inkinen 1999), extraction of buildings (Haala & Brenner 1999, Morgan & Tempfli 2000, Maas & Vosselman 1999, Geibel & Stilla 2000, Vögtle & Steinle 2000), laser data filtering (Vosselman 2000) and use of TIN-models (Axelsson 2000, Maas 2000). Most of the users need a good knowledge of accuracy obtainable using laser scanner. Accuracy is affected by the number of pulses, the measurement geometry (height, field of view), pulse mode used, the sensor and the type of the target, to name but a few main factors. The quality and accuracy of digital surface and target models based on airborne laser scanning data is tackled in this paper.

This paper concentrates on the following aspects:

1. In urban environment, some of the targets are difficult to record with a laser scanner, such as street lamps. We analyzed the visibility percentage of such targets using laser scanner data from two flight altitudes, 400 and 800 m.
2. What is the planimetric accuracy of laser-derived DSM?
3. What is the elevation accuracy of DSM and digital target model in urban environment?
4. How much is the terrain type affecting the obtained elevation accuracy?

2 MATERIAL AND METHODS

2.1 Laserscanning data

The data applied in this study were collected with the German TopoSys-1 laser scanner installed in a local airplane. Properties of the TopoSys laser scanners are in Table 1. Flight 451 was flown on June 15, 2000 at the altitude of 400 m. The swath width was 100 m. Measurements of two strips covered the Otaniemi target area. First pulse mode was used. GPS reference station located in Helsinki less than 10 km from Otaniemi. Flight 454 was flown on June 18, 2000 at the altitude of 800 m and again two strips covered the Otaniemi target area. The swath width was 200 m in this case.

Table 1. Properties of the TopoSys airborne laser scanners (www.toposys.com).

	TopoSys I	TopoSys II
Sensor type	Pulsed fiber scanner	Pulsed fibre scanner
Range	< 1000 m	< 1600 m
Wavelength	1.54 μm	1.55 μm
Pulse length	5 ns	5 ns
Pulse repetition rate	83 kHz	83 kHz
Scan frequency	650 Hz	650 Hz
Scan angle	14 ° (± 7.0 °)	14 °(± 7.0 °)
Width (at max range)	220 m	390 m
Density (at max range)	5 points/m^2	3 points/m^2
Resolution of distance measurements	0.06 m	0.02 m
Measurement possibilities	First or last pulse	First and last pulse simultaneous
Intensity measurements	None	Possible

Each flight track was recorded in its own file. TopoSys delivered the preprocessed data in a packed binary format. After unpacking data were converted from binary to ASCII format (geocentric WGS84

XYZ-coordinates). They were further transformed to the Finnish national coordinate system (KKJ). Analysed heights were orthometric.

Laser scanner survey provided a cloud of points, the x, y and z coordinates of which are known. They form a digital surface model (DSM), which includes terrain points, vegetation points, and points reflected from buildings. For the analysis, the laser measurements were converted into a DSM in regular raster format (a height image). The pixels of the DSM were defined such that they match with pixels of an orthorectified aerial image also used in the study. The pixel size was 30 cm × 30 cm. The DSM was calculated by selecting the maximum height value for each resolution cell from the original laser point measurements. Linear interpolation was used to estimate height values for those pixels that contained no data. Interpolation algorithm based on Delaunay triangulation was applied. Figure 2 depicts part of the laser-derived DSM for the Otaniemi test site.

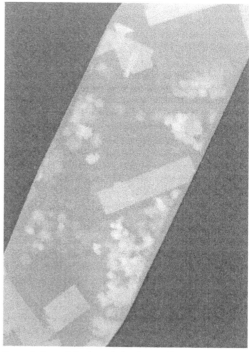

Figure 2. Part of the laser-derived DSM for the Otaniemi test site. The pixel size is 30 cm × 30 cm. Flying height 400 m. Swath width is 100 m.

2.2 Ground truth data

In order to verify the accuracy of the DSM, coordinates of reference points in Otaniemi were measured using Leica SR530 Real-Time Kinematic

(RTK) GPS system. Real-Time Kinematic GPS is the type of GPS measurement where coordinates are measured in real-time with accuracy at centimetre level. RTK is based on the concept of relative GPS where two receivers, one at a reference point with known coordinates and one at a new point, are measuring simultaneously signals from the same satellites. Horizontal accuracy of the RTK measurements was verified to be about 0.015 m and vertical accuracy 0.02 m in another study (Bilker & Kaartinen 2001). The distance from the measured points to the RTK reference point that was in the northern part of Otaniemi was less than 3 km.

Altogether over 700 RTK points were measured, 2/3 of them were ground points and 1/3 were points on buildings or other constructions like walls. The distribution of measured points is presented in Figure 3.

In addition to RTK measurements, colour aerial photos were obtained. By field visit, the street lamps were recorded.

Figure 3. The distribution of measured RTK points (black dots).

2.3 Verification methods

The laser-derived locations and elevations were compared with RTK-derived field measurements. A comparison between the RTK-derived and laser-derived heights was carried out as follows: First a 1x1 m square polygon was generated with the measured RTK point as centre. Then heights in the DSM (the height image) within this polygon were observed deriving maximum, minimum and mean height values and standard deviation of heights. Depending on the target these heights were then compared with RTK-derived heights. For example, for flat roof corners the maximum height value was used, and in the case of flat ground areas (tarmac, grass, etc.) the mean height value was used.

The results of field checks and digital colour orthophotos were used for visual comparison of DSM in regular raster format (height images) and original laser points from 400 m and 800 m altitude. The TerraScan software from the Terrasolid Ltd (www.terrasolid.fi) was used for visual observations of laser points on a workstation.

We compared the visibility of three types of street lamps. Their poles were T-, Γ- or I-shaped. T-shaped lamps are usually used in the central area between lanes in a street and Γ-shaped lamps are at the edges of a street. I-shaped lamps can be along the pavements or walking routes. Totally 114 lamps were reviewed.

3 RESULTS

Small objects such as street lamps are visible from the lower altitude measurements but omissions occur in the 800 m case. This is due to gaps between the across track measurements. The distance between the measured points on the ground perpendicular to the flight direction is 0.8 m from the flying height of 400 m. The across track point spacing is 1.6 m from the flying height of 800 m, accordingly. Laser footprints from 400 m and 800 m altitude are 0.2 m and 0.4 m in diameter. When the lamps are near the trees it is difficult to separate the echoes from the lamp and from the tree branches. The results are in Table 2.

Table 2. Visibility of street lamps in airborne laser scanner measurements.

Lamps		H = 400 m		H = 800 m	
Type	Amount	Visible (%)	Not vis. (%)	Visible (%)	Not vis. (%)
T	23	100	0	96	4
Γ	64	91	9	52	48
I	27	63	37	7	93

To estimate the plane accuracy of laser scanning altogether 44 targets with RTK coordinates were measured. Only points that could be clearly identified visually on the DSM were measured manually using the Erdas Imagine software. Errors in the east and north direction were determined as well as the total error. Derived mean error was 26 cm. There was no significant offset and the error was equal in both directions.

177

Table 3. Differences between the measured plane coordinates in the north and east direction (RTK – DSM) and the total error. 44 measured targets. H = 400m.

	North	East	Total
Mean (m)	0.00	0.01	0.26
Standard deviation (m)	0.24	0.21	0.18
Minimum (m)	-0.82	-0.60	
Maximum (m)	0.36	0.48	0.85

The differences between the RTK and laser-derived DSM heights (height errors) were computed, and the results are shown in Table 4 for all points used in the analysis and for the three most typical ground surfaces measured. There was an offset of 23 cm in laser heights (below the reference heights), but standard deviation was only 6 cm. When comparing height errors for different ground surface materials errors in grass-covered ground are slightly bigger than in tarmac or concrete surfaces, as expected.

Table 4. Height errors (RTK – DSM) for different ground surface types. H = 400 m.

	All	Tarmac	Concrete	Grass
Targets	361	167	42	45
Mean (m)	0.23	0.25	0.19	0.19
Std. dev. (m)	0.06	0.04	0.05	0.07
Min. (m)	0.02	0.16	0.12	0.03
Max. (m)	0.46	0.46	0.34	0.32

The heights of some small buildings (heights 2.5 – 3 m) were computed using the height of the roof corner and the height of a ground point nearby. As shown in Table 5, the height of a building derived from the DSM data is typically 5 cm taller than the reference.

Table 5. Differences in building heights (RTK – DSM). H = 400 m.

Heights	14
Mean (m)	-0.05
Standard deviation (m)	0.06
Minimum (m)	-0.16
Maximum (m)	0.05

4 CONCLUSIONS

The planimetric accuracy of the laser-derived DSM was 26 cm that is in the scope of the DSM grid size. The vertical offset was 23 cm and the standard deviation 6 cm. One reason for this offset could be the inaccuracies of the positioning of the airplane.

A DSM with a regular grid of 30 cm × 30 cm explains some part of the errors in the results. If the laser pulses do not hit the edges of the target then the planimetric and vertical accuracy is affected by interpolation. Also the size and the orientation of the target relative to the flight line have effects on the accuracy. A solution is the ground coverage dense enough but this is also an economic question of flying more strips over the area. Mapping details in urban area is suitable from 400 m altitude because of the relatively dense ground coverage of the laser points.

REFERENCES

Axelsson, P. 2000. DEM generation from laserscanner data using adaptive TIN models. International Archives of Photogrammetry and Remote Sensing. Vol XXXIII, Part B4, Amsterdam 2000, pp. 110 - 117.

Bilker, M. & Kaartinen, H. 2001. The Quality of Real-Time Kinematic (RTK) GPS Positioning. Reports of the Finnish Geodetic Institute.

Geibel, R. & Stilla, U. 2000. Segmentation of laser altimeter data for building reconstruction; different prodedures and comparison. International Archives of Photogrammetry and Remote Sensing. Vol XXXIII, Part B3, Amsterdam 2000, pp. 326-334.

Haala, N. & Brenner, C. 1999. Extraction of buildings and trees in urban environment. ISPRS Journal of Photogrammetry and Remote Sensing, 54, pp. 130-137.

Hyyppä, H. & Hyyppä, J. 2000. Quality of 3-dimensional infrastructure models using airborne laserscanning. The Photogrammetric Journal of Finland, 17: 43:53.

Hyyppä, J. & Inkinen, M. 1999. Detecting and estimating attributes for single trees using laser scanner. The Photogrammetric Journal of Finland, 16: 27-42.

Krauss, K. & Pfeifer, N. 1998. Determination of terrain models in wooded areas with airborne laserscanner data. ISPRS Journal of Photogrammetry and Remote Sensing, 53, pp.193-203.

Maas, H-G. 2000. Least-squares matching with airborne laserscanning data in a TIN structure. International Archives of Photogrammetry and Remote Sensing. Vol XXXIII, Amsterdam 2000, pp. 548-555.

Maas, H-G. & Vosselmann, G. 1999. Two algorithms for extracting building models from raw laser altimetry data. ISPRS Journal of Photogrammetry & Remote Sensing, 54, pp. 153–163.

Morgan, M. & Tempfli, K. 2000. Automatic building extraction from airborne laserscanning data. International Archives of Photogrammetry and Remote Sensing. Vol XXXIII, Part B3, Amsterdam 2000, pp. 616 - 623.

Pyysalo, U. 2000. Derivation of digital elevation models in wooded areas with three dimensional point cloud measured with laserscanner (in Finnish). Helsinki University of Technology, Institute of Photogrammetry and Remote Sensing. 68 p.

Vosselman, G. 2000. Slope based filtering of laser altimetry data. International Archives of Photogrammetry and Remote Sensing, Vol. XXXIII, Part B3, Amsterdam 2000, pp. 935-942.

Vögtle, T. & Steinle, E. 2000. 3D Modelling of buildings using laserscanning and spectral information. International Archives of Photogrammetry and Remote Sensing, Vol. XXXIII, Part B3, Amsterdam 2000, pp. 927-934.

www.terrasolid.fi

www.toposys.com April 24, 2001

Observing our environment from Space: New solutions for a new millennium, Bégni (Ed.)

MERIS for terrestrial remote sensing

P.J.Curran & C.M.Steele
Department of Geography, University of Southampton, Southampton, United Kingdom

ABSTRACT: Principally designed and promoted for ocean applications, the potential of the MEdium Resolution Imaging Spectrometer (MERIS) for terrestrial applications is now recognised. In this paper, we review MERIS's development history, tracking key adaptations to the mission objectives of the sensor that emphasise its potential for terrestrial remote sensing. The product specification of MERIS is examined and we stress how sensor spectral, geometric and radiometric requirements for ocean remote sensing can benefit terrestrial applications. We also review the decision-making processes that resulted in specific features of MERIS, such as the wide swath of the sensor, its programmability and facilities for on-board calibration. We conclude with the suggestion that within a few years from now, MERIS will be seen as one of the more important sensors for the remote sensing of our terrestrial environment

1 THE MEDIUM RESOLUTION IMAGING SPECTROMETER

MERIS is a high spectral and medium spatial resolution satellite sensor and is part of the core instrument payload of Envisat, the European Space Agency's (ESA) environmental research satellite scheduled for launch in October 2001. The primary objective of MERIS is the measurement of sea colour in both ocean and coastal areas. Secondary objectives include atmospheric and terrestrial applications.

Although MERIS was optimised for ocean applications (i.e., it is a sensor with a large dynamic range, high radiometric sensitivity, narrow bands in optical wavelengths, large areal sampling units) it also includes features that are useful for terrestrial applications. Principally, these are MERIS's radiometric and spectral capabilities, both of which have seen several revisions over the span of development of this sensor. In addition, the mission requirement governing the geometric operation of MERIS has been revised recently, further enhancing MERIS's potential for terrestrial remote sensing.

In the first part of this paper, we examine the various revisions of MERIS's original mission requirements. In the second section, we discuss the specific details of MERIS's radiometric, spectral and geometric resolution that have ensured its status as a sensor with a broad environmental remit. In writing this paper, we have relied upon several sources of information, including official ESA documents promoting MERIS, the ESA Envisat website, meeting minutes of the MERIS Science Advisory Group (SAG) and the first report of the International Ocean-Colour Coordinating Group (IOCCG), as well as journal papers.

2 MERIS'S HISTORY

Operational between 1978 and 1986, the pioneer of ocean colour observation was the Coastal Zone Colour Scanner (CZCS) on board NASA's Nimbus-7 satellite. As a proof-of-concept sensor, the CZCS confirmed the value of ocean remote sensing for estimating near-surface pigment fields despite its limited spectral capabilities and radiometric accuracy (Bricaud *et al.*, 1999). The CZCS had few spectral bands and low radiometric sensitivity. Nevertheless, the CZCS inspired a new generation of ocean sensors (Table 1) and its success contributed to the formulation of Earth observation (EO) policy within ESA. During the 1980s and early 1990s, the primary objectives of the ESA Earth observation policy were meteorological/ocean applications and the CZCS mission was widely cited in official ESA documents as justification of these objectives. In 1984, the idea of a European imaging spectrometer with a fast repeat time, high spectral and radiometric resolution that was capable of providing low spatial resolution global data was conceived by the ESA Ocean Colour Working Group. The guidelines laid down by this group formed the basis for MERIS's development.

Table 1. Basic radiometric, spectral and geometric characteristics of the CZCS and subsequent ocean colour sensors (adapted from IOCCG, 1998), resolution is spatial resolution at nadir.

Sensor	Characteristics	Values
CZCS	NEΔL	0.04 (NIR) – 0.21 (visible)
	Spectral bands: 5	4 x visible, 1 x NIR
	Resolution	0.820 km
	Swath	1566 km
OCTS	NEΔL	0.03 (NIR) – 0.19 (visible)
	Spectral bands: 8	6 x visible, 2 x NIR
	Resolution	0.7 km
	Swath	1400 km
MOS	NEΔL	0.03 (NIR) – 0.19 (visible)
	Spectral bands: 11	8 x visible, 3 x NIR
	Resolution	0.5 km
	Swath	200 km
SeaWiFS	NEΔL	0.02 (NIR) – 0.09 (visible)
	Spectral bands: 8	6 x visible, 2 x NIR
	Resolution	1.1 km
	Swath	2800 km
OCM	NEΔL	0.05 (NIR) – 0.26 (visible)
	Spectral bands: 8	6 x visible, 2 x NIR
	Resolution	0.7 km
	Swath	1400 km
MODIS	NEΔL	0.006 (NIR) – 0.05 (visible)
	Spectral bands: 9	7 x visible, 2 x NIR
	Resolution	1.0 km
	Swath	2330 km
GLI	NEΔL	0.007 (NIR) – 0.08 (visible)
	Spectral bands: 15	12 x visible, 3 x NIR
	Resolution	1.0/0.25 km
	Swath	1600 km
POLDER-2	NEΔL	0.03 (NIR) – 0.15 (visible)
	Spectral bands: 8	4 x visible, 4 x NIR
	Resolution	6 x 7 km
	Swath	2400 km
MERIS	NEΔL	0.007 (NIR) – 0.03 (visible)
	Spectral bands: 15	10 x visible, 5 x NIR
	Resolution	1.2/0.3 km
	Swath	1150 km

The MERIS SAG was formed in 1988 to provide scientific expertise and guidance during the development of MERIS. In line with the original remit of this sensor as an ocean instrument, the SAG was composed mainly of scientists with an interest in ocean remote sensing.

The first MERIS technical workshop was held in Villefranche-sur-Mer, France in 1991. The workshop was an important milestone in MERIS's development because it marked the beginnings of the sensor's promotion for more than ocean remote sensing. Three groups (representing ocean, atmosphere and terrestrial applications), reviewed MERIS's development status and mission requirements. The discussions of the ocean group concentrated on technical specifications: spectral requirements, calibration, radiometric resolution and radiometric accuracy.

Therefore, the sensor design and mission requirements remained focussed on ocean applications. The atmosphere group discussed atmospheric variables that could be estimated from MERIS data, and identified wavebands of importance for atmospheric remote sensing.

The land group (terrestrial applications) made several technical comments on the instrument design. For example, they suggested shifts in the location of two wavebands, inclusion of an extra waveband at 960 nm and a requirement for accurate localisation for terrestrial applications. More significantly the land group emphasised the potential of MERIS for terrestrial applications, making nine recommendations for testing the feasibility of producing useful information over land. The report of the land group prepared the foundations upon which MERIS for terrestrial remote sensing could progress. In the introduction to the official MERIS document (Rast, 1996), ESA referred to the origin of MERIS in 1984 and 1988 as an ocean instrument but stressed the refinement and extension of its mission to include atmospheric and terrestrial objectives.

This refinement of the MERIS mission was reiterated during the second MERIS technical workshop held in Villefranche-sur-Mer in 1997. By then, the primary objective was bio-optical oceanography with secondary mission objectives stated as atmospheric monitoring and the study of land surface processes.

The configuration of this imaging spectrometer is now optimised for ocean remote sensing. However, some concessions to this optimisation have been essential in order to preserve the usefulness of MERIS data for terrestrial applications. These concessions are included in the next section, which discusses the radiometric, spectral and geometric characteristics of MERIS.

3 SENSOR CHARACTERISTICS

3.1 Radiometric resolution

Two features of MERIS that are unmatched by current satellite sensors are its radiometric accuracy and dynamic range. Relative to incoming solar radiation, the radiometric error of MERIS is less than 2% of the detected signal between 400 - 900 nm and less than 5% of the detected signal between 900 - 1050 nm (Rast, 1996). A large signal-to-noise ratio (SNR) is necessary for ocean applications because the ocean surface reflects little radiation. Further, a high degree of sensitivity is required to differentiate pigment concentrations in surface waters. MERIS's radiometric performance is beyond that required for typical terrestrial applications but is of value for spectral unmixing of imagery acquired over spatially complex terrestrial surfaces and for the estimation of biophysical variables over vegetated terrain.

Ocean applications also require a sensor with a high dynamic range that can register the radiometric difference between dark ocean and bright clouds (Rast *et al.*, 1999). ESA has defined four sizing radiance conditions for the instrument dynamic range (ESA, 1994): minimum radiance over open ocean (L1), radiance in open ocean (L2), maximum radiance over open ocean (L3) and maximum cloud radiance (L4). The maximum expected radiance of L4 targets was calculated using the assumption that reflectance was greatest over bright cloud at the top of the atmosphere, which were reasoned to be equivalent to a Lambertian target with 108% reflectance illuminated with the smallest zenith angle between Sun and sensor.

Originally the L1 to L4 requirement was argued for all spectral bands, regardless of location. However, at the 16th SAG meeting in 1994, this decision was reviewed and a change was proposed in the gain amplification to L1 radiances. Calibration exercises had indicated that non-linearities in analogue-to-digital conversion would result in a signal error of around 2.5% for L1 radiances, thus reducing the accuracy of radiance measurements over ocean. Gain amplification of L1 radiances so that low reflectance values would be better represented by the 4096 radiance levels was suggested as a solution to this problem. This suggestion contradicted recommendations made by the land group during the 1991 technical workshop. The land group had stressed that should MERIS's dynamic range be decreased to *"somewhat less than 100% in reflectance"* (sic) then the remote sensing of ice and snow environments would be limited. However, the saturation problem is waveband specific. For example, the report of the land group observed that saturation in the blue/green part of the spectrum would be acceptable over land (because bands in these wavelengths contain minimal information concerning land), provided that it would not affect imagery in red and NIR spectral regions as these are useful for terrestrial applications.

The gain amplification debate will continue beyond launch. Band 13 (865 nm) is not programmed to saturate at L4, thus posing a problem for atmospheric correction over ocean. Ocean applications require the capability to saturate over clouds at 865 nm to ensure the quality of atmospheric correction over ocean and therefore the quality of the ocean data itself. It was decided at the 34th meeting of the SAG that band 13 should be set at the threshold preferred for terrestrial applications (70%). This was justified because setting the threshold at the level preferred for ocean applications (42%) would affect the accuracy of radiance measurements over bright (e.g. arid) terrestrial surfaces and the MERIS global vegetation index (MGVI) over dense, dark vegetation. However, this decision may change subject to an analysis of the global distribution of surfaces whose reflectance exceeds 42% and 70% in the 865 nm band.

Some of the problems of gain amplification can be addressed by gain programmability. Gain amplification for each of the 15 wavebands can be set to optimise the saturation level for specific targets. However, the flexibility of this is limited and MERIS's primary mission for continuous data collection over terrestrial and ocean surfaces requires some compromise between the demands of user groups.

3.2 *Spectral resolution*

MERIS will record visible and NIR radiation in fifteen narrow wavebands (Table 2).

Table 2. MERIS default wavebands

Wavebands			Environmental Variables of Interest
No.	Centre	Width	
	nm	nm	
1	412.5	10	Yellow substance, turbidity
2	442.5	10	Chlorophyll absorption
3	490	10	Chlorophyll, other pigments
4	510	10	Turbidity, suspended sediment, red tides
5	560	10	Chlorophyll reference, suspended sediment
6	620	10	Suspended sediment [a]
7	665	10	Chlorophyll absorption [a]
8	681.25	7.5	Chlorophyll fluorescence [a]
9	708.75	10	Atmospheric correction [a]
10	753.75	7.5	Oxygen absorption reference [a]
11	760.625	3.75	Oxygen absorption R-branch
12	778.75	15	Aerosols, vegetation
13	865	20	Aerosols correction over ocean
14	890	10	Water vapour absorption reference
15	900	10	Water vapour absorption, vegetation

[a] Indicates wavebands for calculation of red edge position

Ocean applications require narrow bands that correspond with absorption features specific to oceanographic variables. Broad wavebands, such as those of AVHRR or TM average out the changes in reflectance associated with key spectral features, such as the absorption feature caused by chlorophyll (Huete *et al.*, 1997). Some of the narrow spectral bands required for ocean applications are also valuable for terrestrial applications, e.g. maximum chlorophyll absorption occurs around 665 nm for all photosynthetically-active plant material. AVHRR band 1 and TM band 3 completely obscure this sensitive absorption feature.

A unique feature of MERIS is the programmability of the position and width of its wavebands. While this function allows compensation for spectral drift, it also permits the selection of user-defined wavebands for short periods during each orbit. Therefore, MERIS can be optimised spectrally for very specific applications. For example, a user may choose to select wavebands that are related causally to the phenomena of interest (e.g. canopy pigment content, canopy moisture content). In these cases, wavebands could be located in chlorophyll absorption regions, along the red edge and across the NIR plateau (Verstraete *et al.*, 1999). Using spectral interpolation, a continuous spectrum could be produced and the red edge position (REP) calculated using MERIS's default wavebands (Table 2). Vegetation canopy chlorophyll content and LAI are known to be related causally to REP (Dawson, 2000). Further, moving more wavebands to the red edge can reduce error in spectral interpolation and thus in estimation of these vegetation canopy variables.

The history of optical remote sensing and its use for vegetation quantification is dominated by the development of numerous vegetation indices. Some of these efforts have demonstrated that narrow band indices are more accurate indicators of biophysical and biochemical variables than broad band indices (Yoder and Waring, 1994). One such example is the physiological reflectance index (PRI) (Gamon *et al.*, 1992). The PRI is formulated using data in two narrow red bands (531 nm and 570 nm) and has been shown by several studies to be sensitive to photosynthetic activity at the leaf and canopy levels. Nichol *et al.*, (2000) have demonstrated the applicability of the PRI for monitoring the light use efficiency (LUE) of photosynthesis with airborne spectroradiometer data. The results of this study imply that the PRI may function as an estimator of LUE if calculated from satellite sensor data. Through its programmability, MERIS will be the only satellite sensor capable of providing data for the calculation of the PRI.

3.3 *Geometric resolution*

MERIS will operate in two spatial modes: reduced spatial resolution (1.2 km at nadir) for the primary mission and full spatial resolution (FR: 300 m at nadir) for the background mission (Rast *et al.*, 1999). The FR ground resolution element is invariant at 300 m along track, but varies across track from 260 m at nadir to 420 m at the swath edge. The spatial resolution of the MERIS FR data therefore lies between TM (30 m) and both MODIS (500 m) and AVHRR (1100 m) data. While operation at reduced spatial resolution will facilitate compilation of global data sets, the background FR mode will allow data acquisition over land at regional scales (Merheim-Kealy *et al.*, 1999).

Originally, the operation of MERIS in FR mode was intended for 20 minutes per orbit. However, the time spent over land would vary from nearly 0 minutes (e.g. over the Pacific) to 20 minutes (e.g. Europe/Africa). At the 34th meeting of the SAG, it was recommended that the 20 minutes of FR data per orbit would be treated as an average, thus ensuring the maximum coverage of land and coastal areas at full spatial resolution.

MERIS was originally designed with six contiguous optical modules, each with a field-of-view (FOV) of 14°. The module overlap was to provide an instrument FOV of 81.5° and a swath width of 1500 km from an altitude of 824 km. In 1994 (as part of an Envisat re-evaluation exercise) the number of MERIS's optical modules was reduced to five providing a reduced instrument FOV of 68.5° resulting in a reduced swath width of 1150 km from a revised altitude of 799 km. As a result, the time taken for global coverage has been increased from two to three days with repeat coverage at 35 days. Despite these modifications, the spatial coverage of MERIS data is sufficient to enable viable monitoring of the global terrestrial environment. Further, the combination of the spatial coverage with both medium and coarse spatial resolution will facilitate data collection over terrestrial environments from *region* to *continent* to *globe*, thereby increasing our knowledge of land cover change and ecosystem processes at different scales (Verstraete *et al.*, 1999).

4 CONCLUSION

This paper has provided a brief review of the heritage and history of MERIS. In a discussion of its radiometric, spectral and geometric characteristics MERIS is shown to be a sensor with great potential for terrestrial application. A particular advantage is its unique facility for post-launch programming of waveband gain, width and position. It has been demonstrated that the design requirements of MERIS for ocean applications have resulted in a sensor that is most appropriate for terrestrial applications (Verstraete *et al.*, 1999), particularly at regional to global scales. The MGVI exemplifies this – it will be available as an atmospherically-corrected, global, level 2 data product (Gobron *et al.*, 1999). Some modifications to the MERIS are expected post-launch but it is unlikely that any changes to the default radiometric settings will compromise seriously MERIS's terrestrial mission. The use of MERIS data for land applications is likely to be large given not only the suitability of the sensor but the observation that the terrestrial remote sensing community is much larger than the ocean remote sensing community and the majority of data sales for previous ocean sensors, such as ERS 1 SAR, have been for terrestrial applications. This paper concludes with the prediction that within a few years from now, MERIS will be seen as one of the more important sensors for the remote sensing of our terrestrial environment.

5 REFERENCES

Bricaud A., Morel A. & Barale V., 1999. MERIS potential for ocean colour studies in the open ocean. *International Journal of Remote Sensing* 20:1757-1770.

Dawson T.P., 2000. The potential for estimating chlorophyll content from a vegetation canopy using the Medium Resolution Imaging Spectrometer (MERIS). *International Journal of Remote Sensing* 21: 2043-2051.

European Space Agency, 1994. *The MEdium Resolution Imaging Spectrometer MERIS.* Report of the MERIS Science Advisory Group. Noordwijk: ESA Publications Division.

European Space Agency, 1999. MERIS *The MEdium Resolution Imaging Spectrometer Instrument.* http://envisat.esa.int, (16 April, 2001).

Gamon J.A., Peñuelas J. & Field C.B., 1992. A narrow-waveband spectral index that tracks diurnal changes in photosynthetic efficiency. *Remote Sensing of Environment* 41: 35-44.

Gobron N., Pinty B., Verstraete M.M. & Govaerts Y., 1999. The MERIS Global Vegetation Index (MGVI): description and preliminary application. *International Journal of Remote Sensing* 20: 1917-1927.

Huete A.R., Liu Q., Batchily K. & van Leeuwen W., 1997. A comparison of vegetation indices over a global set of TM images for EOS-MODIS. *Remote Sensing of Environment* 59: 440-451.

IOCCG, 1998. *Minimum Requirements for an Operational Ocean Colour Sensor for the Open Ocean.* Nova Scotia: International Ocean-Colour Working Group.

Merheim-Kealy P., Huot J.P. & Delwart S., 1999. The MERIS ground segment. *International Journal of Remote Sensing* 20: 1703-1712.

Nichol C.J., Huemmrich K.F., Black T.A., Jarvis P.G., Walthall C.L., Grace J. & Hall F.G., 2000. Remote sensing of photosynthetic-light-use efficiency of boreal forest. *Agricultural and Forest Meteorology* 101: 131-142.

Rast M., 1996. *MERIS: The Medium Resolution Imaging Spectrometer.* Noordwijk: ESA Publications Division.

Rast M., Bezy J.L. & Bruzzi S., 1999. The ESA Medium Resolution Imaging Spectrometer MERIS - a review of the instrument and its mission. *International Journal of Remote Sensing* 20: 1681-1702.

Verstraete M.M., Pinty B. & Curran P.J., 1999. MERIS potential for land applications. *International Journal of Remote Sensing* 20: 1747-1756.

Yoder B.J. & Waring R.H., 1994. The Normalised Difference Vegetation Index of small Douglas-Fir canopies with varying chlorophyll concentrations. *Remote Sensing of Environment* 49: 81-91.

Observing our environment from Space: New solutions for a new millennium, Bégni (Ed.)
© 2002 Swets & Zeitlinger, Lisse, ISBN 90 5809 254 2

Forest stand parameters from SPOT-XS data by use of artificial neural networks

A.Öztopal
Meteorology Department, Remote Sensing Group, Istanbul Technical University, Maslak, Turkey

H.Gonca Coşkun
Remote Sensing Division, Civil Engineering Faculty, Istanbul Technical University, Maslak, Turkey

ABSTRACT: This study reports on Artificial Neural Networks (ANN) modelling for the forest parameters and the reflectance values SPOT-XS satellite data on the ground control points in the study area. The study area is chosen as dense forest cover in the west of Istanbul. The city is negatively affected by the residential areas as a result of population growth and industrial development. This situation is assessed by satellite digital data and forest parameter measurements, such as the number of trees, basal area, volume, mean diameter, and the stand density of forest. In order to find possible relationships between remote sensing variables such as the reflectance values and forest parameters, an Artificial Neural Networks (ANN) approach is used for refined determination of regression coefficients. Adaptive parameter estimation in the ANN provides efficient computation. These observed reflectances show strong relationships with the all chosen forest parameters. The necessary values are provided in single pixel and average values for each band at the control point in the study area. As the reflectances (in the forest area) in the longer red, near and middle IR increase faster than the reflectances in shorter blue and green wavelengths, forest parameters become positively related to reflectance.

1 INTRODUCTION

Forests are the major carbon dioxide balance sources for atmospheric components and their survival is of prime importance to ecological life in the world. Due to their irregular, extensive and variety of different species, intensity as well as distribution, it is not possible to make assessments through the classical surveying methods so as to control the balance situation. At this point, satellite images and their treatments in the digital computer provide a rapid, economic and straightforward way of decision making for engineers. Initially, these satellite images furnish a vast amount of qualitative information for the image enhancement and classification of different aspects from the forestry point view. To this end, Artificial Neural Networks (ANN) are currently used in extracting quantitative information for rational decision making. Especially, contrasting of ground measurements with the reflectance of various bands in the satellite image processing yields a quantitative basis for scientific interpretations and conclusions.

Forest areas of the earth surface play an important role for both the exchange of mass, energy and climatic change. The rapid increase in population affects not only the large city areas but also the planet Earth and its natural resources, especially in forest areas. Research on remote sensing applications now provides global synoptic measurements on climatic and environmental parameters relating to major processes in forest areas. In addition to other natural events such as the Mediterranean type of climate, with warm and dry summers throughout most of the country which is partly responsible for the extent and severity of the fire problem. Demographic and socio-economic changes in rural areas of the country, especially during the last 20-30 years represent another set of influential factors (Pereira *et al.* 1995).

Nowadays the monitoring of forest ecosystems and their biodiversity are among the most important areas using satellite remote sensing techniques. The opportunity exists to use operationally multispectral high-resolution satellite data from advanced sensing systems for visible, infrared and microwave parts of the electromagnetic spectrum (Wilkinson, *et al.* 1995; Hame, *et al.* 1984-1995). Satellite remote sensing has been used for forest and land use inventories, mainly at regional levels. The recent advance in the spectral range of satellite images has made the production of a stable and accurate series of multi temporal images possible. By combining satellite images with accurate but old-field information, it seems possible to create control systems for the continuous updating of forest information. The idea is to combine the advantages of both satellite imagery and field inventory (Varjo, 1995; McCormick *et al.* 1995).

Forest destruction and subsequent desertification are unfortunate consequences of the environmental deteriorations. However, remote sensing technology has stimulated remarkable advances in forest management skills. The use of image processing techniques provides rational assessment for forest management, which is the key activity for preservation of forest resources in an area. Remotely sensed data acquired by satellites provide a common basis for many quantitative results both map-wise and correlation-wise between the reflectance value of each band and the forest stand parameters or ground truth measurements.

It is the main purpose of this paper to apply the ANN procedure in order to deduce the quantitative conclusions for forest assessment in the region of Istanbul. Furthermore, a comparison of the already applied genetic algorithm procedure with the ANN approach is also given in this paper for the study area. The Science and Technical Research Council of Turkey (STRCT) supported this latter study. The main aim of this project was to present the possible extent of stand type mapping preparation for forest management planning and a national base forest inventory with satellite data. This project was implemented by co-operation between the Istanbul Technical University Remote Sensing Division and the Forest Management Planning Department of Istanbul University. Hereby, two sets of data as the ground truth measurements such as the tree species, mean diameter, stand density and stand volume are related to reflectance values of each band on the control points. The correlation between each pair of variables in these two sets is sought by the ANN approach, which started appear in the remote sensing data analysis recently. It has been observed that some significant relations become adaptively apparent by the ANN. Such relations help to manage future forest situations by using satellite data only.

2 STUDY AREA AND DATA SETS

The study area is located north-west of Istanbul with mixed characteristics of the Black Sea and Mediterranean Sea forest ecosystems. The primary forest type is composed of oak, coniferous, pines, and deciduous trees. The location of the study area is given in Figure 1, and it covers almost 8000 hectares. In this area 300-measurement control points have been measured to determine the necessary forest parameters. The plots and measurements on the test area used are from 1992 for the forest management plan preparations. Some stand characteristics, · for example basal area, volume, number of trees, mean stand diameter and tree species on each plot have been determined. Each plot's geographical coordinates are laid down on the satellite imagery to get reflectance values. (Coşkun et al. 1998-2000).

Figure 1. Colour composition of study area, using SPOT-XS data with 3-2-1 bands

2.1 Satellite data

Images for the study area are obtained from SPOT-XS satellite data that are used to extract images on 13th June 1993 corresponding to field survey forest parameter measurements in order to evaluate forest mapping. The data was obtained as geo-coded format and analysed for geometric registration using UTM projection. On the other hand, the standard techniques are used with ground control points and a first degree polynomial equation resampling of the multi-spectral datasets to 20m resolution using cubic convolution technique (Welch and Ehlers, 1987; Chaves, 1991). The accuracy of the registration is evaluated on test points and yielded root mean error (RMSE) of ±0.5 pixel for the SPOT XS image subset. The data processing ERDAS package is employed and the ARC/INFO GIS. SPOT-XS with all bands are considered for unsupervised and supervised classifications. 200 test pixels are chosen and compared with the classification results and field truth data. Consequently, overall classification accuracy is obtained as 80 %.(Coşkun et al. 1998).

2.2 Ground Truth Data

300 sample plots are evaluated in this study in 1992 for the purpose of forest management plan preparations. In order to obtain good results for the study sample plots should ideally have been acquired nearer to the time of acquired satellite data. Since one year is not too long a time for changes in forestry to occur, this consideration is not accepted as an important factor affecting the results of our research. Some stand characteristics have been determined such as the basal area, volume, number of trees, mean stand diameter and tree species on each plot.

After determining the geographical coordinates of each plot, all the data listed above are loaded on the computer. Typical natural tree species in the study area are broad-leaved trees. These are oak, beech, chestnut and hornbill trees. Some pine trees such as *P.brutia, P.nigra and P.maritima* have been introduced into the study area in 1970's as a plantation. The planning groups collected required data in plots distributed with 300 x 300 meter spacing systematically. The diameter of all trees, height, site quality and crown closure are determined on the 154 of 300 plots which are chosen reason of confidence according to field survey (Coşkun, *et al.1998*).

3 METHOD

3.1 *Artificial Neural Network (ANN)*

An Artificial Neural Network can be thought of as a black box which produces outputs from given inputs. The system includes layers, which are connected to each other with enough nodes in each layer. A classical ANN model is shown in Figure 2. In general, a neurone has n inputs as X_j, in the form of input signals. Each input is weighted before reaching the main body of processing element (artificial neural) by the connecting strength of the weight factor, W_{ij}. Hence, the signal transferred through the connection strength is equal to a portion of the original signal as $X_j W_{ij}$. On the other hand, for the neurone to produce a signal, the input signal to a neurone must exceed a threshold value, T.

Equation 1 represents total information coming to a node, and this information is transformed through Equation 2, which is called Sigmoid. This function is very popular, because it is bounded, monotonic, and non-linear, and has a simple derivative.

$$Y_j = \Sigma \, X_j \cdot W_{ij} \tag{1}$$

$$F(Y_j) = 1 / [\, 1+ \exp(-Y_j)\,] \tag{2}$$

Y (j) : Output value
X (i) : Input value
Wij   : Weight factor
F( )   : Activation function (sigmoid)

After giving all inputs to the system, the total error (least square) is calculated between outputs and observed data. Subsequently, weight factors are changed with back propagation of the total error (Equation 3 and 4). After determining new weight factors, all inputs are given to the system with the total error calculation and the newer weight factors are changed repeatedly. This procedure continues until a minimum error is reached (Haykin, 1994; Lipman, 1987; Simpson, 1992; Smith and Eli, 1995; Sönmez and Şen, 1998).

$$\mu = \sum_{j=1}^{n} [\, B(j) - Y(j)\,]\,(-1)\,Y(j)\,[(1 - Y(j)]\,X(i) \tag{3}$$

$$W_{ij} = W_{ij} - 0.5\,\alpha\,\mu \tag{4}$$
$$\text{new} \quad \text{old}$$

B (j) : Observed value
Y (j) : Estimated value
X (i) : Input value
Wij   : Weight factor
α   : Learning ratio

4 APPLICATION

In this study, the ANN Model has four input parameters (Bands 2, 3, 4 and 5) and one output parameter (Stand parameter). The model architecture is presented in Figure 3. For training purposes, 143 data values are used with back scattering of the errors. This stage determines connecting parameters as weights between the nodes. In the model, the ANN learning ratio is chosen as 0.04 and 10 000 iterations are applied for the minimum error. After the training stage, the same set of weight parameters are employed for prediction purpose.

Correlation coefficients are seen in Tables 1 and 2 and they are high for stand parameters. Therefore, the ANN model developed in this paper is successful in practical applications.

5 CONCLUSIONS

The investigation in this paper concentrates on the Artificial Neural Network Model with procedures

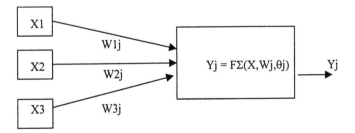

Figure 2. Artificial Neural

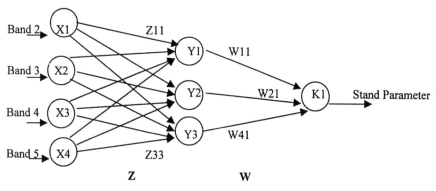

Figure 3. Artificial Neural Network System.

Table 1. Correlation coefficients for training data (143 values of data)

	Number of trees	Basal area	Volume	Mean diameter	Stand density
Correlation	0.81	0.63	0.70	0.81	0.77

Table 2. Correlation coefficients for control data (20 values of data)

	Number of trees	Basal area	Volume	Mean diameter	Stand density
Correlation	0.82	0.73	0.83	0.79	0.80

to assess information for forest stand parameters from SPOT-XS data using remote sensing techniques. Remote sensing image processing techniques can be connected relatively easily with general forest inventory to get help for forestry to prepare stand mapping or national forest inventory. The advantage in this research is that it tries to solve by using maximum training ground truth data containing more specific classes to make decisions based on near time data available with the satellite data and 154 of 300 plots are measured on the study area for the purpose of forest management plan preparations. Some stand characteristics such as basal area, volume, number of trees, mean stand diameter and tree species on each plot are determined with geographical coordinates of each plot loaded onto satellite imagery. The acquisition of forest stand mapping is fundamental to forest management and for the updating of forest stands periodically for National Forest Services. In such an inventory, the main use of remotely sensed satellite data is the transformation of the field plot into stand map information. The results show the possibilities of the presented approach for decreasing the need for field work in the test area. This illustrated experiment can also provide an interesting alternative to solving control and updating problems using satellite data in different conditions from the present situation prevailing in Turkey.

The ANN model has been developed to examine the forest area from imagery for stand parameter measurements and the reflectance values of each band. ANN correlation results are given according to the selected five forest parameters as 0.81 for number of trees, 0.63 for basal area, 0 .70 for volume, 0.81 for mean diameter and 0.77 for stand density of forest as correlation coefficients for 143 values of training data. On the other hand, for the last 20 data values the correlations are 0.82 for number of trees, 0.73 for basal area, 0.83 for volume, 0.79 for mean diameter and 0.80 for stand density in this study. When the model learning ratio is chosen as 0.04 and 10 000 iterations then a minimum error occurs. After the training stage, the same set of weight parameters are employed for prediction purposes.

REFERENCES

Coşkun H. G, Öztopal A, and Şen Z., 2000: *Forest Assessment From Satellite Data by Use of Genetic Algorithms*, EARSeL 7th Symposium on a Decade of Trans-European Remote Sensing Cooperation 14-16 June , Dresden-Germany.

Coşkun, H. G., Örmeci, C., Asan, U., Yesil, A., Musaoğlu, N., and Kaya, Ş., 1998: *Stand Type Classification from Landsat-TM and SPOT Satellite Data, The Science and Technical Research Council of Turkey Project Report, No:1677,pp 1-40.*

Hame, T., 1984: *Landsat aided forest site type mapping*, Photogrammetric Engineering and Remote Sensing, Vol, 50, 1175-1183.

Hame, T., Rauste, Y., 1995: *Multitemporal satellite data in forest mapping and fire monitoring*, EARSeL Advances in Remote Sensing, Forest Mapping and Fire Management, Vol. 4, 93-101.

Haykin, S., 1994: *Neural networks*, Macmillan College Publishing Company, 696.

Lippman, R., 1987: *An introduction to computing with neural nets.*, IEEE ASSP Man., Vol. 4, 4-22.

Mc Cormick, P., Kenedy, P., Folving, S., 1995: *An integrated methodology for mapping European forest ecosystem using satellite remote sensing*, EARSeL Advances in Remote Sensing, Forest Mapping and Fire Management, Vol. 4, 87-92.

Pereira, J.M.C., Olivetra, T.M., Paul, J.C.P., 1995: *Satellite-Based estimation of Mediterranean shrupland structural parameters*, EARSeL Advances in Remote Sensing, Vol. 4, pp.14-20.

Simpson, P., "Artificial neural networks", Paradigms, Applications and hardware implimentations, IEEE Press, 1992.

Smith, J., ve Eli, R.N., 1995: *Neural-network models of rainfall-runoffprocess*, ASCE, J. Water Resour. Plan. Man., Vol. 121, No. 6, 499-508.

Sönmez, İ. ve Şen, Z., 1998: *Artificial neural network approach for natural atmospheric event dynamics and application in meteorology*, 2nd International Symposiumon İntelligent Manufacturing System, IMS'98, 6-7 Ağustos, Sakarya, Türkiye.

Varjo, J., 1995: *Forest change detection by satellite remote sensing in Eastern Finland*, EARSeL Advances in Remote Sensing, Vol 4, pp.102-106.

Welch, R., Ehlers, M., 1987: *Merging Multi-resolution SPOT HRV and Landsat TM Data*, Photogrammetric Engineering And Remote Sensing, Vol. 53, No. 3, March, pp. 301-303.

Wilkinson, G. G., Folving, S., Kanellopoulus, I., Mc Cormick, N., Fullerton, K., Meguro, J., 1995: *Forest mapping from multi-source satellite data using neural network classifiers an experiment in Portugal*, Remote Sensing Reviews, Vol 12, pp.83-106.

Observing our environment from Space: New solutions for a new millennium, Bégni (Ed.)
© 2002 Swets & Zeitlinger, Lisse, ISBN 90 5809 254 2

Processing techniques for CORONA satellite images in order to generate high-resolution digital Elevation models (DEM)

M.Schmidt & G.Menz
University of Bonn, Germany

R.Goossens
University of Gent, Belgium

ABSTRACT: The satellites used on the CORONA missions carried 2 cameras on board, which enabled the recording of stereo images of the earth's surface. During the operational phase of the satellites, panchromatic images were recorded from many regions of the earth at a flight height of 150 km. The resulting data have been available to the public since 1995 and cost 18 US$ per strip. Thus these data are not only affordable for projects with limited budgets, but also are a very useful source in regions where surface information is hard to get or even not available. In this paper two methods are described which transform photographic information into digital format and handle the data with software tools of digital photogrammetry. The first method is applied by photographic enlargement with a scan of the photos; the second approach is to perform a high quality scan of the strips directly. CORONA image strips have no fiducial marks, so this leads to a non metric approach. Techniques of modern photogrammetry allow, nevertheless handling the data and to derive a high resolution digital elevation model. Therefore, during a field campaign it is essential to take Ground Control Points (GCPs) with high accuracy. A LEICA 300 Differential Global Positioning System (DGPS) is used. These points are needed to calculate the external orientation of the Digital Elevation Model (DEM). The achieved accuracy in z direction is in the order of 20 meters relatively, and in x and y direction in the order of 9 metres.

1 INTRODUCTION

While CORONA satellite data have been available since 1995, yet within standard remote sensing software, it is still not possible to handle the data properly (Goossens *et al.*, 2001). The proposed method tries to offer a solution in obtaining topographical information of areas with no, or rear, maps, which are not accessible with aerial photography. Therefore the photogrammetric software package VirtuoZo 3.1 was used in order to handle the stereoscopic images to generate a digital elevation model. The data inherit high image distortions due to the fact that the stereo-images are recorded in a converging manner. The purpose of this study is to benefit from this rather inexpensive data-source available for many parts of the world, and to deliver a methodology to create a DEM that is suitable for environmental and remote sensing applications e.g. LANDSAT terrain corrections. But it is still not possible to obtain SRTM-derived DEMs for remote areas; the best available DEM source is the GTOPO30 from the U.S. Geological Survey with a grid resolution of approximately 1 km. An alternative is always to digitise topographic maps. For many regions of the world these maps are of poor quality and give no detailed information. Hence the purpose was to establish a different methodology to derive DEMs with less effort and costs and with improved accuracy.

2 DATA

The CORONA satellite program was initiated on the demand of Dwight Eisenhower at the end of the 1950's, after the U2 espionage aircrafts of the US Air Force was shot down by the Soviets. In the hope of avoiding further international conflicts, the US observational strategy had to be reorganised, hence the necessity for a new program. CORONA program is a cooperative effort among the US Central Intelligence Agency (CIA), the US Air Force, and private companies. For the construction of the satellite, only the most invented camera-systems were used, and kept top secret. Written notes about Corona were not permitted. (Ruffner, 1995).

In June 1959 the first launches of the rockets

Figure 1: Example of CORONA capsule capture (Jensen, 2000).

carrying the CORONA satellites took place, but the first 9 missions failed due to differing technical problems. The first successful capture of a data capsule, that was parachuted from a CORONA satellite was on the 18-th of August, 1960. (MacDonald, 1995).

Aircraft of the US Air Force had the incredibly difficult, fine-tuned task of capturing the ejected CORONA film capsules before they could plunge into the ocean (figure 1). The capsules were constructed to remain viable for approximately 24 hours, after which they would self-destruct , thereby leaving no chance of being captured by other parties. (Peebles, 1997). This procedure obviously resulted in a loss of much mission data (Ruffner, 1995).

The ground resolution of the first mission data was 40 ft. During the operational phase of the CORONA satellite between 1960 and 1972, the camera systems were improved to a best ground resolution of 6 feet (1.8 metres), which was archived by the KH 4B camera (table 1). This system had the additional advantage of the facility to take stereo-pictures with one camera looking "forward" and another camera looking "afterwards" depending on the flight direction (fig. 2).

Table 1: Data properties of the KH 4B cameras.

System	KH-4B
Camera type	Panchromatic
Format of the frame	5.54x 75.69 cm
Best ground resolution	1.83 m (6 ft)
Flight height	150 km
Size of the observed area	14 km x 188 km
Focal length	60.69 cm

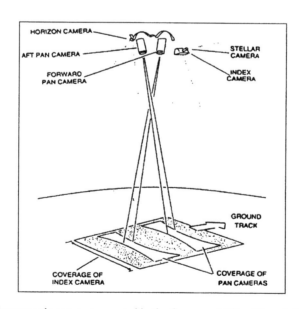

Figure 2: Stereoscopic cameras mounted in the Corona satellites (Campbell J.B, 1996).

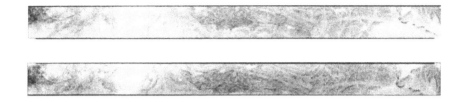

Figure 3: Stereo-couple of the CORONA 1117 mission 26th May 1972.

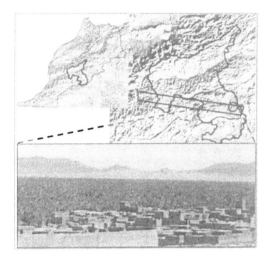

Figure 4: Location of the study region in the Drâa valley, Morocco.

The area that is covered by one strip has a size of approximately 14 km x 188 km. In this study two strips were used from the mission DS-1117-1, the forward strip F-107 and the afterward strip A-113 (figure 3). An example of a stereo-couple is given in Figure 3, where a different image geometry is already obvious.

3 STUDY REGION

The valley of the river Drâa is located on the southern slopes of the Atlas mountains in the south of the Kingdom of Morocco. The river Drâa is a river oasis at the beginning of the Sahara desert, where the hydrological system depends on snow-melt runoff in spring from the High Atlas mountain chain.

In the background of figure 4 is the GTOPO30 elevation model; the outlined areas are the River Drâa catchment, and the location of the Forward CORONA strip. On the right end of the strip is the location indicated from which the terrestrial

photograph was taken. The background of the picture shows the quartzite mountain chain of the Jebel Bani with a wind gap in between, and sand accumulation. A digital model was established around this area.

4 FIELD WORK

During field work, ground control and check points in x,y and z direction were recorded with a LEICA 300 GPS. The points have a relative accuracy of 5cm to each other; the absolute position of the base station was measured with 30 cm precision. This field campaign was conducted in the context of the IMPETUS project. In total, 15 points were measured in the frame of the photogrametric restitution of the images and 86 points were measured from the validation of the generated DEM.

5 METHODS

The photographically stored information of the CORONA strips have to be transformed into digital information, in order to benefit from digital photogrammetric methods. This was done in two different ways (fig. 5).

Within the software VirtuoZo 3.1 it is possible to work on stereo-image data with a non metric approach with no further input parameters other than ground control points. After inputting the images into the VirtuZo 3.1, the software through pattern recognition, enables several matching points in both images, and calculate out of these, a relative orientation including the parameters kappa (κ), phi (ϕ) and omega (ω) of the relatively oriented images. Table 2 shows the results of the calculations of both approaches.

For the calculation of the absolute orientation it was necessary to give at a minimum 6 GCPs to solve the orientation equation system (Chester, 1980). 8 GCP's were used for the exterior information. Afterwards the y-parallax was removed (Mikhail, et

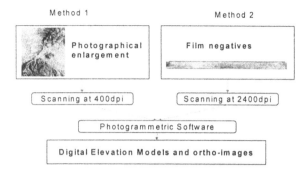

Figure 5: General methodology.

al., 2001) and lines of equal parallax difference and contour-lines were calculated, resulting in a DEM (Maune, 1996). During the processing it is also possible within this software to perform a visual quality check with 3D glasses.

6 RESULTS

Based on the DEM, the ortho-image was created and superimposed with contour lines. The residuals on the GCP's and the mean errors are presented in table 2. As table 2 shows method 2 (the direct film scanning) gave more reliable results, than method 1. Different reasons could be pointed out as follows:

1. It is obvious that the phi error is much greater using method 1 than method 2. This can be due to the fact that during the photo enlargement, the projection table of the enlarger was not completely horizontal. This problem is not faced with method 2.

2. The residuals in x,y and z of the GCP's are lower with method 2. The residuals in x,y are a factor of 1.3 better than with method 2 and in z direction even improved with a factor of more than 2. This can be due to the fact that the scanning resolution allows a finer indication of the ground control points.

3. In addition to the above, with method 2, more ground control points were found automatically, namely 69 versus 50 with method 1. This can be also due to the finer scanning method used with method 2.

4. Different products were generated after the restitution of the images: a DEM (figure 6), which can be inspected using the Nuvision 3D glasses; a file containing the contour lines (labelled or not); the ortho-image and the ortho-image overlaid with the contour lines (figure 7b). This was done for both methods. All these documents are produced in

Table 2: Parameters of the relative orientation of the image matching procedure. a) for the photo-enlarged approach: method 1, b) for the direct scan: method 2.

GCP	dX	dY	dZ	Orientation Parameters		Parameter	Point accuracy
1	1.897	-2.315	-0.54	Kappa (1)	0.0068	RMS Error	0.02
2	-11.86	12.478	2.362	Kappa (2)	-0.0064	mx	9.83
3	8.753	-3.909	-0.883	Omega (2)	-0.0953	my	9.17
4	-10.54	2.681	4.889	phi (1)	-9.1219	mz	3.38
5	0.404	-1.659	0.495	phi (2)	-0.2618	mxy	13.45
6	3.337	5.996	-1.78				
7	-0.792	-6.571	-0.733				
8	6.545	8.442	-3.352				

GCP	dX	dY	dZ	Orientation Parameters		Parameter	Point accuracy
1	1.918	2.237	0.254	Kappa (1)	0.0245	RMS Error	0.02
2	-5.766	-8.557	-0.577	Kappa (2)	0.0027	mx	6.37
3	4.435	7.94	0.611	Omega (2)	-0.982	my	6.37
4	-7.826	2.658	-0.492	phi (1)	-0.3825	mz	1.52
5	-1.547	-0.331	2.094	phi (2)	-0.1804	mxy	9.00
6	4.073	0.245	-1.581				
7	-1.183	-3.54	-1.065				
8	4.9	1.069	0.464				

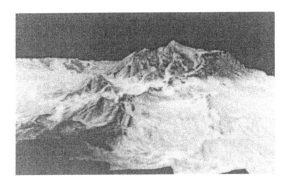

Figure 6: DEM derived from CORONA data .

VirtuoZo 3.1 in a minimum of time. A status report is produced automatically during the process.

A first visual comparison of the contour lines from the topographic map and those once generated by the digital photogrammetric methods show a good resemblance (fig. 7).

The major contour lines indicated on the topographic map can be found back on the created contour map, but moreover the detail is much better on the last one, resulting in a much finer DEM. The contour lines were set up at an equidistance of 10 meters, but they can be generated more finely if wanted. In any case the option of a equidistance of 10 meter was selected to correspond with the final scale chosen for the ortho-image (1:50000).

Besides this visual inspection of the contour lines, a more quantitative analysis of the errors on both created DEM´s was also made using the validation GCP´s measured randomly, distributed in the terrain. For the two DEM's all terrain GCP's

Figure 7:

a) Subset of 1:100.000 Topographic Map – Zagora.

b) Topographic ortho-photomap of the study region derived from CORONA data using the photo-enlargement approach.

Table 3: Height-deviations of terrain measured points and DEM grid-points.

Method	Residuals (mx,my) [m]		Mean height difference (Δz) [m]	Standard deviation [m]	Number of Points
1 (Photo-enlarged)	9.8	9.2	9.54	13.74	49
2 (Scanned)	6.2	6.2	25.13	10.33	30

195

were chosen that lie in the spacing of the mx,my residuals (table 2,3) of the DEM-points. For method 1 these values are 9.8 and 9.2 meters; and for method 2 they are twice 6.2 meters. This was done in order to calculate an average height difference Δz of the z values from the DEM-grid-points and the terrain-GCP's. These findings are displayed in table 3.

With method 1 Δz is less than 10 meters, but the standard deviation is 13.74 meters relatively higher than the standard deviation of method 2, with 10.33 meters. In method 2 the Δz with 25.13 meters is disproportionally high, but this significant Δz error (20 –25 meters) is a systematic underestimation of the real terrain from the generated DEM. While with method 1 no systematic error occurs. A shift of +20 meters to the DEM of method 2 resulted in a very good DEM result. Method 2 also gave better results than method 1 in the flat and sandy plain areas in front of, and behind the mountain ridge.

7 CONCLUSIONS

It can be stated that the proposed methods provide adequate topographical information when topographic maps and aerial pictures are not available for a test area. It is illustrated that the second method with the direct scanning of the films, gives more adequate results with a general under-estimation of the real topography. With the proposed approach of method 1 it is thus possible to generate a DEM with a relative accuracy of approximately 20 metres, which is a reasonable resolution for several applications, e.g. LANDSAT TM/ETM-illumination corrections. This method allows an important reduction of field work because in such a way measurements of the whole terrain with GPS-points is not needed. The proposed method is a cheap way to substitute conventional aerial pictures in case in which they are not available.

8 ACKNOWLEDGEMENTS

The authors would like to thank Dr. Beata De Vliegher, Dennis Devriendt and Frank Benoit (Geography Department – University of Gent) for their help and advice and Angela Altmaier and Christoph Kany (Geography Department – University of Bonn) for their assistance during the field work. Mary Drobin is acknowledged for helping to edit this text.

9 REFERENCES

Campbell J.B. 1996. *Introduction to Remote Sensing*. The Guildford Press.

Chester,C.S. 1980. *Manual of Photogrammetry*. Falls Church: American Society of Photogrammetry. 1056 pp.

Day. A., Logsdon J.M. & Latell, B. (1998). *Eye in the sky: The history of the Corona spy satellites*. Washington D.C.: Smithsonian Institution Press. 303pp.

Goossens,R., De Man,J. & De Dapper,M. 2001. Research to possibilities of Corona-Satellite-data to replace conventional aerial photographs in geo-archaeological studies, practised on Sai, Sudan. In: Buchroithner,M. (Ed.), *A decade of Trans-European Remote Sensing Cooperation*, pp. 257-262. Lisse: A.A. Balkema Publishers. 262 pp.

Jensen,J.R. 2000. *Remote Sensing of the Environment*. London: Prentice-Hall. 544 pp.

Maune D.F. 1996. DEM Extraction, Editing, Matching and Quality Control Techniques. In: *Digital Photogrammetry: An Addendum to the Manual of Photogrammetry*. Bethesda, Maryland, USA. pp.135-141.

MacDonald, R. 1995. CORONA: success for space reconnaissance, a look into the cold war, and a revolution for intelligence. *Photogrammetric Engineering and Remote Sensing*, 61: 689-720.

Mikhail E.M., Bethel J.S. & McGlone J.C. 2001. *Introduction to Modern Photogrammetry*. New York: John Wiley & Sons, Inc. 479pp.

Peebles, C. 1997. *The Corona project: America's first spy satellites*. Anapolis, Maryland, USA: Naval Institute Press. 351pp.

Ruffner,K.C. 1995. *America's first satellite Program*. Washington: CIA History Staff.

Observing our environment from Space: New solutions for a new millennium, Bégni (Ed.)
© 2002 Swets & Zeitlinger, Lisse, ISBN 90 5809 254 2

Development of a solar spectro-irradiometer for the validation of remotely sensed hyperspectral images

A.Barducci, P.Marcoionni, I.Pippi & M.Poggesi
C.N.R.-I.R.O.E. "Nello Carrara", Firenze, Italy

ABSTRACT: The main troubles affecting the application of remote sensing are connected with the radiometric calibration of the sensor signal as well as with the physical parameters to be used for the radiative transfer modelling through the atmosphere. In order to overcome these problems we are developing two new instruments to be used for in-field measurements devoted to the validation of remotely sensed hyperspectral images. The main instrument is a solar spectro-irradiometer, operated in the imaging-spectrometer mode, that samples the impinging irradiance with 7 nm of spectral resolution and a digitalisation accuracy of 10 bit. The 256 independent samples held in the measured spectrum are acquired with a sampling step of about 3.0 nm and cover the range from 200 nm up to 950 nm of wavelength. From this original instrument we have also derived a spectro-radiometer having similar spectral characteristics. In this paper we discuss the development of the main common spectrometer and describe the experimental activity carried out for its calibration. Preliminary experimental data are shown and thoroughly analysed.

1 BACKGROUND

The extended dynamic range of modern detectors (up to 16 bit of digitalisation accuracy) may be fully employed only if the image data can be correctly calibrated and transformed to radiance measures of known physical units, see Barducci & Pippi (1998 and 2001). Moreover, the acquired images must be carefully corrected for atmospheric effects (mainly absorption and scattering, in the visible spectral range), as long as the data would be used for a quantitative assessment of chemical, physical and biological parameters, see Wan et al. (1999) and Barducci & Pippi (1994).

In this field, the main troubles affecting the application of remote sensing are connected with the radiometric calibration of the sensor signal (Nelson et al. 1991) as well as with the physical parameters to be used for the radiative transfer modelling through the Earth's atmosphere. In fact the high digitalisation accuracy of modern detectors makes its radiometric calibration more difficult and the optical properties of the atmosphere often are unknown, since their measurement requires a very complex experimental instrumentation, see Cantella (1993).

In order to overcome these problems we have developed two new instruments to be used for in-field measurements devoted to the validation of remotely sensed hyperspectral images. The main instrument is a solar spectro-irradiometer, operated in the imaging-spectrometer mode, that samples the impinging irradiance with 7 nm of spectral resolution and a digitalisation accuracy of 10 bit. The 256 independent samples held in the measured spectrum are acquired with a sampling step of about 3.0 nm and cover the range from 200 nm up to 950 nm of wavelength. The irradiometer was designed to be operated both on the ground or on board of any airborne platform, under the control of a portable PC which also registers some ancillary data (e.g.: the time of acquisition). The use of a simple optical accessory makes the spectro-irradiometer suitable for the measurement of angularly integrated scattered radiance.

From this instrument we have also derived a spectro-radiometer having similar spectral characteristics (spectral resolution, sampling step, digitalisation accuracy, and so forth).

Both the instruments have being calibrated after performing laboratory measurements concerning the wavelength position, the dark-current signal, the spectral sensitivity and the absolute radiometric response. The two calibrated instruments will provide data about the ground spectral irradiance, the irradiance spectrum of the scattered radiation, the transparency the atmosphere, and the radiometric calibration of the used remote sensing hyperspectral imager.

In this paper we discuss the development of these two instruments as well as the experimental activity carried out for their calibration. Preliminary experimental data are shown and thoroughly analysed. We show that these instruments are useful for controlling in-flight the radiometric calibration factors of any remote sensing imager as well as to capture valuable data for performing the atmospheric correction of the acquired hyperspectral images.

2 HARDWARE ARCHITECTURE

The main aim of our work is to develop an easy-to-use as well compact instrument to be used for in-field measurements during remote sensing campaigns. The developed instrument should serve as a radiometric calibration tool for aerospace sensors and a valuable means to gather in-situ and validation measures for correcting remotely sensed images. Figure 1 details the principal optical characteristics of the chosen monolithic sensor.

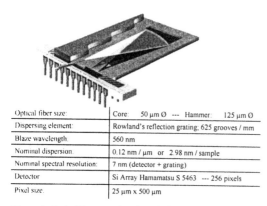

Optical fiber size:	Core: 50 μm Ø --- Hammer: 125 μm Ø
Dispersing element:	Rowland's reflection grating; 625 grooves / mm
Blaze wavelength:	560 nm
Nominal dispersion:	0.12 nm / μm or 2.98 nm / sample
Nominal spectral resolution:	7 nm (detector + grating)
Detector:	Si Array Hamamatsu S 5463 --- 256 pixels
Pixel size:	25 μm x 500 μm

Figure 1. Optical layout and main technical data of the spectro-irradiometer, after microParts (2001). Let us note the compact design of the spectrometer sensor.

Spectral measurements are controlled by the host computer by means of a proprietary serial communication protocol. The computer issues to the micro controller the acquisition command, which contains all the experimental parameters such as shutter status, gain value, integration time, and so forth. The microcontroller then performs the required measurement and conveys the spectral data to the communication port. The spectral acquisition software running on the computer read the incoming data and records it on the hard disk. To this purpose a dedicated software was developed, using the C/C++ programming languages. The software provides the user with all the services for instrument calibration and data pre-processing. Particularly, the software is equipped with algorithms that carry out radiometric

response measurement and calibration, dark-signal characterization and subtraction, channels' spectral position measurement, spectrometer's point spread function (PSF) assessment and spectrum deconvolution..

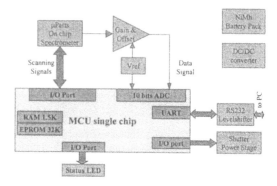

Figure 2. Layout of the readout circuitry of the developed spectro-irradiometer. The electronics is controlled by a Risc CPU running at 20 Mhz of clock frequency. The μC generates the scanning signals for the linear array detector, controls the AD conversion and sends the data to the PC via the RS232 port. Let us note that the data acquisition is triggered by a PC command issued to the μC by the serial interface; this command also contains the requested gain, integration time and shutter status.

We have selected a compact micro-spectrometer, manufactured by the μParts company, which is composed by a concave spectrometer bound to a photodiode array. The spectrometer is fed by a monomode optical fiber that transmits the light collected by the fore optics. The instrument is powered by a NiMh battery pack that allows the user a measurement time greater than 3 hours. The instrument fore optics is a standard imaging objective for the spectroradiometer and a double diffuser for the solar spectro-irradiometer. Figure 2 shows the layout of the detector and the related readout circuitry.

The monolithic sensor is equipped with a 256 photodiodes linear array by Hamamatsu Photonics (model S5461). The output of the array is scanned like a charge coupled device (CCD). The user must provide external shifting clocks (3 lines) and two control lines for reset and noise reduction. At the end of every CCD scan the chip outputs an end of scan signal. The sensor has also a control line that allows the choice between two gain (Hi and Low). The output of the sensor is then sent to a two stages amplifiers circuit with offset control.

A Micro Controller Unit (Arizona microchip model 18C452) running at 20 MHz clock speed sources all the control signals to the linear array. The processor idles until a begin of acquisition string is received via the RS232 communication link.

Inside the micro a 10 bits ADC converts the 256 analogue samples to digital numbers (DN). The DN are stored sequentially in the embedded RAM; when the micro senses the end of scan signal from the sensor stops the acquisition and sends back the samples to the host computer, a 16 bytes tail string that contains housekeeping data follows the data stream. The reception or transmission of data is done by internal serial interface (UART).

The control of integration time is done by means of internal counter timer and can be stepped between 7 milliseconds and about 2 seconds. The micro also performs the control of the shutter used during the measure of the dark signal. An input line of the analogue section of the MCU is reserved to monitor the battery charge status.

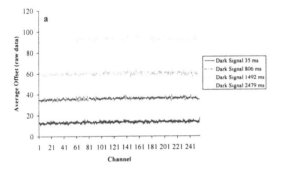

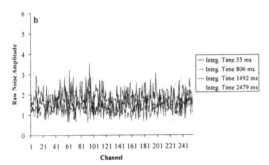

Figure 3. Spectrum of the averaged (ten trials) instrument's offset and noise for different integration times. a: offset, b: noise. Let us note that the offset as well as the experimental noise are spectrally flat; i.e.: they do not change with changing wavelength. (Colour plate, see p. 423)

3 EXPERIMENTAL DATA

In order to test the basic instrument performance and to get common calibration data some laboratory activity was carried out, whose early results are hereinafter discussed. The activity was devoted to measure the instrument dark-signal, the signal-to-noise ratio, the central wavelength of the available spectral channels and the spectrometer's PSF.

A series of dark-signal measurements was performed for the two gain values (high, low) and different integration times. Since the dark-signal contains contributions from the instrument noise and any other source of signal offset, the obtained data permitted to estimate both the noise amplitude (i.e.: the signal-to-noise ratio) and the offset mean spectrum.

Figure 3 shows the instrument overall noise (b) and offset (a) as a function of the channel number; Figure 4, instead shows the wavelength averaged offset and noise versus the imposed integration time. Let us note that both offset and noise have a flat spectrum (i.e.: are uniformly distributed over the wavelength) so that the detector is expected to have a rather uniform sensitivity. Moreover, Figure 4 shows that as expected the offset linearly increases on average with increasing the integration time while the mean noise seems rather insensitive to integration time changes.

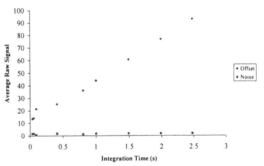

Figure 4. Average instrument's offset and noise amplitude versus the imposed integration time for the high gain. For any point in the plot ten measurements were acquired and the noise amplitude was computed as the signal standard deviation for the single measure. The plotted data have been averaged over repeated trials and over wavelength (different channels) for both the offset and the noise amplitude (Colour plate, see p. 423)

The spectral dispersion of the utilized spectrometer was investigated using a set of Argon, Krypton, Neon and Xenon, low-pressure, spectral calibration lamps. These lamps, manufactured by Oriel, are specifically designed for spectral calibration purposes and their low pressure emitter provides high stability and narrow spectral lines. Figure 5 shows the spectrum measured by our spectro-irradiometer when observing the Krypton lamp. It is to notice that our spectrometer is not able to resolve the different spectral lines emitted by the source, thus adjacent lines are observed as broad blends. We also point out that the SNR obtained for the Krypton lamp measurement was roughly 263 for the single measure.

By means of the measurements performed on the calibrated sources of spectral lines the dispersion curve describing the behaviour of our spectrometer was deduced. The performed analysis allowed us to estimate both the pixel dispersion and the measurement accuracy for the absolute wavelength position of the different photosensitive elements (spectral channels).

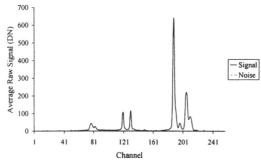

Figure 5. Spectrum of the Krypton lamp together with the estimate noise affecting the measure. The raw data (ten independent spectra) were averaged and the instrument offset, separately measured, was removed. Let us note that the used spectral source really emits a lot of spectral lines, most of which are not resolved by our instrument.

The main result of this activity is shown in Figure 6, where the measured spectral position of 24 different spectral lines are plotted versus their calibrated wavelength.

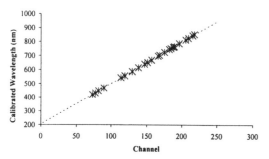

Figure 6. Wavelength position of lines emitted from various calibration spectral lamps versus the channel number of the fitted position of the digitised spectra. Crosses represent the experimental points while the dashed line is the fitted linear dispersion curve.

Let us note that the peak of the lines involved in this calculation was initially fitted with a second order polynomial expression in order to obtain a fair estimate of the true line maximum location. The performed χ^2 fit utilised only a small set of experimental points (e.g.: 7 points) around the apparent maximum (max sample) of the concerned line. Since the fit is sensitive to digitised line asymmetry, it can account for a line central wavelength not exactly centred on the maximum value sample of the measured spectrum. In this way it is possible to obtain

higher accuracy spectral position estimates than simply taking the number of the spectral channel exhibiting the highest value. The plot shown in Figure 6 clearly confirms that the dispersion curve of the spectro-irradiometer is linear, as stated by the following equation,

$$\lambda_k = ak + \lambda_0 \tag{1}$$

where λ_k is the wavelength of the kth spectral channel, λ_0 is the wavelength position of the first channel and a is the pixel dispersion.

The linear dispersion curve of Equation 1 was χ^2 fitted on the experimental points shown in Figure 6, hence producing the regression line herewith plotted. The fitted pixel dispersion as far estimated was 2.96 +/- 0.014 nm / sample, in excellent agreement with the expected value of 2.98 nm / sample (see Figure 1), the difference being easily ascribed to the noise affecting the measurement. As a result of fit we were also able to compute the spectral range explored by the instrument, which resulted to spawn the 205 nm to 959 nm interval. It is worth noting that this result does not necessarily mean that spectral measurement are significant in this broad band. The aforementioned estimate only concerns with the starting and ending wavelengths of the acquired spectra.

Moreover, the executed χ^2 fit allowed us to estimate the accuracy of the absolute wavelength position inferred for the available spectral channels (e.g.: the error on λ_0 estimate). It resulted that the spectral position of every spectral channel is predicted with an overall experimental error of 4.5 nm. We point out that this uncertainty fairly agrees with the expected sensor's spectral resolution of 7 nm (see Figure 1), which is mainly due to beam size output by optical fiber whose core is 50 µm wide.

Finally, we have investigated the PSF characterising the sensor: i.e.: the instrumental cross-talk between adjacent spectral channels. To this purpose we have used a coloured He-Ne laser source, the line centre of which was at 543.5 nm. From a theoretical standpoint we suppose that the laser should emit a very narrow line, whose width should be far below both the spectral resolution and the sampling step of the tested spectrometer. As long as this condition holds true the measured spectrum should directly account for the PSF of the spectro-irradiometer.

One example of the measurements performed is shown in Figure 7, where the digitised spectrum of the laser source is plotted versus the wavelength. As can be seen the central part of the observed laser line corresponds to a reasonably narrow spectral response of the spectrometer. This would be a valuable characteristics, nonetheless very extended line wings are found. Figure 8 shows the same plot of Figure 7 but enhancing the wings of the measured PSF.

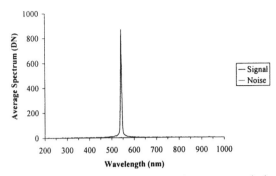

Figure 7. Spectrum of a coloured He-Ne laser source acquired by the spectro-irradiometer. The plotted data were averaged over ten independent measures and wavelength calibrated using the results previously discussed.(Colour plate, see p. 423)

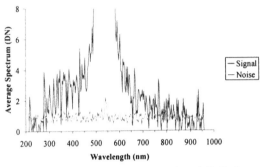

Figure 8. Wings of the spectrum of a coloured He-Ne laser source acquired by the spectro-irradiometer. The plotted data were averaged over ten independent measures and wavelength calibrated using the results previously discussed. (Colour plate, see p. 423)

As shown in Figure 8 the measured PSF extends very far from its centre, hence originating some trouble related to its theoretical interpretation. The first striking remark concerns the origin of these extended wings; i.e.: if they are caused from some unexpected effect affecting the laser source or if really the spectrometer has a broad response. This last option is possible since the wings are extended but have a small value. The investigation of this question requires additional experiments with different light sources working at different wavelengths. These measurements will be performed in the future and will also serve to investigate the wavelength invariant property.

The shape of the response function shown in Figure 7 has been investigated in order to assess its mathematical representation. It was found that no standard exponential function can fit the experimental profile of Figure 8, which instead may be roughly accounted for by a rational function of the following kind:

$$f(\lambda) = \frac{1}{p_2(\lambda - \mu)^2 + p_1|\lambda - \mu| + p_0} + f_0 \quad (2)$$

Here μ, the centre of the response function, is a free parameter and its value needs to be fitted together with that of the other free parameters p_0, p_1, p_2 and f_0. We have tried first ($p_2 = 0$) and second order polynomials in the denominator of Equation 2, thus discovering that second order polynomials gives rise to the better fit. In fact, if imposing $p_2 = 0$ we obtain a worse fit, with a huge relative χ^2 equals to 130. The second order expression provides instead a slightly better as well still unsatisfactory relative χ^2 of 29. It is to notice that this last fit produces a vanishing $p_1 = 0$, which reduces the function of Equation 2 to a standard Lorentz's profile. This result could be important because the pressure broadening of the laser line has this particular expression. Hence, the obtained fit could suggest that the extended wings experimentally found (see Figure 8) would be ascribed to pressure broadening of the emitted line. However, it is very difficult to accept such a large broadening (about 200 nm) as due to a pressure phenomenon and we think that it might be an effect of the optical fiber or the spectrometer. Figure 9 shows the fitted first order and Lorentz's function with the related experimental data. It is evident how the Lorentz's profile is the better approximation to the measured data, particularly in the wings of the response function.

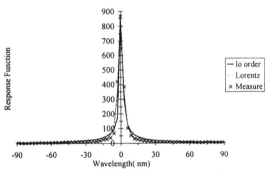

Figure 9. Plot of the fit result and of the experimental data versus the wavelength. It is evident that the Lorentz's profile (red line) is the better approximation to the experimental data (crosses). The response function is not normalized. (Colour plate, see p. 423)

3 CONCLUSIONS

The development of a new spectrometer devoted to in-field measurement for the validation of remotely sensed images has been outlined. The instrument will be the common base for the achievement of a solar spectro-irradiometer and a portable spectrometer for ground reflectance measurement. The main technical result up to now arose has been shown and thoroughly discussed. Particularly we have analysed the instrument dispersion, wavelength calibration, noise amplitude and response to a monochromatic stimulus.

The obtained data show that our spectrometer is characterized by a high signal to noise ratio (roughly 260 for a single measurement) and that its response to a monochromatic source is a well peaked function with extended wings of very low weight. The instrument might noticeably improve the impact of validation campaigns for the processing and final use of remote sensing hyperspectral images.

Open problems concern the investigation of the instrument impulse response function, the pixel sensitivity and radiometric calibration factor.

4 REFERENCES

Barducci, A. & Pippi, I. 2001. Analysis and rejection of systematic disturbances in hyperspectral remotely sensed images of the Earth, *Applied Optics*, vol. 40, No. 9, pp. 1464-1477.

Barducci, A. & Pippi, I. 1998. The airborne VIRS for monitoring of the environment, in Sensors, Systems, and Next-Generation Satellites (EUROPTO '97), H. Fujisada, ed., Proc. SPIE, vol. 3221: pp. 437-446.

Wan, Z. Zhang, Y. Ma, X. King, M.D. Myers, J.S. & Li, X. 1999. Vicarious calibration of the Moderate-Resolution Imaging Spectroradiometer Airborne Simulator thermal-infrared channels, Applied Optics, Vol. 38, No. 30: pp. 6294-6306.

Barducci, A. & Pippi, I. 1994. Atmospheric Effects Evaluation for theAVIRIS Image-Data Correction or Retrieval, in Recent Advances in Remote Sensing and Hyperspectral Remote Sensing (EUROPTO '94, Rome, Italy), Proc. SPIE, EUROPTO Series, vol. 2318: pp. 10-16.

Nelson, M. D. Johnson, J. F. & Lomheim, T. S. 1991. General noise processes in hybrid infrared focal plane arrays, Optical Engineering, Vol. 30, No. 11: pp. 1682-1699.

Cantella, M. 1993. Staring-Sensor Systems, in The Infrared & Electro-Optical Systems Handbook, vol. 5, Passive Electro-Optical Systems, Stephen B. Campana ed, Infrared Information Analysis Center and SPIE Optical Engineering Press: pp. 157-207.

MicroParts GmbH, 2001. VIS spectrometer datasheet.

Observing our environment from Space: New solutions for a new millennium, Bégni (Ed.)
© 2002 Swets & Zeitlinger, Lisse, ISBN 90 5809 254 2

Evaluation of linear mixture modelling of urban areas in Landsat TM satellite imagery

A.D.Hofmann
Institute of Photogrammetry and Remote Sensing, Dresden University of Technology, Germany

ABSTRACT: In this paper, the author presents the experiences made by applying the linear mixture modelling technique to an urban area located in the north of San Diego, Ca. Landsat TM satellite imagery was chosen to perform the analysis on. The paper presents the results of linear unmixing applied to raw and first-order corrected imagery. To evaluate the accuracy of the unmixing method, the results of the sub-pixel classification were compared to a reference, which is represented by high-resolution imagery, called ADAR.

1 INTRODUCTION

Remotely sensed images provide an attractive source of land cover data. However, this source requires a loss in spatial detail. In particular, each pixel within an image provides a single measurement for an area. The general way of classifying satellite data assigns only one material to this single measurement. Especially in areas with a high material diversity, like urban regions, this results in misclassifications. By unmixing a pixel into its component parts, it is possible to enable a more accurate classification of the area. Linear mixture modelling (LMM), as a type of sub-pixel classification can be used to perform the unmixing. It has been used often to analyse vegetation or soil mixtures in multispectral imagery. It has shown very good results. However, it is not very common to use it for classifying urban areas.

The main objectives of this study are the investigation of the potentialities of linear sub-pixel classification of urban areas in Landsat TM satellite imagery as well as the comparison of the sub-pixel classification results in first-order corrected and raw TM scenes.

If linear mixture modelling results in a high classification accuracy, it could be used as an economic way of monitoring land use change. There might be also a possibility of land use class detection.

2 STUDY AREA AND DATA

2.1 Study Area

The study site Del Mar is located in the north of San Diego, California. Del Mar, a young and fast developing city, is a typical urban area at the southern West Coast of the United States.

The climate of this area is Mediterranean. March through October is known to be the dry season of the year. Nevertheless, due to the ocean the humidity is relatively high. It is also important to mention that the atmosphere of this area is often very enriched with dust. This causes a perturbation of the signal penetration, which has to be regarded for the analysis.

The region is marked by low hills and creeks. The assumed ground elevation is about 90m (300ft). The natural vegetation is usually sparse and seared. Only along creeks, Mediterranean vegetation (e.g. palms, succulents, and juniper) grows very densely. Bright green lawn grass, shrubs, and deciduous only grow in irrigated areas. Manmade objects – buildings – are only on plateaus at the top of the hills. Main land use classes of the study area are residential units, schools, commercial and office districts, as well as light industry.

2.2 Data

For the analysis medium resolution imagery was required. Two Landsat-5 TM Scene with acquisition dates in June 1996 and 1998 were used. Ideally, both images were acquired at the same day time. In both scenes, noise removal and geometric registration with terrain correction were performed. Topographic effects were also corrected. Conspicuous are the big reflectance differences of vegetation and soil in the images. This is caused by different climatic backgrounds. 1996 was a year with less precipitation and higher temperatures than usually and 1998, a year dominated by El Niño, was relatively cool and wet. The resulted reflectance variations of especially vegetation might effect the classification results.

Reference values for the sub-pixel classification were obtained by an interpretation of digital airborne photographs taken by an ADAR 5500 system. The ADAR data were acquired in June 1998. The images have a ground solution distance of 1m.

3 LINEAR MIXTURE MODELLING

Linear Mixture Modelling has been discussed often in articles. Therefore here will be only a very brief explanation of the technique. LMM computes the spectral constituents of a pixel by analysing its spectral signature. It assumes a single reflectance of a photon at the object within the given instantaneous field of view. In many cases, it approximates reality. Whenever there is radiation transmission through one of the materials or if there are multiple reflections within a material or between two objects a non-linear mixing occurs. In the LMM, the DN of a pixel in each spectral band is expressed as a linear combination of the characteristic relative radiance of its component endmembers weighted by their respective areal proportions within the pixel. [Ichoku 1996] Thus, the DN_i of a pixel in the i-th band is given by:

$$DN_i = \sum_{k=1}^{n} EM_{ik} \cdot f_{ik} + \varepsilon_i \qquad (1)$$

DN_i relative radiance in band i for each pixel
EM_{ik} relative radiance in band i for each end-member k
f_{ik} fraction of each image endmember k calculated band by band
k each of n endmembers
ε i remainder between measured and modelled DN ("band residuals")

There are several ways to solve equation (1). Generally, the number of unknowns should be less than or equal to the number of equations to get a convenient solution. This implies that the number of endmembers should be less than or equal to the number of bands. Number of Bands plus one (i+1) is possible on condition that a constrained method is used.

4 METHODS

It was already mentioned that the atmosphere of the study area is often enriched with dust. Due to the ocean, haze is also often perturbing the signal penetration in the morning. It is known that a correction of these perturbations in the images will give better classification results. Thus, an atmospheric correction was applied to the data set.

4.1 *Atmospheric Corrections*

Usually, if atmospheric corrections are discussed a full atmospheric correction is understood. But if the atmospheric conditions at the acquisition time are not known, it is very likely that an applied atmospheric correction software does not really improve the data set. At least this is the case for the obtained data set. Therefore the data set was only corrected with an approximate atmospheric correction – the dark subtraction technique.

The dark subtraction technique, a first-order correction, assumes that each band for a given scene contains some pixels at or close to zero brightness value. It is also supposed that atmospheric effects and especially path radiance have added a constant value to each pixel in a band. Subtracting this constant value from the particular spectra will remove the first-order scattering component. A disadvantage of this technique is that problems may encounter in the analysis stage. This is because the DN values selected for the removal may not conform to a realistic relative atmospheric scattering model. It does not account for water vapour and ozone absorption. This lack of conformity may cause the data to be overcorrected or under corrected in some, or all of the spectral bands. The relationship between the bands will not be corrected also. No problems in the analysis stage encountered for the Del Mar data set. The by the dark subtraction technique caused lightening in the image improves the separability of spectral signatures. It is assumed that a better separability of the endmember will increase the LMM results. Along these lines, dark subtraction technique seems to be a valuable method to enhance image data for performing LMM.

4.2 *Software Options of LMM*

Before starting the analysis, it is also necessary to check available software for usefulness and accuracy. The best results were obtained with IPW SMA. IPW is a UNIX-based image processing system developed by the Computer Systems Laboratory, University of California, Santa Barbara. SMA provides three methods of spectral unmixing: modified Graham Schmidt, unconstrained and constrained. Although the unconstrained method is usually recommended the constrained method was used. A main reason is the discrepancy of the big diversity of materials and the limited number of endmember.

SMA obtained on a test image based on reference spectral signatures a classification accuracy of 75%. It was found that concrete, vegetation, and soil can be separate more accurately from each other than asphaltic features and water. The classification accuracy of the software depends on the separability of the spectral signatures and has to be considered in later classification results.

4.3 *Analysis*

First and very important part of the analysis is the endmember selection. The number of endmember is not only limited by the number of available bands, but also by the dimensionality of the data. A feature space plot of the third and fourth band helps to define the dimensionality of the data set.

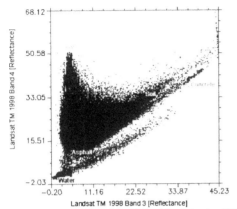

Figure 1 Feature space plot of the third and fourth Landsat TM band.

Figure 1 shows the scattergram of the 1998 TM scene. The five extremities named with Asphalt, Water (Shade), Concrete, Soil, and Green Vegetation (GV) can be recognized easily. The extremities also represent the main material classes later used in the analysis.

Since the analysis was restricted to urban areas the endmember selection had to be adjusted. Concrete, asphalt and vegetation occur most often in the study area. But soil and water need to be included also. Soil represents not only barren areas but also terra cotta roofs, which are seen frequently. Since there has to be a shade endmember to explain brightness variations, shade is assumed to be included in the water extremity. A set of bright and dark pure samples of each main material class were collected and merged. Because of different acquisition geometry and brightness variations the sample areas of 1996 and 1998 differ slightly form each other. Following criteria were required for the sample areas: The pixel representing the main material had to be as pure as possible. The surrounding area of the pixels (at least the eight neighbour pixels) had to be also as pure as possible.

The set of mixed image endmember was used to unmix the images. The spectral signature of the image endmember are graphically represented in Figure 2.

The analysis was restricted to urban landuses, which did not change during 1996 and 1998. The raw (uncorrected) imagery as well as the dark subtracted imagery of both years was analysed.

4.4 Results

The classification results of the linear mixture analysis provides a satisfactory result, what is shown in Table 1.

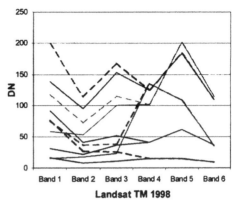

Figure 2. Spectral signatures of the image endmember. (Colour plate, see p. 424)

Table 1 Classification Accuracy of sub-pixel classification

Overall Classification Accuracy [%]	1996	1998
Supervised Classification	47.9	69.0
Sub-Pixel Classification		
Uncorrected Image	75.1	71.5
Dark Subtracted Images	84.2	78.1

For a better idea of usually achieved classification accuracies, the result of a supervised classification was included in the table. The overall classification accuracy percentages were computed by a comparison of the classification result of the Landsat TM scene to those of the ADAR imagery. (The classification of the ADAR imagery has an accuracy of about 98%.) Striking are the relative high differences within the supervised classification results. Since the training sets of both images are alike, only the different climatic backgrounds can be the reason. The differences in the sub-pixel classification accuracy are mainly caused by the different spectral signatures in term of separability of the endmember. Thereby these differences are mainly occur on materials depending on biological cycles.

In the study was found that a better separability of endmember increases the classification accuracy of the LMM. The main problem was that water and asphalt are hard to separate and this ended in misclassifications. Another problem was found in the problematic that darker but still pure pixels of

Vegetation, Concrete, Asphalt, or Soil exist in the image. Environmental effects lower and change the intensity of the reflectivity. To 'explain' the lower reflectivity shade (in this case the water endmember) was assumed to be a part of the pixel. Endmember were chosen from exposed, bright areas, but otherwise, too dark chosen endmembers will give results with too high concrete portions. Another reason for the misclassification is that very mixed pixels have different behaviours than pure exposed pixels. Thus, samples of the endmembers should express the variability of the materials of interest and present a significant variation in mixture portions.

5 CONCLUSIONS AND FURTHER WORK

This paper investigates the usage of spectral mixture analysis on Landsat TM data. Because of the size of the study area, the results might not be portable to other regions.

In general, the classification results of the LMM were motivating. Linear mixture modelling of Landsat TM imagery can be recommended to classify urban and rural areas as long as only few materials are to classify. An application of a first-order correction like the dark subtraction technique improves the classification accuracy significantly.

The accuracy of the LMM method mainly depends on two facts. First, the endmember selection according to the area analysed and second the software used to perform the analysis. The disadvantage of spectral unmixing is the high sensitivity for endmember. Slight differences in the spectral signature end in total different classification results. The method also does very often 'explain' signal attenuation by appearance of Shade. Too high shade fractions are the consequence. The study found that the unmixing method could successfully classify concrete, vegetation, and soil. However, the method did not unmix asphalt and water correctly.

In badly modelled units, a non-linear mixture modelling can be performed to find out if this would increase the classification accuracy. Non-linear mixture modelling has shown progress in vegetation and soil mapping. Regarding the surface structure, it might improve mapping of urban materials as well.

It is assumed that involving additional data can increase the acquisition of information. For instance, a trend involving non-linear or linear mixture modelling in neuronal networks is on the road of being established. Neuronal networks can select endmember automatically and optimise the mixture modelling itself. They only have the disadvantage of being time-consuming in the developing stage. Today, time exposure is a main criterion of image analysis techniques. Another possibility is to combine spectral mixture analysis with already available data. Involving an expert systems (DEM, texture analysis, landuse data, supervised classification) or a GIS can increase the information extraction. Included decision tress could reduce misinterpretation.

6 ACKNOWLEDGEMENTS

Landsat TM and ADAR data were provided by SDSU Department of Geography. I thank Lloyd Coulter as well as Prof. Csaplovics for their support. I also like to thank the Gesellschaft von Freunden und Förderern der TU-Dresden for giving grant.

7 REFERENCES

Adams, J.B., Sabol, D.E., Kapos, V., Filho, R.A., Roberts, D.A., Smith, M.O., Gillespie, A. 1995 Classification of Multispectral Images Based on Fractions of Endmembers: Application to Land-Cover Change in the Brazilian Amazon *Remote Sensing of Environment* 52: 137-154

Bateson A., Curiss, B. 1996 A Method for Manual Endmember Selection and Spectral Unmixing *Remote Sensing of Environment* 55: 229-243

Chavez, P.S. Jr. 1988 An Improved Dark-Object Subtraction Technique for Atmospheric Scattering Correction of Multispectral Data *Remote Sensing of Environment* 24: 459-479

Chavez, P.S. Jr. 1989 Radiometric Calibration of Landsat Thematic Mapper Multispectral Images *Photogrammetric Engineering and Remote Sensing* 55:1285-1294

García-Haro, F.J., Gilabert, M.A., Meliá, J. 1999 Extraction of Endmembers from Spectral Mixtures *Remote Sensing of Environment* 68:237-253

Ichoku, C., Karnieli, A. 1996 A Review of Mixture Modelling Techniques for Sub-Pixel Land Cover Estimation *Remote Sensing Review* 13: 161-186

Jasinski, M.F. 1996 Estimation of Subpixel Vegetation Density of Natural Regions Using Satellite Multispectral Imagery *IEEE Transactions on Geoscience and Remote Sensing* 34(3)

Kressler, F. and Steinnocher, K. 1996 Change Detection in Urban Areas Using Satellite Images and Spectral Mixture Analysis *International Archives of Photogrammetry and Remote Sensing*

Meer F. van der 1995 Spectral unmixing of Landsat Thematic Mapper data *International Journal of Remote Sensing* 16(16) 3189-3194

Oleson, K.W., Sarlin, S., Garrison, J.m Smith, S., Priivette, J.L., Emery, W.J. 1995 Unmixing Multiple Land-Cover Type Reflectances from Coarse Spatial Resolution Satellite Data *Remote Sensing of Environment* 54: 98-112

Price, J.C. 1994 How Unique Are Spectral Signatures? *Remote Sensing of Environment* 49: 181-186

Ouaidrari, H.; Vermote, E.F. 1999 Operational Atmospheric Correction of Landsat TM Data *Remote Sensing of Environment* 70: 4-15

Ray, T.W., Murray, B.C. 1996 Nonlinear Spectral Mixing in Desert Vegetation *Remote Sensing of Environment* 55: 59-64

Smith M.O., Ustin, S.L., Adams, J.B, Gillespie, A.R. 1990 Vegetation in Deserts: I . A Regional Measure of Abundance from Multispectral Images *Remote Sensing of Environment* 31:1-26 and http://cstars.ucdavis.edu/

Strahler, A.H., Woodcock, C.E., Smith, J.A.1986 On the Nature of Models in Remote Sensing *Remote Sensing of Environment* 20: 121-139

Ustin, S.L., Smith, M.O., Adam, J.B. 1993 Remote Sensing of Ecological Process: A Strategy for Developing and Testing Ecological Models Using Spectral Mixture Analysis *19. Remote Sensing of Ecological Process*

http://loasys.univ-lille1.fr/informatique/sixsgb-.html

Observing our environment from Space: New solutions for a new millennium, Bégni (Ed.)
© *2002 Swets & Zeitlinger, Lisse, ISBN 90 5809 254 2*

Spectral signatures of cultivated crops for GIS-supported applications in precision farming

S.Erasmi & M.Kappas
University of Göttingen, Inst. of Geography, Cartography GIS & Remote Sensing Sect., Göttingen, Germany

ABSTRACT: In this contribution the usefulness of spectral reflectance data of canopy cover for applications in site specific crop management (ssm) is investigated. Field spectrometry and high resolution airborne spectrometer data (HyMap®) of different test sites in northern and north eastern Germany are evaluated with primary regard to the biochemical constituents and other geo- and biophysical parameters relevant for the yield potential of winter wheat stands.
The spatial patterns of the spectral features are compared with other site specific characteristics to determine the accuracy of the measurements and to evaluate the relationships between canopy status, soil status and topographic parameters.
Spectral information and auxiliary data form thematic layers for the implementation in a geographic information system specified to model the status of agricultural plots. The resulting data could provide useful and reliable information for inventory management of crop production.

1 INTRODUCTION

Precision farming *(precision agriculture; site specific crop management, ssm)* is the generic term for a large variety of technological concepts that aim at improving the farmers productivity by maximizing the yield and at the same time minimizing the input of means of production. A pleasant side-effect of these techniques could be a sustainable conserving cultivation of arable land.

Uniform fertilisation of arable land generally contributes to a loss of yield at sites with a high yield potential, as well as increasing the waste of fertilizer in sites with a low yield potential. If, however, plots are treated differentially depending on their site specific heterogeneity, the optimal yield can be achieved whilst reducing fertilizer requirements.

Precision farming concepts are based on this in-field-heterogeneity of agricultural plots. The farming methods applied to a field during a vegetation period (seed, fertilizer, herbicide, etc.) are implemented in small homogenous management zones according to their unique mean characteristics.

Such patches are determined using distribution maps displaying such factors as the spatial and temporal variability of nutrient content, soil types and the target yield as well as other site specific factors.

These attributes may be obtained from soil / plant tests, crop yield monitors, topographical maps and remote sensing sensors.

Remote sensing data offer the advantage of a complete coverage of an agricultural area. Unfortunately they still lack spatial, spectral and temporal resolution for most precision farming applications (especially inventory management of crops). Latest approaches use truck-mounted real-time sensors to estimate quantitative plant-features (e.g. nitrogen).

The aim of this investigation is to evaluate the different methods of remote sensing and other data acquisition used to monitor the heterogeneity in winter wheat plots and to suggest a model for the integration of spectral information in the operational processes of precision farming management.

2 DATA

The remote sensing data set comprises a hyperspectral image set (HyMap®, Integrated Spectronics, NSW, Australia) of an agricultural region in N-E-Germany. Data were recorded for two non-adjacent flight tracks during two flight periods (05th/06th May 1999 and 20th June 1999). These two data takes correspond with two major growth stages, namely the beginning of the stem elongation stage (until mid May) and the beginning of the flowering stage (mid June). At these two stages, farmers need up-to-date information about the plant status for site-specific fertilizer application.

HyMap ® data acquisition was accompanied by field spectrometry and ground truth measurements.

Field spectrometry measurements were also conducted on two test sites in central Germany accompanied by an intensive ground data collection of crop parameters. These variables can be estimated by analysing the spectral response of plants and are of major importance for crop management. The following factors were evaluated:
- biomass
- canopy water content
- carbon / nitrogen status
- other major biochemical constituents (Mg, Na, Ca, P)

The comparability of spectral measurements is influenced by the quality of absolute spectral calibration. Berger et al. (2001) point out the influence of the viewing direction and the directional anisotropic behaviour of land surfaces on the measured solar radiation. They stress the use of inverted coupled radiative and atmospheric transfer models that enable to reliably consider the interactions of radiation with the target and on its way through the atmosphere. Nevertheless, most of the required input parameters for those models were not available for the present investigations. The HyMap ® data were corrected by an inflight-calibration using estimated atmospheric parameters and system calibration data which yields acceptable results for homogenous atmospheric areas.

The atmospheric correction of the field spectrometry measurements have at least to take into account the main distortions of the signal that are caused by changes in sun azimuth angle, sun zenith angle and topographic displacement. A reliable simple linear correction algorithm considering these main factors is currently under construction.

Spectral plant data are added by a number of other site-specific data.

Soil type information have been digitised from topographic maps (DGK5-Bo). These soil data provide a good overview of soil patterns but are only of marginal interest for site specific information in plots that are smaller than 20 ha which applies to most sites in central and southern Germany. This fact led to the attempt to valuate alternative sources for the retrieval of high precision site data for site specific farming applications.

In addition, soil moisture has been recorded with a time-domain-reflectivity probe (TDR, IMKO Trime®-P3). The TDR points out an integral value of the absolute volumetric moisture content of the top soil layer (0-15 cm). This method enables the time effective estimation of soil moisture patterns in large scales. These patterns can often be related to changes in soil type and/or topographic effects as well as to supposed changes in crop condition.

Electric conductivity (ECa) measurements were carried out in cooperation with the Agriculture Technology Centre, Bornim, Germany (EM38®, ATB-Bornim e.V). They were used to investigate relations between spectral and spatial plant features and soil conductivity patterns that are mainly caused by soil moisture content and concentration of the clay fraction in the upper 100 cm soil profile.

All data sources were geocoded, then converted into thematic layers and implemented in a GIS (ArcView).

3 METHODOLOGY

The HyMap®-Data were evaluated with regard to the spectral response of biochemical and biophysical components of winter wheat crops. Diverse spectral indices were calculated (e.g. RedEdge, NDVI, see Fig. 1) and spectral transformations were performed to point out the correlations between spectral features and crop characteristics.

The first step was to correlate well known spectral indices (NDVI; REIP "red edge inflection point") with crop features using empirical relationships between the indices and the features. Relations between spectral reflectance and total nitrogen content [% of dry mass], total water content [% of fresh mass] and yield [dt/ha] of winter wheat were established.

In a second step, commonly used semi empirical techniques for retrieving quantitative variables from spectral signatures were evaluated (normalized reflectance, derivative analysis techniques). The outcome of these methods was a dataset consisting of several significant spectral features of the VNIR and especially the SWIR-region. These features act as indicators for the state of the crop at significant growth stages. A multiple regression model was then performed for the prediction of the biochemical constitution (here: nitrogen) of wheat crops.

This model is based on former attempts to describe the state of vegetation by combining various dependent spectral variables. Table 1 shows the results of the authors' model compared to the different efforts in this matter.

The findings of the HyMap®-campaign are currently being validated in a field study on two test sites in Central Germany using high resolution field spectrometry data (ASD FieldSpec Pro®). Spectral crop data are combined with soil and other site specific data (e.g. soil type, moisture, electric conductivity, slope, aspect) in a GIS.

The GIS serves as environment for the development of a quasi dynamic model for precision farming applications. This model takes into account static parameters as well as equations like fertilizer requirements depending on soil type and is supplied with up-to-date crop information.

Table 1. Comparison of methods for plant nitrogen detection

Reference	this study	KOKALY & CLARK (1999)	WESSMAN et al. (1988)	WHITE et al. (2000)	YODER & PETTI-GREW-CROSBY (1995)
Wavelength (nm)	1043 1089 1949 2132 2305	2036 2050 2078 2152 2180	2053 2063 2129 2181 2293	1484 1876 2268	980 1194 1644 1676 2274
data transformation	normalized reflectance (continuum removal)	normalized reflectance (continuum removal)	log(1/R)	1st derivative log(1/R)	log(1/R)
investigated vegetation	agricultural species	forest foliage	forest and grassland species	forest foliage	forest foliage

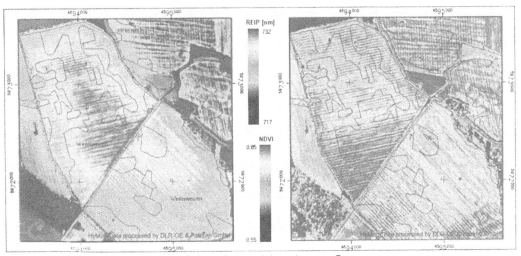

Figure 1. Spatial patterns of spectral indices of winter wheat stands in north-eastern Germany

This information is extracted from spectral signatures of airborne, spaceborne or terrestrial sensors.

4 RESULTS

Spectral data of high resolution can provide useful time-sensitive information about the crop condition which is needed to implement site specific applications during specific growth stages (e.g. 2nd and 3rd N-application in winter wheat). Spectral indices (e.g. NDVI and REIP) are correlated to crop parameters. Generally, NDVI and REIP show different spatial patterns (Figure 1) and weak relations to plant parameters except for a single linear correlation with plant water content (see Table 2).

Table 2. Correlation coefficients (R^2) of crop features and spectral indices

	NDVI	REIP
nitrogen	0,355	-0,302
H_2O	-0,871*	-0,765*
yield	-0,116	-0,537*

*correlation is significant at 0,01 level

Baret & Fourty (1997) also state a lack of robust estimates of biochemical plant components of fresh matter, except for plant water content. The weak relation between spectral response and biochemical constituents of fresh biomass is supposed to be based on strong influences of plant and atmospheric water absorption. This superimposes most of the absorption or reflectance features that are caused by crop factors.

An acceptable solution to minimize the disturbing influence of water absorption is the application of differential absorption methods that emphasize the effect of changes in crop condition on distinct spectral absorption features especially in the NIR and SWIR region.

Differential absorption methods like continuum removal or other rationing methods (see Schläpfer 1998 for details) are used to detect the relative strength of absorption features which is correlated with the total or relative amount of plant constituents. Normalized reflectance data (continuum removal) in the SWIR-region were found to be of greatest interest for deriving chemical plant constituents (particularly nitrogen; see Figure 2). The SWIR-reflectance data (especially the SWIR2-region of 2000 to 2500 nm) face the problem of a lower signal-to-noise ratio (SNR) compared to the VNIR-region. This has to be considered when analysing spectral data in this region of the electromagnetic spectrum.

Improved results in the SWIR-region can be obtained if a minimum of 5 absorption bands are combined in multiple regression model. The result of the regression model for the prediction of nitrogen with spectral information is shown in Figure 3. The models' input data set consists of normalized reflectance data in those wavelength bands stated earlier in Table 1 of this paper. This method compensates the potential source of error (SNR) as well as the influence of overlaying absorption effects.

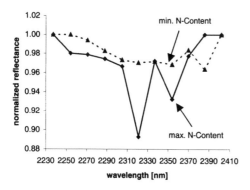

Figure 2. Normalized reflectance of winter wheat plots in the SWIR2-region

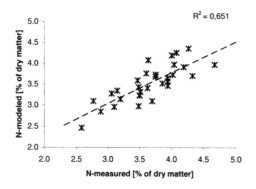

Figure 3. Correlation between modelled (multiple regression) and measured nitrogen status of winter wheat crops

Spectral data provide information about the status of the canopy layer of plants and show up-to-date patterns of the inhomogeneity within a field. These data can only be used effectively if a minimum of other site characteristics that determine the optimal

growth is taken into account. The investigation of electric conductivity (ECa) of soils shows promising results for mirroring the limiting soil parameters (soil water; soil type) for agricultural applications (see Dobers et al. 2001 for details). The different patterns of ECa-measurements and spectral field measurements did not show any correlation (Figure 4). This fact underlines the necessity of integrating as much field and supplementary data as possible to describe the state of an agricultural plot at a specific scale and time.

Besides, digital terrain data (DTM) are also useful to detect accumulation and erosion spots within a field. These geodata are almost time stable. At least a limited number of site factors is essential to reliably estimate the site specific soil status. Such data can be provided as base layers in a GIS to model the environment of agricultural areas.

Up-to-date spectral data combined with suplementary ground data provide the base layers to generate site specific information on soil and plant status for the integration in precision farming procedures. An example of a nitrogen-map based on high precision remote sensing and ground truth data is shown in Figure 5.

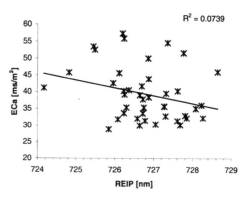

Figure 4. Correlation between crop reflectance data and soil status of winter wheat plots

5 CONCLUSIONS

The investigation emphasises the fact that spectral data have a potential to derive qualitative and (relative) quantitative parameters for site-specific crop management.

Existing GIS & RS-based precision farming applications have to face constraints that lie in the variety of input data sources as well as in data quality. Geodata provide useful base layers but unfortunately they are quite often unavailable or too expensive. Remote sensing data offer the opportunity to establish up-to-date quantitative parameters if calibration and analysing steps are well conducted. The results

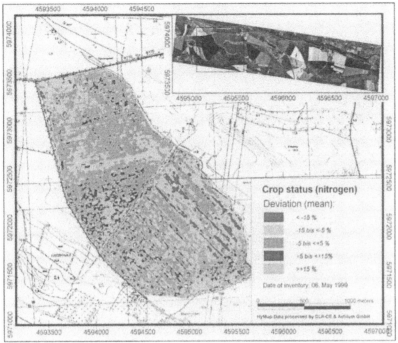

Figure 5. Pattern of nitrogen-distribution within a winter wheat field based on HyMap® and ancillary ground data

of the spectral analysis of crops are promising but as long as absolute calibration is missing they can not be transferred to other agricultural plots. Despite this fact, the provided information about the relative inhomogeneity within fields in general meets the farmers requirements.

Future remote sensing systems will be able to contribute information on site specific characteristics of agricultural land, assuming high geometrical, spectral and temporal resolution (e.g. Rapideye, OrbView4, QuickBird).

Both, Earth Observation (E/O)-data and other geodata need expert-knowledge to retrieve the desired quantitative variables for site specific applications in agriculture. The combination of the different approaches in a GIS provides a reliable practicable technique which can be used as an input into precision farming management systems.

Further research aims at the evaluation of the known and suspected correlations between spectral reflectance and crop data and at the integration of remote sensing data into an operational system for precision agriculture.

6 ACKNOWLEDGEMENTS

HyMap ® airborne scanner data were acquired and processed by German Aerospace Center (DLR) and Astrium Space Inc.

REFERENCES

Baret, F. & Fourty, T. 1997. The limits of a robust estimation of canopy biochemistry. In: G. Guyot & T. Phulpin (eds.) *Physical Measurements and Signatures in Remote Sensing* (2): 413-420. Rotterdam: Balkema.

Berger, M., Rast, M., Wursteisen, P., Attema, E., Moreno, J., Müller, A., Beisl, U., Richter, R., Schaepman, M., Strub, G., Stoll, M. P., Nezry, F. & Leroy, M. (2001). The DAISEX campaigns in support of a future land-surface-processes mission. *ESA Bulletin* 105: 101-111

Dobers, E.S., Meyer, B. & Roth, R. (2001). Comparison of Wheat Yields and Soil Apparent Electric Conductivity of Sandloess-Soils in Brandenburg/Germany. *European Journal of Agronomy* in press.

Kokaly, R. F. & Clark, R. N 1999. Spectroscopic determination of leaf biochemistry using band-depth analysis of absorption features and stepwise multiple linear regression. *Remote Sensing of Environment* 67: 267-287.

Schläpfer, D. 1998. Differential absorption methodology for imaging spectroscopy of atmospheric water vapor. Remote Sensing Series: 32. Zurich.

Wessman, C. A., Aber, J. D., Peterson, Dl. & Mellilo, J. M. 1988. Foliar analysis using near infrared spectroscopy. *Canadian Journal of Forest Research* 18: 6-11.

White, J. D., Trotter, C. M., Brown, L. J. & Scott, N. 2000. Nitrogen concentration in New Zealand vegetation foliage derived from laboratory and field spectrometry. *International Journal of Remote Sensing* 21(12): 2525-2531

Yoder, B. J. & Pettigrew-Crosby, R. E. 1995. Predicting Nitrogen and Chlorophyll content and concentrations from reflectance spectra (400-2500 nm) at leaf and canopy scales. *Remote Sensing of Environment* 53: 199-211

Observing our environment from Space: New solutions for a new millennium, Bégni (Ed.)
© 2002 Swets & Zeitlinger, Lisse, ISBN 90 5809 254 2

Integration of land topography for the improvement of the spatial images classification. Application on the steppic area of Aflou (North of Laghouat, Algeria)

Z.Smahi & A.Bensaid
Arzew, Algeria

ABSTRACT: The principal aim of this study is the combination of the image with information given by the files issued from a digital terrain model so as to correct the image for topographic and atmospheric effects. A test area has been chosen in the region of Aflou containing a great variety of vegetation species and presenting a relief with a variation of slopes and the presence of shadow effects. In a first step, geometrical correction was applied to the image with respect to the map. Secondly, the image was corrected by two absolute calibration procedures and, then, it was classified by the maximum of likelihood procedure. Finally, the results indicate clearly when taking into account the relief parameter that land classification has been improved by 7.8%. Thus, the topography plays an important role in the process of multisource classification improvement when the land is hilly relief.

RESUME : Cette étude vise principalement la combinaison de l'image avec les informations apportées par les fichiers dérivés du MNT afin de corriger l'image des effets topographiques et atmosphériques, pour attribuer à chaque objet sa réflectance réelle. Pour ce faire, un territoire – témoin a été choisi dans la région d'Aflou étendue sur une superficie d'environ (1925,0075 km²), comprenant une grande variété d'espèces végétales et présentant un relief dont les pentes sont variables et les effets d'ombre sont bien présents. Cette région d'étude est couverte par une fenêtre extraite de l'image TM de Landsat de la scène 196/36 du 08 Avril 1995.
Dans un premier temps de cette étude, une correction géométrique a été appliquée sur l'image afin de la superposer par rapport au MNT. Or, le but de la réalisation de ce dernier ainsi que ses dérivées (la pente, l'exposition, l'ombre) est d'intégrer les effets topographiques hors du processus de classification afin d'améliorer la précision des résultats. Dans un second temps, nous avons corrigé l'image par deux méthodes de correction absolue. L'une considère la région comme un terrain plat et l'autre intègre l'effet du relief dans le processus de la correction absolue. Ensuite, afin de pouvoir réaliser une comparaison entre les deux méthodes de correction, une classification par le maximum de vraisemblance a été appliquée sur les trois images :
- l'image brute corrigée géométriquement,
- l'image brute corrigée géométriquement et radiométriquement en terrain plat,
-et l'image brute corrigée géométriquement et radiométriquement en terrain de relief.
Enfin, les premiers résultats indiquent clairement que la correction absolue en terrain plat n'a pas amélioré notre classification. Par contre, après élimination de l'effet du relief la classification a été améliorée de 7.8 %. Ceci, nous conduit à dire que la topographie joue un rôle important dans le processus de la classification multisource lorsque la région d'étude présente un terrain de relief.
Mots clés : MNT, Pente, Exposition, Ombre, Correction absolue, TM de Landsat.

1 INTRODUCTION

In application and research studies, we often need to obtain information from land use. Moreover, these subjects require sometimes that this information be up to date and at the same time as accurate as possible. The realisation of a land use map is traditionally made by the interpretation of aerial photographs. With the development of remote sensing, spatial im-ages have been used as pertinent tools for the establishment of a land use map by using different classification methods. However, most of these methods depend only on the numeric of the image's pixels.

Unfortunately, several studies have revealed many difficulties in the use of these on mountainous areas with slope variations.

However, the use of other data sources such as a Digital Terrain Model and its derived files for the improvement of the classification process will prove to be necessary. This additional information plays an important role in the classification due to the fact that the relief can change the radiometric value of the pixel. So, the objective of this study is the use of a methodological approach in which satellite images are classified by the integration of the DTM and its derived files by considering the topographic and atmospheric effects in the processes of the absolute correction. In this research, Landsat TM of 1999 which cover the region of Aflou are used to evaluate the model applied.

2 METHODOLOGY

2.1 Case study

The test area is situated in northern latitude between 34° and 34°20' and, on the East of Longitude between 2° East and 2°30' East (see figure 1). It is located at 400 Km to the south of Algiers with an area of 1925 km^2. This region is characterised by hilly relief and is constituted by the following main classes of land use :

- brushwood and mountain grassland,
- grassland of Esparto on plains called Alfa,
- irrigated culture along the river and dry culture on plains,
- high rocky structures,
- sand veils and wet land called daya.

The characteristics of the Landsat TM used in the study are illustrated in table 1.

2.2 Data issued from a Digital Terrain Model(DTM)
Due to the fact that the study area is situated in a region characterised by hilly relief, the use of a model

Table 1. Characteristics of the image TM.

Image	TM CCT Format
Path Row	196-36
Total dimension (row – column)	5760 – 6100
Date of acquisition	08 April 1995
Hour of acquisition	9h 35'03"
Latitude at the centre	34°.6116
Longitude at the centre	2°,1569
Sun zenith angle	41°,970
Sun elevation	58°,024
Sun azimuth angle	122°,8719
Satellite declination	98°,1999
Satellite zenith angle	8°,1999

of atmospheric effect correction imposes the use of a Digital Terrain Model. This model is obtained by digitalisation of level curbs issued from topographic map of Aflou N°: J-9. Then, these level curbs were interpolated into a grid of 30 metres resolution. After which this grid served to extract the following information:

- slope map,
- aspect map,
- and the shadow map.

2.3 Satellite data processing

After the computation of the correlation matrix, only three bands TM1, TM3 and TM4 were selected. Then, the TM bands were geometrically precision corrected to the topographic map using 24 ground control points and the third degree of the polynomial model with the near neighbour interpolation. The pooled RMS geometric error compared to map coordinates was 26.7 metres that is less then a pixel of TM (30m) and then, the image was well corrected. In order to validate the method of resampling used in the geometric correction, some parcels have been selected and their numeric values were compared before and after correction.

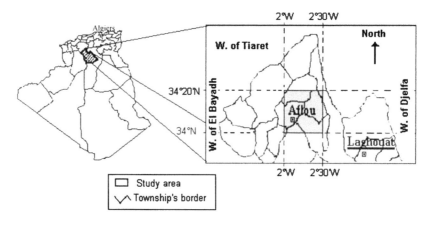

Figure 1. Localization of the study area

216

The results of the statistics computed on the three samples (vegetation, sand, daya) selected in the two images (before and after correction) were reported in the table 2.

Table2. Statistics on the samples for validation.

	Before correction			After correction		
	TM1	TM3	TM4	TM1	TM3	TM4
Sample1: Vegetation						
Min	54	22	93	54	22	93
Max	66	43	135	66	43	135
Average	58.78	27.04	117.98	58.95	28.05	117.31
Standard Deviation	2.85	4.57	1.53	2.82	4.80	1.01
Number of pixels	35			38		
Sample2: Sand						
Min	83	93	103	83	93	103
Max	100	111	115	100	111	115
Average	88,05	100,88	107,76	88,47	101,62	108,16
Standard Deviation	2,71	2,79	1,57	2,70	2,97	1,76
Number of pixels	90			114		
Sample3: daya						
Min	68	52	58	68	52	55
Max	75	61	66	75	61	60
Average	71,45	56,03	60,39	71,38	56,03	39,26
Standard Deviation	1,88	2,18	1,45	1,88	2,17	1,44
Number of pixels	34			34		

According to table2, we remark an equalisation of the numeric values in which the Maximum and Minimum have been conserved for all the samples with the exception of small changes in the standard deviation and the average caused by the variation of the size of the samples selected. So, the image after correction has not altered its original radiometric values.

2.4 Radiometric and Atmospheric corrections

2.4.1. First Absolute Correction Method
This method considers that the area is horizontal or flat in which the presence of the relief was neglected. This method was applied on the image geometrically corrected but without integration of data issued from DTM.

2.4.2. Second Absolute Correction Method
This method has required some input parameters which were obtained from the header of the image. In addition to these data, we needed others data from other sources (see table 3.)

The atmospheric radiance was computed from the Lowtran 7 while the values of the minimal and maximal radiances were given by the constructor (Olsonn, H. 1994).

Table 3. The Input Parameters Of The Model.

Band	TM1	TM3	TM4
L (µm)	0.485	0.660	0.830
Sun irradiance (W.m^{-2}.µm^{-1})	2056.21	1032.99	233.49
Minimal radiance (W.m^{-2}.sr^{-1}.µm^{-1})	-1.5	-1.2	-1.5
Maximal radiance (W.m^{-2}.sr^{-1}.µm^{-1})	152.1	204.3	206.8
Atmospheric radiance (W.m^{-2}.sr^{-1}.µm^{-1})	19	05	03
Visibility (Km)	40		
Number of days	98		

In this method, we have integrated the slopes, the aspect and the shadow maps and a shadow threshold S. The value of this threshold was fixed at 0.28 on the base of the visual interpretation of the shadow grid taking into account the altitude variation of the DTM and the observation conditions (sun position).

2.4.3. Validation of the Absolute Correction
In order to validate the method of absolute correction, we have compared in the three bands the real reflectances computed of the sample sand which was taken as reference with its corresponding theoretic values. Table 5 shows that the real values of this sample were included in the theoretic values interval.

Table 4. Theoretic reflectances of the sand (BECKER, F & Al).

Band	TM1	TM3	TM4
Reflectance	10 – 12 %	23 – 26 %	33 – 37 %

Table 5. Reel reflectances computed of the sand.

	Min	Max	Average	Standard deviation
TM1	0.09	0.11	0.10	0.003
TM3	0.25	0.27	0.26	0.005
TM4	0.37	0.38	0.37	0.005

2.5 Classification of images

In our study area ten (10) classes have been used for the supervised classification based on the maximum likelihood method. On the other hand, the classes of grasslands are composed by species of Alfa and other species of vegetation described as follows:
- grassland1 : *stipa tenacissima* and *schismus barbatus*.
- grassland2 : *stipa tenacissima* and *stipa parviflora*.

The classification was applied on three kinds of images resulting from two methods of absolute correction and on an initial image.

2.5.1 Classification of the initial image

The classification was applied on the image geometrically corrected in which the radiometric

217

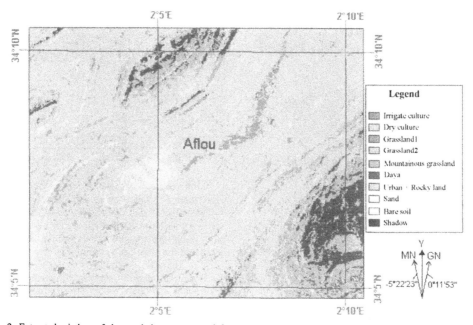

Figure 2. Extracted window of the result image corrected from atmospheric and topographic effects. (Colour plate, see p. 425)

correction was not applied so as to compare it to other images.

2.5.2 Classification of the image resulting from the first method

This classification was applied on the image corrected for atmospheric effects by the application of the first absolute correction method (see table 6).

2.5.3 Classification of the image resulting from the second method

In this case, we have classified the image (see figure 2) resulting from the second method of the topographic effects correction where the digital terrain model (DTM) and its derived files were combined with the image (see table 7).

Tables 6 and 7 contain the surface statistics of the different classes and their improvement rates after the two methods of calibration in comparison with the initial classified image.

3 RESULTS AND DISCUSSION

After analysis of table 6, the first atmospheric correction method has acted on the precision of the classification. Hence, the classified areas (irrigated vegetation, urban and rocky terrain, dry culture) were increased with the exception where the surface of bare soil was lightly decreased. This can be due to the approximations in delineating atmospheric effects. In general, the absolute atmospheric

Table 6. Surface rates improvement of Image's classes without considering the relief.

Classes	Surface (ha)		% of improvement
	Initial image	Without relief	
Irrigate Vegetation	1546.11	1546.38	0.02
Urban + Rocky Terrain	41793.21	41799.42	0.01
Sand	1041.66	1041.66	0.0
Bare Soil	6151.32	6139.08	-0. 002
Grassland1	9590.13	9590.13	0.0
Dry Culture	38598.48	38604.24	0.01
Mountainous Grassland	3967.29	3967.29	0.0
Shadow	20632.14	20632.14	0.0
Grassland2	35969.13	35969.13	0.0
Daya	1400.13	1400.13	0.0
Total	160689.6	160689.6	
Mean rate of improvement			0.042

Table 7. Surface rates improvement of image's classes introducing the relief parameters

Classification	Surface (ha)		% of improvement
Classes	Initial image	Hilly relief	
Irrigate Vegetation	1546.11	1537.56	-0.01
Urban + Rocky Terrain	41793.21	41720.40	-0.01
Sand	1041.66	1090.8	4.7
Bare Soil	6151.32	6561.72	6.6
Grassland1	9590.13	11001.15	14.7
Dry Culture	38598.48	38618.46	0.05
Mountainous Grassland	3967.29	4947.48	24.0
Shadow	20632.14	17915.67	-24.0
Grassland2	35969.13	35947.71	-0.001
Daya	1400.13	1348.65	-4.0
Total	160689.6	160689.6	
Mean rate of improvement			7.807

correction without considering the relief has rectified the radiometry of some elements belonging to a defined class and where they join other new classes. On the other hand, the classes situated on the relief area have not been subjected to any modification such the shadow classes and the grasslands in the mountain. This means that the method used of correction has not taken into account their topographic position. However, at the time of the introduction of topographic effects in the absolute correction processes, the surface of the classes (sand, bare soil, grassland1) has increased, seeing that these last were situated in illuminated areas. Also, the increase of the surface of grassland on the mountain, situated on hilly areas, is explained simply by the presence of high difference in level. Thus, the decrease of the surface of the classes (daya, urban + rocky terrain) is caused by their presence in regions at low difference in level. It is very clear that the first absolute correction method has not improved our classification.

However, after elimination of the relief effects the classification has been improved by 7.8% in which all classes situated on hilly regions have known an augmentation of surface except the shadow classes. The decrease of the surface of this last class has been caused by the fact that the reflectance of some of its elements has been increased and then, has been affected to other new classes.

4 CONCLUSION

The classification applied on the image radiometrically corrected without considering the relief has allowed the elimination of the atmospheric effects, but the image stays less perturbed seeing that our study region is hilly over a large area. Elsewhere, the classification with the integration of the files issued from DTM has given best results in the areas situated on slopes relatively higher. On the other hand, areas of low relief have not changed. Also, the areas situated on the opposite side (shadowed) from the sun have been increased and equalised with those which are illuminated by the sun. Thus, the application of the second model has given good and better results. Finally, the method used in this study proved to be promising for a diachronic study on a region containing a varied morphology and needing the introduction of different layers of information issued from DTM and observation condition of images taken on different dates.

5 REFERENCES

STRUM, B. 1980. The atmospheric correction of remotely sensed data and the quantitative determination of suspended matter in marine water surface layers, in *Remote sensing in meteorology, oceanography and hydrology*. Carnegie Laboratory of Physics, University of Dundee, Scotland. p.163-197

BECKER, F et al. 1978. Principes physiques et mathématiques de la télédétection. Course notes at the summer school of space physics of the Centre National d'Etudes Spatiales, 671 pages, Strasbourg.

OLSSON, H. 1994. Reflectance calibration of Thematic Mapper data for forest change detection. Accepted for publication in International Journal of Remote Sensing, p.1 :22 chap IV in the report 7 *Monitoring of local reflectance changes in boreal forests using satellite data* . Department of Biometry and Forest management, Swedish University of Agricultural Sciences, Umeå, Sweden.

BENHANIFIA, K. 1998. Change detection on mountainous land by two dates images; TM of landsat-5 Corrected for relative atmospheric effects. Case of the forest of M'sila (Oran), *Magister Thesis*, CNTS, 119 pages.

ROUQUET, M. C. 1985\86. *Modélisation atmosphérique* GDTA/ENSG, 15 pages.

DURRIE, S. 1994. Utilisation de la télédétection satellitaire pour la mise à jour de la carte des types de peuplement de l'inventaire forestière national. Applications à une région forestière diversifiée de moyenne montagne. *Doctorate thesis*. The National School of Hydrology and Forestry. 202 pages.

SMAHI, Z. BENHANIFIA, K & BENSAID, A. 1999. Development of an algorithm of Absolute Atmospheric Correction for Multitemporal Satellite images. Application to the region of Oran (West of Algeria).Accepted for publication by the IEEE Geoscience and Remote Sensing Society in International Geoscience and Remote Sensing Symposium (IGARSS).

Observing our environment from Space: New solutions for a new millennium, Bégni (Ed.)
© 2002 Swets & Zeitlinger, Lisse, ISBN 90 5809 254 2

Development of an algorithm of absolute atmospheric correction for multitemporal satellite images. Application to the region of Oran (West of Algeria)

Z.Smahi & K.Benhanifia
Arzew, Algeria

ABSTRACT: This paper presents an algorithm for modelling the different parameters which influence the reflectance at the satellite's level in order to estimate the reel reflectance of the objects at the ground surface. However, this model would compensate the atmospheric effects, the irradiance and the observation conditions, and characterise the objects by their unique and reel values. Thus, it makes directly two images comparable by the compute of the reel reflectance value of the objects from image's digital number. The results show that, in the illuminated area as sea water, sand, vegetal covert, the difference between the two date's reflectance is nearly null. Except in the mountain's areas where the problem of shadow still exists and it has not been resolved. On the other hand, in the areas where the changes occurred (New road of Misserghine and the trails effected into the forest of M'sila), the reflectance is highly different between the two dates.

RESUME : Ce papier présente un algorithme pour modéliser les différents paramètres qui influencent la réflectance au niveau du satellite dans le but d'estimer la réflectance réelle des objets à la surface du sol.
Cependant, ce modèle corrigerait les effets atmosphériques, l'éclairement et les conditions d'observation, et caractériserait les objets par leur unique et réelle valeur. En effet, il compare deux images directement par le calcul de la valeur réelle de la reflectance des objets à partir du compte numérique de l'image.
Dans cette étude, deux images TM de Landsat (1984,1993) ont été utilisées pour tester le modèle de correction atmosphérique où la reflectance spectrale de quatre échantillons tests ont été comparés entre les deux dates pour valider le modèle. Les résultats montrent que, dans la zone éclairée comme l'eau de mer, le sable, et le couvert végétal, la différence entre la reflectance des deux dates est presque nulle sauf dans des zones en montagne où le problème d'ombre existe encore et il n'a pas été résolu. Par contre, dans les zones où les changements survenaient (nouvelle route de Misserghine et les pistes opérés dans la forêt de M'sila), la reflectance est visiblement différente entre les deux dates.
Mots Clés : Eclairement, effets Atmosphériques, ombre, reflectance réelle, TM de Landsat.

1 INTRODUCTION

The reflectance constitutes the characteristic value of an object that is the ratio of the reflected and incident energy and depends on the wavelength, object state and observation conditions (solar incident angle, view observation angle, and relative azimuth between the radiation and observation directions).

At the satellite, the measured values are, for each spectral band, the mean spectral radiance outside the atmosphere. So, from the image, we can calculate the radiance in the upper atmosphere, and then, we deduce the reflectance by algorithms. However, this reflectance is not the same at ground level because of the presence of the atmosphere. This atmosphere,

spatially varied, makes differences in the path affected by the solar radiation. So, the sensor measurements are very delicate to interpret due to the interaction of the radiation proprieties with the atmosphere through the double path sun/earth and earth/sensor.

So, the objective of this study is to model the different parameters influencing the reflectance at the satellite in order to estimate the real reflectance of the surface's object. This modelling, constituting atmospheric correction, must in theory characterise an object by its unique value at the ground surface. In this research, the Landsat TM of the two years 1984 and 1993 are used to evaluate the atmospheric correction model and the spectral reflectance of four

(04) data sets are compared between the two dates and the real value of the reflectance given by laboratory reflectance measurements.

2 METHODOLOGY

The model developed estimates the factor of absolute reflectance of the surface that is assumed lambertian and homogeneous. In the following sections, we describe the different mathematical expressions and equations which are used in the model.

First of all, the total irradiance at the ground level is described by the summation of the direct (Edr) and diffuse irradiance (Edf) and is given by:

$$Et = Edr + Edf \qquad (1)$$

The direct illumination on a horizontal surface at the ground is described by:

$$Edr = Es.\mu s.Tdr(\vartheta s) \qquad (2)$$

where μs is the cosines of the solar zenith angle, Es the solar constant irradiance outside the atmosphere and $Tdr(\vartheta s)$ is the direct transmission factor.

And respectively, the diffuse is given by:

$$Edf = Es.\mu s.Tdf(\vartheta s) \qquad (3)$$

which $Tdf(\vartheta s)$ is the diffuse transmission factor. Then, the total irradiance at ground level becomes:

$$Et = Es.\mu s.T(\vartheta s) \qquad (4)$$

where $T(\vartheta s)$ is the total transmission factor which represents the summation of the two transmission factors (direct and diffuse).

On the other hand, the atmospheric scattering is due to the interaction of photons with the molecules of the Rayleigh scattering and the aerosols of the Mie scattering. However, a percent of photons are transmitted into the direction of radiation at the ground, and this, defines the direct transmission factor:

$$tdr(\vartheta s) = Exp(-\tau/\mu s) \qquad (5)$$

where τ is the optical thickness of the atmosphere. Also, the ground receives a diffuse irradiance which corresponds to the photons scattered by the atmosphere towards the ground. So, the percent of these photons defines the diffuse transmission factor noted $tdf(\vartheta s)$, and the total transmission factor is defined by:

$$T(\vartheta s) = tdr(\vartheta s) + tdf(\vartheta s) \qquad (6)$$

The atmospheric upward direct transmission factor at view angle ϑv of observation is noted by:

$$tdr(\vartheta v) = Exp(-\tau/\mu v) \qquad (7)$$

with μv is the cosines of the zenith observation angle and respectively, the total transmission factor at view angle ϑv is :

$$T(\vartheta v) = tdr(\vartheta v) + tdf(\vartheta v) \qquad (8)$$

Thus, the total transmission factors $T(\vartheta s)$ and $T(\vartheta v)$, at view angle ϑs of radiation and respectively at view angle ϑv of observation, are given by the following analytic expressions :

$$T(\vartheta s) = 1/(1 + b\tau/\mu s) \qquad (9)$$

$$T(\vartheta v) = 1/(1 + b\tau/\mu v) \qquad (10)$$

where $b\tau$ is the retrodiffusion coefficient which corresponds to the percent of diffuse photons. It is noted by:

$$b\tau = br.\tau r + bp.\tau p \qquad (11)$$

In the standard model of continental aerosol, br and bp equal respectively to 0.5 and 0.16. So, the absolute reflectance value of the target is estimated by:

$$\rho = \pi.Lsat - Latm/Et.tdr(\vartheta v) \qquad (12)$$

and the spectral observed at the satellite for each digital number DN is given by :

$$Lsat = (Lmax-Lmin)/255 + Lmin \qquad (13)$$

where Lmin, Lmax represent the spectral radiance for 0 and 255 DN, and Latm is the atmospheric diffuse radiance towards the sensor. It is given (J. Perbos, 1982) by the following analytic expression:

$$Latm = Es.\tau P(\xi)/4.\pi.\mu v \qquad (14)$$

which $\tau P(\xi)$ is the diffusion phase function. It represents the diffuse relative probability of photons defining the angle ξ with the incidence direction. This angle ξ is between the direction of observation and radiation and it is estimated (M. C. Rouquet, 1986) by:

$$\cos(\xi) = -\mu s.\mu v - sqrt(1-\mu s2).sqrt(1-\mu v2).\cos(\varphi) \qquad (15)$$

and

$$\tau P(\xi) = \tau r.Pr(\xi) + \tau pPp(\xi) \qquad (16)$$

which φ is the summation of the two azimuth angles of radiation and observation.

with : $Pr(\xi) = 0.7552 + 0.7345.\cos(\xi)$ (17)

where $Pp(\xi)$ is variable with the aerosol's type and it is estimated by interpolation between values calculated for four (04) wavelength (0.45 µm, 0.55 µm, 0.65 µm and 0.85 µm) and fifty (50) values of $\cos(\xi)$. Also, the atmospheric optical thickness τ which is defined as the summation of Rayleigh and Mie optical thickness, is given by :

$$\tau = \tau r + \tau p \qquad (18)$$

where τr and τp represent respectively the Rayleigh (index r) and Mie (index p) optical thickness.

Thus, the atmospheric expressions of these optical thicknesses are estimated (J. Perbos, 1982) and given by:

$$\tau r = (84.35.\lambda^{-4} - 1.225.\lambda^{-5} + 1.4.\lambda^{-6}).10^{-4} \qquad (19)$$

and

$$\tau p = 0.632.\lambda^{-1} - 0.0194.\lambda^{-2}).Exp(-V/15) \qquad (20)$$

with V and λ are respectively the visibility in Km and the wavelength in µm.

2.1 Satellite data used

The study site is a 36 km by 27 km area located to the west of Oran in the Northwest of Algeria. The area contains a diversity of cover types including agricultural sites, sand, sea water, salty lakes and forest in the mountainous terrain. The area is characterised by relief covering the forest and by horizontal surfaces for most of the rest of the area. Two Landsat 5 Thematic Mapper images were used, scene 198/35 from April 07, 1984 and from March 15, 1993. The bands used are TM1(Blue), TM3 (Red) and TM4 (near IR), (Table 1).

Table 1. Data used for the Landsat 5 TM

	1984			1993		
	TM1	TM3	TM4	TM1	TM3	TM4
Lmin	0	0	0	-1.5	-1.2	-1.5
Lmax	153.6	205.5	207.7	152.1	204.3	206.8
Latm	19	5	3	17	5	3
V (Km)	40			23		
ϑs	39°.76			48°.75		
ϑv	8°.24			8°.22		
φs	129.78			134.21		
Es	2056.21			2084.24		

The values of Lmin and Lmax are given by SSC-Satellitebilt in Kiruna in (w.m^{-2}.sr^{-1}.µm^{-1}), and Esun is estimated from graph (Neckel and laboratory, 1984) in (w.m^{-2}.µm^{-1}). The values of Latm are calculated from Terravue software. The two images are

used just to validate the method and for comparing the values of the absolute reflectance of the features.

3 RESULTS AND DISCUSSION

First, the optical thickness, transmission factors and irradiance are calculated for each canal and date (Table 2). This table 2 shows that the optical thickness estimated in 1984 for the TM1 canal is smaller than in 1993. This is due to the value of visibility that indicates the presence of aerosols and molecules in the short wavelength.

Table 2. Atmospheric parameters (Irradiance in W.m^{-2}.µm^{-1})

	1984			1993		
	TM1	TM3	TM4	TM1	TM3	TM4
τ	0.243	0.109	0.06	0.422	0.242	0.176
tdf(ϑs)	0.163	0.090	0.064	0.316	0.230	0.184
tdr(ϑs)	0.729	0.868	0.914	0.527	0.693	0.766
Edf(ϑs)	257.8	142.9	100.4	433.5	316.1	252.5
Edr(ϑs)	1151	1372	1445	724.5	951.7	1052

So, the diffuse irradiance in the TM1 of 1984 is greater than in 1993, but, the direct irradiance is more greater then in the 1993. This indicates that the image in 1984 is more illuminated than in 1993.

Secondly, in order to valid the method, we have taken four (04) data samples: sea water, sand, wheat and forest in the two images. However, for each image, we have calculated the absolute reflectance value which is extracted for each data sample (Table 3). We remark that the values for the two dates are very similar with a small error. Except in the forest sample where we see a small difference in the TM4 of 1.7%. So, this is due, maybe, to the relief effects and also to the presence of trails which are due to the forest cutting done between the years. The difference of 3.4%, between to two dates in the wheat sample in the TM4, is due that to the fact that the wheat is in maximum chlorophyll activity in April more than in March.

Table 3. Mean values of the DN and reflectance (1984,1993)

	April 7, 1984		March 15, 1993		
Band	DN	ρ	DN	ρ	Difference (%)
		Sea water			
TM1	64.24	0.056	54.07	0.058	0.3
TM3	12.56	0.016	12.42	0.017	0.1
TM4	6.99	0.012	7.77	0.011	0.1
		Sand			
TM1	138.38	0.183	98.17	0.169	1.4
TM3	114.14	0.281	83.83	0.271	1.0
TM4	107.76	0.368	84.50	0.369	0.1
		Wheat			
TM1	67.79	0.062	56.29	0.064	0.2
TM3	25.37	0.050	20.71	0.046	0.4
TM4	146.77	0.505	105.43	0.467	3.8
		Forest			
TM1	).68	0.062	54.31	0.060	0.2
TM3	7.41	0.047	21.16	0.048	0.1
TM4	).04	0.198	44.19	0.181	1.7

In general, the values of absolute reflectance of the data samples, which are unchanged on the two dates, are very near and comparable to the real value of absolute reflectance given by the laboratory measurements.

Finally, this method has made the two images in the same referential in which the atmospheric effects are supposed removed. Then, by this treatment, we can make a multitemporal study in which the changed area may be detected and estimated very easily. So, in our study, after having corrected the two images, we have detected multiple changes that have occurred between the two dates and especially, the new road of Misserghine between Oran and Misserghine cities. Also, we have seen the changes made in the forest of M'sila by the presence of diverse trails cut through this forest.

4 CONCLUSION

The results have demonstrated that the proposed correction algorithm is successful in determining the absolute reflectance values of objects. But, it is very difficult to calculate the real reflectance value of all objects and especially those that are influenced by both the slope, aspect, shadow effects and environments. So, in some future work, we will try to develop a method to take into account these parameters in order to remove theirs effects for calculating and improving the values of objects reflectance situated in mountainous areas.

5 REFERENCES

H. Olsson. 1994. Reflectance calibration of thematic Mapper data for forest change detection. Swedish Univ. of agricultural sciences, dept. of biometry and forest management, remote sensing laboratory, S-901 83 Umea, Sweden, pp. 1-22.

J. Perbos. 1982. Modèle radiométrique d'effets atmosphériques et logiciel associé (ATMLIB). Version I, Division traitement de l'image, CNES, Toulouse, p. 52

K. Benhanifia. 1998. Télédétection et Forêt: Détection des changements dans un terrain montagneux à partir d'images bidates; TM de Landsat-5; corrigées des effets atmosphériques relatifs. Thesis of Magister, C.N.T.S, Arzew, Algeria, p. 119.

M. C. Rouquet. 1985/1986. Modélisation Atmosphérique. GDTA /ENSG Training courses p. 15, CETEL, 1985/1986.

S. Durrieu. 1994. Utilisation de la télédétection pour la mise à jour de la carte des types de peuplement de l'inventaire forestier national. CEMAGREF/ENGREF, Montpellier, p. 202. Thesis.

6 *Natural Hazards*

Observing our environment from Space: New solutions for a new millennium, Bégni (Ed.)
© 2002 Swets & Zeitlinger, Lisse, ISBN 90 5809 254 2

Space technologies for flood risk management: from images to products and services

Th.Rabaute & Gh.Gonzales
Scot, Toulouse, France

J.P.Dupouyet & J.J.Vidal
DIREN Midi-Pyrénées, Service d'Annonce des Crues, Toulouse, France

Y.Colin & B.Denave
CIRCOSC, Bordeaux, France

ABSTRACT: Many studies carried out during the last few years have shown that space technologies efficiently contribute to crisis prevention and management actions. Nevertheless, the operational use of what is still for the risk management community a very new technology remains rather rare. In the same time, facing an increase in losses of human lives and physical damage, citizens expect a lot from science and new technologies in helping to reduce risks and their impact. Time has therefore come to propose a full set of services to end-users in an homogeneous and consistent way, as it is already available in other domains. Aware of this urgent and essential need, some initiatives have been launched. This is the case of the SIREN project, which aims at defining, specifying and testing a dedicated service to provide Earth Observation based information for flood risk management.

1 INTRODUCTION

The end of the XXth century unfortunately gives too many examples of natural disasters all around the world. Floods, earthquakes, fires, hurricanes have caused hundreds of deaths and billions of Euros of damage even in industrialised countries. Whatever their origin (human-made or climatic phenomenon), these events and their consequences on human beings, economic activities and environment are less and less accepted by citizens every day better informed and concerned about the state of our planet.

In many cases, new information and communication technologies as well as improved systems (weather forecast, warning systems...) already help reduce the death toll and mitigate damage. This is confirmed by the growing dependency of emergency management organisations on communication services, some of them based on wireless LANs. This is not the case of the Earth observation (EO) component, which is not yet used in an operational context despite many studies and project pilots demonstrating its usefulness in this domain.

If few years ago the major reason had to be found in the very low awareness of potential customers, this is now less and less the case as many research laboratories, private organisations and public entities (especially space agencies) try to associate final users in their projects. Today, the lack of sensors and satellites specifically designed to meet the requirements expressed by the risk community still remains a problem. But another reason, at least of equal im-

portance, may lie in the difficulty for potential users to find a clear, well defined, secured and reproducible offer of services and products on which they can rely to carry out their every-day tasks.

This pragmatic analysis is at the origin of the SIREN project, funded by the European Union in the context of the DG XII 4th Framework Programme. This project mainly aims at working on "user services" aspects in order to imagine, define and validate specific concepts and solutions to make the use of Earth observation derived information easier and more effective for potential users (disaster management services, public services at national and local level, environmental agencies...).

The general idea is to propose a network organisation to serve national needs at the level of the various European countries (risk management generally follows the principle of subsidiarity), but also to define an architecture that makes the necessary co-operation between these national components possible. Clearly considered since the beginning as a user-oriented initiative, the originality of this project relies both on strong involvement of users throughout the project and on the validation of an integrated approach (Scot 2000b).

This paper describes the methodology used to capture user requirements, analyse their needs and define products and associated services. This approach is illustrated by the example of two French potential customers involved in the Project: the Bordeaux Civil Protection Inter-regional Emergency

Table 1 : The main data and information types

Type	Description
EO raw data	Raw images as obtained from the satellite sensor at the level of the receiving station (no correction)
Image product	EO images resulting from the pre-processing of EO raw data (system corrected, radiometric correction) including various levels of geometric correction (from basic geocoded images to ortho images) to be compatible with user cartographic systems
Information product	Value-added products resulting from the analysis of image product(s) including the use of non-space/ancillary data and which are directly meaningful to user and generally customer/market specific, tailor made and ready-to-use. They are of two types: – Intermediate products: products which may be submitted to additional processing to get more elaborated information (e.g. land use map), – Self-consistent products: full comprehensive products which can be directly used (e.g. flood map).

Control Centre (CIRCOSC) and the Garonne Flood Forecasting Service (SAC).

2 EARTH OBSERVATION IMAGES AND INFORMATION PRODUCTS

Discussions held during this project with users from France, Italy, Germany and Spain confirm, if necessary, the need to deliver high level information products (in opposition to EO images) to users. This question is exactly at the core of the project: definition and progressive implementation of a set of services to generate and make available such information products.

A preliminary remark has to be made concerning the various data and information types to handle. As it is very common to put under the same word different meanings depending on the context or the field of expertise (risk management or space community), it is important to better explain the content of the various data and information exchanged between the key players.

Table 1 gives a short description of the main data and product types used in the SIREN project. Obviously, self-consistent information products are not the only source of information at the level of the end user. The purpose is to give the possibility to the user to combine, during the decision making process, this new source of information with conventional data within the existing decision support system (new value adding step). This is the project final objective: make EO derived information as common and simple to use as standard data.

3 UNDERSTANDING OF USERS' NEEDS AND OPERATIONAL CONTEXT: A NECESSITY

When trying to define a product, it is first necessary to obtain precise (i.e. quantitative) information about the main elements characterising such a product. Many previous studies carried out in the past few years provided only generic information about user needs, as their major objective was to provide a synthetic overview of the needs of the various user "families" (Civil Protection, environment agencies...). In a next step, delivery of a given product to a specific customer involved in one of the previously identified sectors requires to refine the description, as some specific parameters may vary from one user to another (e.g. projection, scale...).

Table 2 describes the items used for the identification and technical characterisation of information products. These guidelines were used during the user requirement capture phase of the SIREN project (Scot 1999).

Table 2: Items related to the product characteristics

	Service/product name	Name of the product or service
Description générale	Purpose	Direct use of the product/services at the user premises
	Concerned risk management phase	Risk management phase to which the product/services applies
	Product level	Classification of the product with respect to its further use by the user (Intermediate product or Self-consistent product)
	Contribution to decision making process	Objectives of the use of this product/service with respect to the user decision making process
Spécifications du produit	Expected information	Main type of information expected by the user from the product/service
	Type	Type of presentation adopted for the product
	Reference system	Reference system to be used for the provision of expected information
	Update frequency	Required frequency for updating delivered information
	Scale	Scale to be used for delivering information (expressed in relevant unit)
	Accuracy	Expected accuracy for the parameters representing the information to be delivered (expressed in relevant unit)
	Delivery time	Expected delivery time for the product/service
	Information support	Type of support used to distribute the information

Table 3. Comparison of Flood Monitoring Product for the Bordeaux CIRCOSC and the Garonne SAC

		CIRCOSC specifications	SAC specifications
General description	Service/product name	Flood Monitoring Product (FMP)	Flood Monitoring Product (FMP)
	Purpose	Identification and follow-up of flooded areas in order to give a general overview of the situation which can be used to make decision	Calibration of flood forecasting models Validation of flood forecasting results Validation/updating of risk prevention plans
	Concerned risk management phase	Crisis	Post-crisis/Prevention (essential use) Possible use in crisis phase
	Product level	Self-consistent product: can be used by the CIRCOSC risk manager and the prefect to take decision	Intermediate product: can be integrated/combined with other data and used as input data for hydrologic models.
	Contribution to decision making process	Operational management of the crisis: update of the intervention plan, optimisation of the localisation of rescue means, alert of concerned organisations and citizens...	Lesson learning (understanding of a major hazard, improvement of existing tools...) and prevention activities (updating and refining risk maps)
Product specifications	Expected information	Location (limits of flooded areas with reference to cartographic information)	Location of flooded areas and monitoring of flow dynamic
	Type	Line type	Line type
	Reference system	French reference system (Lambert II)	French reference system (Lambert II)
	Update frequency	Daily basis with the possibility to update this information every 12 hrs	Hourly basis (including the possibility to cover key dates such as the peak water flow)
	Scale	1:100,000 to 1:50,000	1:25,000
	Accuracy	$u_X = u_Y = 500$ m	$u_X = u_Y < 50$ m
	Delivery time	Near real-time	Near real-time for the "crisis" component (validation/adjustment of the forecasting models), Not critical for the use related to prevention activities
	Information support	Digital information layer (compatible with the GIS and command & control system) or paper map (ready to use document)	Digital information layer (compatible with GIS and model formats)

When detailing the requirements of the Bordeaux CIRCOSC and the Garonne SAC, one of the needed information appears to be related to the observation and analysis of the spatio-temporal development of the flood event. Table 3 summarises the main characteristics of a product, called "Flood Monitoring Product" (FMP), resulting from the analysis of these requirements. It is clear from this table that even for a same class of products, intrinsic characteristics may vary from one user to another.

For CIRCOSC, the objective is to obtain an objective and exhaustive overview of the current situation during the crisis, so that rescue operations can be optimised by the State representative working at the local level. The actual need, in addition to near-real time delivery, is more in terms of synthetic information covering the whole area and highlighting the vulnerable areas at a given period, than of detailed information provided at the level of each property. For the Garonne SAC, the included information will be used to calibrate and validate the output of flood forecasting models, but may also represent objective data that can be used to update risk maps. The level of details as well as the accuracy of the flood limits provided have therefore to be very high.

All these "technical" points have then to be clearly identified as they govern not only the analysis process (which may vary from one user to another), but also the data used to generate the required product. It is not possible to generate a product with a 1:25 000 scale and an accuracy of less than 50 m with the same EO images as those needed to develop a 1:100 000 scale product. In the last case, high (and even medium) resolution images can be used when in the first one only very high resolution images are recommended.

The second aspect of major importance is related to the context in which the product will be used (i.e. integrated in the decision making process). Defining a service is not limited to the provision of an appropriate product (e.g. technical characteristics). It also requires to set up all the processing chain necessary to make this product available to the user at the right time, place, format and even with a cost in relation with the service provided, taking into account competitive offers (other data sources, services...).

When trying to develop an offer of EO derived information for risk management, the service component is of major importance. The best technically designed product is of no use if it is not distributed to people who make critical decisions (emergency managers and decision-makers).

Table 4 gives an example of the criteria used during discussions with users in order to define the expected level of service associated with each identified product.

To go on with the same example, Table 5 gives the results of the analysis of the service required by CIRCOSC and SAC with regard to FMP. A quick comparison between these two kinds of service requirements clearly illustrates the fact that fulfilling user needs requires to set-up and/or adapt the general methodology (generation process), but also the means necessary to perform the task. If for the Garonne SAC a complex analysis process including a detailed verification procedure can be performed with no strong time constraint, this is not possible for CIRCOSC. In that emergency context, it is necessary to propose a fast-analysis approach and take into account all the aspects related to near-real time delivery (including communication aspects).

4 PRODUCT GENERATION: A COMPLEX PROCESS

When defining a service, with the objective to provide users with appropriate information which can be easily accessible and usable in their daily task, as it is the case in the SIREN project, two main problems have to be tackled:
- how to produce the requested products in a reproducible, secured and cost-efficient way?
- who is the most appropriate actor to perform this value adding process (to go from data to information added value)?

After an overview of the state of the art in the field of image processing as well as the analysis of previous pilot projects, it becomes clear that:
- even if many improvements have been made for few years to facilitate and automate tasks (algorithms, software), the analysis of both optical and radar images remains a complex process requiring a solid experience. Moreover, another domain (hydrology) has to be taken into account in this specific field of expertise.
- all the users we are working with do not intend, at least in the short term, to invest in this specific domain in terms of tools and people. It is thus necessary to set-up specific organisation in charge of the generation and distribution of final products.

Concerning the technical aspects linked to the generation of the identified products, it is necessary to define a set of guidelines that can be used during the analysis process. Such guidelines aim at describing the key steps of the selected methodology (specific to a given product) as well as the various modules (geo-referencing, filtering...) that are made available to the operator to perform a specific task (part of the overall analysis). The final objective is to ensure a certain level of quality and homogeneity of the resulting products. These guidelines and support packages are particularly important in a context where the processing of each case is different (not exactly the same conditions, not the same data available, the quality and information content of EO images are never the same...) , giving a crucial role to the "human factor".

Table 4. Items related to the service characteristics

Service specifications	Critical parameter	Most important parameter conditioning the integration of the product/service in the user decision making process
	Service duration	Expected duration during which the user will need information
	Stability of the request	Expression of the stability of the request during service provision (a request may be stable or progressively refined when new information becomes available, more particularly during a crisis)
	Acquisition of near real-time information	Expression of the need to get real-time information to fulfil user requirements

Table 5. Comparison of the service associated to the Flood Monitoring Product for the Bordeaux CIRCOSC and the Garonne SAC

		CIRCOSC specifications	SAC specifications
Service specifications	Critical parameter	Delivery time and updating frequency	Coverage of the whole area during the complete duration of the crisis
	Service duration	Entire duration of the crisis	Entire duration of the crisis
	Stability of the request	No. When new information becomes available (location, duration), the initial request may be refined in order to make adjustments for the monitoring the event	Yes for a request made after the crisis for post crisis and prevention activities (hypothesis of images already acquired during the crisis)
	Acquisition of near real-time information	Yes. Essential to envisage an effective use during the crisis management	Yes even if information products are used for post crisis and prevention purposes

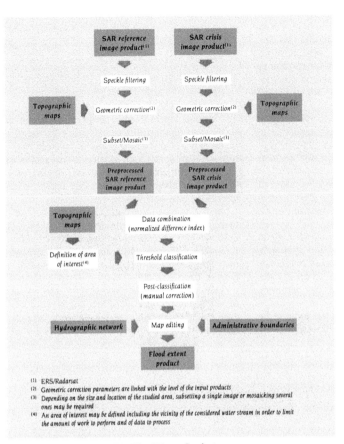

(1) ERS/Radarsat
(2) Geometric correction parameters are linked with the level of the input products
(3) Depending on the size and location of the studied area, subsetting a single image or mosaicking several ones may be required
(4) An area of interest may be defined including the vicinity of the considered water stream in order to limit the amount of work to perform and of data to process

Figure 1. Generation of the Flood Extent Product

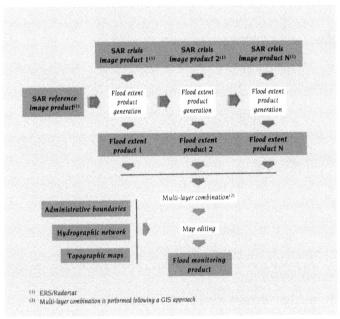

(1) ERS/Radarsat
(2) Multi-layer combination is performed following a GIS approach

Figure 2. Generation of the Flood Monitoring Product

This standardisation effort, which is a very new concern in the Earth observation community (the main objectives of previous projects were more related to the definition of new methodologies than to their operation), appears now as a basic and necessary step in the implementation of a user dedicated service.

Figure 1 shows an example of activity diagram describing the sub-processes and their sequence to apply in order to generate the "Flood Extent Product" (FEP), which provides a "photo" of the situation at a given date.

Figure 2 presents the next step: the integration of several FEP for monitoring the water flow over a given period, resulting in a Flood Monitoring Product.

These two figures describe the methodology at high level only. When implementing the approach, each specific module of these technical guidelines has to be described in detail in order to clearly define the processing to be applied by operators working in an operational context.

In the same way, the selection of the tools to be used is also driven by pragmatic and operational aspects (i.e. only proven techniques and fully validated software are used).

5 VALUE ADDING: A SPECIFIC TASK WITH SPECIFIC FACILITIES

The second aspect to consider is related to the practical organisation of the service itself. In the context of the SIREN project, a possible network architecture supporting this kind of service was proposed resulting from the analysis of the user requirements of the four countries involved in this project (France, Germany, Italy and Spain).

In a first step, a description of the context was carried out, including both the analysis of the actors, external to the system, with the characteristics of their role with respect to the system (Table 6) and the relationships between them and the system itself (Figure 3).

In a second step, the main functions to implement in order to fulfil the identified needs were defined and specified as well as the network architecture necessary to support the proposed SIREN system.

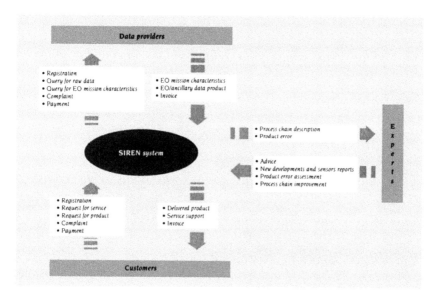

Figure 3. The SIREN context diagram

Table 6. The key actors of the SIREN system

Customer	Generic actor who interacts with the SIREN system to ask for products and/or services
Satellite EO data provider	Generic actor who provides a set of satellite EO standard image products involved in the SIREN product generation process
Non space data provider	Generic actor who provides a set of data involved in the SIREN product generation process. This actor may be the customer himself according to the requested product.
Expert	Generic actor who may give scientific and methodological advice, improve the SIREN product generation processing chain and report on new developments and sensors

232

This design was driven by three main pragmatic considerations:
- all the functions have to be defined with the same "philosophy": implementation of a purely user driven approach thanks to the possibilities offered by the use of new technologies,
- imagine a service and its running compatible with the current user "working style" and responsibilities (especially security and confidentiality aspects),
- design a system with a clearly stated initial objective of setting up a service which provide users, in the short term, with a cost effective level of service.

6 THE PROPOSED NETWORK ORGANISATION: A SUBTLE BALANCE BETWEEN A CENTRALISED SYSTEM AND A SET OF INDEPENDENT NODES

The resulting network organisation, shown in Figure 4, is based on two major components (Scot 2000a):
- a **User Support Service** (USS) in charge of starting relationships with users as well as of organising exchanges between the various components of the system and the other actors working in the field of flood risk management. The functional analysis shows that the activities falling under the responsibility of the USS can be split in three main types:
  - Interface with customers (**Customer Interface Unit** - CIU): this group gathers user oriented support activities (mainly the processing and follow-up of user requests),

- Interface with EO data providers (**Multi-satellite Access and Ordering Unit** - MAOU): this group includes EO data provider oriented support activities (contact with EO data providers, selection and access to EO image products),
- Co-ordination and management activities (Network Coordination Unit - NCU): this includes general activities necessary for an actual and efficient operation of the network (implementation and maintenance of the various nodes, development of the network, management, advertising...).
- **Regional Thematic Units** (RTU), distributed entities implemented at the regional level (i.e. the level of the countries involved in the project) and in charge of answering user needs in terms of tailored information products (such as the FMP presented in this paper) and appropriate technical support.

This organisation is derived from:
- the analysis of user requirements collected during the first stage of the project, which was the preliminary step to extract SIREN service specifications,
- the experience gained by the project partners in previous studies related to the use of EO derived information applied to flood risk management,
- the discussions with data providers and analysis of their current offer, not only in terms of image products but also of related services.

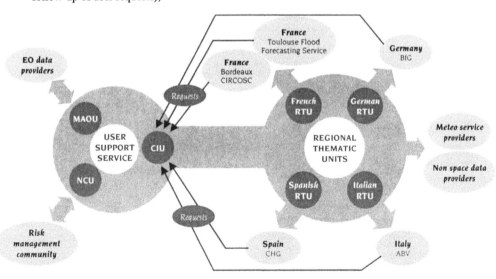

Figure 4. The proposed SIREN network architecture

It also represents a satisfying compromise between a purely centralised architecture, which would be less efficient in terms of relationships with users from different countries, and a fully distributed one, which would result in an increase in implementation and operating costs. In the context of SIREN, the proposed architecture focuses on flood management. However, it can be adapted to any kind of natural hazards.

Finally, it is also important to define the level of service offered to the customer and its positioning within an end-to-end remote sensing information service, which not only includes value adding aspects but also tackles all the other components (space and ground segments) (Alenia Spazio 2000).

Figure 5 illustrates a possible positioning of the SIREN service offer inside the EO value adding chain. The first part is related to the on-the-shelf offer (based on a pre-defined product catalogue where FMP represents one of the proposed products) that can be proposed to any interested user/organisation. The second one is more a tailor-made offer that can be envisaged on demand for users who need a more complete support and/or very specific products/services.

7 CONCLUSIONS

At the difference of the space telecommunications domain, which is now mature and ruled by commercial laws, the Earth observation market is just emerging from the scientific and institutional world.

After many studies and projects demonstrating the potential interest of such a new technology for the risk community, it is now time to define a complete service offer with the appropriate structure taking into account operational and economic constraints.

Considering the current situation, the SIREN project represents a concrete attempt, in a clear user/application oriented approach, to specify and define a full set of services providing information that can be used in existing decision support systems operated at user level.

In complement to necessary R&D activities, it is indeed important to think about a possible service organisation. The results obtained during this project, especially the description of the main functions as well as the presentation of a possible system architecture, represent the basic bricks from which it is possible to go a step further: detailed specifications, system design and progressive implementation of the full system.

However, the development of this market not only includes the definition and generation of standard tools and products. It also strongly depends on the availability of new data/sensors (very-high resolution, constellation...) in order to better fulfil user requirements which are not yet fully covered by existing satellites.

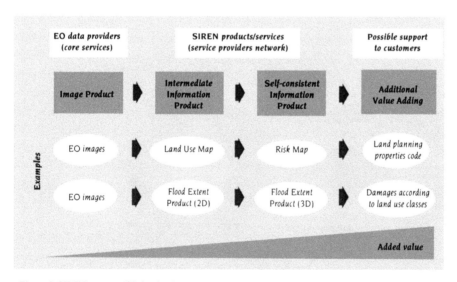

Figure 5. SIREN: two possible levels of services

REFERENCES

Scot, 1999, *SIREN User Requirement Document*, Project report, SIREN/SRD/DT/PPL-TR99224, April 1999

Scot, 2000a, *SIREN System Requirement Document*, Project report, SIREN/SRD/DT/TR99284, Febr. 2000

Scot, 2000b, *SIREN Web server*, http://scot-sa.com/siren

Alenia Spazio, 2000, *European Remote Sensing Information Services (ERSIS)*, Final dossier, LI/ESS/0004/EIT, July 2000

Observing our environment from Space: New solutions for a new millennium, Bégni (Ed.)
© *2002 Swets & Zeitlinger, Lisse, ISBN 90 5809 254 2*

FORMIDABLE: An advanced tool for the information management of natural disasters

F.Lamberti, M. Folino & F.Rossi
DATAMAT Ingegneria dei Sistemi S.p.A., Rome, Italy

ABSTRACT: In the Emergency Management of natural disasters, the effectiveness of interventions and relief actions is affected by the difficult circulation of reliable information among involved organisations and citizens, whenever a disaster occurs. Within this context, the FORMIDABLE advanced tool has been conceived as a joint European enterprise, focused on disasters affecting the Mediterranean area, aimed at standardising operational procedures and information systems. The FORMIDABLE tool is the repository of the information required for natural disasters management, by using suitable technology to maintain and exchange relevant data. In fact the authorised organisations will receive near-real time standardised operational tasks and scenario description, while at the same time useful and friendly information will be provided to involved people. High level of friendliness and flexibility will be achieved by means of the coherent integration of space techniques.

1 INTRODUCTION

The Friendly Operational Risk Management through Interoperable Decision Aid Based on Local Events (FORMIDABLE) System represents the standardised access to data and knowledge required in the emergency management context, with all means to access, maintain and exchange information for disasters, in particular those with a fast evolution time typical of the Mediterranean countries.

Figure 1: The FORMIDABLE Logo

The FORMIDABLE Project has the main objective to propose a European Standard Methodology for Emergency Management based on the consensus of major Mediterranean operational actors, and develop an interoperable support system prototype which integrates the resulting guidelines with all data and tools to operate during any emergency, in line with Council Decision 98/22/EC of 19 December 1997.

The paper will provide an overview of the FORMIDABLE project, starting from the analysis of requirements for Civil Protections. Then the paper will describe how to derive a standard methodology to be translated by each Civil Protection into operational procedures and information requirements. Furthermore, the FORMIDABLE system architecture building blocks will be presented. Finally, the paper will introduce the strategies foreseen for the integration of such a tool with EO systems, and its verification within operational contexts. In particular, due to the early involvement of operational organisations, scientific and industrial entities from the major EU southern countries, concepts and resources will be shared with the Mediterranean neighbours, to demonstrate applications to earthquakes and flood and to improve EuroMed co-operation.

2 THE CONTEXT

In the Emergency Management of natural disasters, the difficult circulation of reliable information among involved organisations and citizens, on occurrence of any hazardous event represents one of the major constraints to the effectiveness of interventions and relief actions. This could have a dramatic impact on the planning and execution of safety and relief procedures, such as people evacuation, thus increasing losses of lives and goods.

A critical analysis of where emergency management operations encountered major problems, causing even partial failure or delay in recovery action, highlights a generic lack of co-ordination, and a poorly unified and homogeneous approach. Often actions have been planned and executed with a limited field of view about the real size of the events and the involved responsibilities for emergency management. Moreover, this "local management" has produced conflicts between the Central Civil Protection authorities and the local administrators, thus hindering the efficient execution of interventions.

Generally the emergency scenarios are quite complex, although in some cases the cyclical occurrence of natural events might help to improve emergency management. Even if the repetition of an event can be regarded as a constant factor, the damage extent and the type of required intervention are variable aspects, also due to the changing features (geomorphologic, administrative) of the affected areas.

From this context, some requirements on the "Emergency Management" of Natural Disasters arise, reflecting the following aspects:

- the need to perform flexible and easy tasks to ensure efficient interventions and immediate relief to affected citizens;
- the need for all involved national and local authorities to share a common knowledge and methods to face the management of natural disasters, from emergency planning to post-event assessment phase;
- the need of quick response times to receive accurate and homogeneous information containing detailed description of both expected crisis scenarios and real ones.

These requirements have been the major drivers in the definition of the project objectives.

3 FORMIDABLE OBJECTIVES

According to the above context, the activities that have been identified to carry out the project will pursue the following major objectives:

1. the contribution to the definition of a **Standard Methodology** at European level for disaster management, initially conceived for the Mediterranean countries partners in the projects but expandable at European level, able to cover complex operational scenarios as products of different features as:
   - Types of natural disasters
   - Civil Protection Organisations
   - Operational Phases
   - Geographic Extension

The resulting Methodology consists of a set of guidelines, recommendations and templates to be used in the emergency management as support to:

- a more rational and efficient organisation of CPAs activities,
- the preparation of relevant documentation and reports according to common template and data presentation,
- the exchange of information between all involved actors following well established criteria to distribute the information related to any natural disaster, tuned according to the event size and to the relevant national organisation.

2. The specification and the development of a **Standard Interoperable Emergency Management System prototype**, compliant with the Methodology and capable to support the decision making process of the authorities responsible for Emergency Management. At the same time the proposed development aims at integrating the new advances in technology in order to provide better answers to both:
   - users requirements, as flexibility, user-friendliness, simple and homogeneous access to specific data will be the drivers to the system specification, taking into account a wide range of natural disasters and operational scenarios;
   - system requirements, for which interoperability, portability, scalability, reliability, performances and maintenance will be the major drivers.

3. The **validation and promotion** of the above results within the Emergency Management community, through the development of specific applications, able to represent suitable test beds with respect to the methodology procedures and the operational tasks to be performed in front of realistic emergency management scenarios.

4 FORMIDABLE DEVELOPMENT LOGIC

The following leading steps have been identified:

a. proposition of a Europe-standard methodology for the emergency management of natural disasters;

b. definition and design of the Emergency Management Support System in terms of system requirements and architecture;

c. development of a prototype representing a significant proof-of-concept, and its application to selected emergency scenarios addressing two different types of natural risk: predictable and unpredictable, and field trials of the applications with different classes of users;

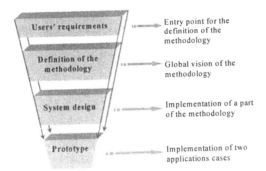

Figure 2: FORMIDABLE Development Logic

d. validation of the objectives with users' involve-
    ment to ensure the validity of the approach at
    different levels (CP authorities, local authorities
    and operational organisations, citizens), the dis-
    semination of the achieved results and the
    evaluation of the potential market for the system.

4.1 Methodology Definition

The main objective of the FORMIDABLE Methodology is to contribute to the definition of a standardised European approach for the Emergency
Management of natural disasters. To this aim, the
activity carried out within the project is focused on
the definition of the guidelines and the criteria to be
used as drivers to decision support in the different
phases of the emergency.

Therefore, the FORMIDABLE Methodology has
been conceived through two main steps:

− the specification of an engineering Functional
   Model of the Emergency Management to be used
   as a reference for the activities and the data to be
   exchanged within the context;
− the extraction of sets of guidelines, recommendations and templates, derived from the model,
   that support the CPAs in their activity organisation and execution during each operational
   phase.

The approach used to derive the Standard Methodology for the CPAs within the Emergency Management of natural disasters has the starting point in the
outcome of the analysis of Civil Protection activities
performed during the User Requirements analysis
phase and reported in the User Requirements Document.

The main result consisted of a high level Functional
Model of the Civil Protection activities for the
Emergency Management of Natural Disasters, i.e.
covering from emergency planning to post-event assessment. The Functional Model provides an engineering view of CPA tasks, by relating each activity
to an input/output process and by establishing the
relationships of each activity with the others as well
as with the external entities.

The major objective of this model was to describe
the complex scenario of emergency management
that includes many actors and many heterogeneous
activities through the schematic diagrams obtained
by applying an engineering analysis methodology.
This allowed to:

− define a general layout applicable to any event
   and/or to any phase,
− refine and optimise the process flows, removing
   redundancy or duplication;
− reach a standard functional model, e.g. applicable to any Civil Protection representing a theoretical model to be used as a reference.

In order to derive from this baseline a Standard
Methodology addressing guidelines for all emergency management processes and information exchanges, the following steps have been performed:

1. Specification of an Emergency Management
    Functional Model, obtained through the refine-
    ment and the detailed analysis of the Civil Pro-
    tection activities in all operational phases. This
    analysis is aimed at providing:
    − a more detailed knowledge of the activities
       required within the Emergency Management
       context;
    − the identification of the complete interaction
       between them.

The operational phases included in the emergency
management context to which the methodology
applies are the following:

− *Emergency Planning*, when all activities are
   performed, along with relevant data and
   means, that are required to prepare and maintain emergency plans.
− *Emergency Control*, which starts from the
   raising of any level alarm up to the implementation of all activities required in the aftermath of a disaster.
− *Post-Event Assessment*, when all activities are
   performed that are required to calculate resulting losses, (both consequential and secondary) and related costs as sources of information for prevention activities, usually taking
   place few weeks after a disaster occurrence.

The activities performed during these three operational phases have been grouped into 12 Auxiliary
Functions (AF), which are the basic elements of
the Functional Model. They are listed in the following:

AF1. Technical & Scientific Support for the sci-
      entific analysis and physical interpretation
      of the event and related data,
AF2. Health, Social Assistance and Veterinary
      Services for the management of all people
      and means involved in the sanitary area,
AF3. Mass-media and Information for the distri-
      bution of specific information to mass-
      media and citizens,

AF4. Resources for estimation of all resources necessary during emergency, their location and availability, including also the co-ordination and training of specific volunteers organisations,

AF5. Transport and Viability for transfer of material and people, to optimise evacuation paths and regulate intervention flows,

AF6. Telecommunications for the provision of TLC networks, as back up and support to operations,

AF7. Lifelines for the co-ordination, restoration and maintenance of all necessary services (e.g. water, electricity, gas),

AF8. Damage Assessment for the estimation/evaluation of damages (e.g. to people, buildings, agriculture),

AF9. Search and Rescue Operational Organisations for the co-ordination of all entities involved in S&R operations,

AF10. Dangerous Materials for the management of storage location and material census with respect to the impact on affected areas,

AF11. Relief Provision for co-ordination of assistance to population, in terms of identification and set-up of suitable areas to provide assistance and necessary services,

AF12. Operational Co-ordination for the management of the auxiliary functions and the rational interventions of means and people.

2. Specification of the FORMIDABLE Methodology, by using the Emergency Management Functional Model as the backbone for deriving a standard approach for homogeneous sets of activities, addressing:
   - criteria and guidelines for activities execution;
   - recommendations for communication exchange;
   - related documentation templates and management criteria.

This process has enabled the identification of suitable areas for standardisation, even in complex scenarios as those including different National CPAs, different types of natural disasters, different emergency phases and different organisation level.

In addition, the identification of all functions and data classes, both common and specific to each functional area, provides an optimised organisation of data exchange within an operational system, able to translate the methodology guidelines into a working tool.

The resulting FORMIDABLE Methodology and the associated Engineering Model herein presented have to be considered as preliminary results towards standardisation. As a matter of fact, even though it is not possible to generate a complete methodology for Emergency Management

within the frame of this project, the results achieved so far show that it is possible to standardise activities and information flows in very complex scenario. In addition to these main outputs, the activities carried out so far also bring other significant achievements, as by-products of the two major ones:
   - the deep knowledge of CPAs responsibilities and specific roles, and their impact on social, administrative and political issues.
   - the definition of a strategy for building up a Standard Methodology, that could be applied to other context or other similar analyses.
   - the definition of terminology and glossary shared between CPAS partners.

However the approach used throughout the analysis performed and reported in this document, aims at maintaining the resulting methodology guidelines as an open issue. This means that the methodology guidelines presented herein are a first important step towards a process of standardisation.

In fact the specification of a Methodology covers a wide scenario, a four-dimension space whose exhaustive exploration is started thanks to Formidable project, performing a "first run" of the Methodology Specification.

In addition, before entering an operational context, the specification of a methodology requires interactions and feedback from the operational people in the Civil Protection. Since the Methodology is conceived for applying to so many different scenarios, (each scenario resulting by the product of a phase by a disaster type by country by responsibility level) it is very important to verify the adherence of the proposed methodology to real operational contexts. A long process of interactions and validation could be necessary to make the FORMIDABLE Methodology a stable and exhaustive operational solution to face natural disasters.

Finally it is worth underlining that the value of the results reached so far, within the Formidable Methodology Specification Document, lay in the fact that they represent the building blocks of the Standardisation Process for the European Methodology; to be used as drivers for a validation and assimilation process.

The major conclusion of this work can be represented by a recommendation, since this wide analysis emphasises the relevance of the deep awareness of duties and responsibilities on one hand and of the acquisition of a certain degree of sensitive attention to disaster management and environmental impacts on the other.

This leads to identify a twofold assimilation process:
   - on the top, CPAs managers have to share principles and the concepts behind the Meth-

odology in order to transfer them to the operational personnel.
- on the bottom, citizens have to be considered as part of the Emergency Management and be involved in the relevant activities.

The merging of these two points is felt as necessary to proceed towards the transfer of a theoretical model into the real life.

4.2 Emergency Management Support System Design

The emergency management support system is specified according to the proposed methodology, through detailed system requirements analysis and operational concept definition. A preliminary architecture foresees that each auxiliary function be implemented and operated on a dedicated workstation. Static (e.g. topography, risk maps, land use, event history) and dynamic (e.g. meteorology, positioning, in-situ measurements) data are used. State-of-the Art technologies such as high performance Internet GIS, multimedia user interface tailored to different operators, data fusion, mobile communications are also taken into account.

The main aim of the FORMIDABLE system design activity is to provide the basic design of a Decision Support System (DSS) able to incorporate both User Requirements and the resulting Methodology criteria, in order to be able to support operational bodies in any operational phase of the Emergency Management. In particular, the FORMIDABLE system has to be conceived as the support tools to the Auxiliary Function Managers that are responsible of activity preparation and activation within their relevant intervention area. More in detail it addresses the coordination of the activities and operations and makes easier the communications among the AF Managers, both internally and with external bodies/organisations.

Then, the FORMIDABLE system aims at supporting the Emergency Management according to the severity of a natural disaster for a given geographical area, this means the system could be configured at several geographical levels:
a) National level
b) Regional level
c) Local level

4.2.1 Operational Concepts.

The definition of the system in terms of functions and interfaces has been carried out in parallel to the preparation of the operation concepts.

This activity aims at identifying how and when the system will be used, the staffing required to operate the system, the type of human interface and the interactions the system has to support. In particular it will be necessary to identify the operational phases

of the system, corresponding to different disaster scenarios.

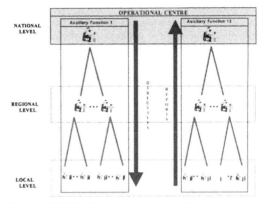

Figure 3: FORMIDABLE Application levels (hierarchical scalability concept)

This activity is carried in close relationship with users, in particular CPs and Local authorities, that could be the potential operators of such a system.

A part of this activity is dedicated to the failure analysis of the system. It seems mandatory to consider an Emergency Management System characterised by a very high availability, since the system shall be robust and easy to maintain at the same time. This analysis provides drivers for the Architectural Design, in order to identify the system critical parts for which redundant elements, and the necessary type of hardware and software have to be considered.

4.2.2 The System Architecture.

The FORMIDABLE System Architecture is strongly dependent on the main critical aspects to be faced in order to develop such a complex system:
- The System has to be independent from the country organisation: this implies allowing the configuration of the system at AF and related responsibility level, it will be possible to assign more than one function to a unique manager. In this way it is necessary to have a configurable Man Machine Interface (MMI) that, when the user login in the system, depending on his/her AF responsibility, is presented in order to allow easy configuration of AFs and specific data. Another geographical problem is about data details, increasing from local level to national level. This problem could be avoided by means of RDBMS utilisation, allowing a Data Base population of the proper geographical data only.
- The system has to optimise information flow, this implies avoiding data duplication inside an Operational Centre. To avoid this problem, all the data to be managed in the Operational Centre will be analysed in order to organise them de-

signing a common Data Base accessible from all the FORMIDABLE system components.
- The System has to manage the Communication between Operational Centre and external/internal entities: the AFs interact each other by means of directives, reports, messages, at the same time the Operational Centre interacts also with external entities.

Then after assessing system requirements and the technical solutions that can be used to guarantee the functional, performance and reliability requirements, the specification of the system architecture provides the basic components of the FORMIDABLE DSS.

Each Auxiliary Function will be implemented and operated on a dedicated workstation. The purpose of this conceptual architecture is to present the feasibility of such a complex system by means of coherent integration of different technological means ranging from Internet to fault tolerant servers up to satellites for both communication and earth observation. The proposed project covers the specification of the system architecture of the Operational Centre, where the auxiliary Functions will be operated. The Emergency Management System can be allocated both at CP's premises and at local Authorities (e.g. Prefects, Mayors) sites, and tailored in size and functions according to the entity of the event.

In the Operational Centre all 12 AFs are connected to the main servers:
- *Application Server*: to manage all the processing foreseen to the Operational Centre
- *Data Base Server*: to manage the storing and all the operations with the common Data Base
- *Web Server*: to manage the citizens interaction (Internet – Operational Centre Web Site)
- *Data Communication Server*: to manage all the external interactions with the data provider
- *Fax & Telephone Server*: to manage all external interaction by means of Fax and Telephone
- *Firewall*: to filter external accesses towards the Operational Centre
- *Clients workstations*: it could be a client for each AF manager in order to have a workstation from which all the activities should be performed.

The System is also able to collect heterogeneous dynamic data coming from different sources such as: in-situ sensors, meteorological radar & satellites, EO satellites, external data-bases.

Once collected, data are stored, together with static information (geographical information, available resources, etc.), into an interoperable "Emergency Management" database, located on the Main Server.

4.3 *The System Prototype*

Within the scope of the EU co-funded project, it is foreseen to provide a reduced implementation of the specified system functions, in order to supply a prototype as a test bed for the methodology on specific application cases.

Within this frame, a selection of subsets of functions and interfaces is developed. The prototype specification is performed in order to guarantee that the methodology guidelines can be successfully translated into an advanced tool. Moreover some technological innovations such as, interoperable Data Base, dynamic GIS, specific Man Machine Interface both for operators and for external users and citizens are taken into account.

With respect to the system architecture, the prototype includes:
- the two main auxiliary functions needed for system co-ordination (AF1, AF12)
- specific auxiliary functions with reference to the required operation context (technical & scientific support, health, volunteers, mass-media & info, etc.)
- an external data provider emulator, to verify the interface behaviour
- the Web server, to analyse the system behaviour with respect to its capability to grant access to citizens.
- A Videoconference module.

4.4 *Applications*

Within the scope of the EC funded project, two applications will be implemented, the first addressing seismic risks, with the second one addressing floods. The selection of these two applications is justified by the need to analyse the prototype behaviour in different operational scenarios due to the occurrence of different natural disasters in two different countries. In particular the two types of disasters reflect two different operational tasks, being one unpredictable as earthquakes and the other predictable as floods. Moreover, these two sites provide a better coverage of a European scenario. The Civil Protections of the two countries, Italy and Spain respectively, will be involved bringing their operational experience, and related organisation, as specific set of test data.

Each application will consist in the definition of the scenario (e. g. event type, extent, affected area, involved organisation).

To this aim, the prototype will be configured to fit the application scenarios by:
- specification of the application to identify the functions, the interfaces and the data required
- application development by using the FORMIDABLE prototype for Data Base population and for ingesting specific information
- Field Trials execution, to test prototype main features e.g., performances, user friendliness and data access interfaces during different operational contexts.

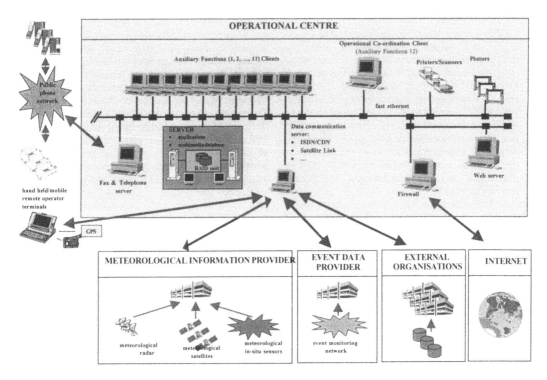

OPERATIONAL CENTRE

Figure 4: FORMIDABLE System Architecture for the Operation Centre

5 INTEGRATION OF EO DATA WITHIN THE FORMIDABLE EMERGENCY MANAGEMENT SYSTEM

The capability of state-of-art technologies in some specific areas, as telecom, interoperable databases, Earth Observation (EO) leads to apply their advantages to the Emergency management needs. Currently, a series of Information Systems is under development in Europe addressing specific environmental risks (e.g. floods) or specific technology (e.g. new EO products).

In the following, some of these projects are recalled from which FORMIDABLE intends to take benefits, in terms of information exchange and synchronisation of developments:

– DECIDE: EO Technologies for Decision Support Demonstrations. These five projects, involving most EU countries and different types of natural hazards, have been launched by the European Space Agency in early 1999 to demonstrate the potential application of EO derived information in support to decision making processes of disaster managers. The demonstration has a pre-operational character and will be conducted with a clear perspective of continued use and service provision.

– SIREN: Service d'Information sur les RisquEs Naturels d'inondation. The main objective of SIREN is to define and specify an Earth Observation based Flood Risk Information Management Service, to implement a service demonstrator and finally to test and demonstrate its feasibility during a pilot operation phase. The proposed service will be the interface between the end-users and the satellite data and information products provider for prevention, mitigation, response and recovery from flood hazards. This project has been funded by EC DG XII as part of the Shared Cost Actions.

– PERSIA Permanent Scatterers Information System Analysis: The main objective of the PERSIA project is to upgrade the use of the Permanent Scatterers technique to a large-scale and daily method for monitoring land surface stability and instability. The PS technique derived from SAR images differential interferometry has already been demonstrated as suitable for users, with different (lower) costs with respect to other techniques, based over e.g. GPS monitoring campaigns. It has been also demonstrated, in case of low data availability, the possibility to increase the Permanent Scatterers network over the interested areas, still at low cost and with high return. Considering this basis, the PERSIA Project is aimed at analysing the possibility of a large scale application, achieving the design of a PS Service Chain.

- CLIFF: Cluster Initiatives for Floods and Fires. CLIFF is an accompanying measure within the IST programme of the European Commission (EC), dedicated to review and analyse selected Flood and Fire projects supported by ESA and the EC. The objective of the project is to identify guidelines, recommendations and common characteristics of operational information and tools (including services and applications). These results will be discussed with an as wide as possible scope of civil protection organisations and operational disaster management authorities across Europe.

The above mentioned projects provide useful references about the use of EO data and products within an operational context.

6 CONCLUSIONS

As repository of the information related to natural disasters, FORMIDABLE will be used as the main source of knowledge towards citizens. This is not only due to the capability to access the information, but to the ease and flexible management of information flow. In particular FORMIDABLE will emphasise:

- the synoptic capabilities of EO data;
- the accuracy of DEM in order to define the extent of disasters both during event and post-event assessment;
- the imaging capabilities of remote sensing sensors in connection with GIS for a straightforward integration with territorial information, during any phase of emergency management.

There will be the possibility to define and extract information according to the emergency phase and event scenarios. As an example the system could be queried to provide information about:

- risk maps for a given geographical region;
- evacuation plans;
- event description;
- contact points for Emergency Management.

The availability of information towards citizens is the cornerstone for the creation and diffusion of an European culture related to natural disasters management, to increase:

- the awareness of how to react during emergency: what must be done, what has to be avoided;
- their involvement in the participation to the preparation of procedures, provision of data;
- the knowledge of the environment where they live.

This should enhance the co-operation between citizens and Local Authorities in the protection of the Environment, since a major knowledge of natural hazards will reflect in better interventions at any level to maximise health and safety.

In addition, FORMIDABLE can contribute to broaden education and then major attention to the overall management of environment and natural resources, make citizens more sensitive to the management of environmental resources, as the only way to prevent risks and reduce losses during events. The possibility to use the FORMIDABLE prototype for simulation and training could help to divulging environmental issues in many different contexts. FORMIDABLE will also allow to prepare training sessions that can be configured according to the purposes and the audience to be involved each time.

7 CREDITS AND PARTNERSHIP

The initiative for the definition of a standard methodology and its translation into an information system is due to a consortium of Mediterranean industries (developers, integrators, service company), research and educational institutes, and customer organisations in charge of operational emergency management in their respective countries, with the particular involvement of the Italian, Greek and Spanish Civil Protection Authorities in charge of representing the users' point of view during any phase of design and development. The FORMIDABLE project consortium is co-ordinated by an Italian company, DATAMAT, with the project kicked-off last January 2000, foreseen to be completed in mid 2002 with an overall duration of 2 years and a half.

All FORMIDABLE Partners are presented in the following table. The authors wish to thank them all for the contributions during this first phase of the project, along with the European Commission DG Information Society that is co-funding this initiative in the context of 5th Framework Programme for Research and Technological Development - IST.

PARTNER	Country
DATAMAT Ingegneria dei Sistemi S.p.A. (co-ordinator)	IT
Presidenza del Consiglio dei Ministri Dipartimento per la Protezione Civile	IT
Fondazione per la Meteorologia Applicata	IT
VITROCISET S.p.A.	IT
Dipartimento per i Servizi Tecnici Nazionali Servizio Sismico Nazionale	IT
Provincia di Modena	IT
Universidad Complutense de Madrid	ES
Direccion General de Proteccion Civil	ES
Centro de Investigacion y Desarrollo Agroalimentario. Region de Murcia	ES
TEUCHOS P. A. C. A.	FR
National and Kapodistrian University of Athens	GR
National Observatory of Athens	GR
General Secretariat for Civil Protection	GR
Earthquake Planning and Protection Organisation	GR

Observing our environment from Space: New solutions for a new millennium, Bégni (Ed.)
© 2002 Swets & Zeitlinger, Lisse, ISBN 90 5809 254 2

High spatial resolution image and digital elevation model data for watershed analysis

A.C.Correa, C.H.Davis, L.Peyton, H.Jiang & J.Adhityawarma
University of Missouri-Columbia, Institute for Commercialization of Remote Sensing Technology (ICREST), Columbia, Missouri, USA

ABSTRACT: The new generation of land observation satellites produces panchromatic image data with 5- and 1 meter spatial resolution that will promote the development of new remote sensing applications and will be used also to create high-resolution digital elevation models (DEMs). The Institute for Commercialization of Remote Sensing Technologies (ICREST) at the University of Missouri-Columbia evaluated some potential applications of the new datasets using simulated and actual Ikonos satellite data in the highly urbanized Jordan Creek watershed in Missouri. This paper describes the characteristics and applications of high spatial resolution land use/land cover information, DEMs and merged products used for watershed analysis.

1 INTRODUCTION

The concept of watershed analysis provides a common framework for the evaluation, use planning and management of watersheds. An important application of this concept is closely associated with the monitoring of land use activities and their interaction with ecological and hydrologic processes. In this paper the concept is applied to the evaluation and understanding of some of the important variables in flood risk analysis.

The work reported on this paper was carried out in the highly urbanized Jordan Creek watershed in the City of Springfield, State of Missouri, USA (Fig. 1).

2 WATERSHED AND FLOOD RISK ANALYSIS

Mathematical simulation models are the main tools in flood risk analysis in a watershed. Two types of models are required: hydrologic and hydraulic. Hydrologic models generate flow rates (volume of water per time) at points throughout the watershed. Hydraulic models use the flow rates as input and generate water surface elevations throughout the flood plain. Each type of model requires different spatial information.

Hydrologic models generally fall into one of two categories: deterministic or stochastic. Deterministic models rely on the spatial and physical characteristics of the watershed to simulate flow rates for various return periods, whereas stochastic models rely on statistical analysis of stream flows. It is for deterministic models that high spatial resolution image and DEM data offer enormous potential.

Hydrologic deterministic models are classified as lumped or distributed. Lumped models use spatially averaged watershed variables, generally ignoring spatial changes within sub-basins. Distributed models use watershed variables that are more closely a function of space. Both of these models can greatly benefit from remote sensing images and digital elevation models (DEMs).

Both lumped and distributed models require similar input variables but at significantly different scales. Commonly required input variables include slope, percent impervious area, vegetation, land use, soil type, travel times, as well as infiltration and storage characteristics. However, lumped models require averages of these variables over scales ranging from hundreds to tens of thousands of hectares, whereas distributed models require these variables over scales ranging from hundreds to tens of thousands of square meters—scales that differ by orders of magnitude.

Figure 1. Location of the City of Springfield, Missouri.

Distributed models have the capability of describing unique features of the watershed in more detail, leading one to hope that the results are more accurate. It is a challenge, however, to measure input variables efficiently and accurately at the required level of detail and to calibrate successfully the model with a larger number of degrees of freedom (larger number of input variables compared to lumped models).

High spatial resolution images can provide measurements of input variables for both lumped and distributed models. However, the potential impact is greater for the distributed models that require measurements over much smaller areas. Of equal significance to scale is the ability for high-resolution multispectral images to distinguish between impervious areas, land use types, and vegetation cover to a much fine degree than before. For example, we now have the potential to distinguish between types of impervious pavements, vegetation species, density of vegetation, and to improve soil maps.

DEMs also play an important role in the measurement of input variables for lumped and distributed hydrologic models. Terrain slopes, flow paths, travel times, storage characteristics and sub-basin boundaries can all be rapidly determined using DEMs.

The potential impact of DEMs can be equally or even more significant for hydraulic models, where land surface elevations throughout the flood plain are the most critical input for accurately simulating flood elevations. Hydraulic models require flood plain cross sections perpendicular to the direction of river flow at frequent intervals along the path of the river. The shapes of these cross sections determine the area available for flow and to some extent the resistance to flow. These have a large impact on the height that the water surface will reach during flood stage and therefore on the total flood inundation area. The use of DEMs acquired from airborne or space images to determine these flood plain cross sections have the potential to significantly reduce the cost of flood plain studies by reducing the field time required for survey crews to measure these cross sections.

Another sensitive variable in hydraulic modeling is the roughness coefficient, a measure of the relative resistance to flow. This coefficient varies from point to point throughout the channel and overbanks. The coefficient is influenced by many variables, most of which can be identified and mapped on high-resolution images. In the channel, type and density of channel vegetation, extent of channel debris, type of channel bed and bank material, and degree of channel meander influence the roughness coefficient. In the overbanks, this coefficient is influenced by land cover type such as vegetation (species and density), paved surfaces (area extent and type), presence of buildings and other obstructions.

3 HIGH RESOLUTION DIGITAL ELEVATION MODELS

Digital elevation models (DEMs) provide three-dimensional representations of the Earth's surface that are becoming widely used for many applications. The United States Geological Survey (USGS) is a major source of DEMs in the US. The highest resolution DEMs available from the USGS are generated from a national database of 1:40,000 aerial photographs available over 90% of the country. These aerial photographs are routinely collected every 5-6 years as part of the National Aerial Photography Program (NAPP). The USGS DEMs have a 30-m spatial resolution (x,y resolution) and a 7-15 m RMS elevation (z) accuracy (U.S. Geological Survey 1997).

The resolution and accuracy of these USGS DEMs are not suitable for the vast majority of urban applications. The spatial resolution is too coarse for urban related projects and more appropriate for natural resource management programs. The accuracy estimates for the elevation data are often representative throughout the entire DEM because only a limited number of ground control points are used in estimating the RMS vertical error.

3.1 *DEMs from airphotos*

The DEM produced for our study in Springfield was derived from digital photogrammetric stereo-processing of the 1:40,000-scale NAPP photos after being digitally scanned at 1200 dpi (pixel resolution of 0.85 m). PCI OrthoEngine™ software was used for DEM extraction from the stereo photo pairs and for production of an orthorectified mosaic of the study area (Fig. 2).

A kinematic GPS (Global Positioning System) survey carried out in the study area generated a checkpoint database with more than 50,000 positions. The GPS data were rigorously analyzed in reference to eight geodetic survey locations and 46 ground control points from a rapid-static differential GPS survey throughout the area to eliminate data of questionable quality. The results of this study indicate that measured GPS elevations are accurate to within ±10 cm (Davis & Wang 2001). The Springfield DEM produced by ICREST had 3-m horizontal resolution and an estimated vertical accuracy of ±1.8 m (Davis & Wang 2001).

A 1-m DEM was also generated for Springfield but the vertical accuracy was worse than the 3-m DEM because there fewer pixels are used in the final iterations of the stereo image matching process.

The ICREST 3-m DEM is a reasonable proxy of future DEMs that will be soon available from 1-m

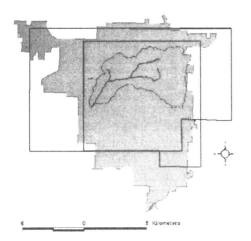

Figure 2. Outline of the Jordan Creek watershed within the limits of the City of Springfield (light gray). The polygons show Ikonos coverage (central part of the city) and the extent of the DEM mapping.

resolution satellite data. Nevertheless, the ICREST 3-m DEM demonstrates that conventional 1:40,000-scale aerial photography can be used to produce DEMs with spatial resolution and vertical accuracy an order of magnitude smaller than the USGS DEMs derived from the same data (Davis & Wang 2001).

The importance of DEMs for floodplain and flood risk analysis is very high. The official maps presently used for floodplain management in the Springfield area, known as FIRMs (Flood Insurance Rate Maps), display the topography digitized from 1:24,000 scale USGS quadrangle maps (Federal Emergency Management Agency 1996).

3.2 DEMs in floodplain mapping

Topographic maps are one of the basic datasets required for watershed analysis. Up to now the Federal Emergency Management Agency (FEMA) produces FIRMs based on USGS 1:24,000 topographic paper maps. Topographic spacing contour in those maps is 10 ft or approximately 3.05 m. Field data collected for hydrologic analysis of watersheds are plotted on these maps as well as the outlines of 100- and 500-year flood boundaries.

The availability in the near future of DEMs with 1-m spatial resolution should create the need for a new methodology to present flood risk data to the public. The work done with 3 m resolution data shows some of the advantages and limitations of the new datasets.

A significant quality improvement introduced by DEMs to the topographic information is the capability to provide elevation data for each ground parcel corresponding to an elevation data cell. A hard copy topographic map provides similar quality information only for locations on a mapped contour line. Elevation values for locations within the area between contour lines are determined by interpolation and have variable accuracy. This is not a significant problem in steep terrain where elevation values change rapidly. However, this situation is a major issue on floodplains, which are nearly flat and may cover large areas.

A limitation of the DEM dataset generated for this project is the lack of a "bare earth" terrain display. The high resolution of the data source and the reliance on automatic mapping generates a display that includes structures (buildings) and tree canopies in the DEM. Figure 3 is a 3-D display of the DEM dataset with 3 m resolution that provides a general overview of the Jordan Creek watershed and surrounding areas within the city limits. In this display isolated peaks correspond to buildings and trees in the central part of Springfield.

Figure 3. Three-dimensional display of the Jordan Creek DEM file where the Jordan Creek stream bed shows up dark gray and most peaks on the image surface correspond to buildings and large trees. Arrow shows North.

Figure 4. DEM and topographic contours (3 m spacing) and the outline of the floodway boundary (white line).

The contour map generated for the DEM dataset reproduces correctly the terrain morphology and the presence of manmade structures and natural features mentioned above. When the floodway boundaries determined by FEMA for Jordan Creek are added to the DEM contour map it is apparent that this outline could have been placed more accurately if the ICREST DEM basemap had been available to FEMA (Fig. 4).

4. HIGH SPATIAL RESOLUTION SATELLITE IMAGES

The Ikonos satellite provides panchromatic and multispectral coverage of the Earth's surface with 1- and 4 m spatial resolution respectively. The Ikonos pan image has a spectral range (0.45 – 0.90 μm) that goes beyond the visible range of standard panchromatic film. The information content of both, however, does not provide as much information for visual or automatic interpretation as color images produced by multispectral datasets.

The Ikonos produces multispectral images in four bands (0.45-0.53, 0.52-0.61, 0.64-0.72 and 0.77-0.88 μm). These bands were combined to produce natural color and infrared color images and as input to an automatic supervised classification of land cover. A maximum likelihood classifier was used with single-pixel training sets containing 100 pixels for each class. The automatic image classification resulted in five classes significant for hydrologic modeling. Vegetated areas were grouped into grasslands and woodlands. Grasslands were further subdivided into those with 50% or less grass cover (including bare ground) and those with more than 50% (average 75%) grass cover that were referred as "poor" and "good" grasslands. The woodland class included areas with a fairly dense tree cover. Water bodies were classified as part of the "water" class. An impervious class included the surfaces with concrete, asphalt and common roofing materials (mostly asphalt shingles).

The supervised classification results were validated with a commonly used accuracy assessment of the results and by doing field verification in sing selected city blocks.

The accuracy assessment of the supervised classification results was based on a simple random sampling of pixels throughout the image. The overall accuracy level obtained from the confusion or error matrix was 79 % from 250 observations. This accuracy level is representative of the confusion introduced in the classification because of the shadows in the image.

The image acquisition information provides important clues for the interpretation of the classification results. The Ikonos image used in this study was acquired on 17 September 2000, in early fall when deciduous trees still had their leaves on. At this time of the year the sun elevation at the time of image acquisition was not too high and the shadows created by buildings and trees masked part of the terrain (Table 1).

Field verification of selected areas was done in combination with the mapping of our selected land cover classes on panchromatic aerial photos using visual interpretation (Fig. 5).

Table 1. Image acquisition parameters

IKONOS	Degrees
Azimuth	215.5
Elevation	69.9
SUN	
Azimuth	151.2
Elevation	51.4

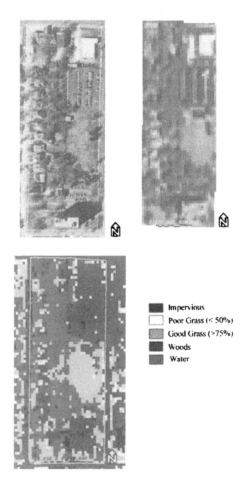

■	Impervious
□	Poor Grass (< 50%)
▨	Good Grass (>75%)
■	Woods
▨	Water

Figure 5. Airphoto (top left), Ikonos multispectral image (top right) and supervised classification result (original figure in color).

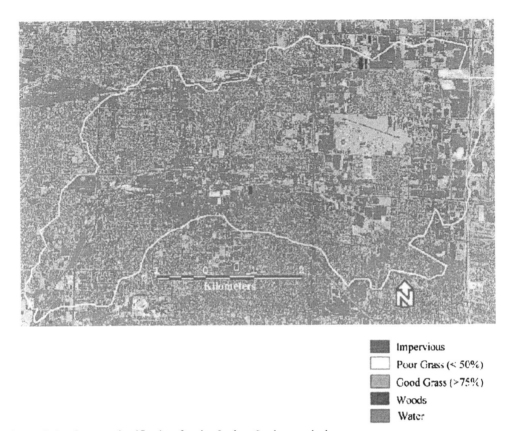

Impervious
Poor Grass (< 50%)
Good Grass (>75%)
Woods
Water

Figure 6. Land cover classification for the Jordan Creek watershed

The aerial photography available for field check had been acquired in April 1999. At that time of the year the deciduous trees had no leaves. This is probably the best situation for land cover mapping because shadows are reduced and grass and impervious surfaces that would be hidden by tree canopies are well exposed. The combined effect of large tree canopies and the shadows cast by building and trees provide the best explanation for the relatively low supervised classification results obtained for Ikonos images. It was also observed that the lack of a "shadows" class in the training dataset causes some data misclassifications.

5. INPUTS FOR HYDROLOGIC MODELS

The conventional approach to provide land cover/land use input to hydrologic models is to obtain the information from available panchromatic aerial photographs or land use maps. This information may be coupled with field observations to establish runoff curve numbers (CNs) according to the widely adopted the Soil Conservation Service (SCS 1986) procedure.

The high spatial resolution images collected by the Ikonos satellite are optional data sources with the advantages of multispectral characteristics and digital format conducive to automatic cover type classification and more up-to-date information because of the repetitive satellite coverage.

A supervised maximum likelihood classifier was used to map five land cover classes in the Jordan Creek watershed. The five land cover classes: impervious surfaces, poor-to-well and well-developed grass, mixed grass and woodland and water are representative of land cover in urban areas and significant to hydrologic studies (Fig. 6).

To specify the SCS Curve Numbers required by hydrologic models it is necessary to have the land cover estimates and the hydrologic soil groups present in the area of interest. The land cover classification and the soil map for the watershed were referenced to the same basemap using a geographic information system (GIS). This approach allows for the determination of CNs for each grid cell and accurate results of an average CN for each subcatchment area or for the entire watershed that is used as input in the hydrologic model (Fig. 7).

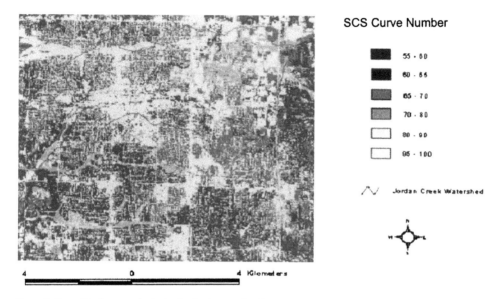

SCS Curve Number

- 55 - 60
- 60 - 65
- 65 - 70
- 70 - 80
- 80 - 90
- 95 - 100

Jordan Creek Watershed

Figure 7. Curve Number map obtained using land cover classification from Ikonos multispectral image and soil map for the Jordan Creek watershed and surrounding areas.

6. CONCLUSIONS

Work is in progress to evaluate in detail the results of CN determinations obtained with the approach described above and the usual methodology. One of the results with immediate application is the knowledge that in some situations, already available airphoto coverage may be used as an alternative to produce high quality digital DEMs to improve existing official digital topographic maps. This may represent savings in time and money for some users.

7. ACKNOWLEDGMENTS

This study was funded by a grant from the National Aeronautics and Space Administration (NASA). The Space Technology Applications and Research laboratory of the Center for Environmental Technology, Institute for Commercialization of Remote Sensing Technologies (ICREST) at the University of Missouri-Columbia performed the work.

Thanks to Harold E. Johnson III and to D. Scott Adams for providing support work to our research.

8. REFERENCES

Davis, C.H. & Wang, X 2001. High resolution DEMs for urban applications from NAPP photography. Photogrammetric Engineering and Remote Sensing 67(5): 585-592

Federal Emergency Management Agency 1996. Compliant metadata for Q3 flood data coverage. Federal Emergency Management Agency, Washington, DC.

Soil Conservation Service (SCS) 1986. Urban Hydrology for Small Watersheds. Water Resources Publications, Technical Release No.55 (TR-55).

United States Geological Survey 1997. Standards for digital elevation models, Part 1: General; Part 2: Specifications; Part 3: Quality Control. Department of the Interior, Washington, DC.

Observing our environment from Space: New solutions for a new millennium, Bégni (Ed.)
© 2002 Swets & Zeitlinger, Lisse, ISBN 90 5809 254 2

Detection of storm losses in Alpine forest areas by different methodical approaches using high-resolution satellite data

M.Schwarz, Ch.Steinmeier & L.Waser
Swiss Federal Research Institute WSL, Birmensdorf, Switzerland

ABSTRACT: Based on the detection of storm losses in Swiss alpine forest areas, two different digital classification approaches were compared. In contrast to the pixel based classification we investigated an object-oriented classification procedure. The eCognition software package of Definiens offers this possibility. The comparison was performed for images with different spatial resolution - very high resolution images of IKONOS, and images of SPOT in the sharpened mode. The evaluation of the IKONOS image indicated a significantly higher accuracy for the object-oriented classification approach than for the pixel-based method. The eCognition software handles the high level of detail and the associated high texture better than the pixel-based parallelepiped-algorithm. The quality of the pixel-based approach, which takes into account only the spectral information and some derived data-products is limited for very high resolution images. The classification of the SPOT presented approximately the same results for both methods.

1 INTRODUCTION

In Switzerland thirty-one percent of the area are covered by forests which are of vital importance and play multifunctional roles. Forest management as well as forest monitoring with remote sensing data have a long tradition in Swiss research work (Kellenberger, 1996). Especially as a result of the hurricane Lothar on 26.12.1999 the public interest in forest management has increased.

During the last decades the classification of forest was the focus of many remote sensing supported investigations. The spatial resolution of the available systems (Landsat, SPOT, IRS) ranged between thirty and twenty metres for multispectral data. The new generation of satellite-systems, such as the spaceborne IKONOS system, offers a spatial resolution of between one and four meters which allows on the one hand new perspectives in many applications but on the other needs more basic investigations (Manakos *et al.*, 2000).

Most classification approaches are based exclusively on the digital number of the pixel itself. Thereby only the spectral information is used for the classification. Limitations arise through this way of thinking, especially for the new sensor generation (Steinocher, 1999). The images show a very high level of detail and are very strongly textured (Manakos *et al.*, 2000). A single tree in a Landsat image for example appears as a homogeneous object. The same object in an IKONOS image is represented as several pixels with different reflections. Due to this, homogeneous objects are not only characterised by their spectral signature but also by the texture and their local context.

In order to overcome these limitations, an object oriented way of data assessment is adopted in new image analysis methods. The eCognition software promotes this new perspective (de Kok *et al.*, 2000). With this new software product not single pixels are classified but homogeneous image objects are extracted during a previous segmentation step. This segmentation can be done in multiple resolutions, thus allowing to differentiate several levels of object categories eCognition, 2000).

The objective of this study was to test a pixel-based approach against the object-based classification approach by eCognition for the detection of storm losses in alpine forest regions. Classifications were performed for two different satellite systems representing different spatial resolutions, SPOT and IKONOS. Another objective of this project was to extract rules and needs for a successful classification of storm losses.

2 MATERIALS AND METHODS

2.1 Testsite

The testsite is located in the area of Bern in Switzerland between 7°42'03"E and 7°50'48"E longitude and 46°40'55"N and 46°52'15"N. The main part belongs to both the alpine and the prealpine region. The altitude elevates between 555 m (Thunersee) and 2060m (Gemmenalphorn). According to this difference in elevation different plant societies exist, corresponding the altitude.

2.2 Satellite data

One IKONOS image was at our disposal collected at August, 12th 2000. The spatial resolution is 4 m for the multispectral channels and one meter for the panchromatic band. For the classification only the multispectral bands blue, green, red, and near infrared were used. Furthermore we could get a SPOT image, recorded at July, 2nd 2000. The processing of this SPOT image was done as "pansharpened" which results in an image with a spatial resolution of 10m. Before the merge with the panchromatic band, the original spatial resolution of the multispectral bands were twenty meters. The image was collected by the SPOT4 serie including green, red, near infrared, and middle infrared bands.

2.3 GIS-Data

In addition to the satellite data the forest areas of the national map 1:25'000 were available. This thematic layer was an important input into the classification process, and was resampled on the spatial resolution of the satellite data.

2.4 Groundtruth

The groundtruth is based on the visual interpretation of aerial images, collected shortly after the hazard. The interpretation of the damages was performed by several independent engineering companies. The groundtruth was resampled on the spatial resolution of the satellite data.

2.5 Software

The object-oriented classifications were performed with Definien's software product eCognition. On the other hand the ERDAS Imagine was used as well for the pixel-based approach as for the accuracy assessment and all image pre- and processing steps. The classifications were performed with ERDAS Imagine's expert-classifier and the accuracy assessment was realised with Modelmaker.

2.6 Image Processing

All satellite images were corrected geometrically and georeferenced to the national map 1:25'000. The images were resampled using the nearest neighbour algorithm. This georeferencing was necessary to eliminate the topographic effects and to connect the satellite data with the GIS-data described in 2.3 which is used for the classification.

2.7 Pixel-oriented classification method

The measured electromagnetic energy per pixel serves as the base for the pixel-oriented multispectral classification. Each pixel is characterised by a special reflectance in the multidimensional feature space (spectral signature). Due to this spectral signature the pixel will be associated to a certain class (Leiss, 1998). In contrast to a visual image-interpretation it is not possible to take into account contextual or topological information.

In this investigation the parallelepiped classification algorithm has been chosen representing a supervised pixel-oriented classification approach. The parallelepiped algorithm is based on a deterministic approach. The classification of an object-class is set by rectangular "regions of decision" in the multidimensional feature space. Each class is explicitly defined by a minimum and maximum value for one or several bands. The pixel is assigned to a class only as a result of the spectral information (digital number) for the different spectral bands. For the extraction the most suitable bands has been chosen. For further detailed information see also (Kraus et al., 1988). The selection of the bands was based on the maximum separability. As a result of the training samples a histogram was generated for each class. Further the exact value of the separation was computed, by the maximum difference of the relative cumulative frequency of each class (Kellenberger, 1996). This means that only spectral information was taken into account in the process of classification. The entire classification process can be summarised as follows.

1st step
Differentiation of forest areas from the rest on the base of the forest mask (see also 2.3)

2nd step
Selection of training-samples which describe the different object-classes

3rd step
Selection of the most suitable bands and computation of the parallelepiped values

4th step
Building the classification rules

5th step
Classification of the entire image

6th step
Accuracy assessment

2.8 *Object oriented*

The eCognition software allows an object-oriented classification. The basic difference, especially when compared to pixel-based procedures, is that object-oriented classification methods does not classify single pixels but rather image objects, which are extracted in a previous image segmentation step. eCognition allows the segmentation into highly homogenous object primitives in any chosen resolution. These object primitives represent the image information in an abstract manner. Beyond the spectral information many other additional attributes can then be used for classification: shape information, texture information, relations to neighbouring objects and a good deal more. The basic part of the software is the previously mentioned segmentation.. It was developed to extract image objects in optional resolutions and high quality (eCognition, 2000; Baatz et al., 2000).

eCognition supports supervised classification-techniques and provides different methods to train and build up a knowledge base. The frame of eCognition's knowledge based classification of image object is the so called class-hierarchy. This class-hierarchy contains the classification rules to which the image will be classified. Each class is defined by a class-descriptor which offers a lot of different parameters and object attributes. This way the forest-mask can be incorporated into the classification rule for example (eCognition, 2000; Baatz et al., 2000). The entire classification process is summarised in Figure 1.

1st step
All bands including the thematic layer were segmented in different levels representing the level of detail.

2nd step
Definition of the class-hierarchy and class-description. The distinction between forest and non-forest areas is based on the thematic layer containing the forest mask. The object class forest area was subdivided into the classes possible forest and possible forest damage. Training-samples for both subclasses were gathered.

3rd step
Based on the classification hierarchy the entire image was classified. The correlation to one of the subclasses was performed by a nearest-neighbour func-

tion which was calculated on the base of the training-samples.

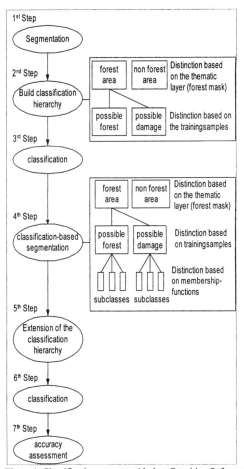

Figure 1. Classification process with the eCognition Software

4th step
Further a classification based segmentation was performed. In many cases image objects of interest cannot be extracted following a relatively general homogeneity criterion. For this reason, eCognition provides techniques for classification-based segmentation of image objects, merging neighbouring objects of the same structure group. The structure groups were built with the desired classes that were defined in the class hierarchy. All contiguous objects belonging to the same object class were merged to super objects. The advantage of this step presents the user driven segmentation. Figure 2 illustrates a part of the original IKONOS Image compared with the same part after a classification based segmentation illustrated in Figure 3. As a result of this step a new segmentation level was obtained.

253

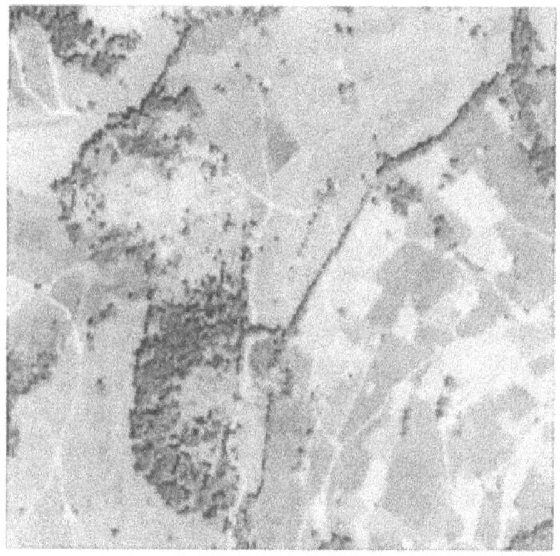

Figure 2. IKONOS Image (4m resolution)

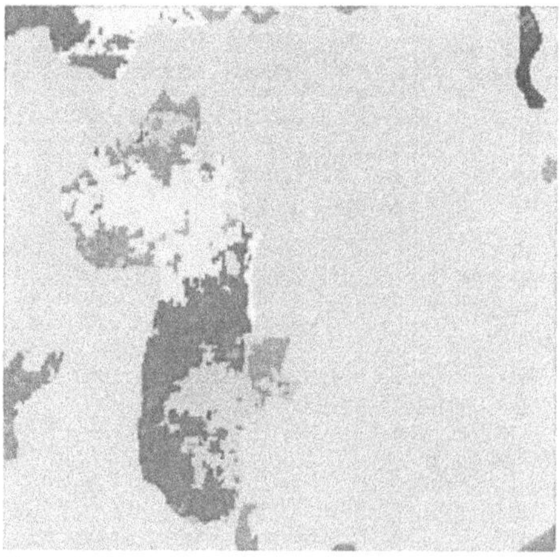

Figure 3. IKONOS image after classification based segmentation

5th step
This new level was reclassified by defining a new class hierarchy. The classification was enhanced by using membership functions. The Membership functions allow the formulation of knowledge and concepts. They offer relationship between feature values and computed fuzzy values. The use of membership functions is recommended if a class can be separated from other classes by one or several features.

6th step
The classified image was imported into ERDAS Imagine where the accuracy assessment was performed.

2.9 *Accuracy assesment*

The classifications were compared to the groundtruth which is described in chapter 2.4 to assess the quality. An error matrix was computed based on the pixel to pixel comparison. Due to the error matrix, the following accuracy-parameters were calculated in order to obtain more reliable comments about the quality of the classification (Kellenberger, 1994).

Overall accuracy:
The sum of all correct classified pixels divided by the sum of pixels in the entire classification.

Producer accuracy:
The sum of all correctly classified pixels that belong to the class (x) divided by the sum of pixels in the groundtruth that belong to the class (x).

User accuracy:
The sum of all correctly classified pixels that belong to the class (x) divided by the sum of pixels in the classification that belong to the class (x).

Inclass accuracy:
The sum of all correctly classified pixels belonging to the class (x) divided by the sum of pixels belonging to the class (x) which are minor or surplus classified.

3 RESULTS

In this chapter, the results of the classifications are described. The creation of a difference-map proved to be a very useful tool for the analysis of the incorrectly classified pixels with respect to their spatial distribution. Therefore a difference map was calculated for each classification. Resulting classes were: correctly classified forest, correctly classified damage, surplus-classified forest, minor-classified forest. In the following figures each class is represented by the color described in Table1.

Table 1. Colortable for the difference-map

class	color
Not forest	white
Correctly classified forest	light grey
Surplus classified forest	dark grey
Minor classified forest	grey
Correctly classified damage	black

3.1 *Pixel-oriented classification results*

Table 2 presents the most important results of the accuracy assessment. For the IKONOS image the accuracy was slightly better as for the SPOT image. The IKONOS classification obtained an inclass-accuracy of 1.49 for the object class damage, compared to 1.31 for the SPOT classification. Within the testsite 43.5ha of damage-areas were found in the IKONOS image and 42.5ha in the SPOT image, compared to 44.1ha in the groundtruth. For both classifications the user accuracy and the producer accuracy were quite the same. This indicates that the classification was satisfying balanced. The best feature extraction was achieved by the green, red and near spectral channels. The derived Normalized Difference Vegetation Index (NDVI) could improve the results slightly but not significantly. The results shown in the table are the results without using the NDVI.

Table 2. Accuracy assessment of the pixel-oriented classification

Setting	SPOT		IKONOS	
	forest	damage	forest	damage
area [ha]	546.9	42.5	542.9	43.5
inclass-accuracy	22.26	1.31	24.20	1.49
user-accuracy	0.98	0.74	0.98	0.76
producer-accuracy	0.98	0.71	0.98	0.74
overall-accuracy	0.96		0.97	

3.2 Object-oriented classification results

In Table 3 the results of the accuracy assessment for the object oriented classification are shown. The accuracy for the IKONOS image was significantly better than for the SPOT image. The calculated inclass-accuracy for the object class damage is 1.29 for the SPOT image compared to 2.16 for the IKONOS image. Within the testsite 44.7ha of damages were found in the IKONOS image and 40.9ha in the SPOT image compared to 44.0ha in the groundtruth. For both images the user-accuracy and the producer-accuracy were quite the same, which means that the classification was satisfying balanced. The best feature extraction was achieved by the green, red and near spectral channels. The derived Normalized Difference Vegetation Index (NDVI) could improve the results slightly but not significantly. The results shown in the table are the results without using the NDVI.

Table 3. Accuracy assessment of the object-oriented classification

Setting	SPOT		IKONOS	
	forest	damage	forest	damage
area [ha]	548.9	40.5	541.8	44.6
inclass-accuracy	22.45	1.29	31.98	2.16
user-accuracy	0.75	0.98	0.81	0.99
producer-accuracy	0.69	0.98	0.81	0.98
overall-accuracy	0.96		0.97	

3.3 Pixel-oriented versus object-oriented approach (SPOT)

Both, the pixel-oriented and the object-oriented classification approaches revealed similar results with comparable accuracy. The inclass-accuracy varied from 1.31 for the pixel-oriented to 1.29 for the ob-

ject-oriented classification method. The sources of errors were well recognised by analysing the difference images. Most popular errors appeared at the borders of the damages. It seems that there still is a problem in the geometric accuracy of the georectification. Although all data were georeferenced a certain inexactness in the geometric position implicates some errors. Further a few small damages were not recognised which is due to the spatial resolution. Other errors are found in steep slopes, west to north-east oriented. These errors are mainly based upon the low solar radiation in those areas. A similar problem was detected by the shadows of the healthy trees which affect the areas located in the south of a damage section. Figure 4–5 show the Difference-Map for a part of the testsite with the object- and the pixel-oriented classification.

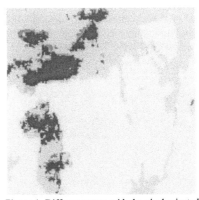

Figure 4. Difference-map with the pixel-oriented classification for the SPOT image (for color explanation see Table 1)

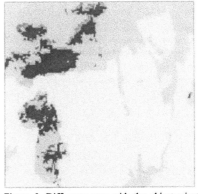

Figure 5. Difference-map with the object-oriented classification for the SPOT image (for color explanation see Table 1)

3.4 Pixel-oriented versus object-oriented approach (IKONOS)

The accuracy-assessment reveals a significant difference. The inclass-accuracy for example varies from 1.49 for the pixel-oriented approach to 2.16 for the object-oriented classification method. In addition to the errors described in the previous section, some misclassifications are implicitly due to the higher spatial resolution of the IKONOS data with its increasing level of detail and texture. Trees or shrubbery are sometimes within obviously damaged areas. The digital number of these pixels are the same values as of the vital vegetation but in fact these objects should be grouped to the damage class. This problem can not be handled by the pixel-oriented approach but by the object-oriented method. A similar problem was recognised for unclenched forest stands: Figure 6 shows the noisy pixel-oriented classification in contrast the homogenous result with the eCognition software and the right hand side shown in Figure 7.

Due to the geometric inexactness some missclassifications along streets or along the border of forest with the pixel-oriented method were found (Figure 8). These missclassifications could avoided with the eCognition software as Figure 9 illustrates.

4 DISCUSSION AND OUTLOOK

In this study two different classification approaches were tested for satellite systems with different spatial resolutions. The investigation was based on the detection of storm losses in alpine forest areas.

The classification of the SPOT image shows that both, the pixel-oriented and the object-oriented classification approach revealed similar results and comparable accuracy. Both methods are available within software packages suitable for operational application in business or research work. As long as the spatial resolution ranges between ten and thirty meters no significant difference can be detected for the overall-accuracy.

Classifications of the IKONOS image show that the usage of eCognition results in a significantly higher accuracy. The pixel-based approach reached its limits of operational use because of the internal variance and the level of texture of the spectral image information. For this reason the pixel-oriented approach is not very helpful not only in terms of detection of recent storm losses but also of land-use

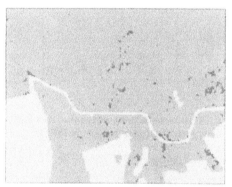

Figure 6. Difference-map performed by the pixel-oriented-approach for the IKONOS image (for color explanation see Table 1)

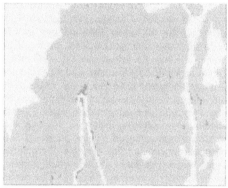

Figure 8. Difference-map performed by the pixel-oriented-approach for the IKONOS image (for color explanation see Table 1)

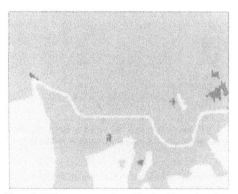

Figure 7. Difference-map performed by the object-oriented-approach for the IKONOS image (for color explanation see Table 1)

Figure 9. Difference-map performed by the object-oriented-approach for the IKONOS image (for color explanation see Table 1)

classification in general. To overcome this problem, an object based analysis should be preferred. Moreover the eCognition software allows good integration of GIS- and remote sensing data. With the possibilities to implement additional information about the geographic position, orientation and the relationship to neighbouring objects a better classification is guaranteed. The expert knowledge can be incorporated and the implicit richness of information can be fully exploited.

Further on a few general perceptions in terms of detection of recent forest losses could have been set up. A satisfying classification is only assessed by including external GIS-data for the rough boundaries of forests.. Areas with recent forest damage may have the same reflectance as parts of urban areas or bare agricultural land. The green, red and NIR are the spectral bands which are well suited for this kind of classification. In some cases the use of the normalised difference vegetation index NDVI could be a helpful additional information. For a proper classification result the satellite images should be collected during the vegetation period from mid May to mid October, otherwise deciduous forest will not be classified well enough.

For a more complex land-use classification the possibilities offered by the eCognition software will surely improve the results. The user friendly way to include spatial, contextual and internal object information will be the standard in image interpretation software for the future.

5 REFERENCES

Baatz, M. & Schäpe, A. 2000. Multiresolution Segmentation - an optimization approach for high quality multi-scale image segmentation. Proceedings, Angewandte Geographische Informationsverarbeitung (AGIT) XII. Salzburg: AGIT

Darvishsefat, A. 1995. Einsatz und Fusion von multisensoralen Satellitenbilddaten zur Erfassung von Waldinventuren. Dissertation, Remote Sensing Series Vol.24. Zürich: Remote Sensing Series

De Kok, R. & Buck, A. & Schneider, T. & Ammer, U. 2000. Analysis of image objects from VHR imagery for forest GIS updating in the Bavarian Alps. ISPRS 2000, Vol. XXXIII, Working Group III/5, Amsterdam: ISPRS

Ecognition, 2000. Ecognition User guide, München: Definiens

Kellenberger, T. 1996. Erfassung der Waldfläche in der Schweiz mit multispektralen Satellitenbilddaten. Dissertation, Remote Sensing Series Vol.28. Zürich: Remote Sensing Series

Kraus, K. & Schneider, W. 1988. Fernerkundung, Band 2, Auswertung photographischer und digitaler Bilder. Bonn:Dümmler Verlag

Manakos, I. & Schneider, T. & Ammer, U. 2000. A comparison between the ISODATA and the eCognition classification methods on basis of field data. ISPRS, Vol. XXXIII, Supplement CD. Amsterdam: ISPRS

Leiss, I. 1998. Landnutzungskartierung mit Hilfe multitemporaler Erdbeobachtungs-Satellitendaten. Dissertation. Zürich: Geographisches Institut Zürich, Switzerland

Steinocher, K. 1997. Texturanalyse zur Detektion von Siedlungsgebieten in hochauflösenden panchromatischen Satellitenbilddaten. In Dollinger, F. & Srobl, J. Proc. of Angewandte Geographsche Informationsverarbeitung, IX, Salzburger Geographische Materialien, No. 26. Salzburg:

Observing our environment from Space: New solutions for a new millennium, Bégni (Ed.)
© 2002 Swets & Zeitlinger, Lisse, ISBN 90 5809 254 2

Natural disaster monitoring by remote sensing in Hungary: waterlogging and floods in the 1998-2001 period

M.Lelkes, G.Csornai & Cs.Wirnhardt
Remote Sensing Centre, Institute of Geodesy, Cartography and Remote Sensing (FÖMI), Budapest, Hungary

ABSTRACT: Natural disasters management can substantially be helped by remote sensing in many areas. Not only the objective documentation of the disasters can be done through the utilization of satellite data in many cases, but this technology also provides information for the planning, during the prevention, preparation period and also in the most serious time of the mitigation. Excessive and lacking water, that is flood, waterlog and draught are quite frequent in Hungary, but in the past decade their occurrence was remarkably higher. This paper shows examples, how remote sensing techniques can operationally be used in the combat, mitigation management at local and national levels.

1 INTRODUCTION

In the past decade weather extremes occurred more frequently in Hungary. After the years of draught (1992-93) waterlog and flood (1998-2001) damages caused severe losses to the whole national economy. The serious damages caused and the disasters themselves highlighted a few well-known facts. One of these is that the tasks of protection and prevention can only be planned and performed within the framework of an adequate, scientifically and technically sound multidisciplinary system. The other is that in accurately, rapidly and objectively recording and documenting the current status and the processes of the Earth surface at local and at the same time at regional level, remote sensing is an essential, unique tool which is much more efficient than traditional methods.

2 BACKGROUND

During the Hungarian Agricultural Remote Sensing Program (HARSP 1980-) 300 man years R&D was invested by FÖMI Remote Sensing Centre (FÖMI RSC) to develop the methodology of crop production forecast by remote sensing. The original final objective of the program was to introduce remote sensing to the operational agro information system in Hungary. The operational system monitors crops in the entire country, providing accurate, timely and reliable information on the area of the major crops, their development quantitatively, plus reliable yield forecast and final yield estimates in the counties (9)

and for the entire country. The HARSP program's two main periods were: the development of the methodology basis, the crop mapping and area assessment methods plus the yield forecast models and validation (1980-96) and the operational period (from 1997 to date) (Csornai 1999).

The operational satellite based National Crop Monitoring and Production Forecast program (CROPMON) makes the implementation of other monitoring programs possible and very cost effective applying the same data, infrastructure and know-how bases at the FÖMI RSC.

3 OPERATIONAL WATERLOG MONITORING AND IMPACT ANALYSIS WITH REMOTE SENSING

For mapping and monitoring waterlog and for supporting the related necessary measures to be taken (planning of waterlog canals, compensation for waterlog losses, etc.) remote sensing is an extremely good tool. The different satellite images that can be used are different and complement each other with respect to spatial, temporal and spectral resolution and price. Their multi-level system can be used in monitoring and assessing waterlog and flood, giving the possibility of rapid, country-level overview and detailed field level survey.

3.1 *Operational remote sensing based waterlog monitoring in 1998 and 1999.*

Three levels of products were developed in waterlog assessment carried out on CROPMON basis in 1998

Figure 1. Spatial coverage of high resolution satellite images used to derive waterlog maps in spring 1999.

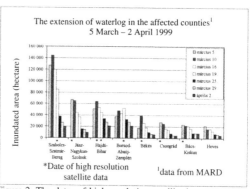

Figure 2. The dates of high resolution satellite images used in waterlog assessment were close to the peak inundation periods in the counties.

and 1999, using the different available satellite data sources:

1 Low resolution (120 hectares) waterlog maps, giving country-wide overview on a daily basis by fast information extraction from NOAA AVHRR satellite images
2 Medium resolution (3.6 hectares) country- and county-level waterlog and waterlog change maps, together with the areas of the different waterlog categories in the counties, derived from IRS-1C/1D WiFS satellite data.
3 High resolution (0.1 hectares) country-, county- and field-level waterlog, waterlog change and impact maps based on Landsat TM, IRS-1C/1D LISS-III satellite images.

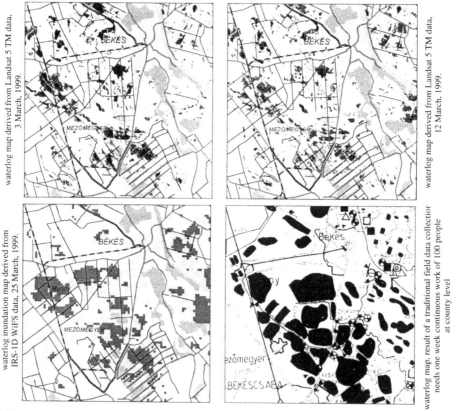

Figure 3. These high-resolution (upper two images) and medium resolution (lower left) waterlog maps illustrate that the satellite images based waterlog maps are more detailed, more accurate than the results of the traditional field data collection (lower right) performed by the local water management authorities.

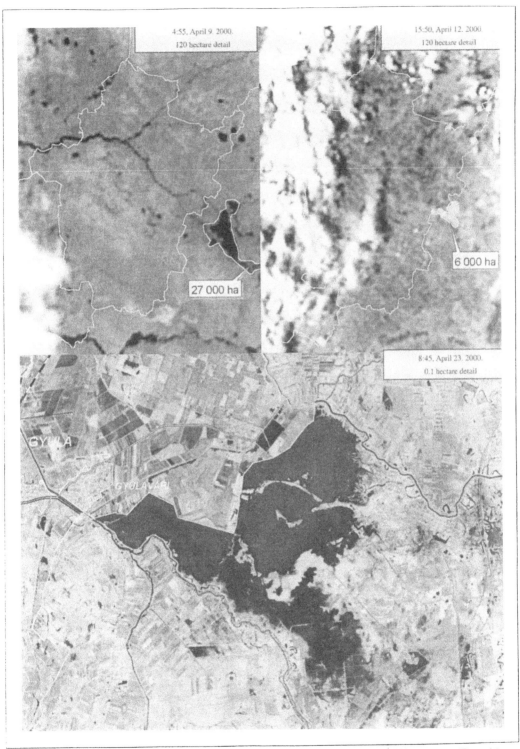

Figure 4. The Water Management of Körös Region augmented information on the flood in the neighbour country (Romania) contacted FÖMI RSC. Using the satellite based flood-maps they could estimate the coming rise or decrease in the flood wave highs. Stilted the old dam on the border they could protect a part of the Hungarian cropland

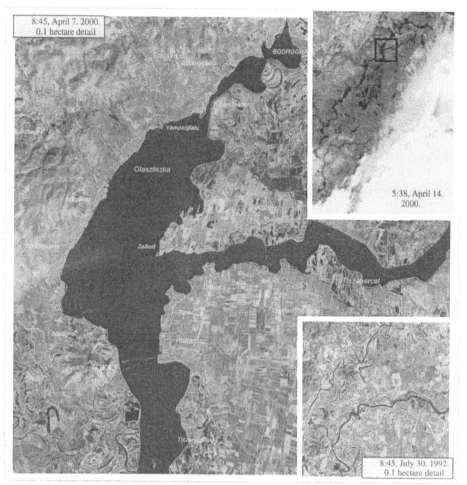

Figure 5. With the daily time series of NOAA AVHRR data based flood maps the Regional Water Management network could follow the changes on the catchment of Tisza river . With the less frequent high resolution satellite data FÖMI Remote Sensing Centre (FÖMI RSC) objectively documented the flooded and waterlogged areas. The contrast between the flooded and dry (lower right) period is obvious.

The operational remote sensing based waterlog mapping was established in 1998. After the heavy precipitation in April-May 1998, waterlog was mapped at 0.1 hectare detail for 4 counties, giving the areas of the different waterlog affected categories at county level. The effect of waterlog on arable crops (crop kills) were also detected and monitored.

The real-time monitoring of spring waterlogs in 1999 was performed for the Ministry of Agriculture and Regional Development (MARD) at all the three levels mentioned above for 7 counties of Hungary, east of river Danube (Fig. 1). The dates of satellite images used fitted the peak inundation period (Fig. 2).

Beyond the static status assessment of the areas under water or having saturated soils, impact analysis on the crops and the dynamism of changes could also be monitored quantitatively. The resulted GIS data base and printed maps were utilized by MARD intensively and proved to provide fast, operational information for decision makers. Figure 3 shows examples of the high- and medium resolution waterlog maps as well as the map that was derived by traditional field inspection methods at the local water management authorities.

The remote sensing technology was applied in two ways: to collect reliable statistics and maps for the affected vast areas, plus at the scale of individual parcels. At one hand, almost half of the country was monitored and on the other hand remote sensing was used as a control tool in the waterlog loss compensation program, launched by the government. This second program required very accurate, parcel and owner/user specific procedures, surveys and documentation (Csornai et al. 1999)

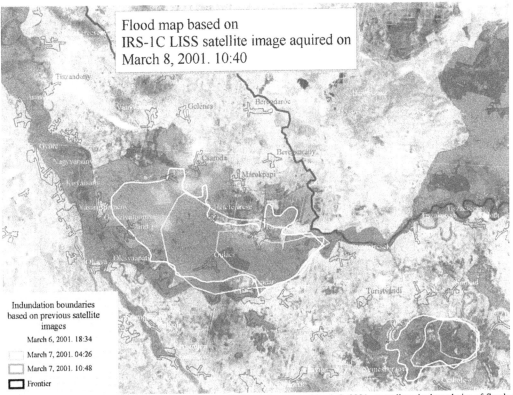

Figure 6. This high resolution flood map shows the actually flooded areas on March 8, 2001, as well as the boundaries of flooded areas based on previous high-, medium- and low resolution IRS LISS, IRS WiFS and NOAA AVHRR data.(Colour plate, see p. 426)

3.2 *Flood monitoring in 2000 and 2001*

On April 9, 2000 the Hungarian government declared parts of eastern Hungary a disaster area. The combat against floods had already begun earlier along Hungary's second largest river, the Tisza, and its several tributaries. Large tracts of land were flooded, destroying many homes and threatening hundreds of others. Hundreds of residents were evacuated. The Minister of Transport, Communications and Water Management was appointed government commissioner at a Cabinet session (April 8) to oversee and coordinate all efforts. About 1,700 miles of dikes had to be reinforced to contain the floods. Hungary's armed forces were placed on alert in order to mobilize quickly if necessary. The water level of the Tisza surpassed previous record highs.

Immediately after the declaration, FÖMI Remote Sensing Centre (FÖMI RSC) began to start operations with its available resources to help the combat providing real time satellite data for the disaster area. Each day as soon as the NOAA data were recorded by the FÖMI RSC's receiving station they got processed, analyzed and transmitted directly to the Ministry of Transport, Communications and

Water Management, to the Regional Water Management network headquarters in the most threatened areas plus to the Ministry of Agriculture and Regional Development. This information and quick estimation of the flooded areas were very much acknowledged by the institutions involved in disaster management and by the experts in their actions and planning the tasks during the prevention. In some cases they could estimate the coming rise or decrease in the flood wave highs (Fig. 4). The reliable and quantitative information helped the flood management in making the necessary intervention steps more adequately and in time. This helped in saving some financial and other resources. FÖMI RSC could electronically transmit these flood-maps within 2 hours from the reception.

Even the slower and less frequent high resolution data were acquired and processed relatively very quickly to monitor and document the flood (Fig. 5).

Spring 2001 unfortunately also started with serious flood situation along the upper part of Tisza river. The sudden snow melt and the heavy rains (more than 120 mm precipitation in two and a half days) caused extreme flood situation by 5th March..

In spite of the efforts of the responsible authorities that were put on alert, at the first time after 53 years, the dike along river Tisza was breached by the water at midday on 6th March. The water flooded the neighboring areas through a 120 m wide gap threatening tens of villages and thousands of people.

On the morning of 6th March 2001, FÖMI started its operational flood monitoring using NOAA AVHRR satellite images acquired by its own receiving station and followed the availability of other medium and high resolution satellite images for the region. The extent of flooded areas was evaluated both on the Ukrainian and the Hungarian side and the high-, low- and medium resolution flood maps were forwarded to the central and local management authorities through electronic transmission.

3.3 Conclusions

These terrible disasters clearly pointed out the potential of the satellite based information. FÖMI RSC could mobilize its operational capabilities that had been developed and utilized in the National Crop Monitoring and Production Forecast Program (CROPMON) that has been operational for the 5th years now. Monitoring the waterlog, its progress and impact on crops can quantitatively be done by remote sensing from local to national and regional scale. Moreover remote sensing can successfully be used at the parcel specific disaster compensation program for the control of claims.

Although the examples at one way prove the usefulness of remote sensing during the mitigation of flood, these techniques have much greater potential in revealing the reasons of these phenomena, related to the changed land use and deforestation in the river catchment. These studies can contribute to the improvement of flood forecast.

REFERENCES

Csornai, G. 1999. Operational crop monitoring by remote sensing in Hungary. In G.J.A. Nieuwenhuis, R.A.Vaughan & M.Molenaar (eds), *Operational Remote Sensing for Sustainable Development*: 89-95. Rotterdam: Balkema

Csornai, G., Wirnhardt, Cs., Suba, Zs & Lelkes, M.1999. Area based subsidy and compensation control by remote sensing in Hungary: as a CROPMON application. Poster presentation on the 5th Conference on Control with Remote Sensing of Area-based Subsidies, 25-26 November 1999, Stresa, Italy.

Observing our environment from Space: New solutions for a new millennium, Bégni (Ed.)
© *2002 Swets & Zeitlinger, Lisse, ISBN 90 5809 254 2*

Seismotectonic activity in Dinaric Alps based on satellite data and geophysical survey

M.Oluic, D.Cvijanovic & S.Romandic
Zagreb, Croatia

ABSTRACT: The area of Dinaric Alps (Dinarides) has been analysed on Landsat TM images. Many of the faults have been registered and categorised into three classes: 1st > 50 km in length, 2nd 10 -50 km and 3rd < 10 km. The majority of the determined faults are from the neotectonic age and they have special significance for seismotectonic activities. The region mentioned has been geophysically researched (deep geoelectric, geomagnetic, gravity), too. Based on both kinds of data, the tectonic complex of the investigated area has been interpreted. The epicentres of stronger earthquakes occurring and registered during the last two centuries have been drawn into the tectonic map. The epicentres of the earthquakes appeared mostly in the tectonically active zones, so it was possible to separate the potential , seismic areas.

1 INTRODUCTION

The area of Dinaric Alps (Dinarides) was analysed on images obtained by the Landsat TM satellite. The analysis was made on images from different spectral bands, by both visual and computer processed techniques. The main objective of the analysis was to obtain important tectonic elements, those of the regional significance in particular, especially the fractures (faults) and ring structures.

Some parts of the mentioned region was geophysically researched (deep geoelectrical sounding, gravity and geomagnetic measurements). In that way some deep faults were recognised and connected to earthquake epicentres, too. Satellite data, i.e. faults and ring structures, and geophysical data were interpreted and connected with seismic activity. The epicentres of stronger earthquakes occurring and registered during the last two centuries were also drawn into a tectonic map. They appeared mostly in the tectonically active zones, which are predominantly linked to regional faults and to the places where the faults of different directions cross each other, or the zones between two or more ring structures. The comparison of many data obtained by different methods of investigation and their interpretation allowed us to discover the relationship between tectonic and seismic activity.

2 TECTONICS BASED ON SATELLITE DATA

The Dinaric Alps represent a part of the Alpine Orogen and the Pannonian massif. There are several large geotectonic units obvious. The most significant are: the Dinaric Alps or Dinarides with high Alpine relief, composed mostly of carbonate rocks -karst area (outer Dinarides); mountains of medium and lesser height, divided by depressions and river valleys, composed of different sediment-magmatic and metamorphic rocks (inner Dinarides); the Pannonian Basin with flat ground and isolated mountains, and finally the Adriatic Basin and its off-shore area. These geotectonic units are completely different in geologic composition and tectonics, but their definite tectonic shape was obtained in the course of Alpine Orogeny.

Neotectonic movements provided for the final structural elements, and had a decisive influence on the contemporary morphological terrain composition. Therefore the evolution and origin of tectonic elements was influenced by the paleo-tectonic movements during a long period of time, with structures shaped in the course of the Alpine Orogeny, and by the neotectonic movements, which took place in the final phase of the Alpine Orogeny (Neogene and Quaternary).

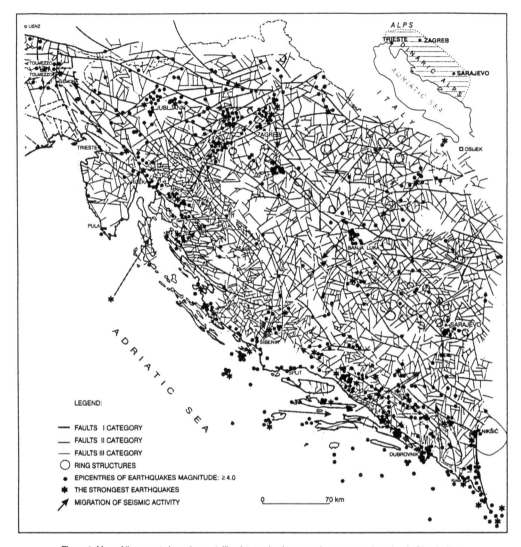

LEGEND:

— FAULTS I CATEGORY
— FAULTS II CATEGORY
— FAULTS III CATEGORY
○ RING STRUCTURES
• EPICENTRES OF EARTHQUAKES MAGNITUDE: ≥4.0
✳ THE STRONGEST EARTHQUAKES
↗ MIGRATION OF SEISMIC ACTIVITY

0 _____ 70 km

Figure 1. Map of lineaments based on satellite data and epicentres of stronger earthquakes in Dinaric Alps

Neotectonic movements are of particular significance for this research, as they conditioned the creation and construction of the recent relief, which was formed by horizontal and vertical faulting. Longitudinal large fractures are of particular importance, because tectonic activities of varying intensities are almost permanently taking place along them. Such younger dislocations are essential for the seismic activity and they are very well shown on the satellite images [1].

Fractures. In explored area numerous fractures have been registered on the Landsat TM images (Fig.1). The majority of them were followed in greater length for the first time, and precisely located. According to the length of stretching and their intensity, they have been classified into several categories [2]:

- faults of the first category stretch from 50 km to a few hundred kilometres,
- faults of the second category are the ones approximately 10–50 km long,
- faults of the third category are not longer than 10 km.

The majority of the first category fractures stretch longitudinally in the NW -SE direction. Those fractures are predominantly old dislocations reactivated by neotectonic movements, while transversal faults which are of the NE -SW and N-S directions, are younger than the previous ones (the paleotectonic fractures can be more easily recognized on satellite images).

The most prominent longitudinal fractures of the first category were registered along the Adriatic coast (outer Dinarides) and on the Sarajevo -Banja Luka -Zagreb -Ljubljana -Tolmezzo (interior Dinarides) line.

Some of these fractures can be traced in a length exceeding 350 km, e.g. Banja Luka - Zagreb - Ljubjana - Tolmezzo fault, or Knin - Gemona fault (Fig. 1). Those faults are intersected by several transversal and diagonal faults, particularly evident in the vicinity of Gemona, Zagreb, Bnaja Luka, Knin, Nikšić, etc. Such faults are usually of second categories. Most frequent are the faults of the third category. They dislocated rocks within individual tectonic units or blocks.

Ring structures. Landsat images are very convenient for discovering ring structures. A number of them have been identified in the Dinaric Alps. The majority are in the inner Dinarides, between Sarajevo and Zagreb, as well as in the Pannonian Basin (Fig. 1). Their diameters range from a few to several tens of kilometres, (e.g. ring structure near Nikšić). Their origin is connected to magmatic activity (Nikšić, Sarajevo), but to tectonic and diapiric activity as well (east of Dubrovnik). Some of them show increased geomagnetic and gravimetric maximums (3). It has also been noticed that the seismic activity is less intensive within the ring structures, but it is most increasing in the zones between two or more ring structures e. g. Dubrovnik-Nikšić [4].

3 SEISMICITY

Territorial and time variations of seismic activities in the area of the Dinaric Alps are shown by:

- the map of stronger earthquakes,
- the strongest earthquakes in more active epicentral areas, and
- time variations and directions of migrations of seismic activity [5].

Instrumental data of earthquakes were mostly used within the period of 1900 to 1980, as well as reliable macroseismic data covering the same period and also the data on the strongest earthquakes in the17th, 18th and 19th centuries. The map of earthquake epicentres (Fig 1)) shows all stronger earthquakes of the M≥ 4.0 magnitude. The map shows, as it has been already ascertained [2], that the model of earthquake occurrence in the Dinarides consists of

two characteristic zones: one of them stretching along the coastal part of the Adriatic and the other in the inner part of the Dinarides.

The *coastal* zone (the outer Dinarides) shows two directions of earthquake epicentres, one under the sea and the other on the land. The examples are: surroundings of Rijeka, Senj with the neighbouring islands (the northern Adriatic), surroundings of Zadar, Benkovac, ~ibenik, Split (Biokovo) with the neighbouring islands (the central Adriatic) and the surroundings of Dubrovnik, Kotor (the Montenegro coastal region) and Skadar (the southern Adriatic).

The *inner* zone (the inner Dinarides) of more intensive seismic activity stretches from Sarajevo over Banja Luka, Zagreb and Ljubljana to Tolmezzo. Besides the mentioned zone some other areas with intensified seismic activities have been registered, such as the one in the Pannonian Basin (SW from Osijek) and the ones Nand NW from Zagreb.

The same map (Fig. 1) also shows the strongest registered earthquakes in more intensive epicentral areas. So, the strongest earthquake in the coastal zone occurred in the vicinity of *Dubrovnik in 1667*, with maximal intensity Imax = X degrees of MCS, then in the Montenegro coastal region (Ulcinj) in 1979 (of the M = 7.1 magnitude), in the vicinity of Skadar in 1905 (M = 6.6), on the NE of Split (the Biokovo mountain) in 1923 (M = 6.2) and in Furlania (Tolmezzo-Gemona) in 1976 (M = 6.4).

The strongest earthquakes of the inner zone were in the vicinity of Ljubljana in 1895 (Imax = IX degrees of MCS), of Zagreb in 1880 (Imax = IX degrees of MCS), then in the vicinity of Banja Luka in 1969 (M = 6.4), in Pokuplje (S ofZagreb) in 1909 (M = 6.0) and in the vicinity of Sarajevo in 1962 (M = 6.0).

Time variations of seismic activity may be identified also on the basis of older data of earthquakes which, although being not quite reliable, indicate activities as follows:

At the beginning of *17th century* (in 1608) an intensified seismic activity took place in the southern part of the Dinaric Alps (Boka Kotorska), then in the northern part of the Dinarides (in 1626) in Istria and in Vinodol (in 1648). In the second half of the century the strongest earthquake of the all investigated area occurred in the vicinity of Dubrovnik in 1667.

In the *18th century* strong earthquakes took place in the northern part of the Dinarides (Vinodol) in 1721, then in their central part (vicinity of Zadar) in 1750 and, by the end of the century (in 1780) in the southern part (area of Boka Kotorska).

The *19th century* was rich in strong earthquakes in the Dinaric Alps. At the beginning of the century (in1823) the seismic activity was strong in the southern part of the Dinarides, then some time later (in1838) in the northern part, then again several times running (in 1843,1849, 1850 and 1855) in the southern part. Towards the end of the century (in 1870) strong earthquakes took place in the northern part of the Dinarides, some time later (in 1890) in the southern part, while the very end of the century (1898 and 1899) registered strong earthquakes in the central part of the Dinarides.

At the beginning of *20th century* (in 1900, 1901 and 1907) the first to become active was the central part of the Dinarides, then (in 1905) the southern one and finally (in 1916 and 1920) the northern one. Some time later (in 1923, 1924, 1925 and 1927) the central part of the Dinarides was submitted to a series of strong earthquakes, which simultaneously and later on (in1926 and 1928) occurred also in the northern part of the Dinarides, their central part being active again in the middle of the century (in 1941, 1942, 1956, 1962 and 1970). After those earthquakes, in 1966 and 1979 a particularly intensified seismic activity was manifested in the southern part of the examined area, as well as the one in 1976 in the northern part.

According to the sequence of activation of individual epicentral areas within the same year or within a shorter period of several years time the main directions of migration of seismic activity (Fig. 1) were ascertained as follows:
Furlania –Cerkničko jezero and back (in 1926 and 1928), the submarine area of the Adriatic -Vinodol (in 1916), Podvelebitski kanal- Lika (in 1949), Zagreb -Ljubljana (in 1880 and 1895), Zagreb - Pokuplje (in 1901, 1906 and 1909), Dilj Gora - Banja Luka- Knin - Šibenik (in 1964, 1969 and 1970), Hvar- Biokovo -Sarajevo -Dilj Gora (in 1956, 1962 and 1964), Biokovo- Ljubinje (in 1927), the Montenegro coastal region -Dubrovnik - Nevesinje (in 1979), Dubrovnik -Mljet -Ston and back (in the middle of 19th century).

There were, however, "simultaneous" seismic activities in mutually very far apart epicentral areas, what may be hardly regarded as a causal relationship. Therefore, the directions shown of migration of seismic activity may serve as a reference to some other causal and consecutive considerations of tectonic processes.

4 GEOPHYSICAL DATA

Geophysical data of the regional and super-regional character were considered within the frame of seismic activities and tectonic data obtained by satellite images. The Landsat images have registered huge tectonic dislocations of the NW -SE direction, while the ones transversal or diagonal to this direction are less expressed. The question was raised about how the mentioned dislocations and seismological data are shown and recognized in various geophysical maps. A partial answer to this, as well as to some other questions related to the seismic activity of the investigated area. was given by the comparison of data resulting from various research methods. So, the comparison was made of residual gravity data with the geomagnetical and seismological ones, as well as with those obtained by satellite images. A similar comparison was also made with the deep geoelectrical sounding data.

Three separate areas were examined within the Dinaric Alps. i.e. the ones in their south-eastern, central and north-western parts.

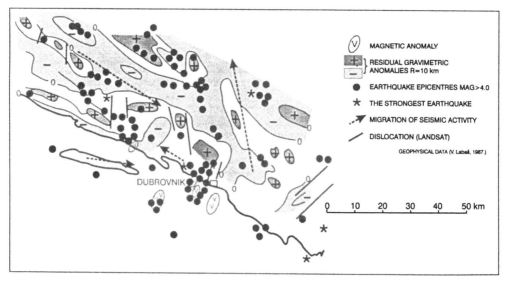

Figure 2. Geophysical measurements in comparsion with seismic and Landsat data (Dubrovnik)

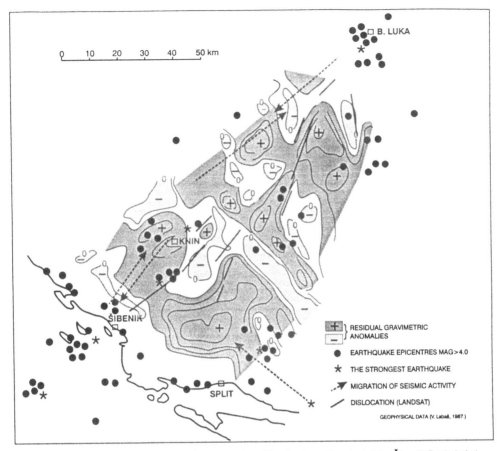

Figure 3. Gravimetric measurement in comparsion with seismic and Landsat data (Šibenik-Banja Luka)

The *south-eastern area* includes wider surroundings of Dubrovnik (Fig. 2). where important residual gravity anomalies (R = 10 km, d = 2670 kg/ m^3) have been obtained. stretching in the NW-SE direction, thus corresponding to the direction of large longitudinal faults (Fig. 1). However. the smaller positive residual anomaly, in the north of Dubrovnik, the concentration of earthquake epicentres and geomagnetic anomaly forms. probably caused by magmatic intrusion in the sea, as well as the faults shown on the Landsat images, have generally orientated in the NE -SW or N -S direction. A similar orientation is shown by an large important residual anomaly in the north of Dubrovnik, which is accompanied by emphasized directions of migration of seismic activity (Fig. 2). The mentioned data lead us to conclude that an increased seismic activity should be connected also to very deep transversal fractures.

The central part of the Dinarides, the area between Sibenik and Banja Luka (Fig. 3) has been correlated from the aspect of residual gravity and seismological data (6), as well as of the fault tectonics interpreted on the basis of the satellite images. Based on the consulted various data the examined terrain shows a wide fracture zone stretching from the Adriatic Sea (Sibenik -Knin) towards Banja Luka and Osijek, in the SW -NE direction. This is an old and deep fracture zone of intensive movements, generating tension, which results in delivery of seismic energy along the neotectonic longitudinal faults, such as the ones in the wider area of Banja Luka.

The northern part of the Dinarides is illustrated by the example given by the area between Rijeka and Senj (Fig. 4). From the apparent resistivity map AB/2 = 15 km it is evident that the areas shown with high resistivities $\geq$ 1000 Wm are made of Mesozoic carbonate rocks, generally stretching in the NW -SE direction.

They are intersected by smaller lower resistivity zones $\geq$ 500 .Ωm corresponding to clastics of

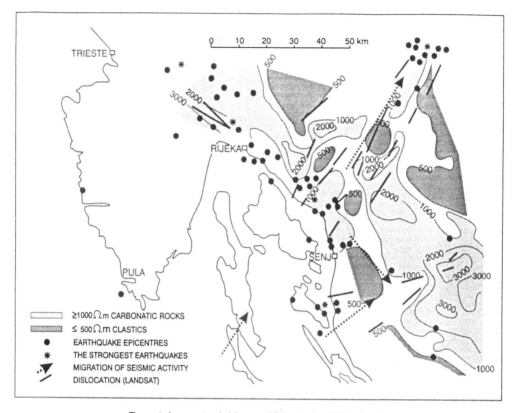

Figure 4. Apparent resistivity map AB/2 = 15 km (Rijeka-Senj)

Paleozoic age. The forms and the orientation of geoelectrical "structures" and contour-resistivity lines generally correspond to the directions of the migration of seismic activity as well as to the transversal fractures shown on the satellite images.

On the basis of the available data it may be ascertained that also this part of the Dinarides includes a wide and deep fracture zone stretching from the Adriatic Sea towards Zagreb, thus crossing the Dinarides massif. Similar as in the preceding fracture zone, this one is also submitted to tectonic movements generating tensions that are also delivered in the form of seismic activity along younger faults.

5 CONCLUSION

The area of Dinaric Alps (Dinarides) has been analysed on Landsat TM images in order to register important tectonic and morphostructural elements.

Based on the Landsat images we registered a large number of faults, which were divided into three categories according to their length and intensity. The most important longitudinal faults are stretching in the NW -SE direction (the first category), whilst the transversal faults are mostly of the second category and they intersect the first ones. On the same images also were registered quite a number of ring structures.

The map of faults also shows the epicentres of stronger earthquakes, which are most frequently connected to neotectonic longitudinal faults or to the places where they are intersected by the transversal ones, as well as to the unstable zones between ring structures.

Paleotectonic deep fractures, which are usually difficult to recognize on satellite images, were registered on several locations by geophysical measurements. They are, most frequently, deep transversal fractures stretching in the SW- NE direction, generating tensions, delivering energy and inducing seismic activity along younger neotectonic faults.

The comparison of satellite provided data and of the ones obtained by geophysical measurements with the data on stronger earthquake epicentres shows a certain regularity in earthquake manifestations. It means that the intensified seismic activity covers two tectonicly expressed very disturbed zones, in the outer and inner Dinarides.

Numerous data on earthquakes and on the activation sequence of individual epicentral areas were the basis for the ascertainment of important directions of migration of seismic activity.

6 REFERENCES

M. Oluic, Remote Sensing in Geology, 1983 pp. 127. *Remote Sensing in Geosciences*: Donassy, Oluic, Tomašegovic), Yugosl. Acad of Science and Arts, pp. 333.Zagreb. [1]

M. Oluic and D. Cvijanovic, 1991. Tectonic Elements Registered on the Landsat TM Imagery and Seismic Activity in the Western Part of Yugoslavia, in Proceedings of the 8th Thematic Conference on Geologic Remote Sensing 2, pp. 1207-1214, Denver, Ca-USA. [2]

M. Oluic and I. Kubat, 1981. The Connection between Ruptural and Ring Structures on Landsat Imagery and Mineral Occurrences in Central Bosnia. Yugosl. Acad. of Science and Arts, 11/2, pp. 45-57, Zagreb [3]

M. Oluic, D. Cvijanovic and E. Prelogovic, 1980. Some New Data on the Tectonic Activity in the Montenegro Coastal Region (Yugoslavia) Based on the Landsat Imagery. Acta Astronautica 9/1, pp. 27-33, Oxford. [4]

D. Cvijanovic, 1981. The Seismic Activity in Croatia. Doctoral dissertation, University of Zagreb, 1981. [5]

D. Cvijanović and V. Labaš, Comparison of Seismic Activity and Residual Gravity Anomalies in the Wider Knin Area (in press). [6]

7 Water

Observing our environment from Space: New solutions for a new millennium, Bégni (Ed.)
© 2002 Swets & Zeitlinger, Lisse, ISBN 90 5809 254 2

Monitoring and water pollution modelling of the Bosphorus by regression analysis using multitemporal Landsat-TM data

H.Gonca Coşkun, S.Ekercin & A.Öztopal
Istanbul Technical University, Maslak, Turkey

ABSTRACT: Environmental pollution is among the most common problems of different nations as a global problem. Especially, water pollution is a continuous threat to surface water resources and transportation routes in the world. Istanbul Straight (Bosphorus) is a model for such a situation due to water passage from the Black Sea to the Mediterranean Sea. In fact, pollution sources of the Bosphorus originate from the wastage of many nations that have costal areas along the Black Sea and Danube River in addition to the population as well as industrial pollutants from the city itself. Pollution surface measurements are carried out concerning total suspended solids (TSS), humic materials (HM), chemical oxygen demand (COD), poliaromatik hydrocarbons (PAH) and hydrodynamic conditions of water. On the other hand, digital multispectral Landsat-5 satellite data were recorded and co-registered for a portion of the Bosphorus. Relationships are sought between the water quality parameters and the reflectance values by the use of regression analysis. Although such an analysis has been carried out in the past during 1986 since then there are signs of significant water pollution increment in the Bosphorus. Therefore, this paper aims at the deduction of the most recent data analysis from 1997 in comparisons with earlier studies so as to make temporal decisions in addition to the spatial features. Adaptive parameter estimation in the regression analyses provides efficient computation in an economic manner. These observed reflectance values show a strong relationship with the water quality observation. The necessary values are provided in single pixel values for each band at the station point in the Bosphorus. Satellite data provide a useful index of TSS, HM and PAH. As the reflectance (in the turbidity area) in the longer red and near IR increases faster than the reflectance in shorter blue and green wavelengths, it can be seen that turbidity levels are positively related to reflectance.

1 INTRODUCTION

The Bosphorus (Istanbul Strait) runs through Istanbul which is one of the most populated and ancient cities of Europe. Not only is this strait a site of incredible beauty, it is also an area steeped in history and myth. Today this beauty is under extreme threat due to the pollution load being dumped along its coasts. Water quality in the Bosphorus is being strongly affected by waste discharges from many residential areas. The Bosphorus receives domestic and industrial waste from about thirty large and small-scale towns. Due to the rapid increase in pollution levels, a monitoring program must be instituted to measure coastal pollution using remote sensing techniques.

Water quality parameters may chance according to type of the study area. For example, reflectance is expected to increase with suspended sediment concentration in visible wavelength. It increases in the algal concentration in the green but decreases reflectance in the blue wavelength. Organic matter has a tendency to decrease reflectance in visible and infrared wavelength. Iron oxide selectively reflects red light but absorbs green light (Bhargava and Mariam, 1991; Lathrop and Lillesand, 1996).

In this investigation water quality measurements have been taken from an earlier study in 1986. For instance as total suspended solids (TSS), humic materials (HM), chemical oxygen demand (COD), poliaromatik hydrocarbons (PH) and the reflectance values of the each bands on the oceanographic station points are taken for the statistical model purpose of the water pollution estimation in the Bosphorus. In order to find possible significant relationships between remote sensing variable as the reflectance values and the surface measurements a regression approach is used for refined determination of regression coefficients. The regression correlation calculated from mentioned two variables by a model, are in good agreement.

The objective of this paper was to determine water pollution changes in the study area using remote sensing and GIS (Geographic Information Systems) techniques. This investigation presents temporal water analysis on the Bosphorus using the results of image processing of two different dated data from

1986 and 1997 with 30 meters resolution on multispectral and 60 meter resolution on thermal band. This study is an outcome of a project that has been supported by Istanbul Water and Sewerage Administration sponsorship in collaboration with Istanbul Technical University Remote Sensing Laboratory. The project title is "Satellite data to describe the differences of pollution on drinking water dams and Bosphorus"

2 STUDY SITE

The Istanbul Strait connects the Marmara Sea and the Black Sea, and it is a natural boundary between the European and Asian continents, like the Dardanelles Strait that runs from the Marmara Sea to Aegean Sea, which is another natural separation between the two continents. This system of straits is approximately 300 km in length and connects the Mediterranean Sea via the Aegean Sea, to the Black Sea. The system has an important influence on oceanographic conditions in the Black Sea and in the Marmara Sea. The Bosphorus is a meandering strait some 31-km in length, with widths varying from 0,7 to 3,5 km averaging at about 1,6 km. Its bed is a drowned river channel, more than 50 m deep, extending beyond a sill of 32 to 34 m depth located at the southern entrance of the Strait between Kabataş and Üsküdar. One of the distinct characteristics of the Bosphorus Strait is a two-layer current and density distribution. Less-saline, Black Sea surface water is carried into the Marmara Sea at the surface of the Strait, while underlying bottom water from the Marmara Sea, which is originally from the Mediterranean, is carried in the opposite direction towards the Black Sea (Bastürk *et al.* 1988). The surface current follows (Fig.1) through the Strait's meandering path and forms numerous small circulations in several bays on both sides. The bottom current follows more closely the windings of the channel. Many researchers have investigated vertical stratification in order to understand the important role of water exchange throughout the Bosphorus Strait in determining the oceanographic regime, especially that of the western Black Sea and the Marmara Sea. English, Russian and German observers first conducted studies of water exchange, vertical stratification and current characteristics in the early 1800s.

The Middle East Technical University's Institute of Marine Sciences (METU/IMS) have collected water quality reference data for the Bosphorus.

The monthly surface and 20 m depths of measurement values of TSS, HM, COD and PAH are taken at the oceanographic stations (Fig.2) in the Bosphorus.

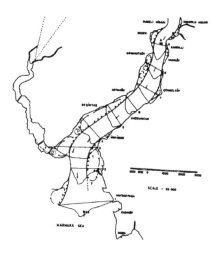

Figure 1. Location map and Strait of Istanbul at a certain cross section of the Bosphorus

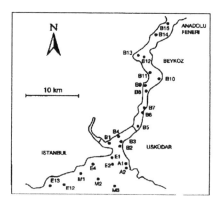

Figure 2. Oceanographic stations in the Bosphorus (Çeçen, et al. 1981)

3 MONİTORING WATER POLLUTION AND HYDRODYNAMIC CONDITIONS IN THE BOSPHORUS USING SATELLITE DATA

In this study, a multispectral digital data set of Landsat-TM data was used for the location of Istanbul Strait in Turkey. The hydrodynamic conditions in the Bosphorus are quite complex. With its synoptic view capabilities and image processing methods, remote sensing data combined with actual ground measurements can be regarded as a useful tool for monitoring the coastal zones (Coşkun *et al.* 2000). This study investigates the water pollutants with hydrodynamic conditions in the Bosphorus using satellite images based on Landsat 5 Thematic Mapper data on visual, near infrared and thermal bands data with multispectral bands which have 30 meter and 60 meter depth for thermal resolution of data that is dated of October 1986 and July 1997. Image analysis is performed for satellite data evaluation related to the Bosphorus for 1986 and 1997 using ERDAS Imagine 8.4. As a first step, the radiomatically corrected satellite data were geometrically corrected. Satellite images have transformed into UTM (Universal Transformation Marcator) projection co-ordinate system. In this investigation used ground truth measurement are used with 1/25000 scale standard topographic maps, various thematic maps and water quality measurements.

The turbidity distribution map (Fig. 3a) of Bosphorus is derived from Landsat 5 TM data of October 1986 formed by controlled classification method. (Colour indicates: yellow= waste discharges from textile factories; orange=organic materials, including products of biologic decomposition (HM) and industrial paints and pigments; blue=Black Sea water (<10°C); pink=Marmara Sea water; red=domestic wastes; black=wastes discharged from heavy industries (including metals)). The turbidity distribution map (Fig. 3b) of Bosphorus is derived from Landsat 5 TM data of July 1997 and formed by controlled classification method. Final images of Bosphorus, as shown in Figure 3 a-b are formed by controlled min-distance classification method using TM data and it is compared with the ground data sets (Fig.1). It is observed that the image in Fig.2 is compatible with actual ground data given in Fig.1. The pollution is caused due to the Istinye Stream and the Shipyard in the Istinye Inlet, the Baltalimani Stream, Bebek Inlet, the Coal Depot in Kurucesme; the registered or unregistered waste extends till Besiktas. The northwestward flowing surface water carries pollution from the Bosphorus which is supplemented by local discharges into the Golden Horn. The shipyard in İstinye Inlet was moved in June 1992. We can see clearly decreasing pollution levels at the İstinye Inlet when comparing 1986 and 1997 Landsat classified images.

The hydrodynamic situation of the Bosphorus can be seen as shown in Figure 4. Bosphorus water enters the Sea of Marmara (Landsat-5 TM thermal band dated 1984(a) and 1997(b)). Light blue=water temperature between 10 and 15°C; medium blue=water temperature between 10 and 15°C; dark blue=water temperature < 10°C. dark blue=water temperature < 10°C. The Thermal Bands have formed both images which are dated from 1984 and 1997 for the Bosphorus. These images clearly display the temperature differences.
As seen in the images, the cold water coming down from the Black Sea mixes with the warmer water of the Marmara Sea causing turbulence. When the images are zoomed, the light blue zones representing the increasing temperature of the wastewater discharge locations of the Bosphorus and the Golden

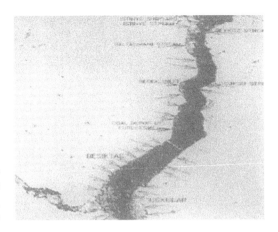

(a)

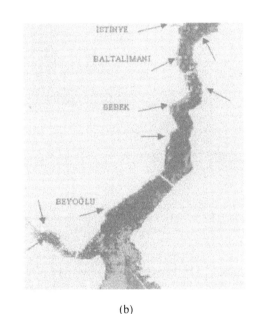

(b)

Figure 3. Turbidity distribution map of Bosphorus derived from Landsat-5 TM of 1986 (a) and 1997 (b), formed by controlled classification method. (Colour plate, see p. 427)

Horn are visible. It can also be observed that Black Sea water is dominant in Bosphorus and that it flows into the Marmara Sea.

4 STATISTICAL MODEL FOR WATER

In order to obtain meaningful relationships, considering water quality parameters that are shown

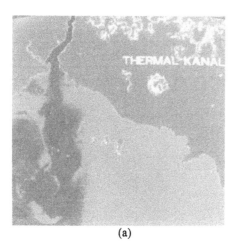

(a)

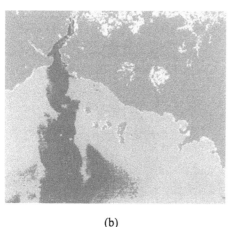

(b)

Figure 4. Thermal images are formed by the thermal bands of Landsat-5 TM dated 1984 (a) and 1997 (b) of the Bosphorus and the Marmara Sea. (Colour plate, see p. 427)

in Table 1 the reflectance values are used for classic regression. The necessary values are provided in single pixel values for each band at the station point in the Bosphorus and the two data sets mean reflectances, water quality measurements are analysed and the coefficient estimates (a,b) are shown in Table 2. During the regression analysis, water quality measurements on hydrologic station point are taken as Y; and the values of reflectance on the same points are denoted by X. As a result, coefficient estimates and coefficient of determination (R) are obtained for TM data (24th October 1986) and the water quality reference data that are collected in the first week of November. The coefficients of determination for the first band are observed as 0.83% for TSS, 0.86% for HM, 0.47% for COD, 0.80% for PAH. On the second band, however, they are observed as 0.86% for TSS, 0.63% for HM, 0.49% for COD, 0.68% for PAH. Finally, band values have shown in Table 2 (Coskun, 1995).

5 CONCLUSIONS

Monitoring of water pollution using remote sensing techniques can help towards the improvement of water quality development and environmental protection. Hydrologists need to have up to date systematic information on the water quality patterns to control water environmental problems. Remote sensing technology with computer-based geographic information systems (GIS) is available as a useful means of supplying up-dated information to check and improve the management of water. This study reports temporal Landsat-TM satellite data using remote sensing techniques for water quality modelling development in the Bosphorus which creates serious environmental problems. Therefore, measuring water quality change is a very important issue for controlling the future development of the Bosphorus.

Temporal Landsat-5 TM satellite data using remote sensing techniques in the Bosphorus is analysed for monitoring water pollution in two steps. The first step includes satellite data classification according to ground truth water quality measurements. The second step is demonstrated by the correlation of Landsat-TM images and ground truth measurements using statistical model. Hydrodynamic-conditions and water pollution of the Bosphorus are quite complex in nature. It may be appropriate to investigate this pollution by processing the digital images taken from data in conjunction with the help of the conceptual methods in remote sensing. Processing and classifying the satellite data can examine the very important environmental problem of water pollution.

As the remote sensing method permits, large regions of the environment should be studied, the pollution problems should be determined both quickly and correctly. Measured reflectance values of each band on the oceanographic stations show a strong relationship with the suspended solids, chemical oxygen demand, and with humic materials and Poliaromatik hydrocarbons. Satellite data provide a useful index of TSS, HM and PAH. As the reflectance (in the turbidity area) in the longer red and near IR wavelength increases faster than the reflectance in shorter blue and green wavelengths, it can be seen that turbidity levels are positively related to reflectance. The relationship between the water quality parameters and the individual band reflectance are supported by regression analysis. According to the theory of remote sensing the established statistical model is compatible. Monitoring of temporal Landsat-TM about water quality can help to manage the collection of water ground truth measurements.

Table 1. The monthly surface water measurement values of TSS, HM, COD, PAH at Bosphorus oceanographic stations (mg/l) (METU/IMS, 1986-1999).

Sta.	TSS		HM		COD		PAH	
	Sept.	Nov.	Sept.	Nov.	Sept.	Nov.	Sept.	Nov.
B9	0.80	1.12	1.40	0.50	0.37	0.27	-	-
B8	1.20	1.34	1.55	0.57	0.24	0.30	0.52	0.09
B7	0.90	1.12	1.10	0.55	0.39	0.32	0.07	1.12
B6	0.95	1.24	1.35	0.75	0.18	0.37	-	-
B5	1.10	1.40	0.30	0.70	0.18	0.28	-	0.09
B4	5.30	1.33	1.40	0.83	0.39	0.56	1.20	0.56
B3	-	1.50	-	0.67	-	0.45	-	8.64
B2	-	-	-	-	-	-	-	-
B1	1.50	1.96	2.25	1.70	0.91	0.42	0.34	5.76

Table 2. Results of Regression Analysis Between Pollutants and Pixel Values Measured at the Stations of the surface Water in Bosphorus.

Band No	TSS			HM			COD			PAH		
	a	b	R	a	b	R	a	b	R	a	b	R
1(p)	-5.61	0.11	0.83	-9.51	0.17	0.86	-1.10	0.02	0.47	-34.63	1.98	0.80
2(p)	-1.05	0.13	0.68	-2.44	0.17	0.63	-0.28	0.03	0.49	-34.63	1.98	0.81
3(p)	-0.02	0.10	0.51	-1.60	0.17	0.60	0.33	0.00	0.04	-15.37	1.26	0.48
4(p)	0.52	0.11	0.24	-1.24	0.25	0.44	0.26	0.01	0.10	-19.97	2.84	0.50
5(p)	1.18	0.05	0.22	0.77	0.00	0.00	0.40	-0.01	0.10	2.27	0.12	0.00
6(p)	11.82	-0.09	0.14	28.18	-0.23	0.32	-1.80	0.02	0.10	-382.47	3.26	0.47
7(p)	1.48	-0.06	0.14	1.16	-0.22	0.32	0.49	-0.07	0.37	-3.50	3.88	0.64

REFERENCES

Bastürk, O.A., Yılmaz, C. Saydam and I.Salihoğlu, 1988: Health of the Turkish Straits Chemical and Environmental Aspects of the Sea of Marmara and Golden Horn. *Institute of marine Sciences Middle East Tech. University.*

Bhargava, D.S. and D.W. Mariam, 1991: Effect of suspended particle size and concentration on reflectance measurement. *Photogrammetric Engineering & Remote Sensing*, 57 (5), 519-529.

Çeçen, K., Güçlüer, S., Sümer, M., Doğusal, M., and Yüce, H., 1981: Oceanographic and Hydrologic Study of Bosphorus. *ITU Civil Engineering Press Turkey.*

Coşkun H.G.: 1995: Monitoring of Bosphorus and the Golden Horn Using Landsat-TM Data. *I T C (International Institute for Aerospace Survey and Earth Sciences) Journal* -1 pp 38-42.

Coşkun H.G, Öztopal, A., Şen, Z., 2000: Genetic algorithm model for the water pollution estimation in the Bosphorus by using satellite data, *Proceedings 20th EARSeL Symposium, Dresden.* pp 311-318.

Lathrop, R.G.Jr. and Lillesand, TM, 1996: Utility of thematic mapper data to asses water quality in Southern Green Bay and West and West Central Lake Michigan, *Photogrammetric Engineering & Remote Sensing*, 52, 671-680.

METU/IMS, 1986-1999: Istanbul Sewerage Project Marina Studies on Observations of Sewage disposal Systems Before and After Operations *1st. Interim Report*, January.

Observing our environment from Space: New solutions for a new millennium, Bégni (Ed.)
© 2002 Swets & Zeitlinger, Lisse, ISBN 90 5809 254 2

Determination of snow water equivalent from passive microwave measurements over boreal forests in Canada

K.Goïta
École de sciences forestières, Université de Moncton, Campus d'Edmundston, Boulevard Hibert, Edmundston, Nouveau-Brunswick, Canada.

V.Roy
CARTEL, Département de Géographie et Télédétection, Université de Sherbrooke, Sherbrooke, Québec, Canada.

A.E.Walker & B.E.Goodison
Climate Processes and Earth Observation Division, Climate Research Branch, Meteorological Service of Canada, Environment Canada, Downsview, Ontario, Canada

ABSTRACT: The objective of this study is to examine the efficiency of the Helsinki University of Technology (HUT) snow emission model to predict snow water equivalent (SWE). The test site corresponds to the former BOREAS study sites in Saskatchewan and Manitoba in Canada. Airborne measurements of passive microwave data at 18 and 37 GHz in both vertical and horizontal polarizations were considered in the study, in addition to ground measurements of snow parameters, meteorological and forest biometric data. The prediction of absolute values of brightness temperatures using the HUT snow model is not very good. However, snow water equivalent can be estimated by inverse modeling based on constrained least square minimizations of spectral and polarizations difference metrics. Some of the metrics considered yield encouraging SWE estimations, with a mean retrieval error less than 20 mm.

1 BACKGROUND

Snow is an important component of high latitudes climate system, global and regional water cycle, agriculture management and water sources for hydroelectricity. Therefore, accurate monitoring of snow properties (aerial extent, depth and water equivalent) is critical. Conventional methods based on snow course data are not sufficient to reach this goal because of the large variabilities in snow cover both in space and time. The integrated nature of remote sensing measurements is an advantage compared to conventional methods when measuring snow parameters at regional scales. Spaceborne passive microwave data has been investigated since the 1970's for the retrieval of snow depth and snow water equivalent (SWE). Several empirical algorithms were developed. A review can be found in Hallikainen et al. (1992). Microwave radiometry of water equivalent of dry snow is based on the decrease of the observed brightness temperature with increasing SWE, due to volume scattering. This decrease is on the order of tens of kelvins, depending on observation frequency and other technical parameters (Hallikainen et al. 2000). Empirical models have extensively used a spectral polarization difference between a high frequency and a low frequency (namely 37 GHz and 19 GHz for SSM/I sensor) as a mean to quantify the volume scattering which is then related to SWE. An algorithm developed by Goodison & Walker (1994), based on vertical polarizations, is operationally used to estimate SWE with good accuracy over the Canadian Prairies, where there is no vegetation. As pointed out by several researchers (Kurnoven & Hallikainen 1997, Goïta et al. 1997), land cover categories, and particularly the vegetation cover, have a major influence on the retrieval of SWE from passive microwave measurements.

In this study the Helsinki University of Technology snow emission model is considered in order to take into account the contributions of the vegetation cover and snow characteristics. This paper presents the results of a validation of the model over the Boreal Ecosystem – Atmosphere Study (BOREAS) project study area in western Canada.

2 TEST SITE

The test site corresponds to the BOREAS project study region, located in Saskatchewan and Manitoba in Western Canada (latitude 50-60° N, longitude 93-111° W). The southern part of the study area is near the southern limit of the boreal forest. Mixed woods composed of aspen and white spruce are common where the sites are well drained. In the poorly drained areas, bogs support black spruce with some tamarack. Jack pine stands also exist especially on dry sites composed of coarse texture soils. The northern part of the study region is typical of extreme northern boreal forest. It is covered primarily with black spruce, scattered birch and some stands of jack pine. Stem volume are relatively low in that area.

3 DATA

3.1 *Microwave data*

The data used in this investigation were collected during the February 1994 BOREAS winter field campaign in the study area. Airborne passive microwave data were collected along a set of flight lines, located in both the northern and the southern study areas. Microwave brightness temperatures were measured by three NASA microwave radiometers with frequencies of 18, 37 and 92 GHz (vertical and horizontal polarizations), and viewing at 45° incidence angle (Chang et al. 1997). Data at 18 and 37 GHz were considered in this study. Few lines acquired over the Canadian Prairies, just south of the BOREAS southern study area, were also considered.

3.2 *Surface data*

Extensive ground surveys were conducted during the experiment along designated flight lines to measure snow characteristics such as snow depth, density, SWE, and snow pack structure, including grain size and temperature profiles, in representative land types. Table 1 shows the average parameters values for the southern and the northern areas. As it can be seen in the Table, the measurements were taken in a very cold period.

Table 1. Average values of microwave brightness temperature, snow and surface data.

Parameters	South	North
TB_{18V} (K)	234.6 ± 5.4	235.2 ± 2.4
TB_{18H} (K)	226.5 ± 4.1	228.3 ± 2.8
TB_{37V} (K)	219.7 ± 7.2	215.5 ± 7.9
TB_{37H} (K)	199.3 ± 11.0	195.8 ± 10.7
Air Temp. (°C)	-32.3 ± 4.2	-27.5 ± 3.2
SWE (mm)	48 ± 8	55 ± 13
Stem volume (m³ha⁻¹)	200 ± 43	113 ± 18

Land cover types were determined during the field surveys. Forest volume information were extracted from Halliwell and Apps (1997a,b) and extrapolated to the nearest flight lines according to vegetation cover type. Forest stem volume are more important in the south, where several patches of dense forest occur. The maximum stem volume estimated from all the sample plots considered in the study site is less than 300 m³ha⁻¹.

4 MODEL DESCRIPTION AND INVERSION

The model used in this study is the HUT snow emission model (Pullinainen et al. 1999). It is a semi-empirical model based on radiative transfer approach. The HUT model describes the emission contribution and attenuation of a snow pack as a function of SWE, snow density, snow grain size and snow temperature. The snow cover is modeled as a single homogeneous layer. The vegetation contribution is modeled by using an empirical forest canopy attenuation which is based on forest stem volume. The basic assumption in the model is that the scattering is mostly concentrated in the forward direction. This assumption gives the brightness temperature of a homogeneous snow layer of thickness d on top of a ground surface in a direction θ as:

$$T_B(d,\theta) = T_{B,g} + T_{B,s} \tag{1}$$

where $T_{B,g}$ is the brightness temperature contribution originating below the snow layer and attenuated by the snow layer, and $T_{B,s}$ is the actual emission of the homogeneous snow layer. The two components of equation (1) can be expressed in terms of snow water equivalent as follows:

$$T_{B,g} = T_{Bgs}(\theta)e^{-(\kappa_{em}-q(\kappa_{em}-\kappa_{am}))\sec\theta\, SWE} \tag{2}$$

$$T_{B,s} = T_s \frac{\kappa_{am}}{\kappa_{em}-q(\kappa_{em}-\kappa_{am})}\left(1-e^{-(\kappa_{em}-q(\kappa_{em}-\kappa_{am})\sec\theta\, SWE)}\right) \tag{3}$$

T_{Bgs} is the ground emission contribution just above the soil-snow boundary. T_s is the snow physical temperature. κ_{am} and κ_{em} are the snow cover mass absorption an extinction coefficients respectively. Coefficient q is an empirical constant that describes the fraction of intensity scattered toward the receiver antenna beam. The total brightness temperature of a forested area with underlying snow-covered ground, in a given direction, can be estimated as:

$$T_{B,surf} = (1-t_{for})T_{for} + t_{for}T_{B,snow} + $$
$$(1 - \varepsilon_{snow})(1 - t_{for})t_{for}T_{for} \tag{4}$$

where $T_{B,snow}$ is the brightness temperature contribution of snow-covered terrain (see equation 1), ε_{snow} is the emissivity of snow-covered terrain, T_{for} is the physical temperature of the forest and t_{for} is the forest transmittivity. In the case of boreal forest, t_{for} is modeled using empirical formulas which depend on forest stem volume.

The simulation of the total brightness temperature in a given frequency and polarization, requires the knowledge of snow temperature, layer thickness, density, grain size diameter, moisture and salinity in addition to vegetation temperature and stem volume. Using the complete data set available over the test site, the brightness temperatures were simulated at 18 and 37 GHz in both vertical and horizontal polarizations. These results were compared to the measured values.

SWE retrieval technique is based on the use of constrained least-squares minimization algorithms. A maximum a posteriori method using a spectral polarization difference is proposed in Pulliainen et al. (1999). Since all parameters were known, we considered several metrics based also on spectral and polarization differences, by optimizing the values of snow depth. The solution of the iterative process is found when the minimum value of the metric is reached. Three of the metrics considered are listed below:

$$d_v = \left(T_{B,simu}^{19V} - T_{B,meas}^{19V}\right)^2 + \left(T_{B,simu}^{37V} - T_{B,meas}^{37V}\right)^2 \qquad (5)$$

$$d_h = \left(T_{B,simu}^{19H} - T_{B,meas}^{19H}\right)^2 + \left(T_{B,simu}^{37H} - T_{B,meas}^{37H}\right)^2 \qquad (6)$$

$$d_{v+h} = d_v + d_h \qquad (7)$$

5 RESULTS AND DISCUSSION

Figure 1 represents the comparison between the snow model brightness temperatures predictions and the airborne measurements at 18 and 37 GHz at both vertical and horizontal polarizations. Overall the model describes the general trends of the microwave observations made during the BOREAS winter field campaign. However, the scatter is very important, meaning that the absolute values of predicted brightness temperatures are not as good.

It can be seen that the model underestimates the brightness temperature at 37 GHz vertical polarization, while there is a slight overestimation at 18 GHz in both vertical and horizontal polarizations. Several factors influence the model behaviour. In this case, predicted values depend straightly on input parameters. Measurements errors on temperature, vegetation and snow parameters have an impact on the model estimation. Since the scattering effect increases with frequency, the errors on the simulated brightness temperatures at 37 GHz become more important than at 18 GHz. This can be linked to the higher sensibility of the simulated temperature at 37 GHz, compared to 18 GHz, to snow grain size and to a lesser degree to the vegetation density. The results may be slightly affected also by atmospheric effects, but these effects are negligible in the frequencies used here. Several assumptions in the model may bias the predictions of absolute brightness temperatures, but their effects are reduced when using spectral and polarization differences in SWE

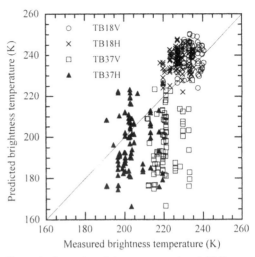

Figure 1. Comparison between measured and HUT snow model predicted brightness temperatures at 18 and 37 GHz.

estimation. The empirical components of the model, for example the forest transmittivity factors developed in Finnish conditions, may not be suited for our study site.

The performance of an inversion scheme in SWE determination depends on the metrics to minimize. In fact, we have tested several metrics including the three metrics given in equations 5 to 7. We found that each metric has its own minimum value which corresponds to a given solution of the inversion procedure. This solution varies from one metric to another. In all the simulations, the metrics which give the closest predictions of SWE values are d_v, d_h and d_{v+h} (equations 5 to 7).

The results of the three metrics predicted SWE are compared to the measured values in Figure 2. The correlation coefficients are statistically significant and approximately similar ($R \approx 0.50$). The mean retrieval error is ±18 mm, ±16 mm, ±15 mm, respectively for d_v, d_h and d_{v+h}.

Even if the model failed to accurately predict absolute values of brightness temperatures, the results of SWE estimations are very encouraging. It's an indication that the model can be used to predict SWE values in the type of environment considered. As in empirical algorithms, spectral polarization difference is a way to reduce the effects of external parameters such as such as atmospheric effects, or errors in the input parameters.

Due to the good quality of the dataset used, the inversion scheme considered is a straightforward method, which is sufficient to have an indication of the potential performance of the model in terms of SWE estimation in a low to medium density boreal forest. In practice, when dealing with satellite data, the ancillary data are not as complete with such a

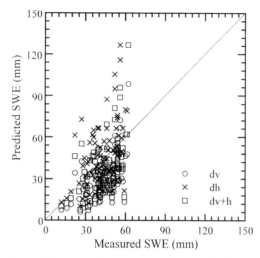

Figure 2. Comparison between measured and predicted snow water equivalent using three different metrics.

good quality. It is therefore necessary to consider more complex inversion schemes. Works are in progress on inversion schemes in which both snow grain size and depth are optimized to estimate SWE.

The test site is a low to medium density boreal forest, including also some Prairies (no vegetation) sites. SWE values were not as high during the experiment period (maximum SWE around 50 mm). It may be interesting to evaluate the performance of the model in more deeper snow conditions and in dense vegetation sites, in which empirical models are not efficient.

6 ACKNOWLEDGEMENTS

This work was supported by the Canadian CRYSYS Program (Environment Canada) and Natural Science and Engineering Research Council (NSERC). J. Pulliainen of the Helsinki University of Technology (Finland) is gratefully acknowledged for providing the HUT snow emission model source codes.

7 REFERENCE

Chang, A.T.C., Foster, J.L., Hall, F.G., Goodison, B.E., Walker, A.E., Metcalfe, J.R. & Harby, A. 1997. Snow Parameters Derived from Microwave Measurements During the BOREAS Winter Field Campaign. *J. Geophys. Res.*, 102 (D24), 29663-29671.

Goïta, K., Walker, A.E., Goodison, B.E. & Chang, A.T.C. 1997. Estimation of snow water equivalent in the boreal forest using passive microwave data. *Proc. Int. Symp. Geomatics in the Era of Radarsat, May 27-30, Ottawa, Canada.*

Goodison. B.E. & Walker, A.E., 1994. Canadian development and use of snow cover information from passive microwave satellite data. In Choudhury et al. (ed.), *ESA/NASA Int. Workshop, 245-262.*

Hallikainen, M., Jääskeläinen, V.S., Pulliainen, J. & Koskinen, J. 2000. Transmissivity of Boreal Forest Canopies for Microwave Radiometry of Snow. *Proc. IGARSS'2000.*

Hallikainen, M.T. & Jolma, P.A. 1992. Comparison of algorithms for retrieval of snow water equivalent from Nimbus-7 SMMR data in Finland. *IEEE Trans. Geosci. Remote Sens.*, 30: 124-131.

Halliwell, D.H. & Apps, M.J. 1997a. BOReal Ecosystem-Atmosphere Study (BOREAS) biometry and auxillary sites: overstory and understory data. *Report, Canadian Forest Service, Northern Forestry Centre, Edmonton, Canada.*

Halliwell, D.H. & Apps, M.J. 1997b. BOReal Ecosystem-Atmosphere Study (BOREAS) biometry and auxillary sites: locations and descriptions. *Report, Canadian Forest Service, Northern Forestry Centre, Edmonton, Canada.*

Kurvonen, L. & Hallikainen, M. 1997. Influence of land-cover category on brightness temperature of snow. *IEEE Trans. Geosci. Remote Sens.*, 35: 367-377.

Pulliainen, J., Grandel, J. & Hallikainen, M. 1999. HUT snow emission model and its applicability to snow water equivalent retrieval. *IEEE Trans. Geosci. Remote Sens.*, 37:1378-1390.

Observing our environment from Space: New solutions for a new millennium, Bégni (Ed.)
© *2002 Swets & Zeitlinger, Lisse, ISBN 90 5809 254 2*

Remote sensing applications for sustainable watershed management and food security

G.L.Rochon & D.Szlag
U.S. Environmental Protection Agency (EPA), Office of Research & Development (ORD), National Risk Management Research Laboratory (NRMRL), Cincinnati, Ohio, USA

F.B.Daniel
U.S. EPA ORD National Exposure Research Laboratory (NERL), Ecological Exposure Research Division, Andrew W. Breidenbach Environmental Research Center, Cincinnati, Ohio, USA

C.Chifos
Assistant Professor, School of Planning, University of Cincinnati, College of Design, Architecture, Art & Planning (DAAP), Cincinnati, Ohio, USA

ABSTRACT: The integration of IKONOS satellite data, airborne color infrared remote sensing, visualization, and decision support tools is discussed, within the contexts of management techniques for minimizing non-point source pollution in inland waterways, such as riparian buffer restoration, and of strategies for food crop production sustainability, in an array of ecosystems. Case studies include the Little Miami River Watershed, Ohio; the Barataria Basin and Bayou Sauvage, Louisiana; and Sudan's Gezira. The investigation's objectives are to ascertain the utility of environmental decision support, informed by archival satellite data visualization and alternative future scenarios generation, and to explore site-specific sensitivity of such technologies, in support of inter-generationally prudent management of watersheds and sustainable regional food security. Follow-on collaborative research with European counterparts (e.g. Foundation for Research & Technology-Hellas (FORTH), Crete, Greece) is planned to include enhancement, hardening, environmental technology verification, and field deployment of the prototype Temporal Analysis, Reconnaissance and Decision Integration System (TARDIS).

1 BASIC HUMAN NEEDS: WATER AND FOOD

1.1 *Sustaining water quality*

Water quality in inland waterways has been under assault from point-source industrial pollutants, passive urban effluent and non-point source agricultural run-off. The impacts have been discernible, in a local context, on drinking and recreational water quality, on fisheries and benthic community habitat and on water-borne disease prevalence and, remotely, on suspected down-stream eutrophication, such as in the Gulf of Mexico's hypoxic zone. Efforts to protect surficial and groundwater resources in the United States have had mixed results, largely mitigating direct industrial contaminants; but less efficacious in deterring urban and agricultural contributions of nutrients, pollutants and pathogens. Man-made earth impoundments and impervious channelization have in some instances contributed to increased sedimentation and pollutant loading.

In the developing areas, dam and irrigation canal construction (Fenwick, 1981) have been associated with inadvertent improvements to the habitat of certain disease vectors and hosts, (e.g. the snails, *Biomphalaria pheifferi* and *Bulinus truncatus*, in the

case of *Schistosoma mansoni* and *S. haematobium;* and *the black fly, Simulium damnosum*, in the case of *Onchocerciasis*).

1.2 *Sustaining food security*

With specific reference to Sahelian Africa and the semi-arid central Sudan, in particular, the prospects for an immediate transition from an emergency relief mode to a sustainable food security mode seems to be overly optimistic. In this regard, the Climate Change 2001 Reports issued by the Intergovernmental Panel on Climate Change (IPCC) (2001) echo the earlier findings of the International Research Partnership for Food Security and Sustainable Agriculture, issued by the Consultative Group on International Agricultural Research (CGIAR) (1998). Climate Change impacts within that region, however dire the projections, still represent but one of many threats to food security in Sudan and certain other African countries, south of the Sahara. Grinding poverty, minimal access to health care services, warfare, reported resumption of slave trading, a virtual HIV pandemic in some areas and an array of endemic infectious diseases, all militate against achievement of nutritional stability at the individual, family, district, national and multi-country regional levels.

Under such circumstances, any proponent of a quick technological fix to such complex and life-threatening disasters would be naive, at best. One must be sensitive to the charge, "Let them eat information!" in response to any overly aggressive push to deploy laptops loaded with satellite data, and software for GIS, visualization and decision support, in lieu of basic human needs and the means to ultimately develop self-sufficient sustainable food security.

It is with great humility, trepidation and caution that planning support tools should be developed and widely disseminated to assist decision-makers in "better serving the people," particularly when certain decision-makers have justifiably earned a reputation for egregious human rights abuses, even including the withholding of food, as a weapon of a war, and the foreign export of food supplies during periods of widespread domestic malnutrition.

In its idealized manifestation, remote sensing can be viewed as an equal opportunity technology, in the sense that underdeveloped countries can access data with the same overflight frequency and spatial resolution as can the developed countries. Moreover, now that remotely sensed data can be downloaded to a PC, with an otherwise isolated region, there is at least the potential for regional empowerment and digital autonomy (Rochon, 2001).

Yet, even if the digital divide slowly becomes narrower, due to advances in satellite telecommunications technology, one must not lose sight of the fact that the gap between rich and poor nations continues to widen, as does distributional inequities within countries, and consequently, unless dramatic international bi-lateral and multi-lateral intervention occurs to ensure food security in the face of climatological perturbations and other etiological factors, not the least of which is income insecurity, the gap between the fed and the unfed may also continue to widen.

2 TECHNOLOGY INTEGRATION

2.1 *Aerial color infrared* and *IKONOS satellite data*

The U.S. EPA Office of Research & Development (ORD) National Exposure Research Laboratory (NERL) commissioned 1m. resolution color infrared overflights of the Little Miami River Watershed in Ohio, which comprises thirty-five sub-watershed study sites, that have been the subject of intensive long-term sample collection and groundtruthing. The color infrared data will form one of the data layers in the spatial database.

Subsequently, investigators from three other research entities within the EPA (i.e. National Risk Management Research Laboratory (NRMRL), National Center for Environmental Assessment (NCEA) and the National Health and Environmental Effects Research Laboratory (NHEERL) joined together with NERL to collaborate on a nascent multi-disciplinary investigation of "Riparian Zone Restoration in Midwestern Watersheds and its Potential for Reducing Nutrient, Sediment, and Pathogen Export to Downstream Ecosystems," for the same long-term research study area within the Little Miami River Watershed.

To facilitate this interdisciplinary study, involving hydrologists, environmental engineers, soil scientists, ecologists, microbiologists, chemists and remote sensing scientists, Space Imaging Corporation was commissioned to acquire approximately 3,500 km^2 of 11 bit IKONOS CARTERRA 1m. resolution panchromatic and 4m. multispectral data in Albers Conic Equal Area projection, during a period of maximum green-up and vegetation species differentiation, namely June 15 to July 15, 2001.

Investigators plan to be in the field conducting groundtruthing and collecting samples for chemical and microbiological analysis during the satellite passes. These multiple data sets will be available as input to a prototype decision support system, in order to develop an array of alternative future scenarios for watershed sustainability status, under differing resource management and intervention regimes. It is hypothesized that an intensive regimen of riparian zone buffer rectification will be propitious in mitigating deleterious impacts on downstream ecosystems.

2.2 *TARDIS*

To address the perceived need to facilitate adequately informed decision-making, under conditions of environmental and social complexity, the concept of a decision support tool, incorporating critical antecedent data, such as site-specific terrain, landcover, landuse and event history, *inter alia*, as well as the capability to represent alternative future scenarios, was developed and first articulated in an interview with the presenter, appearing in Naval Meteorology and Oceanography Command News, in the February/March, 1997 issue (Willis, 1997).

The concept evolved into a prototype Temporal Analysis, Reconnaissance and Decision Integration

System (TARDIS). Components of the TARDIS include virtual exploration of archival satellite data. This feature required value-added modification of familiar virtual overflight software (e.g. ERDAS Virtual GIS, PCI's FLY!, Autometric EDGE Wings), allowing not only user-selected direction, altitude and speed, but also virtual time travel, through reconnaissance of successive georeferenced archival imagery of available earlier time periods for the study area. User-specified parameters for noting feature-specific change detection on-the-fly is an option.

Integration of *in situ* data would typically be required for complex environmental decisions, in addition to remotely sensed data (e.g. demographic data, discrete data associated with precision agriculture or with species-specific habitat constraints). This integration is achieved simply by accepting data layers, independent of which specific GIS package is available to, or preferred by, the user/decision-maker.

It is further assumed that for many decisional applications, historical awareness of what critical events have transpired within the confines of the study site, while frequently necessary, may be insufficient. The consequent sequella of contemporary action, and/or of historical trends, are often of paramount importance, especially when prodromal indicators of environmental degradation or of ecological instability are not clearly manifested.

To this end, the TARDIS future scenario generation module was designed to permit visualization of an array of alternative futures. The visual feature is justified on the basis of potential horizontal extension of decisional input, thereby embracing stakeholders not necessarily conversant with tabular data or statistical representation. Moreover, given situational complexities, even the statistical cognoscenti could conceivably be challenged by the enormity of data volume and/or by data heterogeneity.

Virtual future scenarios also posed challenges for the investigators. Unlike visualization of relatively recent historical periods, wherein one might have access to the archive of satellite scenes and/or airborne imagery, and associated digital elevation models, those hypothetical futuristic terrains are, by definition, not archivable. They could be generated with animation packages (e.g. Alias Wavefront/Maya); but, not in a sufficiently timely fashion for most critical response decisional time constraints. The solution chosen was to select a past scene from the satellite data archive that most closely matched the future parameters and then to utilize false color, vector overlays, and/or masking to depict the future

phenomena in question, as dictated by the results, for example, of one's General Circulation Model (GCM), Integrated Assessment Model (IAM) or crop forecasting model, thereby retaining the look and feel of the satellite imagery.

With respect to the TARDIS alternative future generation module, current options include simple trend extrapolation, model output representation, as well as a wide array of decision-maker determined *what if* scenarios (e.g. changes in labor availability, funding, public policy, climate, biodiversity, international commodity market fluctuations, habitat disturbance, degree of animal integration in farming, improvements in mechanization technology, availability of hybridized or transgenic seeds, upgraded resource management practices, disease prevalence, insect infestation, population pressures, sedimentation rates, discharge fluctuations, meteorological perturbations or other biogenic or anthropogenic disasters). Subsequent future scenario generation capabilities, under consideration, include similarity methods (Holt, 2000), and putative self-organized criticality methods, based upon complexity theory.

The extent to which decisions benefiting from such contemporaneous hindsight and foresight are actually superior to decisions made without such benefit remains to be demonstrated definitively. It would seem to be logically appealing that the optimal decisions are those that are made in an information-rich environment. Moreover, it seems reasonable that awareness of historicity and sensitivity to the potential future implications of one's immediate actions or of one's failure to intervene would be preferable to those decisions made in ignorance of the past and in either innocence with respect to the future or insouciance toward longer-term future outcomes.

On the other hand, it may be discovered that there are acquired perspectives, skills, and behaviors that are necessarily propaedeutic to a decision-maker's ability to synthesize visual data from past and present phenomena, and thereby to arrive at and intelligently select from a plausible array of optional future paths. Moreover, the ideological perspective, class identification and political consciousness level of the decision-maker may prove to be far more indicative of the nature of decisions rendered than either the wealth or paucity of available data, visual or otherwise. Axiological considerations, including aesthetics, ethics, multi-cultural societal valuation, inter-generational eco-sensitivity and distributional justice issues may also be highly relevant factors in environmental, political, economic and commerce-related decision-making processes. Accordingly, further controlled research is needed to ascertain the specific utility of decision support aided by TARDIS-like technologies, as well as the relative site-specific and issue-specific sensitivity of the TARDIS modules.

The modular structure of TARDIS enables the end-user to insert commercial or public-domain packages of choice for spatial database development, image analysis, qualitative or quantitative decision support, and modeling. This feature is beneficial for mobility and dissemination within heterogenous software environments; but it further complicates the environmental technology verification phase.

3 DUAL OBJECTIVES

3.1 *Assessment of TARDIS utility: decision support with hindsight and foresight*

A controlled prospective study of decision-makers' effectiveness, with the aid of conventional data manipulation sources, alternatively with TARDIS decision support and, in the third cohort, without either of the above means to divine hindcasts or forecasts is planned during the environmental technology verification and field deployment phases of the research. It is anticipated that this can be accomplished through comparative accuracy and eco-sensitivity assessment of environmental decisions, based upon short-term planning horizons that seek to achieve similar ecologically benign interventions, without necessarily resorting to closed research setting assessments of decision-makers' responses to TARDIS remote sensing data visualization cues, such as those technological approaches involving media effects laboratory monitoring of heart rate, eye movements and galvanic skin responses.

3.2 *Site-specific sensitivity*

In order to amass a suitable archive of data and procedures to test and refine TARDIS utility for operational deployment, it will be necessary to conduct field tests in widely differing environments. All the case studies described below have lives of their own and were not created as vehicles for TARDIS field testing. The occasions for refinement of TARDIS specifications were, therefore, piggy-backed onto projects developed to address an array of complex issues deemed appropriate for future decision-support, incorporating remote sensing, visualization and the need for voluminous data integration (i.e. sustainable watershed management, wetlands monitoring, and regional food security).

4 CASE STUDIES

4.1 *Little Miami River Watershed - Ohio*

Remote sensing of watersheds has been undertaken over the past twenty years for a wide array of purposes, in many different locations, integrated with numerous models, indexes and ancillary technology, and utilizing both course and high resolution multi-spectral, panchromatic, radar, x ray and microwave sensors.

Of particular relevance to this investigation are those studies that addressed non-point-source pollution/sedimentation control (e.g. Jakubauskas et al., 1992, Kang & Bartholic, 1994, Basnyat et al., 1999).

Analysis of both color infrared data and of satellite imagery will be conducted at the U.S. EPA ORD NRMRL GIS Laboratory (Andrew W. Breidenbach Environmental Research Center in Cincinnati, Ohio). The primary hardware platform is a network of Silicon Graphics workstations, with Windows NT operating systems and ERDAS IMAGINE 8.4 Professional GIS Software (Erdas, Inc., Atlanta, GA). Classification of the scenes will be accomplished with the ERDAS Expert Classifier/Knowledge Engineer module. Classes of particular interest will include riparian buffers, forested areas, scrub/shrub, agricultural row crops (e.g. specific varieties of corn, soybeans, winter wheat, legumes), fallow fields, pasture, lawns/golf courses, impervious surfaces, roadways, agro-industrial and other manufacturing facilities, wholesale and retail facilities, and the urban built environment. Given the availability of high resolution data and scheduled intensive field investigations, it is planned that the researches will obtain species-specificity with respect to grasses, trees, and agricultural crops.

This investigation is a watershed-scale study, utilizing analysis of IKONOS CARTERRA panchromatic and multi-spectral imagery, in combination with 1 m. color infrared aerial data. Moreover, in order to further examine the effects of scale on the efficacy of watershed management and to facilitate detection of inter-annual perturbations, the investigators will obtain multi-date, seasonally matched Landsat Thematic Mapper (TM) and Multispectral Scanner (MSS) data, as well as NOAA/AVHRR High Resolution Picture Transmission (HRPT) Local Area Coverage (LAC) data for the study area. Scalability has emerged as a significant recurrent issue, given the need to determine the appropriateness of extending local/regional model validity to watershed-scale and beyond and, conversely, the utility of global change models to watershed-scale impact assessment.

For the Little Miami River Watershed case study, the emphasis will be directed toward sustainable watershed management, with particular focus on riparian buffer zones.

4.2 *Barataria Basin & Bayou Sauvage National Wildlife Sanctuary- Louisiana*

Over a four year period, with support from NASA/EPSCOR's Carbon Cycling in Shallow Coastal Estuaries Research Project in association with the Louisiana Space Grant (LaSpace) and from the United States Department of Agriculture (USDA) Forest Service for the Urban Ecosystems and Global Change Study, faculty researchers and student research assistants at Dillard University collected soil, water and vegetation samples from twelve wetland study areas surrounding New Orleans, Louisiana, to determine the distribution of urban effluent within peri-urban wetlands. The samples were tested at Tulane University's Coordinated Instrumentation Facility, utilizing atomic absorption, x-ray fluorescence and inductively coupled plasma techniques. Additionally, remotely sensed data was obtained for the study area, including imagery from NASA's calibrated airborne multispectral scanner (CAMS), mounted aboard a Lear Jet, and data from an airborne digital camera.

The initial hypothesis, that wetland pollutant contamination was correlated with urban industrial proximity, was supported. Hence, samples from the Bayou Sauvage National Wildlife Refuge, located within the City of New Orleans, itself, consistently revealed higher concentrations of heavy metal contamination than was found in the Barataria Basin study site, which was furthest removed from the urban/industrial center. Moreover, it was determined that the combination of remote sensing techniques and laboratory analysis of samples collected *in situ* was a more effective means of monitoring wetland sustainability that either method used alone.

As this case study began prior to the development of TARDIS, the decision support tool was not deployed within the wetland study areas; however, lessons learned during the course of the four year study were essential during the design phase of the TARDIS prototype. The sensitivity of the TARDIS modules in facilitating decision-making relating to sustainable wetland ecosystem management is scheduled during the field deployment and technology verification phase, in collaboration with scientists on the faculty of the University of Cincinnati's School of Planning.

4.3 *The Gezira - Sudan*

Approximately six months were devoted to field investigations on location in Sudan, with support from the Massachusetts Institute of Technology (MIT), the United Nations University (UNU) and the Dorothy Danforth Compton Foundation. Initially, the primary subject of inquiry was a comparative analysis of American, British, French, German, Russian, Chinese, Kuwaiti, United Arab Emirates, and Saudi Arabian investment practices in Sudan's agricultural sector and the implications of such modes of investment for nutritionally vulnerable populations in Sudan. Due to a variety of exigencies, including the intensification of the civil war and rapid diminution of foreign investment capital, the study evolved into an investigation of remote sensing applications for monitoring drought-related famine conditions in Sudan's Gezira area.

The Gezira was the venue for the first proof of concept for the TARDIS. The ultimate theoretically-intended end-users/decision-makers for this initial design were the team members of the Annual Crop Forecasting and Nutritional Needs Assessment Mission to Sudan, conducted by the United Nations Food and Agriculture Organization (UN FAO) and the World Food Programme (WFP). In view of the anticipated severe regional impact of global climate change on the semi-arid zone of central Sudan, where the Gezira is located, the study recommended an inclusion to the multi-disciplinary U.N. team of a representative from the newly established United Nations Global Terrestrial Observing System (GTOS), so as to facilitate regular monitoring of the longer term regional climate impact, in addition to the annual crop forecasting and nutritional assessments. As reported by Hulme (1989), this region of Sudan has experienced a 20 percent decline in rainfall since 1921, a contraction of its wet season by approximately three weeks, a southward displacement of its rainfall zones by 50 km to 100 km, representing 25 percent of the of the total zonal rainfall fluctuation for the past 20,000 years, according to paleo-climatological analysis (Rognon, 1987, Hulme, 1989, Hulme & Kelly, 1993).

Deploying TARDIS for the task of supporting decisions relating to the achievement of sustainable food security in the Gezira was considered to be an acid test of the planning/decision support technology, given the degree of inherent regional complexities. In addition to the civil war and the likelihood of decreased precipitation and increased temperature, which may constrain even drought-resistant varieties of the primary food crop, Sorghum bicolor, there are other formidable constraints imposed on the food crop by hosts of insect and nematode species and by the periodic advance of the desert locust, Schistocerca gregaria. The cash crops, including long staple cotton (Gossypium barakatensis) and short staple cotton (Gossypium aegypti) are also perennially plagued by various infestations during their growth cycle, including Pectinophora gossypiella and Helicoverpa armigera (Madden, et al., 1993).

Moreover, the local Gezira Scheme tenant farming population and surrounding communities are subjected to endemic schistosomiasis (*Schistosoma mansoni* and *S. haematobium*) and malaria (*Plasmodium falciparum*) and periodic outbreaks of cholera (*Vibrio cholerae*) and leishmaniasis (*leishmania spp.*), associated with the sand fly vector (*Phlebotomus orientalis* and *P. paptasi*) (Elnaiem, et al., 1999; Thompson, et al., 1999). The instability of produce markets under current economic and political conditions in Sudan adds yet another layer of complexity.

Lest one become too caught up in the advocacy of state-of-the-science technological solutions to age-old problems, it appears that the development of immunities among local insect populations to commercial pesticide applications may soon necessitate a major return to cultural practices for insect control, largely abandoned in favor of the less labor-intensive imported pesticides.

Current diplomatic tensions between the USA and the Government of Sudan as well as the conditions of military insecurity make it unlikely that a verified and hardened TARDIS deployment will soon occur in the Gezira; however, the lessons learned there will hopefully facilitate deployment in other areas of Africa, and potentially in Afghanistan and North Korea, where widespread malnutrition underscores the need for informed planning and decision-making related to sustainable food security.

5 CONCLUSIONS

Resolution of the broader questions relating to TARDIS ultimate generic utility to decision-makers, and both site-specific and issue-specific sensitivity, will necessarily await the results of the technology verification and field deployment phases of the research in Crete, Greece and subsequent collaborative test sites. However, preliminary observations, based upon feedback from developers during the beta test phase, would seem to indicate that TARDIS utility would be enhanced by user-friendly improvements to its graphical user interface (GUI), that multi-lingual interface options are essential to successful dissemination to applications within Europe and within the developing areas, that an array of public domain modules for basic operations would make the product more accessible and cost-effective to a larger potential user group, rather than by sole reliance on end users' providing their own GIS software. The option for users to easily substitute for the provided generic modules, their own spatial database, decision support or visualization software would, however, be retained.

The investigators, however, are confident that, despite the constraints imposed by the complexities of the externalities, there are substantive benefits that potentially can be achieved by involvement of a broader array of stakeholders within environmental decision-making processes and by those environmental planners and decision-makers who opt for a new paradigm, based upon a *modus operandi* that dynamically incorporates knowledge of historical progression, awareness of contemporary inter-disciplinary and interstitial complexity, and evocative visualization of the likely future outcomes of existing trends and of both the intended consequences and unintended sequella of interventionist strategies.

6 FOLLOW-ON RESEARCH

In addition to continued deployment within the context of the case studies discussed, it is intended that the TARDIS prototype undergo hardening, technology verification, documentation develop-ment, model integration and further testing and enhancement based upon feedback from beta users and the results of the prospective decision-maker effectiveness research described earlier. It is intended that the EPA investigators will be assisted in this process by collaborators in Greece and that other interested parties with appropriate multi-temporal satellite data and ancillary data sources are invited to participate in the field deployment and testing phases.

7 ACKNOWLEDGMENTS

The investigators are indebted to a wide array of agencies that contributed financial and/or institutional resources to support various stages of this initiative. These include the USDA Forest Service's Global Change Program and the NASA/ EPSCOR Louisiana Space Grant Consortium (LA SPACE) Global Change Cluster for support of the research on the impact of urban effluent on peri-urban wetlands in Louisiana, NASA/JOVE and the Jet Propulsion Laboratory (JPL) at CalTech for access to synoptic satellite coverage of Sudan's Gezira, the United Nations University (UNU), UN

FAO, UNICEF, the Dorothy Danforth Compton Foundation, Tanmiah, Inc., the Department of Geography at the University of Khartoum, and three former Ministers of Agriculture (Drs. Ali El Tom, Abas Abdel Magid and Tewfik Hachem) for facilitating field investigations in Sudan, the U.S. DOD Naval Oceanographic Office (NAVOCEANO) Major Shared Resource Center Programming Environment & Training (PET) Program for underwriting preliminary development of the TARDIS, the three co-investigators during the PET phase of research (Dr. Mohamed A. Mohamed, Dr. Sidney Fauria and Mr. Marseyas Fernandez), the on-going collaboration of scientists affiliated with four research entities within the U.S. EPA's Office of Research and Development (ORD) (i.e. NRMRL, NCEA, NERL and NHEERL) for interdisciplinary investigations and for acquisition of IKONOS and aerial color infrared data for the Little Miami River Watershed. Research facilities were also provided by the International Nutrition Foundation in Boston, the Environmental Change Unit at Oxford University, and, in New Orleans, at Dillard University's Remote Sensing & GIS Laboratory and at Tulane University's Inorganic Laboratory, Coordinated Instrumentation Facility.

8. REFERENCES

Avery, Dennis. 1998. The promise of high yield agriculture. *Forum for Applied Research and Public Policy. 13 (2), Summer, 1998, 70-76.*

Basnyat, P., L.D. Teeter, K.M. Flynn & B.G. Lockaby. 1999. Relationships between landscape characteristics and non-point source pollution inputs to coastal estuaries. Environmental Management.23 (4), 539-549.

Fenwick, A., A.K. Cheesmond & M.A.Amin. 1981. The role of irrigation canals in the transmission of Schistosoma mansoni in Gezira Scheme, Sudan. Bulletin of the World Health Organization. Vol. 59, No. 5, 777-786.

Holt, A. 2000. Understanding environmental and geographical complexities through similarity matching. *Complexity International. 107 (1)*
http://life.csu.edu.au/ci/vol107/holt01/html.

Hulme, M. 1989. The Changing Rainfall Resources of Sudan. Transactions of the Institute of Geography. New Series. 15, 21-34.

Hulme, Mike and Mick Kelley. 1993. Exploring the links between desertification and climate change. Environment. 35 (6) July,/August, 5-45.

IPCC Intergovernmental Panel on Climate Change. 2001. Climate Change 2001: Impacts, Adaptation and Vulnerability. Geneva, Switzerland: IPCC Feb. 13-16, 2001.

Jakubauskos, Mark E., Jerry L. Whistler, Mary E Dillworth, and Edward A. Martinko. 1992. Classifying remotely sensed data for use in an agricultural non-point-source pollution model. EPA Region VII grant # X-007-331-001. Journal of Soil and Water Conservation. 47 (2), March/April, 1992, 179-183.

Kang, Yung-Tsung & Jon Bartholic. 1994. A GIS-Based Agricultural Non-point Source Pollution Management System at the Watershed Level. American Association for Photogrammetry and Remote Sensing Conference Proceedings. ASPRS ACSM. Orono: University of Maine, 1994.

Rochon, Gilbert L. 2001, in prep. Sustainable Food Security: Nutritional, Socio-Economic and Climatic Vulnerability in Sudan's Gezira. Lewiston, N.Y.: Edwin Mellen Press, anticipated 2001.

Rognon, Pierre. 1987. Aridification and Abrupt Climate Change on the Sahara Northern and Southern Margins, 20,000 Y BP to Present, in W. H. Berger and L.D. Labeyrie. Abrupt Climate Changes: Evidences and Implications;Dordrecht: Reidel Publishing Co.

Willis, Cathy L. 1997. NAVOCEANO's Scientific Visualization Center: Creating a Virtual Environment for DOD. Naval Meteorological and Oceanography Command News 17 (2) February/March, 1997, 8-10.

Observing our environment from Space: New solutions for a new millennium, Bégni (Ed.)
© *2002 Swets & Zeitlinger, Lisse, ISBN 90 5809 254 2*

Contribution of GIS to the planning and management of irrigation. Application to Zriga (Western Algeria)

A.Mendas, M.A.Trache & O.Talbi
Laboratoire de Géomatique, CNTS, Oran, Algeria

Editor's note: This paper was presented by a francophone author and we are publishing this exceptionally in French, since publishing delays have not allowed us to make a translation.

RESUME: S'appuyant sur les données physico - climatiques relatives à un périmètre d'irrigation, ce travail a pour objectif d'apprécier l'apport des Systèmes d'Information Géographique, outil d'aide à la gestion et à la prise de décision, à la planification et la gestion de l'irrigation à l'échelle du périmètre. En utilisant les capacités de stockage, traitement, analyse et de visualisation des SIG, il s'agit, au premier lieu, de pouvoir situer les zones les plus favorables à la mise en place de l'irrigation et déterminer l'amplitude de ces projets grâce à la connaissance des superficies irrigables en fonction de l'eau disponible. Le système mis en place permet de répondre aux différentes requêtes des utilisateurs potentiels à l'échelle d'un périmètre d'irrigation. Une application au périmètre d'irrigation de Zriga (ouest de l'Algérie) est présentée.

Mots clés: Irrigation, SIG, Base de Données, Planification, Gestion.

ABSTRACT: Relying on the physico-climatic data relative to an irrigation perimeter, the objective of this work is to appreciate the contribution of GIS, a powerful tool for management and decision-making, for the planning and management of irrigation at a perimeter scale. Using the GIS capacities of stocking, processing, analysis and visualization, we aim to situate the most favorable zones for irrigation and to determine the amplitude of these projects knowing the irrigable areas according to the water availability. The more accurate and up to date the information about the environment, the better the manager will be able to take decisions and determine policies. An application to the perimeter of Zriga (Tlemcen) in west of Algeria is presented.

Key words: Irrigation, GIS, Data base, Planning, Management.

1 INTRODUCTION

L'irrigation représente un facteur principal de développement agricole. Compte tenu de la place qu'occupe l'agriculture et du poids de la facture alimentaire, la recherche des terres aptes à l'irrigation, des eaux nécessaires à cette opération et leur meilleure gestion restera une préoccupation constante des services techniques. De ce fait, les gestionnaires du secteur ont besoin de disposer d'informations abondantes et fiables, notamment sur le sol, l'eau, la plante et le climat. Ces études nécessitent une approche multidisciplinaire, faisant intervenir la topographie, la pédologie, la géologie, la géomorphologie, l'agronomie, l'écologie, l'économie, etc...

La diversité et la quantité de ces données imposent de devoir les organiser en un système approprié pour garantir leur gestion et leur traitement, et les méthodes classiques de gestion deviennent de plus en plus inadaptées. De nouvelles méthodes, en cours de développement et basées sur la technologie des SIG s'avèrent particulièrement bien adaptées aux problèmes de l'irrigation puisqu'elles consistent notamment en:

- la prise en compte des données multiples,
- la mise en évidence des connexions entre les différentes données,
- la détermination des conséquences des différentes opérations d'aménagement.

Les principales opérations que subiront les données thématiques et spatiales sont de type agrégation, discrétisation et analyse spatiale. Cela nécessite l'utilisation d'un outil parfaitement adapté, comme peut l'être un Système d'Information Géographique (SIG). Derrière cet acronyme, se cache un ensemble de données géoréférencées et descriptives organisées et traitées par une structure humaine à l'aide d'outils informatiques, de manière à produire des informations visant à répondre à des objectifs précisément définis.

La première utilisation du SIG concernera ses potentialités analytiques permettant de différencier les processus spatiaux et de les modéliser. Dans un deuxième temps, nous utiliserons une autre potentialité qu'offre un S.I.G. pour la gestion des ressources naturelles et qui concerne les capacités de cet outil pour l'organisation des données aboutissant à la gestion d'un projet d'irrigation.

2. PRESENTATION DE LA ZONE D'ETUDE

Le périmètre d'irrigation de Zriga se trouve à 20 km de la ville de Maghnia et à 150 km de l'Ouest de la ville d'Oran. Il constitue le prolongement Nord-Ouest de la plaine de Maghnia, longeant ainsi la frontière Algéro-Marocaine. D'altitude moyenne voisine de 420 m, la zone de Zriga est limitée au Nord par les monts de Beni-Snassen, à l'Est par la plaine de Maghnia, au Sud par la plaine des Angads (Maroc) et à l'Ouest par la frontière Algéro - Marocaine.

Il occupe une superficie de 2050 ha où transitent l'Oued El Aouedj et l'Oued Bou Naïm qui confluent au niveau de la route nationale n° 7 (Maghnia - Ghazaouet) pour constituer l'Oued Mouilah qui se jette dans la Tafna à l'amont de Hammam Boughrara. Le climat est de type méditerranéen, rangé dans l'étage semi-aride, avec une pluviométrie moyenne annuelle de 391 mm.

3 METHODE ET MATERIELS INFORMATIQUES

A partir des données météorologiques, d'une carte numérisée des sols, des données topographiques et des informations régionales sur l'agriculture, on vise à établir un plan d'irrigation local en passant par :

• la délimitation des zones irrigables, selon les facteurs climatiques et physiques,
• l'estimation des ressources en eau disponibles, superficielles et souterraines (Henriksen C. & Hall L., 1992),
• la détermination des besoins en eau du périmètre en passant par la détermination des périodes au cours desquelles l'irrigation est nécessaire,
• l'estimation des besoins en eau des cultures (Billlaux, P., 1978) puis les besoins en eau du périmètre et, enfin,
• la confrontation entre les besoins et les disponibilités.

La méthodologie générale adoptée dans cette étude repose sur l'utilisation d'un logiciel de détermination des Besoins en Eau des Cultures (BEC) et d'un SIG pour la manipulation des différentes informations relatives au périmètre d'irrigation, organisées en base de données.

Au regard de la problématique de notre travail, il apparaît que l'un des critères fondamentaux de choix de l'outil SIG approprié doit être la richesse des opérations d'analyse spatiale qu'il peut permettre. Il doit également pouvoir supporter une base de données relationnelles.

Dans notre application, nous avons utilisé le logiciel SIG Arc/Info (ESRI, 1995).

Pour la détermination des BEC, le logiciel CROPWAT - version 5.7 (1991) a été utilisé. Ses programmes sont réalisés par AGLW / F.A.O. et fonctionnent sur PC (Smith M., 1992).

L'intégration, des données obtenues, dans le SIG choisi a nécessité:

• la digitalisation des données géométriques préalablement collectées,
• la correction des erreurs résiduelles de numérisation pour chaque couche et l'établissement de la cohérence entre couches,
• la construction de la topologie,
• la création de liens entre les données géométriques et les tables d'attributs correspondantes.

Pour le cas qui nous concerne, nous avons identifié les composantes de bases de l'information nécessaire à l'étude du périmètre d'irrigation de Zriga (Mendas A., 1997). Sept couvertures de base ont été choisies:

a) Une couverture des limites de la zone pilote;
b) Une couverture du réseau hydrographique;
c) Une couverture des puits situés à l'intérieur du périmètre;
d) Une couverture des courbes de niveau et un semis de points cotés;
e) Une couverture du réseau routier, comprenant les routes et les voies ferrées;
f) Une couverture contenant le contour des différentes agglomérations situées à l'intérieur de la plaine;
g) Une couverture des sols comprenant les caractéristiques pédologiques des sols de la zone. Elle permet d'établir une carte d'aptitude culturale, d'indiquer la nature et l'importance des aménagements hydroagricoles souhaitables pour que les sols soient aptes aux cultures indiquées et de définir sommairement des zones de priorité de mise en valeur;

L'information climatique fait certainement partie des composantes de base et devrait figurer dans cette liste. Mais compte tenu de la dimension de la zone d'étude et du fait que celle-ci n'est couverte que par une seule station météorologique, nous avons la même valeur pour chaque paramètre climatique pour toute la zone.

Au niveau de la zone d'étude, les différentes couvertures du SIG peuvent être complétées par des données locales plus détaillées.

Pour chacune des couvertures de base a été créée la topologie. Celle-ci a généré des tables attributaires de points pour les puits et le semis de points cotés, d'arcs pour le réseau routier, les courbes de niveau et le réseau hydrographique et de polygones pour les unités des sols.

Le système de codage permet de mettre en relation les identifiants des entités graphiques (points, arcs et polygones) avec les données descriptives:
– pour les puits: qualité et quantité d'eau;
– pour le réseau hydrographique: régime et qualité d'eau;
– pour le réseau routier: catégorie de la route;
– pour les sols: caractéristiques pédologiques du sol;
– pour les courbes de niveau et le semis de points : cotes;
– Pour les agglomérations: nom de l'agglomération.

4 TRAITEMENTS DES DONNEES ET RESULTATS

La délimitation des zones à irriguer se fait à partir des conditions climatiques et des conditions physiques. Les types de paramètres les plus importants sont relatifs aux aspects pédologiques (éléments agronomiques, (voir Fig. 1) et topographique (voir Fig. 2).

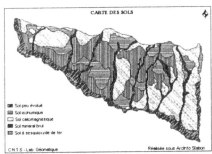
Figure 1. Carte des sols

Suivant ces facteurs, le sol irrigable peut être hiérarchisé en plusieurs classes, en tenant compte des contraintes dues à la topographie, au sol et à l'eau, et normalisées en fonction des cultures, des modes d'irrigation et des aménagements foncier

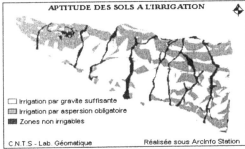
Figure 3. Carte d'aptitude des sols à l'irrigation

La quantification *des besoins en eau* peut être réalisée en deux étapes :

La synthèse des analyses utilise à la fois les données spatiales, les données descriptives et les données concernant les normes à respecter.

Le MNT est calculé à partir des couvertures des courbes de niveau, et d'un semis de points, auxquelles est appliqué le module TIN. Dans la présente application, la constitution d'un MNT a pour objectif d'une part, de disposer d'une valeur altimétrique en n'importe quel point du domaine d'étude, et d'autre part d'en dériver les paramètres de la morphométrie nécessaires pour les étapes ultérieures de ce travail. C'est en particulier le cas pour les deux sous produits les plus intéressants, le gradient de la pente et l'orientation de la pente. Les produits dérivés du MNT réalisés dans le cadre de cette application sont les suivants:
• une carte des pentes par classe
• une carte d'orientation par secteurs de 90°;
• et une vue en perspective du terrain.

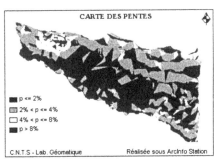
Figure 2. Carte des pentes

A ce stade, nous disposons d'une carte des sols, obtenue par digitalisation de la carte thématique pédologique et complétée par des informations sur les caractéristiques physiques et chimiques. Nous disposons aussi d'un MNT et de ses produits (carte des pentes, carte d'orientation) relatifs à la même zone.

La confrontation de la carte des sols avec celle des pentes permet de dégager une quantité importante d'informations sur la répartition des terres. Dans ce sens, disposant des données nécessaires et bénéficiant des facilités offertes par l'outil SIG, nous avons pu réaliser plusieurs types d'opérations (Mendas A., 1997) et particulièrement l'étude de la répartition des terres irrigables selon les facteurs sol - pentes.

Ainsi, en considérant la carte des sols et celles des pentes, l'opération d'intersection consiste à ressortir l'utilisation du sol couvrant certaines pentes et certaines caractéristiques des sols spécifiés (voir Fig. 3).

a - Calcul des besoins en eau des cultures (BEC) :

Une fois délimitées les zones potentiellement aptes à une mise en valeur agricole et hydriquement déficitaires, il s'agit de déterminer les périodes où les apports sont nécessaires, quantifier les besoins

295

Mois	J	F	M	A	M	J	J	A	S	O	N	D	Annuel
P (mm)	51	49	48	51	35	12	2	3	14	36	35	55	391
ETP (mm)	57	81	105	138	171	219	237	222	159	102	75	60	1626
P - ETP (mm)	-6	-32	-57	-87	-136	-207	-235	-219	-145	-66	-40	-5	-1235
P+RFU-ETP (mm)	44	18	-7	-37	-86	-207	-235	-219	-145	-66	-40	45	-935

annuels en eau d'irrigation puis connaître la quantité d'eau requise dans le cas théoriquement optimal d'une mise en culture intégrale de ces zones.

Sachant que K_c est le coefficient cultural, l'évapotranspiration de la culture Et_c est le produit entre l'ET_0 et K_c ($Et_c = K_c * ET_0$). Les besoins nets d'irrigation (I_n) sont déterminés par : $I_n = Et_c - P_{efficace}$ (Smith M., 1992).

Connaissant la période d'irrigation, la valeur I_n peut être transformée en débit moyen (l/s) puis en débit continu (l/s/ha). Cette période correspond à la somme des mois agronomiquement secs au cours desquels l'irrigation est nécessaire.

Quelque soit le type de culture envisagé, un mois est agronomiquement sec lorsque:
• P + RFU < ETP si le mois antérieur est excédentaire,
• P < ETP si le mois antérieur est déficitaire.

La confrontation des totaux pluviométriques et des ETP mensuelles (ces dernières calculées suivant la formule la plus adoptée au climat local, voir tableau 1), permet d'évaluer le nombre des mois secs au cours desquels il faut irriguer (Durand J. H. 1983).

Le choix des cultures s'effectue en fonction des conditions du milieu naturel (pédologie - climat), des objectifs généraux de la planification algérienne et selon que les cultures se font en sec ou en irrigué.

La collecte des informations sur les cultures comprend la durée du cycle de croissance, les coefficients culturaux et la profondeur d'enracinement, etc.

Ainsi, les besoins en eau mensuels de chaque culture peuvent être déterminés.

Si le planificateur dispose de la localisation des zones irrigables, de la période au cours de laquelle il faut les irriguer et des quantités d'eau requises à leur mise en valeur agricole, il pense à quantifier ou à découvrir les sources d'alimentation en eau susceptibles de couvrir les besoins. La première option, à long terme, nécessite un investissement financier très important et la deuxième, à court terme, permet une intervention rapide, nécessite de faibles investissements.

b – Calcul des besoins en eau du périmètre (BEP)

Connaissant les besoins unitaires de chaque culture, il convient de déterminer les besoins globaux du périmètre. Dans cette estimation interviennent l'efficience du système d'irrigation, les superficies

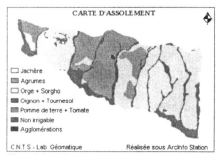

Figure 4. Carte d'assolement

Tableau 2. Besoins réels en eau d'irrigation par mois et par culture dans le périmètre.

Type de culture	Besoins réels en eau (m^3/ha)											
	J	F	M	A	M	J	J	A	S	O	N	D
Pomme de terre	-	-	100	660	1590	2190	860	-	-	-	-	-
Tomate	-	-	-	280	1050	2130	2580	1980	55	-	-	-
Oignon	-	-	-	280	1020	1860	2220	1950	150	-	-	-
Tournesol	-	-	-	80	810	2190	2520	680	-	-	-	-
Sorgho	-	-	-	-	-	360	1500	2190	1380	50	-	-
Orge	60	510	810	400	-	-	-	-	-	-	-	-
Agrumes	30	210	420	660	990	1380	1560	1380	930	420	210	30

mises en jeu par l'irrigation et la répartition relative des différentes cultures (Mendas A.,1999).

Les débits continus des différentes cultures, leur répartition et leur superficies sont présentés sur la figure 4.

Basée sur les travaux de (Kilborn K. *et al.*, 1992), l'application de la méthode de THIESSEN (voir Fig. 5) disponible dans le logiciel Arc/Info, a permis l'estimation des réserves en eau souterraine de la région.

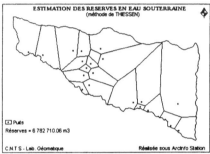

Figure 5. Estimation des réserves en eau souterraine

Connaissant la localisation, les besoins en eau des zones irrigables et la quantité des eaux disponibles, le planificateur peut faire une confrontation, une sélection par priorité des zones à mettre en irrigation et une mise en place des projets particuliers. Les résultats des besoins en eau du projet pour le plan de culture donné peuvent être comparés à la réserve en eau disponible à la dérivation, au barrage de prise ou au réservoir.

En effet, avec un débit moyen de 0.5 l/s/ha, les besoins en eau de tout le périmètre est de 7.88 hm^3/an. Par contre, avec les débits de chaque culture sélectionnée, les besoins en eau du périmètre cultivé (estimé à 71% du périmètre irrigable) est de 7.03 hm^3/an.

Or, les ressources en eau disponibles sont estimées à 97.7 hm^3/an (barrage de Beni-Bahdel : 63 hm^3/an, eaux souterraines: 7 hm^3/an, eaux superficielles "Oued Mouilah": 27.7 hm^3/an). Sans tenir compte des autres exploitations de ces ressources, le besoin est nettement inférieur au disponible..

5 CONCLUSION

L'utilisation des SIG pour la planification de l'irrigation requiert dès le début une méthode où les objectifs doivent être définis avec précision, avant

toute manipulation de logiciel. En raison de la puissance de l'analyse spatiale conçue pour un SIG, ce dernier doit s'intégrer dans une démarche géographique, afin que la réponse attendue ne reste pas seulement descriptive des phénomènes mais devienne analytique et spatiale.

La confrontation entre ressources potentielles en eau et en sols et leur usage actuel permet d'ores et déjà au planificateur de préparer les diverses étapes chronologiques d'un plan d'irrigation orienté vers la mise en valeur des terres irrigables

Le traitement numérique des données dans les SIG pourrait aider à maintenir à jour les informations concernant les terres irrigables et/ou irriguées et par conséquent, la précision du bilan hydrique. En effet, l'utilisateur disposant des données estimatives suffisantes peut aboutir à une meilleure aptitude culturale puis réaliser une planification harmonieuse et rationnelle de l'irrigation.

Par l'utilisation du logiciel CROPWAT muni d'une banque de données climatiques et culturales réalisé par la F.A.O, et du SIG Arc/Info permettant des opérations spatiales puissantes pouvant prendre en compte plusieurs facteurs limitants, les résultats quantitatives sont plus réalistes par rapport aux résultats obtenus par mode traditionnel, et peuvent être extrêmement variés.

Les systèmes d'information géographiques, techniques devant être maîtrisées et adaptées à la réalité locale par valorisation des données existantes, peuvent apporter une réelle réponse à la planification de d'irrigation. Cependant, leur utilisation requiert une méthode où les objectifs doivent être définis avec précision, avant toute manipulation de logiciel. Il est également nécessaire de réfléchir à la structuration de l'analyse relationnelle, entre entité, et spatiale pour assurer d'une part une sélection précise des variables constitutives du système et d'autre part, un choix des échelles d'entrée, d'analyse et de restitution.

Bien entendu, plus les données sont disponibles, plus fine sera l'analyse et par la suite les résultats. Les couches d'information présentées s'interprètent comme un test de faisabilité suivant les fonctionnalités offertes par les SIG. Il est important de souligner qu'elles ne constituent pas un document définitif mais restent expérimentales.

La mise en place d'une base de données fiable et de qualité, sa mise à jour régulière grâce à un circuit d'information bien organisé sont donc les prémices d'une véritable montée en charge du SIG. Les gains liés à la disponibilité des informations sont cependant réels, au point de rendre un SIG bien intégré dans la vie d'une entreprise, un outil à ne pas passer.

Cette expérience a montré que le SIG ARC/INFO est d'un apport majeur sur plusieurs aspects. L'automatisation, la rapidité de manipulation et la mise en mémoire des données avec capacité de réactualisation ont donné pleine satisfaction.

Enfin, à une époque où la technologie offre des possibilités sans cesses croissantes dans ce domaine, il est essentiel que les décideurs prennent conscience et financent convenablement ces outils d'aide à la gestion et à la prise de décision en matière de ressources naturelles.

REFERENCES

Durand J. H. (1983) *Les sols irrigables, étude pédologique. Agence de coopération culturelle et technique.* France : 322 p.

ESRI. 1995. GIS by ESRI. *Understanding GIS, The Arc/Info method.*

Henriksen C. & HALL L. (1992) GIS measures water use in the arid west, USA : Water Resources, *Geo Info Systems.* July/August : 63-67.

Kilborn K. & Rifai H. S. & Bedient P. B. (1992) Connecting Groundwater Models and GIS solutions *: Geo Info Systems.* February : 26-30.

Mendas A. (1997) Thèse de magister : *Apport des systèmes d'information géographique à la planification de l'irrigation .* CNTS, 112 p.

Mendas A. (1999) Estimation de l'eau nécessaire à un périmètre d'irrigation par l'utilisation d'un SIG. *Journées techniques et scientifiques de l'eau.* Blida les 22 et 23 février 1999, 13 p.

Smith M. (1992) Cropwat : un logiciel pour la planification et la gestion des systèmes d'irrigation. *Bulletin F.A.O d'irrigation et de drainage* n° 46, 133 p.

Observing our environment from Space: New solutions for a new millennium, Bégni (Ed.)
© 2002 Swets & Zeitlinger, Lisse, ISBN 90 5809 254 2

Using satellite data for land use change detection: a case study for Terkos water basin, Istanbul

C.Goksel, S.Kaya & N.Musaoglu
Istanbul Technical University, Remote Sensing Division, Istanbul, Turkey

ABSTRACT: This study covers the analysis of temporal change of the land use in Terkos basin which is among the basins, actually the most important one, that were studied within the project of "Analysis of the Land-Use Changes by Using the Satellite Images of the Water Basins of Istanbul" carried out by a group from Istanbul Technical University (ITU), Remote Sensing Division for Istanbul Water Board Authority (ISKI). As the spatial resolution is high, IRS 1C-1D and LISS III satellite images belonging to three different years were used in the project. In the first stage of the study, geometric rectification process of the images were made. In the second stage, land-use classes were determined by applying classification algorithms on the basis of outer border of the long-term conservation area of Terkos basin. From the classification process and its results, area values were calculated and temporal changes were ascertained. In the last stage of the study, accuracy evaluations has been carried out.

1 INTRODUCTION

Being the largest metropolis of Turkey, Istanbul is face to face with environmental problems brought about by fast population growth. Uncontrolled building-up and related destruction of the forests poses a great threat for Istanbul. In order to maintain its life, mankind has to use the available natural resources in the most economic way and protect such resources so as to leave the best for the generations to come. Parallel to the developing technology and the increased scientific studies today, developments in the remote sensing science appear to be of great help to humanity.

Superficial water resources providing water for Istanbul are facing the problem of pollution due to change in land use. Physical, chemical and bacteriological deteriorations in the quality of water generally arise out of the settlements and therefore human effects in the water collection basins. Nowadays, vegetation cover of the water production basins in Istanbul is ending up as mine pits planned or unplanned industrial premises. The pollutants and the waste produced by these facilities are not only obliterating the vegetation but also directly pollute the water basins (Goksel, 1998).

As a result of the urban developments in the water basins, agricultural land and forests are disappearing, superficial drinking water is polluted, animal kinds are decreasing, and air pollution and garbage become a serious concern. Protecting these areas which are the source of our most essential need, that is water, is important not only for the present but also for the future generations. Change of the forest areas in the water basins which have to be entirely surrounded by a green belt and failure to control this change lead to a great deal of problems.

It is necessary to monitor and know, at certain intervals, the amount and temporal change of the parameters such as area, tree wealth, growing sites and type of trees of the forest resources As the forest inventory to be produced by means of terrestrial studies is a time consuming, expensive and difficult task, it is shown by the studies made that satellite images yield sufficient and reliable information in a short time for this purpose (Musaoglu, 1999). Over the past decades, land management and regulatory agencies, earth system scientists and non-governmental organisations have mounted a growing number of campaigns to develop large-area digital maps of vegetation and land cover (Botkin, 1984; Franklin et al., 2000). Satellite images widely used in the fields of national forest inventory and monitoring the periodical changes in forest lands are combined with aerial photographs to prepare plans of forest management (Hame and Rauste, 1995).

This paper, by using satellite images of various dates, compares the temporal changes in the forest areas of the Terkos Lake Basin in terms of visual statistics and areas.

2 STUDY AREA

Terkos (Durusu) Lake is a freshwater lake located on the European Side of Istanbul by the Black Sea coast covering .688km^2 being nearly 70 kms away from the city center (Figure 1) Since the late last century, this lake has been being used as a water depot providing for most of the water requirement of the city of Istanbul. It is separated from the sea by a low threshold whose width ranges between 0. 5 – 2 kms, covered with sand and sand dunes. Terkos Lake is basically an old bay with inlets and outlets formed as a result of a network of valleys being inundated by a final sea tide encased within the formations. Its water has become fresh because of the streams feeding this bay. The most important one of these rivers is the River Istranca in the northwest.

There are seven drinking water basins in Istanbul, each of which is separated into conservation bands according to distances within itself. Rules concerning constructions for each of these conservation bands are set by the Basins Directive updated in 1998 by the Istanbul Water Board Authority. Figure 1 illustrated Terkos boundries.

3 USED DATA

3. 1. Maps

A number of maps (topographic maps, thematic maps and plans) were used in the study. For geometric correction, 13 pieces of 1/25000 standard topographic maps of the Terkos basin have been used and 1/ 50000 scale and 1/5000 scale orthophoto maps covering the definite conservation (0 – 300 m) of the basin were utilised.

Figure 1. Terkos Water basin Borders.
(Colour plate, see p. 428)

3. 2. Satellite Images

19 satellite images were used in the project. Due to high resolution of Pan image, the preferred images were IRS 1C, 1D and LISS III. IRS 1C has been launched into polar orbit on the 28th December, 1995. In Table 1, characteristics and in Table 2 dates of the images used of the Indian satellite which has been utilised are displayed.

Table1. Indian Remote Sensing Characteristics

SENSOR	PAN	LISS-III
Spatial Resolution	5.8 m	23m (VIS and NIR)
		70 m (SWIR)
Swath-width	70 km	142 km
Radiometric resolution	6 bit	7 bit
Spectral Covarege	500-750 nm	520-590nm (Band 2)
		620-680nm (Band 3)
		770-860nm (Band 4)
		1550-1570nm(Band 5)

Table 2 . Images Dates

Sensor / Path-Row		Date
1996		
PAN 1C	46 / 40	30 June 1996
PAN 1C	46 / 40	24 July 1996
LISS III	46 / 40	
2000		
PAN 1 D	45 / 40	09 May 2000
LISS III	45/40	09 May 2000

4 METHOD

4. 1. Geometric Correction of the Satellite Images

Original satellite images obtained from the remote sensing systems contain systematic errors such as scanning errors and slope of earth's surface as well as non-systematic errors such as position of the satellite and height of the sensor platform. Systematic errors can be removed by means of corrections made according to error sources. Non-systematic errors, on the other hand, can be removed by means of physical features (such as crossroads, sharp details) that can be definitely distinguished on the image or by means of mathematical connections set among the point co-ordinates established by GPS (Berstein 1983, Richards 1986, Jensen 1996, Musaoglu 1999). Moreover, for the integration of satellite images with other data groups, it is necessary to define all the data groups within the same projection system.

In geometrical correction of satellite images,

- 1/ 25000 scale standard topographic maps with JPG extension transferred to computer media after having been scanned and 1/ 5 000 scale orthophoto maps of the area have been rectified during the first stage.

- In the second stage, an average of 20 ground control points distributed homogenously and which can be sharply distinguished in panchromatic satellite images at 5.8m spatial resolution and on maps, have been selected.

- The images have been referenced to UTM (Universal Transverse Mercator) by using average 0.5–0.7 pixel root mean square error and cubic convolution resampling.

- Multispectral images have been rectified by using *image to image* and *map to image* methods with average 15 points and 0. 5 pixel rms.

4.2 . Merge and Classification

In order to enhance the visual interpretability of the satellite images and differentiability of the surface cover features, images obtained in two platforms can be collected within one set of data. In the merging process, the most known technique is IHS conversion. In this study, it is intended to obtain more detailed image of the basin by merging the single-channel IRS-1C Pan image processing 5.8m Ground resolution with the image of 3-channel LISS III possessing 23.5 resolution. In order to carry out the process of merging, number of lines and columns as well as data file co-ordinates of both images have to be the same. Green, Red and Near Infrared channels on the LISS III image were executed as RGB, and IHS conversion was applied by means of the algorithm in the available ERDAS Imagine software. The Intensity Channel that was obtained later has been replaced with IRS 1C Pan channel, applying the reverse process to obtain RGB mode again. The resulting image obtained is a multispectral image enhanced as a spectral one with 5.8m ground resolution. Figure 2 Indicates the merged image thus formed.

Different feature types in digital images constitute combinations, which contain different digital values depending on the natural spectral reflection and emittance characteristics. The purpose in classification is to group objects bearing same spectral features (Musaoglu, 1999).

In the classification process ISODATA (Iterative-Self Organizing Data Analysis Technique) and method were used. In ISODATA method, image elements are classified iteratively. Attributing certain mean values to the random groups starts the

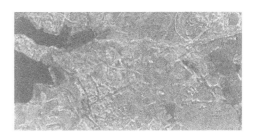

Figure2. Merge Image (IRS IC + LISS 111). (Colour plate, see p. 428)

process and this process is repeated until these arbitrary mean values are classified to the group values in the data. In this method, the groups can be obtained like sampling areas in controlled classification and the groups that are found appropriate can be used in other controlled classification methods too.

LISS III images have been classified with ISODATA method. As a result of the classification where number of groups is entered as 40, the groups have been analysed by utilizing topographic and thematic maps and some groups were deleted/combined to decrease the number of groups to 6.

- *Water*
- *Forest*
- *Bare soil*
- *Rangeland*
- *Agricultural land*
- *Urban*

The obtained results were once again analysed with the ancillary data belonging to the image area and a legend was prepared for the formed classes Figure 3. Indicates the LISS images obtained as a result of this classification. It has been established that the classified image is generally consistent with the landuse map but as the northern parts of the studied area is on the Black Sea coast there are forests and evidently decreased forest lands due to attraction for agricultural purposes of the fertile Thrace Territory. In the southern parts of the basin there appears to be differences in the fertile forest formation. Particularly in the southeastern part of

Table3: Classification Results

Class	1996 (ha)	2000 (ha)
Water	3670.19	3470.44
Bare Soil	1851.50	1556.88
Forest	53227	51421
Rangeland	2100.19	5445.81
Agricultural land	10692.90	3647.25
Urban	2303.81	3210.56

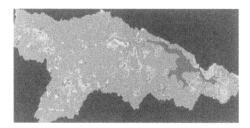

Figure 3a. Classified 1996 LISS image. (Colour plate, see p. 428)

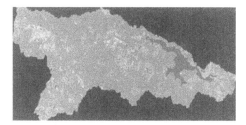

Figure 3b. Classified 2000 LISS image. (Colour plate, see p. 428)

the area the formation of the forest cover is noted to have been spoiled (thickness decreased) in the course of time.

When the basic land use groups obtained by means of the classification results are calculated areawise according to the years, changes have been observed in land use within a 4-year period. Although the Terkos basin attracts less population for human settlement compared to other water basins of Istanbul, it is seen that a change has started also in this region.

4.2. Evaluation of Accuracy

The quality of an image classification and its appropriateness for use is evaluated typically by its accuracy (Foody, 1992). The most common way to express the accuracy of such images is by a statement of the percentage of the map area that has been correctly classified when compared with reference data.

Accuracy of classification has been determined by using error matrices formed with random sampling. In an evaluation carried out with 200 randomly samples. Error matrices and statistics were produced for each classified images. In this study for classified images of 1996, overall classification accuracy for six classes was 77 % and kappa coefficient was 0.74. For the 2000 images overall classification accuracy for seven classes was 81 % and kappa coefficient was 0.79.

5 CONCLUSION

Impact of the urbanisation in Istanbul can be determined through satellite images in the light of visual and statistical results also in the Terkos basin. It is facing problems of human settlement and rising population. Destruction of the forestlands is seen especially in the north and west of the basin. Urbanisation leads to increased speed the flow of surface water as well as the quantity and variety of the objects being accumulated on the ground, causing a major pollution of resources. It is essential that satellite images are to be used in monitoring such intensive changes. By using especially the satellite images with high spatial resolution, one must determine the purpose for which the forest areas that are subject to changes in the water basins, which have vital importance for the cities, are used. Thus, these areas are to be analysed temporally in order to take the necessary measures on time.

REFERENCES

Botkin, D.B., Estes, J.E , MacDonald, R.M., Wilson, M.V., 1984. Studying The Earth Vegetation From Space, Bio Science, 34181:508-514.

Elijah, W., Ramsey, M., Sensen, J.R., 1996. Remote Sensing and Mangrove Wetlands Relating Canopy Spectra to Site - Specific Data,Photogrammetric Engineering&Remote Sensing , Vol. 62, No: 8, pp. 939-948.

Foody, M. G. 1992.Classification Accuracy Assessment: Some Alternatives to the Kappa Coefficient for Nominal and Ordinal Level Classifications. Proceedings of the 18th Annual Conference of the Remote Sensing Society, Dundee pp: 529-538.

Franklin J.,Woodcock C., and Warbington R., 2000. Multi-Attribute Vegetation maps of Forest Sevice Lands in California Supporting Resource management Decisions. Photogrammetric Engineering and Remote Sensing Vol,66, No. 10,pp.1209-1217

Goksel, C., 1998. Monitoring of a water basin area in Istanbul using Remote Sensing Data. Water Science and Technology, 38 (11), 209-216

Hame, T., Rauste, Y., 1995 . Multi Temporal Satellite Data in Forest Mappind and Fire Monitoring, EARSel Advances in Remote Sensing, Vol.4, No: 3, pp. 93-101.

Musaoğlu, 1999. The Integration of Different Data Groups with Satellite Images, International Symposium on Remote Sensing and Integrated Technologies, Proceedimg, 391-396.

Varjo, J., 1995. Forest Change Detection by Satellite Remote Sensing in Eastern Finland, EARSel Advances in Remote Sensing , Vol.4, No.3.

Observing our environment from Space: New solutions for a new millennium, Bégni (Ed.)
© 2002 Swets & Zeitlinger, Lisse, ISBN 90 5809 254 2

The use of LANDSAT7 ETM + - and SeaWiFS-data to detect sediment plumes in the Rio de la Plata

T.Vanderstraete
Department of Geography, Ghent University, Gent; Research Assistant, Fund for Scientific Research-Flanders, Brussel, Belgium

R.Goossens & M.De Dapper
Department of Geography, Ghent University, Gent, Belgium

ABSTRACT: The Rio de la Plata is economically and ecologically very important for South-America. Using data derived from LANDSAT7 ETM+- and SeaWiFS- images it is possible to detect sediment plumes in the estuary. This information is of special interest for the maintenance of the man-made channels that assure access to e.g. the harbor of Buenos Aires, as for the monitoring of the quality of the water. Part of the study has also explored the possibilities to extrapolate information derived from the Landsat7-image to a coarser SeaWiFS-image. A Resurs-01-image is used to link both images. Several conditions have to be fulfilled before this can be done. If these prerequisites are met, it is possible to develop a less expensive and more global monitoring system for the Rio de la Plata based on remote sensing, as is possible with conventional monitoring techniques.

1 PHYSICAL GEOGRAPHICAL SETTINGS

The Rio de la Plata-estuary (Fig. 1) is situated on the east coast of South America between 33°50'S - 58°20'W and 36°10'S - 56°30'W. It is nearly 250 km in length and covers an area of about 38,800 km² (López Laborde 1997).

Figure 1. Localisation of the Rio de la Plata (Source: Encyclopedia Britannica, Inc. 1994).

The funnel-shaped estuary becomes broader toward the southeast from 32 km between Colonia de Sacramento and La Plata, to 100 km between Montevideo and Punta Pietras and to 230 km between Punta del Este and Cabo San Antonio (López Laborde 1997).

The Rio de la Plata is the collector of the second largest drainage basin of the continent, covering 3,170,000 km² (Tossini 1959 in: López Laborde 1997), formed by the Uruguay and Paraná-Paraguay rivers. These two river systems are responsible for 97 % of the total water input into the Rio de la Plata (López Laborde 1997). The Rio Paraná is by far the more important of these two rivers (Table 1). According to Nagy et al. (1997) the flood of the Paraná River occurs in late summer and autumn, the minimum in winter and spring. The Uruguay River flood occurs in winter, with a secondary maximum in October, and a minimum flow from summer to autumn (November-May). Because the seasonal cycles of the Paraná and Uruguay Rivers are partly opposed and mutually compensated and the variation of the discharge is low (Table 1), there is no clear seasonal cycle of the water in the Rio de la Plata (Nagy et al. 1997). The suspended sediment load of the Paraná River is greater than that of the Uruguay River (Table 1), because the Uruguay River basin is partially developed in crystalline rocks (Nagy et al. 1997).

In the west, the Rio de la Plata is boarded by the 14,100 km² (López Laborde 1997) large delta of the Rio Paraná, which has an average growth speed of 219.4 m² a year (Cavallotto 1987, Parker & López

Laborde 1990 in: López Laborde 1997). Surficial bottom sediments (Urien 1966, 1967, 1972, Parker et al. 1985, López Laborde 1987a, 1987b, Parker & López Laborde 1990 in: López Laborde 1997) in the Rio de la Plata exhibit a graded distribution due to selective sedimentation with sands predominating where the rivers enter the Rio de la Plata, mainly silts in the middle region and clayey silts towards the mouth.

Table 1. Hydrological characteristics of the Rio Paraná, the Rio Uruguay and the Rio de la Plata.

	Rio Paraná	Rio Uruguay	Rio de la Plata
Extreme maximum discharge	22,000 m³/s[*]	14,300 m³/s[**]	30,600 m³/s[**]
Mean discharge	17,000 m³/s[*]	4700 m³/s[**]	20,000 m³/s[**]
Extreme minimum discharge	8000 m³/s[*]	800 m³/s[**]	10,800 m³/s[**]
Variability of the discharge	4 %[**]	9 %[**]	5 %[**]
Mean sediment load	121 mg/l[***]	52 mg/l[****]	40–225 mg/l[*****]

[*] Measurements between 1884-1975 at Rosario (source: CARP-SINH-SOHMA 1990 in: Nagy et al. 1997)
[**] Measurements between 1916-1975 at El Hervidero (source: Nagy et al. 1997)
[***] Tossini 1959 in: Nagy et al. 1997
[****] Nagy 1989 in: Nagy et al. 1997
[*****] Measurements in the upper tidal river, values increase near the middle of the river (source: Nagy 1989 in: Nagy et al. 1997)

The area of investigation specified for this study is restricted to the westernmost part of the Rio de la Plata and is roughly boarded by the virtual line between Colonia de Sacramento (Uruguay) and La Plata (Argentina) in the east. The study area can be divided into 4 different 'morphological units' (Cavallotto 1987, Parker & López Laborde 1988, 1990 in: López Laborde 1997) (Fig. 2). The largest unit, the "Playa Honda", is a widespread shallow area that forms the subaqueous extension of the Paraná River delta, limited approximately by the 6.0 m isobath. It is formed by the deposition of the fluvial load from the river as the water velocity decreases. The deposited sediments form bars across the tributary outlets. In this way the outflowing water is blocked and starts to form new outlets. As these bars become joined, islands are formed which eventually connect to each other, allowing the delta to enlarge. The discharge of the Uruguay River causes a redistribution of the sediments furnished by the Paraná River tributaries and is, in this way, responsible for the differential lineal growth of the Paraná delta. The Playa Honda is also responsible

for the separation of the sediment transport along the north and the south border of the Rio de la Plata (Parker et al. 1985 in: López Laborde 1997). The channels in the north, as can be seen on the DTM (Fig. 2) are formed by the erosive action of the Uruguay and Paraná Bravo Rivers and are known as the "Sistema Fluvial Norte". It includes channels, longitudinal benches and subaqueous, asymmetrical sandy banks. This unstable morphology has the characteristics of a braided river, as it is the result of the dynamic balance between the sedimentation process of the Paraná River delta and the erosive forces of the Uruguay River. In the south, a westernmost spur of the "Gran Hoya del Canal Intermedio" can be detected. This Canal is formed by the influences of the tidal currents in the Rio de la Plata. In general it is deeper than the Playa Honda. The subaquatic slope offshore the coast of Argentina is known as the "Franja Costera Sur". It is limited to a depth of 6 to 9 m and is subject to sedimentation.

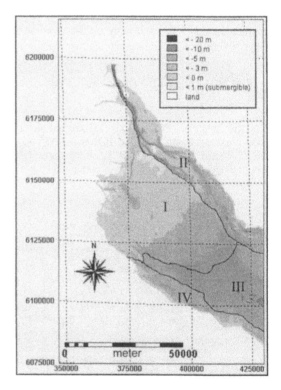

Figure 2. Digital Elevation Model of the study area (Source: Goossens et al. 2000) with delimitation of 4 'morphological units' (Source: Cavallotto 1987, Parker & López Laborde 1990 in: López Laborde 1997):
I Playa Honda
II Sistema Fluvial de Norte
III Gran Hoja del Canal Intermedio
IV Franja Costera Sur

2 DETECTION OF SEDIMENT PLUMES USING REMOTE SENSING

The Rio de la Plata is, economically and ecologically, very important for Argentina and Uruguay. It gives access to the harbor of Buenos Aires and La Plata, to the Rio Paraná and the Rio Uruguay. It is important for tourism, for fishing, as supplier of water for domestic use, waste disposal and so on (Wells & Daborn 1997). As the contribuary rivers, especially the Rio Paraná, discharge large amounts of suspended material into the Rio de la Plata -mean load greater than 75% (Nagy et al. 1997)-, the estuary is subject to sedimentation. The supply of suspended material, combined with the relative shallow depth of the Playa Honda, makes it necessary to dredge channels in order to allow modern shipping to frequent the harbor of Buenos Aires or other ports more upstream. Because of the importing sedimentation taking place, maintenance is also very important. For this purpose it is necessary to have insight into the patterns of sediment transport in the Rio de la Plata. In this way the dredging activities can be carried out more efficiently.

Table 2. Technical information concerning the LANDSAT7 ETM+-image used (Source: Goossens et al. 2000).

Reference number	E1SC:L70RWRS.002:2000705192
Processing level	LANDSAT-7 LEVEL-1 WRS-SCENE V002
Cloud Cover	0%
Coordinates centre	34.6118°S 58.7611°W
Coordinates upper left corner	33.6483°S 59.5631°W
Coordinates upper right corner	33.9487°S 57.5033°W
Coordinates lower left corner	35.5689°S 57.9392°W
Coordinates lower right corner	35.2626°S 60.0400°W
Path/Row	225/084
Date	19/01/2000
Time	13:37:24 UTC

Remote sensing can play a major role in the study of the transportation of the sediment throughout the Rio de la Plata. In comparison with conventional ship based monitoring techniques, some advantages can be noted (Goossens et al. 2000). First of all, remote sensing gives the opportunity to get a synoptic, global overview of the situation in the area of investigation. The conventional methods only produce point information connected with the sampling transect. In order to get the same coverage as with remote sensing, the sampling density has to be higher, which has repercussions on the costs. It is also impossible to map an equally large area with the conventional techniques in one short campaign. In this way, the remote sensing technique is less expensive and less labour intensive. Remote sensing also allows to make a multi-temporal study of the sediment plumes and to track seasonal changes.

Because information is required on materials in suspension, passive sensors have been chosen which detect in the visual range, more precisely in the blue and the green range as these wavelengths are less absorbed by the water. For this study a LANDSAT 7 ETM+ – image (Table 2) and a SeaWiFS – image (Table 3) are used. Both images are recorded on the same day, namely January 19th, 2000. The LANDSAT7 ETM+-image, with a spatial resolution of 30 m, is used for a detailed investigation of the sediment plumes in the study area, while the coarser SeaWiFS-image, with a spatial resolution of 1.1 km, is used to obtain an overview of the entire Rio de la Plata and adjoining part of the southern Atlantic Ocean (Goossens et al. 2000).

Table 3. Technical information concerning the SeaWiFS–image (Source: Goossens et al. 2000).

Reference number	S2000019153816.L1A_HARG_BRS
Processing Level	LEVEL-1A
Date	19/01/2000

3 IMAGE PROCESSING

The images are processed using the 'ILWIS 2.2 for Windows' - software, developed by ITC and distributed by PCI Geomatics.

3.1 *LANDSAT 7 ETM +*

Only the part of the LANDSAT7 ETM+-image covering the study area is used in the following image processing methodology. First of all the image is georeferenced to a UTM-coordinate system for this study specifically (Table 4) using ground control points. The coordinates of these ground control points are derived from 'An International travel map of Uruguay (1st edition), 847' on a 1/800,000 scale. This small-scaled map is appropriate for this purpose but is responsible for a sigma-error of 3.221 when the 'full second order' transformation method is applied.

Table 4. Properties of the study specific UTM-coordinate system (Source: Goossens et al. 2000).

Min X, Y: -59.00° / -40.00°
Max X, Y: -40.00°/ -27.00°
Projection: UTM
Datum: WGS 1984
Ellipsoid: WGS 84
Southern Hemisphere
UTM-Zone 21

Because only the Rio de la Plata is in the interest of this study, a mask is developed to cover the surrounding land. The mask is created using the infrared band (nr. 5). As this spectral range is nearly totally absorbed by the water, a clear distinction can be made between the spectral values of the water and the spectral values of the land. By masking the land in black, more contrast can be put into the spectral range of the water in the estuary.

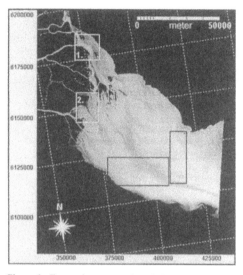

Figure 3. True colour composite of the LANDSAT7 ETM+ showing different sediment structures (white boxes) and man-made channels (black boxes) (Source: Goossens et al. 2000).

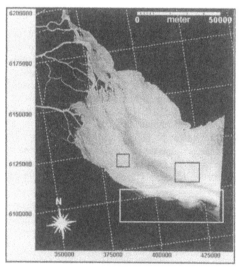

Figure 4. Conventional False Colour Composite of the LANDSAT7 ETM+ showing the eutrophic sewage of Buenos Aires (white box) and several spoil-grounds (black boxes) (Source: Goossens et al. 2000).

In order to detect the sediment plumes in the Rio de la Plata, two colour composites are created, namely a True Colour Composite (TCC), a combination of bands 1, 2 and 3, (Fig. 3) and a Conventional False Colour Composite (FCC), a combination of bands 2, 3 and 4 (Fig. 4). In theory the near infrared band nr. 4 would not give important information concerning the quality of the water, as it is almost totally absorbed. But, as can be seen on Figure 4, in this case it does give important insight into the composition of the suspended material (Goossens et al. 2000). Mid-infrared bands 5 and 7 are excluded because they are too much absorbed to give any additional information about the properties of the water. The thermal band 62 (Fig. 5) is used separately, due to the coarser resolution, to help explain the hydrological situation in the Rio de la Plata. The panchromatic band 8 is not used in this study because, although the much finer resolution of 10 m, it doesn't give any significant extra information. In making these colour composites a linear stretch is applied in order to get the most contrast, equally distributed over the spectral range, into the images.

As can be seen on the TCC (Fig. 3) the man-made channels can be detected on the satellite image. In order to get a better view on them, a Y-directional (3 × 3) filter (Table 5) is applied on the first ETM + -band (Fig. 6a) (Goossens et al. 2000).

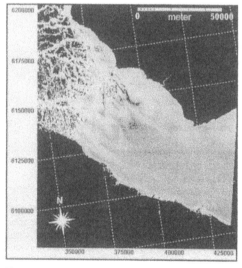

Figure 5. Thermal LANDSAT7 ETM+ - band 62 showing contrast between the salty water and the warmer fluvial water (Source: Goossens et al. 2000).

Table 5. Y-directional filter (3 × 3) used to enhance the view on the channels (Source: Goossens et al. 2000).

-20	-20	-20
0	0	0
20	20	20

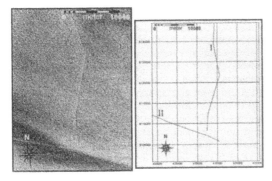

Figure 6a. Image of the LANDSAT7 ETM+ band nr.1 after applying the Y-directional filter (Source: Goossens et al. 2000).
Figure 6b. Vectorisation of the man-made channels:
I: Canal Martin Garcia;
II: Canal de Acceso al Puerto de Buenos Aires
(Source: Goossens et al. 2000, Admiralty Chart 3561 1995).

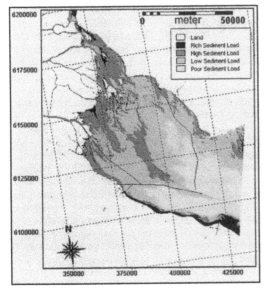

Figure 7. Density Slicing on LANDSAT7 ETM+ - band nr.1 showing 5 classes of sediment load with an overlay of the detected channels (black lines) (Source: Goossens et al. 2000).

The result of this operation makes it possible to vectorise the channels into a different layer (Fig. 6b) that can be combined with other raster maps (Fig. 7).

Due to lack of information concerning the properties of the sediment at the time the image was recorded, it is not possible to make a quantitative classification of the image. This problem can partly be managed by making a density slicing on one band in order to get a relative classification (Goossens et al. 2000). The density slicing is executed on the first spectral band and a distinction is made between 5 different classes (Table 6). This 'Density slicing' is based on a subjective separation but is still appropriate to distinguish different sediment loads. The result can be seen on Figure 7.

Table 6. Threshold values and distinguished classes used in the density slicing of LANDSAT7 ETM+ - band nr.1 (Source: Goossens et al. 2000).

Threshold value	Class name
0	Land
90	Rich sediment load
94	High sediment load
98	Low sediment load
110	Poor sediment load

3.2 SeaWiFS

As one SeaWiFS-image covers the entire southern part of South-America, a selection is cut out of the different bands covering the Rio de la Plata and an adjoining part of the Atlantic Ocean. To geocode this image, the coordinates of the ground control points will be acquired from the georeferenced LANDSAT 7 ETM + - image. As the difference in resolution between these two images, respectively 30 m for the LANDSAT7 ETM+ and 1.1 km for SeaWiFS, is large, a RESURS-01-archive image, with a spatial resolution of 160 m (Table 7), is used as an intermediate to link these two images geometrically to each other (Goossens et al. 2000). This is done using the slave/master - method whereby first the RESURS-01-image is registered to the LANDSAT7 ETM+-image (sigma: 0.375) and afterwards the SeaWiFS-image to the RESURS-01-image (sigma: 0.419).

Table 7. Technical information concerning the RESURS-01 SK – image (Source: Goossens et al. 2000).

Reference number	SK951220S34W055
Coordinates center	33.85°S 54.68°W
Coordinates upper left corner	30.97°S 56.28°W
Coordinates upper right corner	30.08°S 50.52°W
Coordinates lower left corner	35.73°S 58.65°W
Coordinates lower right corner	36.78°S 51.25°W
Date	20/12/1995
Time	12:46 UTC

Also for these bands a mask for the land is made, based on an infrared band (nr. 8). 2 colour composites are produced. The first one is similar to the conventional false colour composite of the LANDSAT7 ETM+-image. Whereas the SeaWiFS-sensor has 2 green and 2 near-infrared bands, several combinations can be made. The combination of bands 8 (IR), 6 (Red) and 5 (Green) has proven out to have the best resemblance to Figure 4 (Fig. 8) (Goossens et al. 2000). Also a colour composite (Fig. 9) is produced based on the Optimal Index Factor (O.I.F.).

This factor calculates the optimal combination of three spectral bands so the abundancy is minimal (van Westen & Farifteh 1997). The best combination based on this O.I.F. is a combination of bands 3 (blue), 2 (blue) and 1 (blue). This image gives more information concerning the sediment transport in the Atlantic Ocean (Goossens et al. 2000).

Figure 8. Colour Composite (band8/6/5) of the SeaWiFS with the position of the detailed image (Fig. 10) in white (Goossens et al. 2000).

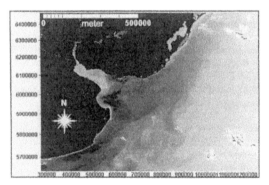

Figure 9. O.I.F.-colour composite (band3/2/1) of the SeaWiFS showing the sediment plumes in the Atlantic Ocean (Goossens et al. 2000).

4 RESULTS

Although the true (TCC) and conventional false (FCC) colour composites used for the LANDSAT7 ETM+-image are not the result of complex calculations or image processing techniques, lots of information can be derived out of them. Several patterns of transport and sedimentation can be detected besides small vessels and parts of the dredged channels. Remarkable is the contact zone between the water supplied by the Rio Uruguay and the Paraná Bravo River (Fig. 3, marked in white). As in these zones, the speed of the water is normally lower and

more turbulent, more suspended sediment is concentrated there (Goossens et al. 2000). Several other contact zones can be noticed.

Also remarkable is the possibility to detect the sewage water from Buenos Aires. Two important subaquatic sewage outfalls, called Berazategui and Berisso, release industrial and domestic effluents from Buenos Aires and La Plata City respectively (AGOSBA-OSN-SIHN 1994 in: Kurucz et al. 1997). Based on the spectral characteristics of the sewage on the FCC (Fig. 4, marked in white) it can be concluded that this water is very eutrophic and has a high phytoplankton concentration. The environmental quality of the water in this part of the Rio de la Plata is of major significance because it constitutes the main source of potable water (Kurucz et al. 1997). There are three main water intakes: the main one is located 1050 m from the coast off the "Aeroparque Metropolitano Jorge Newbery"-airport (2.5 million m³ per day); the Bernal intake located 2500 m from the coast (1.1 million m³ per day) and one located at Piria (0.4 million m³ per day) (AGOSBA-OSN-SIHN 1994 in: Kurucz et al. 1997). Important interferences have not yet been detected with conventional methods (AGOSBA-OSN-SIHN 1994 in: Kurucz et al. 1997). In future, the quality of the sewage and risks of interferences with the drinking water intakes can thus be monitored by remote sensing (Goossens et al. 2000).

On these colour composites, also some spoil-grounds of the dredgers can be delimited (Fig. 4, marked black). A remarkable example can be detected on the FCC of the LANDSAT7 ETM+-image as an elongated spot. This spoil zone has a cross section of approximately 10 km from northwest to southeast and of nearly 4 km southwest – northeast (Goossens et al. 2000). This information is of special interest as these spoil-grounds are major obstacles for the navigation on the Rio de la Plata. As they are subject to erosion and displacement over time, it is also important to follow up the evolution of these deposits. Based on remote sensing the updating of the nautical maps can be done much easier and much faster as could be possible with conventional surveys (Goossens et al. 2000).

In combination with the thermal data, the difference between the river water and the intrusion of the more saltish water from the Atlantic Ocean can be seen (Goossens et al. 2000). On the thermal band (Fig. 5) the colder intruding ocean water is shown in dark grey, while the relatively warmer fluvial water is shown in lighter grey tones. These observations are backed up by information derived from the tides predicting program 'WXTide 32' (Hopper 1998) (Table 8). As the Landsat7 ETM+ - image was taken at 13:37:24 GMT (Table 2) or, put other ways, at 10:37:24 local time, it can be derived from the information in Table 8 that, globally, in the upper part of the Rio de la Plata, the tide is changing from high to low. In the extreme northwest of the estuary, near Isla Martín García, it is nearly high tide. This mean that, at the moment of image recording, the water coming from the Rio Uruguay and the northern tributaries of the Paraná River is partially blocked by the upcoming water. As a consequence, the water in the northwest of the estuary is more turbulent than in the southwestern part where the water from the Paraná de Las Palmas can stream directly into the Rio de la Plata (Goossens et al. 2000).

Table 8. Tidal information for the stations of Colonia de Sacramento, Buenos Aires and Isla Martín García on 19/01/2000 (source: Hopper 1998).

	Time of high tide	Time of low tide
Colonia de Sacramento	8:51	15:12
Buenos Aires	8:34	15:44
Isla Martín García	10:44	17:08

On the TCC and the FCC of the LANDSAT7 ETM+ - image, the man-made channels of the Rio de la Plata can be seen (Fig. 3, marked in black). This is of special importance for the planning of upcoming dredging activities. As the position of the channels is known, as well as the changes in spatial distribution of the sediment transport, future dredging activities can be more restricted in time and space (Goossens et al. 2000). This will lead to a more cost-effective maintenance the channels.

On the SeaWiFS O.I.F.-composite (Fig. 9) it is possible to follow the sediment plumes into the Atlantic Ocean. These sediment plumes extend for several kilometres and are affected by two ocean currents. On the image the influence of the cold Falkland Current coming from the southwest and the warmer Brazilian Current coming from the northeast can be noticed (Goossens et al. 2000).

As can be seen on the false colour composite made of the SeaWiFS – image (Fig. 8) and more precisely on the detail for the study area (Fig. 10), almost the same structures in the Rio de la Plata can be distinguished, even though less detailed due to the coarser resolution of the SeaWiFS-sensor. An example is the eutrophic sewage coming out of Buenos Aires that can be seen in both images (Fig. 4 and Fig. 10). Based on these resemblances, it seems possible to extrapolate information derived from the detailed LANDSAT7 ETM+ - image to the more synoptic SeaWiFS-image in order to get a general idea of the amount of suspended material in the entire Rio de la Plata. To realize this it is also necessary to gather a large amount of information concerning the composition of the suspended material based on conventional surveying methods in order to quantify a classification based on the LANDSAT7 ETM+ - image. This is, though, not so evident. Several problems can be encountered (Goossens et al. 2000). First of all it is difficult to tune the moment of pass-

ing over of the satellite to the moment the samples are taken in the Rio de la Plata. A second important problem is the difficulty to find a cloud free image over the area. This is a particular problem in winter and especially for SeaWiFS. The annual average cloud coverage percentage is of the order of 40 to 50 % of the total sky (Nagy et al. 1997). Relatively strong winter cloudiness starts mainly in May, with maxima in June and October. A weak but characteristic maximum is found in August, and an ubiquitous one-to-two-weeks "out of season" maximum often occurs in April (Rivero, pers. comm. in: Nagy et al. 1997). Besides, a large number of seaborn data is necessary to be statistical relevant which makes the basic quantification also very expensive.

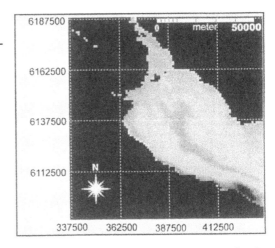

Figure 10. Detail of Fig. 8 with the same area covered as the FCC of the LANDSAT7 ETM+-image (Source: Goossens et al., 2000).

5 CONCLUSIONS

It can be concluded that the information gathered with the LANDSAT7 ETM+- and the SeaWiFS-sensor is useful for the study of sediment plumes in the Rio de la Plata. Globally, remote sensing is less expensive and has several advantages in comparison with conventional monitoring techniques. As both the sediment transport and the man-made channels can be detected on the LANDSAT7 ETM+-image, it is possible to ameliorate the dredging plan in order to maintain the channels. By doing this, the dredging activities can be carried out more effectively and less cost-intensive. Remote sensing can also be used to monitor the quality of the sewage coming out of the Argentinean capital district, Buenos Aires, and to warn for possible harmful interferences with the potable water intakes. Remote sensing gives also the opportunity to update the navigation maps more frequently which is of special interest for the dynamic spoil-grounds. By linking the LANDSAT7 ETM+-image to the SeaWiFS-image, using a Resurs-01-image as an intermediate, it is also possible to extrapolate the information derived from a detailed study area to the entire Rio de la Plata and part of the adjoining southern Atlantic Ocean. Though several conditions has to be fulfilled before this can be done. When combining the remote sensing information with other data resources in a GIS a more comprehensive monitoring system can be developed for the Rio de la Plata.

ACKNOWLEDGMENTS

This research was funded by grant # T4/02/53 of the Belgian Federal Office for Scientific, Technical and Cultural Affairs (OSTC). Special thanks also to ir. B. Malherbe of HAECON N.V., Harbour and Engineering Consultants for the technical support.

REFERENCES

Admiralty Charts and publications, 3561, Rio de la Plata – La Plata to Nueva Palmira, 1995 (corrected 2000), Taunton (UK), scale 1/100000 at lat 35°00'

An International Travel Map of Uruguay (1st edition), 847, 1997, ITMB Publishing LTD/ Volker Schniepp, Vancouver (CA), scale: 1/800000

Encyclopedia Britannica, Inc. (1994). http://www.brittanica.com. 23/04/2001.

Goossens R., De Dapper M. en Vanderstraete T. (2000). *Onderzoeksovereenkomst nr. T4/02/53. Studie van Sedimentpluimen in grote deltasystemen aan de hand van terreinopmetingen en teledetectiegegevens – Eindverslag.* Rapport: Universiteit Gent: Gent (BE). 57 p.

Hopper M. (1998). *WXTide32.* http://www.wxtide32.com. 20/03/2000.

ILWIS (The Integrated Land and Water Information System), version 2.2 for Windows, 2000, ITC (The Netherlands) and PCI Geomatics (Canada).

Kurucz, A., Masello, A., Méndez, S., Cranston, R. & Wells, P.G. (1997). Environmental Quality of the Río de la Plata. In: Wells, P.G. & Daborn, G.R. (eds). *The Rio de la Plata. An Environmental Overview. An EcoPlata Project Background Report.* Halifax: Dalhousie University. p. 69 – 84.

López Laborde, J. (1997). Geomorphological and geological setting of the Rio de la Plata. In: Wells, P.G. & Daborn, G.R. (eds). *The Rio de la Plata. An Environmental Overview. An EcoPlata Project Background Report.* Halifax: Dalhousie University. p. 1-16.

Nagy, G.J., Martínez, C.M., Caffera, R.M., Pedrosa, G., Forbes ,E.A., Perdomo, A.C. & López Laborde, J. (1997). The Hydrological and climatic setting of the Rio de la Plata. In: Wells, P.G. & Daborn, G.R. (eds). *The Rio de la Plata. An Environmental Overview. An EcoPlata Project Background Report.* Halifax: Dalhousie University. p. 17 – 68.

van Westen, C. & Farifteh, J. (1997). *ILWIS 2.1 for Windows – User's Guide.* Enschede: ITC.

Wells, P.G. & Daborn, G.R. (eds). (1997). *The Rio de la Plata. An Environmental Overview. An EcoPlata Project Background Report.* Halifax: Dalhousie University. 248 p.

8 *Health and Risk Management Strategies*

Observing our environment from Space: New solutions for a new millennium, Bégni (Ed.)
© 2002 Swets & Zeitlinger, Lisse, ISBN 90 5809 254 2

Disaster mitigation via telecommunications: the Tampere Convention

M.Harbi
Senior Policy Advisor, United Nations/OCHA, Place des Nations, Geneva, Switzerland

ABSTRACT: Disasters, both natural and man-made, cause tremendous human sufferings and property damage. Since the early 1990's, the international community has come to realize that modern telecommunications can be of substantive help in humanitarian assistance. Consequently, the Tampere Convention on Emergency Telecommunications was adopted to facilitate these activities. This international effort was endorsed by various international conferences. In the future, the cooperation between public and private sector will still be crucial in properly and effectively implementing the Tampere Convention.
Disasters happen: they won't change.
In 1999, about 105.000 people died in 326 major disasters, total property losses reached about US $ 100 billion, the second highest level ever recorded; 1999 was not an anomaly. Compared to the 1960s, the 1990s has seen the disasters triple, costing 9 times as much (World disaster Report)

1 INTRODUCTION

A sudden and towering tsunami inundated a coastal country, resulting in huge casualties and loss of shelters. Communications with the rest of the world were totally severed. Within hours, international search and rescue teams rushed to the capital, as requested by the national government. The team leaders were stunned beyond belief when warned by the custom officials that hefty import duties must be paid for their telecommunication equipment, while their telecom operators must obtain operating licenses before commencing sorely needed telecommunication with the outside world. Consequently, precious time vital for saving human lives was wasted in the ensuing bureaucratic haggling.

Scenarios such as that described above are encountered time and again by various humanitarian relief agencies, both public and private, around the world. This is most unfortunate in view of the fact that modern telecommunication equipment has proved itself indispensable in humanitarian relief and disaster mitigation. In fact, mobile and satellite technology has found usage in a wide variety of humanitarian-related fields, ranging from remote sensing for disaster mapping to global positioning system (GPS) for exact locationing of relief operations to real-time voice and text relays between headquarters and fields.

2 HISTORY OF THE TAMPERE CONVENTION

The first of these gatherings was the Tampere Conference of 1991, which resulted in the adoption of the first Declaration on Disaster Communications. This landmark declaration is a statement of experts from major humanitarian organizations, regulatory authorities and the private sector. It became the foundation of all subsequent work towards an international treaty to allow the unhindered use of telecommunication technology in international disaster response.

In particular, the Tampere Declaration called on ITU Member States to take all practical steps to facilitate the rapid deployment and effective use of telecommunications equipment for disaster mitigation and disaster relief by reducing, and where possible, removing regulatory barriers and strengthening transborder cooperation between countries.

The combined efforts of the ITU and the Office for the Coordination of Humanitarian Affairs (OCHA) of the United Nations led to the adoption of the Tampere Convention on the Provision of Telecommunications for Disaster Mitigation and

Relief Operations, an international treaty, deposited with the United Nations Secretary-General. As part of its name suggests, this treaty was adopted in Tampere at the Intergovernmental Conference on Emergency Telecommunications in 1998 (ICET-98). The Tampere Convention was opened for signature in New York on 22 June 1998, and will remain open until 21 June 2003. Under international law, the Convention will officially come into force 30 days after an official ratification or "consent to be bound" has been received from 30 countries. As at 14 May 2001, the Convention had been signed by 50 Member States and ratified by eight. This treaty is particularly important for humanitarian organizations, which rely heavily on telecommunications equipment to coordinate the complicated logistics of rescue operations.

Very often, the transborder use of telecommunication equipment is impeded by regulatory barriers that make it extremely difficult for humanitarian organizations to import and rapidly deploy such equipment without the prior consent and approval of the relevant local authorities. These organizations include OCHA, the United Nations High Commissioner for Refugees (UNHCR), the International Committee of the Red Cross (ICRC), the International Federation of Red Cross and Red Crescent Societies (IFRC) and other organizations. The Convention, among other things, aims to reduce administrative formalities of all kinds (licences, customs, charging, etc.) and to promote the drawing up of national plans and inventories of telecommunication resources to be deployed in the event of a disaster or emergency.

3 THE INTERNATIONAL TELECOMMUNICATION UNION (ITU): A decade of involvement in disaster communications

1991: ITU took an active part in the First Conference on Disasters Communications which took place in Tampere in 1991 and which recommended that an intergovernmental conference should be convened to prepare for the negotiation of an International Convention on Disaster Communications no later than 1993.

1994: ITU participated in the Annual Meeting of the International Institute of Communications (also held in Tampere), which included a special session on disaster communications.

1994: The ITU World Telecommunication Development Conference (WTDC-94) held in Buenos Aires passed Resolution 7, inviting ITU to study the technical, operational and regulatory aspects of emergency telecommunications; study charging and accounting in domestic and international disaster communications; and help developing countries, particularly the least developed ones, to prepare their telecommunication services in the event of disaster. The resolution also requests ITU to work closely with OCHA with a view to increasing the Union's involvement in disaster communications

1994: The ITU Plenipotentiary Conference (Kyoto, 1994), endorsed the Tampere Declaration and passed Resolution 36 instructing the Council to address the issues raised in Resolution 7 of WTDC-94 and take the action necessary to implement this Resolution.

1994–1997: The Working Group on Emergency Telecommunications (WGET) drafted the Tampere Convention. WGET is an open forum that facilitates the use of telecommunications in the service of humanitarian assistance. It comprises United Nations entities, major non-governmental organizations (NGO), ICRC, and experts from the private sector and academia.

1996: The annual session of the ITU Council gave the Secretary-General the go-ahead to enable ITU to play an important role in the implementation of the future Convention. In October 1996, the ITU Secretary-General consulted all ITU Members on the draft Convention.

1997: The ITU World Radiocommunication Conference (WRC-97), held in Geneva in October-November, passed Resolution 644 urging administrations to give their full support to the adoption of the Convention.

1998: The Valletta Declaration, adopted at the second ITU World Telecommunication Development Conference (WTDC-98), underlined the importance of emergency telecommunications and the need for an international Convention.

1998: The ITU Plenipotentiary Conference (Minneapolis, 1998) instructed the ITU Secretary-General to study the possibilities of increasing the use of telecommunications for the safety and security of humanitarian personnel in the field and to report to the 1999 session of the Council for appropriate action to improve that use. Furthermore, this conference urged Member States to ensure that humanitarian personnel have unhindered and uninterrupted use of telecommunication resources required for their safety and security, in accordance with the national rules and regulations of the States concerned. To this end, the conference called on Member States to work towards the earliest possible ratification, acceptance, approval or final signature of the Tampere Convention.

2000: The ITU World Radiocommunication Conference (WRC-2000), held in Istanbul (May 2000), approved revisions to two major resolutions. Resolution 644 invites ITU to continue to study, as a matter of urgency, those aspects of radio-communications that are relevant to disaster mitigation and relief operations, including amateur radio facilities, mobile and portable satellite terminals.

Resolution 10 urges ITU Member States to take account of the possible needs of the International Red Cross and Red Crescent Movement for two-way wireless telecommunication means when normal communication facilities are interrupted or not available. The Resolution also urges countries to assign to these organizations the minimum number of necessary working frequencies in accordance with the Radio Regulations, while taking all practicable steps to protect such communications from harmful interference. The two organizations heavily rely on two-way wireless telecommunication facilities (particularly HF and VHF radio networks) for the efficient and safe conduct of their humanitarian operations.

4 AN INTERNATIONAL FRAMEWORK FOR EMERGENCY TELECOMMUNICATION ASSISTANCE

The Tampere Convention creates an international framework for the provision of telecommunication resources for disaster mitigation and relief between States and between a State and a non-State entity. Under this framework, a State which perceives the need for disaster telecommunication assistance in its territory will request such assistance through the UN Emergency Relief Coordinator, who is the Operational Coordinator under the Convention and who will then channel the requests to other concerned entities. On the other hand, a providing State party is obliged to set down in writing the fees it expects to receive or have reimbursed. The fees, if any, will be based on an agreed model of payment and reimbursement, as well as on other factors such as the nature of the disaster and the particular needs of developing countries. Procedures are also set

forth in the Convention for termination of telecommunication assistance and for dispute settlement. Nevertheless, this framework does not preclude the existing or future arrangements between States and between a State and a non-State entity in emergency telecommunication assistance.

Indeed, the Tampere Convention urged States and non-State entities to cooperate in facilitating the use of telecommunication resources in disaster mitigation and relief. This include the deployment of terrestrial and satellite equipment to predict and monitor hazards and disasters; the sharing and dissemination of information about hazards and disasters; and the installation and operation of reliable and flexible telecommunication resources for humanitarian relief and assistance organizations.

The States and non-state entities should also develop and make available model agreements and best practices which provide a foundation for bilateral or multilateral agreements facilitating telecommunication assistance. They should also establish mechanisms for training and instruction in equipment handling and design and construction of emergency telecommunication facilities.

The Convention also recommends States to reduce or remove regulatory barriers that currently impede the use of telecommunications resources for disaster mitigation and relief operations. It further safeguards the privileges, immunities and facilities accorded to persons providing disaster assistance by granting them immunity from arrest and detention and exempting them from taxation and duties.

Under the Tampere Convention, the International Telecommunication Union (ITU) will work closely with the Operational Coordinator on several provisions of the Convention. These include maintaining contact with focal points within States that are authorized to request, offer, accept and terminate telecommunications assistance. States will also compile a telecommunication assistance information inventory listing, among others, competent authorities and points of contact. This inventory will be maintained and updated by the Operational Coordinator with the help of the ITU.

As hinted above, it is worth noting that many provisions of the Convention are applicable to non-state entities such as intergovernmental organizations and non-governmental organizations. This is evidently important because non-state entities often work in conjunction with governmental organizations in disaster mitigation and relief operations. In addition, States which have not yet signed the Convention may also apply the Convention provisionally.

« Information on what has happened, how much damage there has been and what kind of help is needed, are key questions in disasters situations. To be able to answer them, we need a communications system that can be set up quickly and will guarantee of flow of information even when the local systems are inoperative. This is the focus of the <u>Tampere Convention</u> adopted in 1998. »

Tarja Halonen, President of the Republic of Finland
Patron of the 2nd Tampere Conference on Disasters Communications. 28-30 May 2001

5 THE ROAD AHEAD

The application of the Tampere Convention depends heavily on respective national implementing legislation. Guidance and assistance, in particular to the most disaster-prone countries, is vital for the implementation process. As such, a Plan of Action for the Implementation of the Tampere Convention is being implemented by the different partners in humanitarian relief and telecommunication development. The action items range from a series of regional emergency telecommunications workshops and pilot projects for development of model agreements and best practices in emergency telecommunication provision. An active campaign for the promotion of the Convention will be launched at the 2nd Conference on Disasters Communications which will be held in Tampere on 28 to 30 May 2001.

In summary, the synergy among the public and private partners in disaster relief and in telecommunications is as crucial for the implementation of the Tampere Convention as it has been for its adoption.

About the Author

Mohamed HARBI is former Director of Strategic Planning, External Affairs and Corporate Communication Units at the International Telecommunication Union (ITU). As Special Advisor to the ITU Secretary-General he was responsible for telecommunication assistance to Bosnia and Herzegovina and Palestinian Authority, and for the adoption and implementation of the Tampere Convention on "Emergency Telecommunications". He is currently Senior Policy Advisor to United Nations/OCHA on Disasters Communications.

Observing our environment from Space: New solutions for a new millennium, Bégni (Ed.)
© 2002 Swets & Zeitlinger, Lisse, ISBN 90 5809 254 2

Space surveillance of epidemics: Case of the Rift Valley fever

P.Sabatier
INRA-ENVL, Unité BioMathématiques et Epidémiologie, Marcy l'Etoile, France

M.A.Dubois
CEA -SPEC, CEN Saclay, Orme des Merisiers, Gif sur Yvette Cedex, France

R.Lancelot & P.Hendrickx
CIRAD-EMVT, ISRA -LNERV, Dakar-Hann, Sénégal

J-P.Lacaux
MEDIAS-France, Toulouse Cedex 4, France

ABSTRACT: The unexpected appearance of new infectious diseases, often of zoonotic origin, is a surprising phenomenon at the end of the beginning of the third millennium (TAYLOR, L.H. & WOOLHOUSE M.E.J. 2000). To trigger off epidemics, and not only to entail sporadic forms, these new agents require amplification effects linked to natural or artificial changes in the environment. This project, *CNES* sponsored, develops a predictive model using remote sensing data for monitoring a vector-born disease. The study is based on a thorough modelling of the epidemic process of a zoonotic haemorrhagic fever (Rift Valley Fever, RVF); in a Sahelian area (basin of the Senegal River). The preliminary results of the space modelling: *(i)* express the periodicity of the outbreaks; and *(ii)* display the change of the space behaviour of RVF epidemics when the infection, or transmission probability, rises above a threshold for bond percolation on the network. An increase in the number of moving populations causes faster spreading by increasing the number of vertices in the graph's largest connected component. This integrative approach of the *S2E consortium* ("*Space Surveillance of Epidemics*"), will open a new field for early warning epidemiological systems, able to alert on and to predict the geographical extension of diseases.

1 INTRODUCTION

Rift Valley fever (RVF) is an insect-borne disease of man and animals caused by a member of *Phlebovirus* genre of the family of *Bunyaviridae*. Signs of the disease tend to be non-specific, making it difficult to recognize individual cases of RVF. The main indication of an epidemic is frequently the numerous cases of abortion and neonatal mortality of sheep associated to human deaths. Index cases and sporadic cases are usually misdiagnosed. RVF was recognized first in the Rift Valley of Kenya at the turn of this century (1913), but the agent was not isolated until 1931 (DAUBNEY, R. *et al.* 1931). Until 1977, major outbreaks of RVF have occurred in eastern and southern Africa and the furthest north that the disease occurred was the Sudan. From 1977, occasional outbreaks have been noted in other parts of the continent. In 1977-78, a major epidemic occurred in the Nile delta and valley, in Egypt. Several million people were infected and thousands died during a violent epidemic of this disease (MEEGAN, J. M. *et al.* 1979). A severe epidemic affected the Senegal River Basin in Mauritania and Senegal in 1987 (three hundred died in Rosso) and again in Egypt in 1993 (JOUAN, A. *et al.* 1989).

The natural history of the disease indicates that the potential exists for spread to other regions of the world. Since this time, RVF reached the Middle East in September 2000. A total of 882 individuals have been affected in Saudi Arabia and Yemen, of whom 757 have fully recovered and 124 have died. What are the chances that RVF could follow this route moving on to South Asia, i.e. India, Pakistan, Bangladesh, Nepal ? Were it to do so, the outcome would probably be a disaster on a large scale. So this move by RVF may have much greater implications in the future and should spark a serious international surveillance and response to eradicate it. Increasing evidence linking the human deaths to animal disease as well has led epidemiologists to include RVF to list of emergent zoonoses, like Aids and Ebola that infect thousands of people each year. Recurrent viral activity occurs in localized areas where mosquitoes transmission of RVF virus to animals occurs during most years. This provides one key to understanding virus survival during inter-epidemic periods, and virus intensification after heavy, prolonged and often unseasoned rainfall favoring the breeding of mosquito vectors. Can remote sensing provide information about environment of the mosquitoes that would be useful in preventing these outbreaks?

2 CLIMATE AND SATELLITE INDICATORS

Epidemics, in East Africa, are associated with unusually heavy rainfall and large number of mosquitoes (DAVIES, F. G. *et al.* 1985, LINTHICUM, K. J. *et al.* 1999). The occurrence of mosquitoes, in

Kenya's Rift Valley, is known to follow periods of widespread rainfall which floods mosquitoes breeding habitats. These habitats, called dambos, are shallow depressions, often located near rivers, which fill with water during the dry season.

2.1 NDVI anomalies in East-Africa

Dambos typically have distinctive tall edge grass in middle areas and payrus and several other grasses around edges. This pattern of vegetative growth makes them look quite different from the surrounding area. Even in the dry season, the dambos appear greener than other areas. Vegetation in dambos that are frequently flooded is healthier and more abundant than vegetation in drier dambos. Dambos respond to increased rainfall and can be easily measured by satellite. A dambo can be a kilometer in length and several hundreds of meters in width.

From 1950 to 1998, in Kenya, exceptionally higher than normal rainfall coincided with major RVF epidemics in 1951-53, 1961-63, 1968-69, 1977-79, 1997-98. Vegetation responds to the increased rainfall, and can be measured by the Normalized Difference Vegetation Index (NDVI), data from advanced high resolution radiometer (AVHRR) on National Oceanic and Atmospheric Administration (NOAA) satellite. Linthicum finds a relationship between an increasing of Normalized Difference Vegetation Index (NDVI), caused by abnormally heavy rainfall, and epizootics. In 1998, during a period of RVF activity, elevated NDVI anomalies were observed through the normal dry season. The NDVI anomalies, in East Africa, were significantly correlated with following RVF activity, 1 to 2 months after (p<0.5).

2.2 Climate deviations in East-Africa

Linthicum used indicators of the Global Change Climate to predict RVF outbreaks up to 5 months in advance. Warm ENSO (El Niño Southern Oscillation) events are known to increase precipitation in some regions of East and Southern Africa [5]. In East Africa, above normal rainfall and more green vegetation can be associated with: (1) the negative Southern Oscillation Index (SOI), index comparing atmospheric temperature in Tahiti and Darwin (Australia); (2) the Sea Surface Temperature (SST), measured from an equatorial region in the Pacific Ocean or in the Western Indian Ocean, which are the most commonly used indices for ENSO phenomena.

From 1950 to 1998, the RVF activity followed a period of strong negative deviation of the Southern Oscillation Index (SOI), and Sea Surface Temperature (SST). The major outbreaks starting in 1951, 1961, 1968 occurred after SOI anomalies <-2. Subsequently, RVF activity starting in 1982 and 1997 followed several months of strong concurrent equatorial Pacific (>3°) and Indian (>0.5°) Ocean SST anomalies.

2.3 Ability to forecast RVF epidemics

Strong negative SOI anomalies occurred in 1964, 1969, 1972, 1981 and 1991, and concurrent Pacific and Indian SST anomalies occurred in 1988, 1993, 1995, without detectable RVF activity in Kenya. So Linthicum suggests to combine all these indicators in an ARIMA model (SBC = -106; analysis variance df = 192 ; P < 0.01). He considers that the "best" conditions for an RVF outbreak in East Africa were achieved when abnormal equatorial Pacific and Indian Ocean SST as well as NDVI data were used together. These data could have been used successfully to predict, in East-Africa, the RVF outbreaks in 1982, 1989 and 1997, but unsuccessfully to predict the epidemics of 1999.

Given the sporadic distribution in the past, risks of RVF transmission and epizootics during the next rainy season may appear in areas in which the virus has not been detected.

However, such a relationship has not been established in Sahelian areas. For example, the enzootic maintenance of the RVF virus in Senegal in 1993 coincidentally occurred in Mauritania without any relationship with heavy rainfall or extensive NDVI. And the RVF activity described in Southern Mauritania in 1982-1985 occurred during a period of drought; it was therefore not associated with extensive rainfall. In the Sahelian bioclimatic region, suc as the Ferlo valley, characterized by a short rainy season from July to September with annual rainfall of 350 mm, temporary ground pools filled soon after the first rains, are the unique water resources for up to four months into the dry season.

3 MODELLING THE RVF TRANSMISSION

Understanding the spread of the disease is of vital importance to monitor and to control the epidemics. Why did RFV outbreaks occur in these new areas (Sahelian bioclimatic region)? Why did they disappear during many years (inter-epidemic pause)? The use of epidemiological models has been initiated in 1927 (KERMACK, W.O. & MCKENDRICK, A.G., 1927) and developed later (ANDERSON R.M. & MAY R.M. 1979, MAY R.M. & ANDERSON R.M. 1979) to respond to such questions. Theory and applicable techniques have been developed to study the transmission of infections through populations.

3.1 Parameters and state variables

Consider a population of domestic animals (sheep, and/or cows) distributed in N classes of age, each class containing $z_\alpha I \alpha \in \{1, ..., N\}$ such that the total size of the population is $\sum_{\alpha=1}^{N} z_\alpha$. For simplicity we assume $z_\alpha = 1$ so that the population size is N. Consider three main immunological states: susceptible, infectious and resistant (THIONGANE, Y. *et al.* 1991, THIONGANE, Y. *et al.* 1999). But the infectious state is fleeting, so the sanitary status $p(\alpha, t)$ of an animal of the α class of age, in the course of time t

is : $p(\alpha,t) = \begin{cases} 0; \text{IgM seropositive (resistant)} \\ 1; \text{seronegative (susceptible)} \end{cases}$

Consider a population of infected eggs of endemic vectors of RVF virus, *Ae.vexans*, *Ae. ochraceus*, and *Ae. Dalzieli*. These eggs are accumulated at the edge of temporary Sahelian ground pools (ZELLER, H. G. *et al.* 1997, FONTENILLE, D. *et al.* 1998). They may develop into infected larvae and adults in the four days following the flooding of their habitat. Consider a ground pool of depth L subdivided into L stata (or levels) containing a number $o(s,t)$ of infected eggs in the strata s in the course of time t. For simplicity we assume that the value of $o(s,t)$ lies between 0 and V, i.e. $0 \le o(s,t) \le V$. In addition the ground pool is subject to be filled up to the h status (or level) from water from rainfall ("hivernage"). In order to mimic the stochastic nature of the rainfall abundance, the "hivernage" level $h(t)$ is assumed to be a random function of time uniformly distributed in the interval $[H_{min}, L]$.

Data obtained in the last few years have shown that the pattern of RVF transmission in West Africa is different from the known pattern in East Africa (1). *Ae. vexans* control an enzootic cycle of RVF virus quite different of the epidemic cycle of the vectors of East Africa (*Ae. cumminsii*, *Ae. circumluteolus*, and *Ae. mcintoshi*). There is endemic cycle is caracterized by:
- laying of eggs on the soil of temporary ground pools (in the dried mud surrounding the ground);
- survival of eggs that can (and has to) endure dry condition for long periods;
- hatching of the eggs after the filling of the ground pools at the beginning of the rainy season;
- viral transmission to the susceptible cattle on which they feed selectively, and to the *aedine* mosquitoes descendants (transovarial).

3.2 0-Dimension model of RFV epidemics

The time discrete evolution of the system, with a time increment Δt, is describe by :

$$\begin{cases} p(\alpha,t+\Delta t) = [1-f_1(t)]p(\alpha-1,t); \alpha \in [2...N] \\ o(s,t+\Delta t) = (k\Delta t-1)H[s-H(t)]o(s,t) + [1-H[s-H(t)]]f_1(t)f_2(t)V; s \in [1,...L] \end{cases}$$ (1)(2)

with the boundary condition, $p(1,t) = 1 \ \forall t$, and $k\Delta t > 1$.

We define two functions of time $f_1; f_2 : N \to [0, 1]$:

$$f_1(t) = H\left[\sum_{s=1}^{h(t)} o(s,t) - \theta\right]$$ (3)

$$f_2(t) = H\left[\sum_{\alpha=2}^{N} p(\alpha,t)\right]$$ (4)

where θ is defined as being the threshold for triggering the infection of an animal population, and $H(x)$ is the Heaviside step function defined as $H(x)=0$ for $x < 0$ and $H(x) = 1$ for $x > 0$. Note that $f_1(t)$ is thereby $h(t)$ a random function of time. (5)

The outputs of the dynamical system (1, 2) controlled by the hibernate level $h(t)$, i.e. the filling of the ground pool, are :

$O(t) = \sum_{s=1}^{h} o(s,t)$ for the total number of the infected eggs near of the monitored ground pool 9. (6)

$P(t) = N - \sum_{\alpha=1}^{N} p(\alpha,t)$ the total number of infected (IgM seroprevalence) of the animals in the area 9. (7)

$A(t) = f_1(t) \sum_{\alpha=2}^{N} p(\alpha,t)$ the total number of the abortion of the sheep (clinical prevalence) in the area 9. (8)

3.3 2-Dimension model of RFV epidemics

The Sahelian herds (and breeders) are moving from one accommodation site, made up of a ground pool and a plot of arboreous savanna, to another. These movements of inhabitants (transhumance, trade), destroy partially the spatial correlation of the populations. The spatial contacts, and consequently the epidemic contacts, between herds are affected by these movements. So the movements affect the epidemic process, and we must distinguish the contacts with nearby herds, and the contact with distant herds.

Consider a grid G of herd accommodation sites i, characterised by local resources {ponds, fodders, trees, mosquitoes, etc}, and living inhabitants {herds, humans}(DURETT, R. & NEUHAUSER, C. 1993, HASTINGS, A., 1993). Each site is linked to another by a bond E, which denotes the contact withmoving herds. $Gr = (G, E)$ is a graph of G and E ; Gr is locally finite. Consider a transition rule, applied to each site (mapping). $f_i : P^{|U_i|} \to P$; U_i : is the neighbourhood of the vertex i : $U_i = \{j \in G | \{i,j\} \in E\}$. U_i is defined as a Von Neumann neighbourhood (four neighbours). P is the number of resistant animals (IgM) on a given site.

In this diffusion model, the transmission of the virus to a herd, situated on the site i is controlled by three epidemic rules. The herd becomes infected, when the animals are partially susceptible, and that: (1) they arrive at an infected ground pool ; (2) or they contacts a neighbouring infected herd (with an α probability) ; (3) or they contacts a distant infected herd (with an β probability).

319

$$p(i,a,t+\Delta t) = \left[1 - f_1(i,t) \times f_2(i,t) - \alpha \times f_3(i,t) \times f_2(i,t) - \beta \times f_1(j,t) \times f_2(j,t)\right] \times p(i,a-1,t);$$

(9)

$$f_1(i,t) = H\left[\sum_{s=1}^{h(t)} o(i,s,t) - \theta\right] \qquad (10)$$

$$f_2(i,t) = H\left[\sum_{a=2}^{N} p(i,a,t)\right] \qquad (11)$$

We define a new function of time: $f_3 : N \rightarrow [0, 1]$:

$$f_3(i,t) = H\left[\sum_{|U_i|} f_1(i,t) \times f_2(i,t)\right] \qquad (12)$$

4 PRELIMINARY RESULTS

This is the simplest mean field model of the dynamics of the RVF transmission. We used it to explore the expected patterns of the RVF outbreak, and explore the consequences of one biological uncertainty: transmission rate and filling of the ground ponds. For θ fixed and given initial conditions, $h(0)$, $o(s,0)$ I $s \in \{1, ..., L\}$, and $p(\alpha,0)$ I $\alpha \in \{2, ..., N\}$, the time discrete evolution of the system is computed with a time increment $\Delta t = 1$ year. To study the phenomenology of the system, numerical simulations were carried out varying the "hivernage" level $h(t)$.

4.1 *First result*

We find three main patterns of RVF infection: *(A)* no infection, there is no secondary cases; *(B)* sporadic infection, there is a single outbreaks; *(C)* epidemic infection with periodic oscillations (and eventually with amortized oscillations) (*Fig. 1*). Phase and frequency of the ground pond filling have preponderant influences. This is consistent with the behaviour of the disease, with alternating stages: endemic and epidemic.

Therefore it is possible to compute accurately the epidemic behaviour with the mosquito transmissions. These preliminary results are relevant to the dynamics of the disease. The model reveals the two epidemic patterns: (*1*) endemic period with a low circulation of the virus (persistence), on the one hand; (*2*) epidemic period with amplification of the virus in the cattle, on the other hand. These results confirm the possibility of generation of epidemic waves with simultaneous intensification of vector and viral activity. It is not necessary to speculate about the existence of a cryptic foci and of carrier animals.

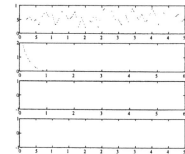

(Pattern A)

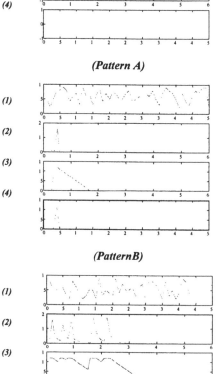

(PatternB)

(Pattern C)

(1) *h(t)*: Temporary ground pool level;
(2) *O(t)*: Mosquitoes infected eggs density near a pool;
(3) *P(t)*: Cattle Sero-prevalence ratio near a pool;
(4) *A(t)*: Abortion prevalence of the cattle near a pool.

$$p(a,\ 0)=1\ \forall a\ ;$$
$$o(s=7,\ t=0)=10;\ o(s,\ t=0)=0;\ \forall s \neq 7$$
$$a \in \{1, ..., 12\}.\ ;\ s \in \{1, ..., 10\}\ ;$$

Figure 1. Temporal epidemic patterns of RVF in sub-Sahelian area.

4.2 *Second result*

The evolution of this 2-Dimension discrete dynamic system is computed with the same time increment Δt

= *1* year, and the same given initial conditions. For α, β, θ fixed, we find three main spatial patterns of the RVF spread: *(A)* no infection: there are no secondary cases; *(B)* classical diffusion from original foci with epidemic wave ($\alpha = 0.3$; $\beta = 0$) ; *(C)* non-linear diffusion with generation of secondary distant foci ($\alpha = 0.3$; $\beta = 0.1$) *(Fig. 2)*.

These results are consistent with the fact that, in sub-Saharan Africa, epidemics can occur simultaneously over geographic areas separated by several hundred kilometres. They confirm that the distant links have a preponderant influence (non-local effects), and give a non-classical diffusion process without epidemic waves. This hypothesis corresponds epidemiologically to the movement of cattle over long distances, and is consistent with the very short duration of the viraemie of the animals (around ten days).

4.3 *Third result*

Therefore it is important for space surveillance of the environmental risk factors of RVF to compute accurately the epidemic behaviour of the disease. Such an epidemic process, spatially distributed, with several spatio-temporal levels of details, cannot be monitored directly from the earth observing system. It is necessary to derive the surveillance from an "aggregated" analytical model using relevant state variables integrating regional and local levels (AUGER, P. & ROUSSARIE, R. 1994).

For example, it is possible to obtain from remote sensing data the surface water resources (hydrography, lakes, dams, ponds, etc) from the red, infrared and medium infrared channels. From the physical point of view, water has a relevant property: the medium reflectance in the visible area falls down suddenly in the near and medium infrared, unlike most other bodies (vegetation, ground) which have a medium or high reflectance in these wave length scales. So the separating « water / no water » is easy in remote sensing when we can work in the near or the medium infrared area. The cartography of the water area and of the irrigable area can be designed by using data of VEGETATION-SPOT 4 in low resolution but high repetitivity: analysis of the channel PIR et MIR; time series analysis of the MIR, NDVI, NDWI.

(NDVI: Normalised Difference Vegetation Index; NDWI: Normalised Difference Water Index; PIR: Proche Infrarouge; MIR: Moyen Infrarouge; R: Rouge; HRVIR Haute Resolution Visible et Infrarouge)

From the NDVI index (calculated from the PIR and R) and the NDWI (calculated from the PIR and MIR) on the one hand, and from the channel MIR itself, on the other hand, we can deduce the water mask. This water mask calculated with the data of VEGETATION-SPOT 4 is well coordinated with the data of the HRVIR-SPOT 4 (high resolution visible and infrared but low repetitivity) *(Fig. 3)*.

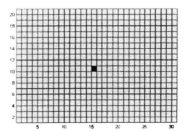

(PatternA)

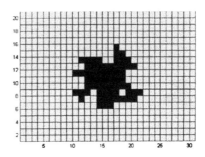

(PatternB)

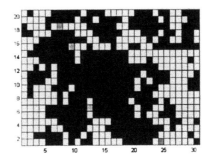

(Pattern C)

$h(t)$: periodical fonction: $T = 3,1$;
$p(x, y, a, 0)=1$ $\forall x$; $\forall y$; $\forall a$;
$o(x = 10, y = 15, s=7, t = 0)=10$;
$\alpha = 0.3$; $\beta \in \{0, 0.1\}$; $\theta = 1$;
$a \in \{1, ..., 12\}$. ; $s \in \{1, ..., 10\}$;
$x \in \{2, ..., 20\}$; $y \in \{2, ..., 30\}$;

Figure 2. Space epidemic patterns of RVF in sub-sahelian area.

Satellites can be used to detect conditions suitable for the earliest stages of an RVF epizootic. But these data have a too low spatial and temporal resolution to monitor the "hivernage" level $h(t)$, i.e. the filling of the temporary ground pools of Sahelian areas, which is an important control parameter of the

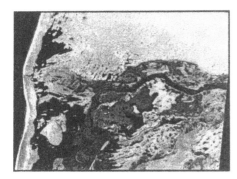

Satellite imagery:
SPOT-4 HRVIR (20 m)
September, 21, 1999

Water mask at local scale:
SPOT-4 HRVIR (20 m)
September, 21, 1999

Water mask at regional scale (Zoom):
SPOT-4 Vegetation (1 Km)
September, 2° decade, 1999

Figure 3. Water masks images generated at two complementary scales: local and regional. (Delta of the Senegal river).

dynamics of the RVF. The hydrologic regime of the Sahelian pool have two different components: the filling up by surface runoff; the emptying through

drainage and evaporation (DESCONNETS, J. C. *et al.* 1997). The filling regime of the pool has a common pattern during a few hours. The water depth increases quickly after rainfall, and diminishes at a strong but decreasing speed (during many weeks). There is a threshold separating a zone of rapid emptying and a zone of slow emptying. But the pattern of the emptying regime is sensitive to the changes in distribution of rainfall. Therefore it is impossible to monitor the spatial distribution of the RVF through the surveillance of the ground pools without an aggregated model.

5 CONCLUSION

Geographic distribution of vector-borne diseases is mostly determined by environmental factors that condition both the pathogenic agent and its vectors. Based on satellite imagery data, some of these factors could be apprehended and quantified. Some projects have studied remote sensing indicators to predict the density of vector density closely associated with disease risks (ROGERS DL. & PACKER M.J. 1993, HAY S.I. *et al.* 1996, ROBERTS D.R. *et al.* 1996, THOMSON M.C. *et al.* 1997, HASSAN, A.N. *et al.* 1997, WASHINO RK, WOOD BL. 1994). Low resolution data such as those from the National Oceanic and Atmospheric Administration satellites (NOAA) have been used for general studies of disease vectors in broad climatic zones or regions (18, 20). Higher resolution data from Landsat or the Système Pour l'Observation de la Terre (SPOT) XS sensors have been used to study disease distribution in association with smaller terrestrial features such as villages, rice fields, etc. (20 21). For interpreting these higher resolution data, it is necessary to have a more complete understanding of vector biology taking into account, climatic, hydrological, agronomical, zootechnical, epidemiological and social factors. This study, sponsored by the CNES, is based on a thorough modelling of the epidemic process of the RVF. This integrative approach of the *S2E consortium* (*"Space Surveillance of Epidemics"*), will open a new field for early warning epidemiological systems, able to alert on and to predict the geographical extension of diseases.

REFERENCES

Taylor, L.H. & Woolhouse M.E.J. 2000. Zoonoses and the risk of disease emergence. *International conference on emerging infectious diseases. 16-19 juin 2000,* Center for diseases control, Atlanta.

Daubney, R., Hudson, J. R., & Garnham, P. C. 1931. Enzootic hepatitis of Rift Valley Fever, An undescriptible virus disease of sheep, cattle and man from East Africa. *J. Pathol. Bacteriol.,* 34, 545-579

Meegan, J. M., Hoogstraal, H., & Moussa, M. I. 1979. An epizootic of Rift Valley fever in Egypt in 1977. *Veterinary Record*, 105(6), 124-5.

Jouan, A., & al. 1989. Analytical study of a Rift Valley fever epidemic. *Research In Virology*, 140(2), 175-86.

Thiongane, Y. & al. 1999. Données récentes de l'épidémiologie de la fièvre de la vallée du Rift (FVR) au Sénégal. *Bull Soc. Med. Afrique Noire de Lang. Franç. Spécial Congrès. 1-5*

Thiongane, Y. et al. 1991. Changes in Rift Valley fever neutralizing antibody prevalence among small domestic ruminants following the 1987 outbreak in the Senegal River basin. *Research In Virology*, 142(1), 67-70.

Linthicum, K. J. & al. 1999. Climate and Satellite Indicators to Forecast Rift Valley Fever Epidemic in Kenya. *Science,* 285(5426), 397-285

Davies, F. G. & al. 1985. Rainfall and epizootic Rift Valley fever. *Bulletin Of The World Health Organization,* 63(5), 941-3

Fontenille, D. & al. 1998. New vectors of Rift Valley fever in West Africa. *Emerging Infectious Diseases*, 4(2), 289-93.

Zeller, H. G. & al. 1997. Enzootic activity of Rift Valley fever virus in Senegal. *Am. J. Trop. Med. Hyg.*, 56(3), 265-72

Kermack, W.O. & McKendrick, A.G., 1927, A Contribution to the Mathematical Theory of Epidemics. *Proc. R. Soc.* A115. 700-21.

Anderson R.M. & May R.M., 1979, Population Biology of infectious diseases : Part I, *Nature*, Vol. 280, 361-367.

May R.M. & Anderson R.M., 1979, Population Biology of infectious diseases : Part II, *Nature*, Vol. 280, 455-461.

Auger, P. & Roussarie, R. 1994. "Complex ecological models with simple dynamics: from individuals to populations" Acta Biotheoretica 42: 11-136.

Durett, R. & Neuhauser, C., 1993. "Particle system and reaction-diffusion equations", *Ann. Probab.*, 22, 289-333.

Hastings, A., 1993. "Complex interactions between dispersal and dynamics: Lessons from coupled logistic equations", *Ecology*, 74, 1362-1372.

Desconnets, J. C. & al. 1997. Hydrology of Hapex-Sahel, Central Super Site: surface water drainage and aquifer recharge through the pool systems. *Journal of Hydrology*, 188-189, 155-178.

Hay S.I. & al. 1996. Remotely sensed surrogates of meteorological data for the study of the distribution and abundance of arthropod vectors of disease. *Ann Trop Med Parasitol*, 90:1-19

Roberts D.R., & al. 1996. Prediction of Malaria vector distribution in Belize based on multispectral satellite data. *Am. J. Trop. Med. Hyg.* 304-308.

Rogers DL. & Packer M.J. 1993. Vector-borne diseases, models, and global change. *Lancet*, 342:1282-1284

Thomson M.C. & al. 1997. Mapping malaria risk in Africa: what can satellite data contribute. *Parasitol Today* ; 313-8.

Hassan, A.N. & al. 1997. Prediction of villages at risk for filariasis transmission in Nile Delta using remote sensing and geographic information system technologies. *J. Egypt. Soc. Parasitol.* ; 28 : 361-74.

Washino RK & Wood BL., 1994: Application of remote sensing to arthropod vector surveillance and control. *Am J Trop Med Hyg,* 50 (suppl): 134-144

9 *Monitoring*

Observing our environment from Space: New solutions for a new millennium, Bégni (Ed.)
© 2002 Swets & Zeitlinger, Lisse, ISBN 90 5809 254 2

A multiresolution modelling approach for semi-automatic extraction of streets: application to high-resolution images from the IKONOS satellite

R.Péteri & T.Ranchin
Groupe Télédétection & Modélisation, Ecole des Mines de Paris, Sophia Antipolis, France

I.Couloigner
Department of Geomatics Engineering, University of Calgary, Calgary, Canada

ABSTRACT: In order to help people working in the urban areas, we are concerned about extracting road networks in from high resolution satellite images. However, due to the urban landscape complexity, the development of reliable and robust algorithms to extract information from city images is difficult.
An existing street extraction method is applied on high-resolution images from the IKONOS satellite. Results of this method are discussed. Then, a methodology is proposed which aims at semi-automatically extracting all types of urban street networks and at managing their topology.

1 INTRODUCTION

With the launch of new satellites with high spatial resolution sensors (less than 1 meter), such as IKONOS, the interest of people working in urban areas (urban planners, cartographers, urban geographers, etc ...) for such a data source is growing. Indeed, the spatial resolution of spaceborne data was till now too coarse for the scale used in urban mapping (1 :200 to 1 :20,000). With new high resolution images, one can expect to reach a scale better than 1 :5,000.

Among possibilities of remote sensing for urban uses, extraction of street networks is an important topic. Indeed, a good knowledge of the street network is necessary for urban planning and map updating. Moreover, an assisted street extraction from remotely sensed data will highly help cities which do not have a cartography of their road network yet.

However, the development of reliable and robust algorithms to extract information from city images is very difficult, because of the urban landscape complexity.

In this paper, a state of the art on road extraction will first be presented. Then a method for extracting urban streets (Couloigner 98), will be applied on high resolution IKONOS satellite images. Results of the method will be discussed. We finally propose a methodology adapted to both urban context and high resolution satellite images, and give our conclusions and perspectives on this subject.

2 STATE OF THE ART

Road extraction from remotely sensed images has been the purpose of many works in image process-

ing field, and because of its complexity, is still a challenging topic. These methods are based on generic tools of image processing, such as mathematical morphology (Destival, 1987; Serendero, 1989; Wang et al., 1996), linear filtering (Wang and Howarth, 1987; Wang et al., 1996), or on more specific tools using Markov fields (Merlet and Zerubia, 1996; Stoica et al., 2000), neural networks (Bhattacharya and Parui, 1997), cooperative algorithms (McKeown and Denlinger, 1988), dynamic programming (Jedynak, 1995; Gruen and Li, 1995), or multiresolution (Baumgartner et al., 1996; Couloigner and Ranchin, 2000 ; Laptev et al., 2000).

Most of the methods are based on simple road models. The main properties of these models are common for all authors, *i.e.* the radiometry along one road is relatively homogeneous and when compared to its background could be highly or poorly contrasted according to the surroundings of the road under consideration. Moreover the width of the road and its curvature are supposed to vary slowly, and the road network is supposed to be connex.

Recent studies try to take the context of the road into account in order to focus the extraction on the most promising regions (Baumgartner et al., 1999 ; Ruskoné, 1996).

In order to overcome all problems raised by an automatic or semi-automatic road extraction, most of those works focused on rural or semi-rural areas, where artefacts are less numerous than in cities. However, streets (term exclusively used in urban areas) are near enough to roads (term used in no urban areas) to take benefit from the proposed methods for our urban context.

The recent possibility to have high resolution satellite images (less than 1 meter) has modified the road extraction issue in satellite images. Indeed, at this current high resolution (less than 1 meter), a road is represented as a surface element which was not the case with lower resolution where streets were seen as linear elements.

Furthermore, on the one hand, when spatial resolution increases, the images of urban areas present higher complexity, and the street extraction becomes more complicated. On the other hand, a higher resolution enable a more accurate extraction.

The image of figure 1 is a good example for showing what kind of artefacts generate both the urban context and the high spatial resolution. Vehicles (circled), the bridge and its shadow disturb the road extraction process. Other artefacts encountered can be trees, tarred areas (parking, airport) or buildings with a radiometry similar to roads and with an important contrast with their environments.

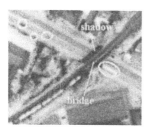

Figure 1. Artefacts encountered in urban context and high resolution images. Copyright 2000 Space Imaging Europe. *Courtesy of GIM - Geographic Information Management*

Baumgartner *and al.* (1996), and Couloigner (1998) have shown the interest of using multiresolution algorithms, especially on high resolution images: small details, such as shadows or vehicles do not disrupt the numerical processing.

The method developed by Isabelle Couloigner (1998), is a good starting point for our work. Indeed, it is one of the only methods working on urban areas, and it gives promising results. We have then chosen to apply it, in order to explore the advantages and the limits of this method.

3 STREET EXTRACTION IN URBAN AREA

3.1 *Application of a method of urban road extraction to high resolution IKONOS images*

In this section, a semi-automatic method for extracting urban street network is presented. This method focuses on urban linear street networks and is generic to a certain extend. An explicit model of streets was developed according to a multiresolution

analysis associated with the wavelet transform (Couloigner and Ranchin, 2000).

Mallat (1989) first introduced the concept of multiresolution algorithm (MRA) for multiscale representation. Figure 2 is a very convenient description of pyramidal algorithms. From the original image (bottom of the pyramid), MRA allows the computation of successive approximations with coarser and coarser spatial resolutions. Climbing the pyramid, the different floors represent successive approximations. To describe difference of information between two successive approximations, Mallat associates MRA with the wavelet transform (WT), in a mathematical framework. In this case, WT describes the difference of information between two consecutive approximations. Hence, when applying a MRA using WT, one can describe, hierarchically, the information content of a remotely sensed image (see *e.g.* Ranchin 1997).

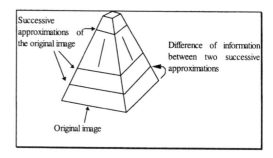

Figure 2. Representation of the successive approximations of an image by the means of a multiresolution algorithm

The algorithm used for this purpose in the proposed method was proposed by Dutilleux (1987). Figure 3 presents this algorithm. From the original image, the "à trous" algorithm allows the computation of successive approximations by smoothing. The wavelet coefficients (results of the application of a wavelet transform on a signal or image) are computed by a pixel-to-pixel difference between two successive approximations. All the approximation and wavelet images have the same size. This is more suitable for multiscale analysis of the characteristic structures of an image, but leads to a redundant representation of information.

In the case of the proposed method, this hierarchical description is used to define a model of roads. The model is based on the three following properties:

- *geometry* : streets are linear and the range of their width is known. Streets can be represented by a quadrilateral.
- *radiometry* : the radiometry is homogeneous along street axis and exhibits contrast in the perpendicular direction.
- *cross-section* : depends on the class of street of

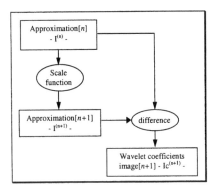

Figure 3. The "à trous" algorithm. The iteration is on the *n* variable (n=0 corresponds to the original image)

interest and on the spatial resolution (Figure 5). Strips are brighter than streets in the studied cases. 4 classes of streets have been defined, independent of spatial resolution.

The method is developed in two steps.

The first step consists in the extraction of the two sides of streets using the approximations at different spatial resolutions by applying the "à trous" algorithm. The processing enables to obtain on the image, the location of the two intersection points of different cross-sections of the studied street. The processing is semi-automatic and iterative. As an initialisation, the user indicates on the original image, the points that locate the beginning and the end of both sides of the street of interest. Then, the algorithm follows the street in two research windows, in a "estimation-registration" process, using the cross-sections provided by multiresolution.

The second step consists in the extraction of street characteristics. A multiresolution processing has been developed to specify characteristics of streets, like central reservation, according to their class. This iterative processing uses the wavelet coefficients images of the original image to locate reservations, if any. *(For more details, see Couloigner 98 and Couloigner and Ranchin 2000).*

Here are presented results of this method applied on images from the region of Hasselt , Belgium with different classes of road as defined in the method. These images are provided by the 1 meter resolution panchromatic band of the IKONOS satellite.

In the image of figure 4, we can see a road with a central reservation, which is called "class 3 street" with the model we use. Figure 5 shows an example of the evolution of a class 3 street cross-section with different spatial resolutions using the "à trous" algorithm.

One can notice that the central reservation does not appear anymore one the coarsest approximation (16 m).

The "à trous" algorithm was applied on the original

Figure 4. The original image of the Hasselt region Copyright 2000 Space Imaging Europe. *Courtesy of GIM - Geographic Information Management*

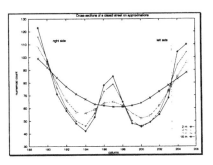

Figure 5. Example of a cross-section of a class 3 street at different spatial resolutions.

image, giving coarser resolution context images and wavelet coefficient images. (see figure 6).

The central reservation does not appear anymore in the 8 meter and 16 meter approximation images (Figure 6a. and 6b.). Figure 7 shows the result of the extraction after applying the method previously described.

3.2 *Discussion on results*

First of all, there is a good extraction of both sides of the road. The algorithm is not stopped by vehicles (circled on the image) and/or by shadows, which are artefacts mostly encountered in urban area. Moreo-

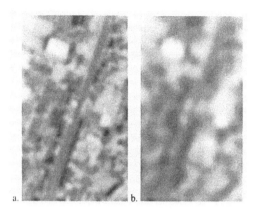

Figure 6. 8 metre (a) and 16 metre (b) approximations of the original image using the "à trous" algorithm.

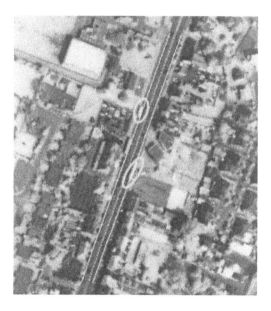

Figure 7. Result of the extraction of a class 3 street.

ver, the central reservation fits correctly with the image, as the good continuity between the dotted line extremities and the reservation in the image shows.

Nevertheless, this method was only developed for linear road networks and should manage roads with curvatures, often encountered in town. Models used were more specifically developed for new cities like Jeddah in Saudi Arabia, with a linear street network and with specific models of streets. However, it is not a real problem as it just requires to modify the models of streets. Another point is that this method does not manage the topology of the street network.

4 FURTHER DEVELOPMENTS

4.1 *The new methodology*

In the context of urban area and high resolution satellite images, our current work aims at extracting both linear and curvilinear streets and also at characterising the street networks. We propose a methodology, which goal is to help operational uses, inspired by the approaches of Airault *et al.* (1995) and Baumgartner *et al.* (1999).

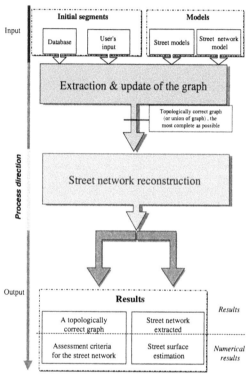

Figure 8. The new algorithm including topology management and street reconstruction

For operational purposes, the main constraint is the reliability of the result : false detection or over detection are banned. To fit to this goal and increase the robustness, a semi-automatic approach has been chosen: the inputs of the algorithm are linear elements coming from a database or manually given by the user. In both cases, it will give an initialisation of the graph representing the road network, which vertices are street intersections and ridges are street axis. Models of streets (using roads properties used by Couloigner 98) and properties of street network (such as connexity) will be introduced to perform the extraction.

Our approach is decomposed in two sequential parts (see Figure 8).

Firstly, a topologically correct graph of the street network will be extracted. This step aims at giving correctly spatial connections between streets as well as an approximation of their length and location. The work to do in this step depends on the inputs of the algorithm:

- If the input is a database, a registration of the road and an update of the network topology will be done.
- If the input is manually given, a registration of the initial road segments and completion of the graph will be performed.

After this step, the network will be represented by a topologically correct graph. A very precise road registration is not needed at this step.

The next step is the actual street reconstruction. Due to the high resolution images, a surface reconstruction has to be performed. This step will use the previous step of graph management as an initialisation for the reconstruction.

Finally, results will be quantitatively characterised, using mathematical criteria (Couloigner and Ranchin, 1998). Assessment criteria for the topological characterisation of the street network will be established as well as a street network surface estimation.

4.2 *Advantages of this approach*

The important point of this methodology is to separate the process in a step of topological management and an other of street reconstruction as surface element. First of all, it will enable to focus separately on each step which are both of them very challenging. An other advantage of separating the algorithm is that it will enable potential user to check if the road network topology is correct during the process.

The first step, which is a topologically correct graph extraction, will benefit from methods which have been developed when satellite sensors had a coarser resolution, and roads were represented as linear elements. Moreover, the fact that a very precise registration of the street is not needed at this step will permit to use an multiresolution approach in order to get rid of numerous artefacts encountered in urban area and increased by high resolution.

The second step will benefit from the results of the first one, by using the correct graph extracted for the reconstruction. The two tasks are then sequential and not parallel.

Finally, this method will establish new quantitative criteria on street network topology and geometrical features.

5 CONCLUSION AND PERSPECTIVES

Automating street extraction is an important issue in order to help people working in the urban environment. Urban street extraction is still almost undeveloped and high resolution images have a raised new prospects in this matter.

A recent method to extract particular model of street network have been applied, and its advantages and drawbacks have been shown.

In a context of extracting all types of urban street networks and managing their topology, we have then proposed the methodology currently under development. To reach the reliability required by operational uses, a semi-automatic approach was chosen.

The first step of this semi-automatic method aims at extracting a topologically correct road network. In the second step, a reconstruction of roads as surface elements will be proceed. The use of active contours, such as snakes, is envisaged in this step for the street extraction. They will be initialised with the graph obtained at the first step of the algorithm. Finally, criteria for characterising both the street network and street elements will be developed.

An important effort will be held to find some characteristics proper to the urban context which will help the street extraction.

This approach fits to the new problematic of street extraction in urban area with the adding of information that bring high resolution images.

ACKNOWLEDGEMENT

We would like to thank the firm G.I.M. (Geographical Information Mapping) for having provided the IKONOS images from the Hasselt area.

REFERENCES

Airault, S. & Jamet, O. 1995. Détection et restitution automatique du réseau routier sur des images aériennes. *Traitement du Signal*, 12(2):189-200.

Baumgartner, A. Steger, C. Wiedemann, C. Mayer, H. Eckstein, W. & Hebner, H. 1996. Verification and multiresolution extraction. *Photogrammetric Engineering and Remote Sensing*, 31(3):53-58.

Baumgartner, A. Steger, C. Mayer, H. Eckstein, W. & Heinrich, E. 1999. Automatic road extraction based on multiscale, grouping, and context. *Photogrammetric Engineering and Remote Sensing*, 65(7):777-785.

Bhattacharya, U. & Parui, S.K. 1997. An improved backpropagation neural network for detection of road-like features in satellite imagery. *International. Journal of Remote Sensing*, volume 18, n°16, pp. 3379-3394.

Couloigner, I. 1998. Reconnaissance de formes dans des images de télédétection du milieu urbain. Ph.D. Thesis, Université de Nice Sophia-Antipolis.

Couloigner, I. & Ranchin, T. 1998. Extraction of urban network from high spatial resolution satellite imagery : assessment of the quality of an automated method, *Proceedings of the EARSeL Symposium "Operational Remote Sensing for Sustainable Development"*, Enschede, The Netherlands, 11-14 May, Balkema, Rotterdam, The Netherlands, pp. 309-314.

Couloigner, I. & Ranchin, T. 2000. Mapping of Urban Areas : A Multiresolution Modeling Approach for Semi-Automatic Extraction of Streets. *Photogrammetric Engineering and Remote Sensing*, 66(7):867-874.

Destival, I. 1987. Recherche automatique de réseaux linéaires sur des images SPOT. *Bulletin de la Société Française de Photogrammétrie et de Télédétection*, n° 105, pp. 5-16.

Dutilleux, P. 1987. An implementation of the "algorithme à trous" to compute the wavelet transform. *Proceedings du congrès ondelettes et méthodes temps-fréquence et espace des phases*, Marseille (France), Springer Verlag Editors, 14-18 décembre 1987, pp. 298-304.

Gruen, A. & Li, H. 1995. Road extraction from aerial and satellite images by dynamic programming. *ISPRS Journal of Photogrammetry and Remote Sensing*, 50, 4, pp. 11-20.

Jedynak, B. 1995. Modèles stochastiques et méthodes déterministes pour extraire les routes des images de la terre vue du ciel. Ph.D Thesis of the University of Paris Sud, January 1995, 186 pages.

Laptev, I. Mayer, H. Lindeberg, T. Eckstein, W. Steger, C. & Baumgatner, A. 2000. Automatic extraction of roads from aerial images based on scale space and snakes. *Machine Vision and Applications*, 12:23-31.

Mc Keown, D.M. & Denlinger, J.L. 1988. Cooperative methods for road tracking in aerial imagery. *In Proceedings of IEEE Computer Society Conference - Computer vision and pattern recognition*, Ann Harbor, Michigan, 1988, pp. 662-672.

Mallat, S.G. 1989. A theory for multiresolution signal decomposition: the wavelet representation. *IEEE Transactions on Pattern Analysis and Machine Intelligence*, 11, 7, pp. 674-693.

Merlet, N. & Zérubia, J. 1996. New prospects in line detection by dynamic programming. *IEEE Transactions on Pattern Analysis and Machine Intelligence*, 18, 4, pp. 426-430.

Ranchin, T. 1997.Wavelets, remote sensing and environmental modelling, *Proceedings of the 15th IMACS World Congress on Scientific Computation, Modelling, and Applied Mathematics*, Berlin, Germany, 24-29 August 1997: *Application in Modelling and Simulation*, Vol.6,edited by Achim Sydow, pp. 27-34.

Ruskoné, R. 1996. Extraction automatique du réseau routier par interprétation locale du contexte : application à la production de données cartographiques. Ph.D thesis of the University of Marne-La-Vallée.

Serendero, M.A. 1989. Extraction d'informations symboliques en imagerie SPOT : réseaux de communications et agglomérations. Ph.D thesis of the University of Nice, December 1989, 160 pages.

Stoica, R. Descombes, X. & Zérubia, J. 2000. A Markov Point Process for Road Extraction in Remote Sensed Images. Rapport de recherche de l'INRIA, n°3923, avril 2000, 38 pages.

Wang, J.F. & Howarth, P.J. 1987. Automated road network extraction from Landsat TM imagery. *In Proceedings. of Annual ASPRS/ACSM Convention* ,Baltimore, MD, vol. 1, pp. 429-438.

Wang, D. He, D.C. Wang, L. & Morin, D. 1996. L'extraction du réseau routier urbain à partir d'images SPOT HRV. *International. Journal of Remote Sensing*, 17, 4, pp. 827-833.

Observing our environment from Space: New solutions for a new millennium, Bégni (Ed.)
© 2002 Swets & Zeitlinger, Lisse, ISBN 90 5809 254 2

Monitoring natural disasters and 'hot spots' of land cover change with SPOT VEGETATION data to assess regions at risks

F.Lupo, I.Reginster & E.F.Lambin
University of Louvain-la-Neuve (UCL), Belgium

ABSTRACT: The impact of global change on land surface attributes will not be uniformly distributed geographically. Changes in land surface attributes have major implications for ecosystem processes, geochemical cycles and society. Assessing the regions «at risk» of rapid land-cover changes and/or natural disasters is therefore a priority for global change research and for policies aimed at mitigating the impact of these changes. The paper describes a change detection technique based on the multi-temporal change-vector to two years of SPOT VEGETATION data. The main processes detected are related to droughts, floods and deforestation. The study was extended to Europe and the whole of Africa over a time period from May 1998 to April 2000. Continental scale maps of change in surface attributes were generated over the growing seasons of every biome. Finally, a validation of the change patches identified allows us to interpret the type of change and their environmental significance.

1 INTRODUCTION

All recent scientific evidence clearly points to the fact that the impact of global change on land surface attributes will not be uniformly distributed geographically. Assessing the regions "at risk" of rapid land-cover changes and/or natural disasters is therefore a priority for global change research and for policies aimed at mitigating the impact of these changes. The objectives of this project are to: (1) Use SPOT VEGETATION data to monitor over large regions the impact on ecosystems of natural disasters such as droughts, fires, floods and vegetation diseases, as well as land-cover change 'hot spots'; (2) Validate and interpret SPOT VEGETATION-based maps of natural disasters and extreme land-cover changes with collateral data on natural disasters and 'hot spots' of land-cover change; (3) Integrate this validated product in the current efforts of the global change scientific community, sponsored by the International Geosphere-Biosphere Programme (IGBP) and International Human Dimensions Programme on Global Environmental Change (IHDP), (4) Assess regions 'at risk' of rapid environmental change in order to focus research on most vulnerable areas and support the design of appropriate mitigation policies.

2 METHOD

The land-cover change detection approach is based on a comparison of the seasonal development curve for successive years of a remotely-sensed land-cover indicator - e.g. a vegetation index or a measure of spatial heterogeneity. When the time trajectory of the indicator over a particular pixel departs from the pixel's reference time trajectory, a change in land-cover is detected. This multitemporal approach is very sensitive to changes in seasonality and ecosystem dynamics, in addition to more abrupt landscape disturbances. It is quite insensitive to atmospheric and sensor noises that only affect isolated periods. The seasonal dynamic of a land-cover indicator can be represented by a multitemporal vector $p(i, y)$ for pixel i and year y. Any change in accumulated value and/or in seasonal dynamic of the indicator between the current and reference years can be measured as: $c(i) = p(i, ref) - p(i,y)$ where $c(i)$ is the change vector for pixel i between the reference year ref and the year y. The magnitude of the change vector $|c|$ measures the intensity of the change in land cover. It is calculated by the Euclidean distance between the pixel position, for the current and reference years, in the multi-temporal space of the observations (i.e. each dimension of that space represents the land-cover indicator for one observation period; the coordinates of a pixel position are the values for that pixels, for that year, of the land-cover indicator for the observation periods). Note that this arithmetic calculation assumes that the land-cover indicator I is a quantitative measure which is linearly related to some land-cover attributes. Through this vector difference, all the input images (decadal or monthly composites for two years) are reduced to a single land-cover change magnitude map. Tests and validations of this methodology have already been published: (see references [1], [2]), [3], [4]), [5]), [6], [7], [8]).

The methodology has been applied to the SPOT VEGETATION data. The change image based on monthly composites is developed by temporal aggregation (maximum value compositing method) of the decadal images.

3 RESULTS

3.1 *Description of data sets*

3.1.1 *Remote sensing data*

We used the ten-day synthesis product (VGT-S10, see reference [15]) which corresponds to the highest value of top-of-atmosphere Normalized Difference Vegetation Index (NDVI) for every pixel during a ten days period. The radiance data were atmospherically corrected with the Simplified Method for the Atmospheric Correction (SMAC) procedure. The VGT-S10 products were delivered in the Plate carrée 1km map projection. The data cover growing seasons of different agro-ecological regions (from May 1998 to April 2000). Over this period, one expects to detect mostly the influence of inter-annual climatic variability and natural hazards.

An additional cloud screening was necessary as some clouds remained, particularly along the Gulf of Guinea. The cloud mask provided with the data was expanded by a neighbourhood analysis. Firstly, all pixels in a 5x5 moving window were added to the cloud mask when one pixel of the window was a cloud. Then, a 10x10 majority filter was applied. Thirdly, the cloud mask was widened to include the pixels with a high change magnitude that were neighbours of cloudy pixels.

3.1.2 *Reference data*

A database on natural disasters was assembled to develop an independent estimate of change. Decadal rainfall data were extracted from the Africa Data Dissemination Service (see reference [10]). Daily data on fires detected from NOAA AVHRR were taken from the World Fire Web (see reference [11]).

Information on the major natural disasters in the region was assembled from FAO's Global Information and Early Warning System (see reference [12]), the Famine Early Warning System programme (see reference [13]) and the Contributions for Natural Disasters (see reference [14]).

3.1.3 *Vegetation indices*

Three vegetation indices were tested: the NDVI, the Soil-Adjusted Vegetation Index (SAVI) and the Normalised Difference Infrared Index (NDII) computed by replacing the red band by the short-wave infrared band (centred on 1.65 μm) in the NDVI formula. In this band, reflectance is related to water content of the canopy. It is also less sensitive to aerosols than the red band.

3.2 *Presentation of the results*

The results are:

1. a digital map representing the impact of the natural disasters and rapid land-cover changes which occurred during the growing seasons in the different agro-ecological regions of Africa and Europe between May 1998 and April 2000;

2. a detailed validation of this map, with collateral data on natural disasters (droughts, floods), rainfall, and information on the process of change and their environmental significance.

In Africa, the regions were defined from the Ecosystem Map of White (1983). The period corresponding to the growing season for each region was defined from rainfall and spectral vegetation index profiles. In Europe, the limits of the agro-ecological regions and their growing season are based on rainfall and air temperature distributions. The limits of the northern Europe region are based on the Agro-ecological Zones of the FAO (see reference [9]).

A region of severe moisture constraints is also defined in Asia. It is also based on the FAO maps.

4 DISCUSSION AND CONCLUSIONS

This new land surface product represents the impact on surface attributes of the major natural disasters which are mainly caused by rainfall anomalies. Most of the validation information is related to severe floods, heavy rains, cyclones or important difference in rainfall distribution between two growing seasons. High change magnitudes detected over known deforestation 'hot spots' are too close from clouds to

Fig. 1. Map of land-cover changes in Europe and Africa according to the growing seasons of different agro-ecological regions from May 1998 to April 2000. The change intensity is measured by the change vector magnitude. The clouds are represented in white (mask) and the red circles delimitate area where a detailed validation has been done (see http://www.geo.ucl.ac.be/Disasters.htm). (Colour plate, see p. 429)

be validated with confidence. The validation highlights the good performance of the method. There is a need for improvements in the pre-processing of SPOT VEGETATION data as the incomplete cloud mask and sensor noises forced us to work with monthly composites, even though droughts and floods would be better detected on decadal composites.

REFERENCES

E.F. Lambin and A. Strahler. 1994. Multitemporal change-vector analysis: A tool to detect and categorise land-cover change processes using high temporal resolution satellite data. *Remote Sensing of Environment.* Vol.48: pp.231-244. [1]

E.F. Lambin and A. Strahler. 1994. Remotely-sensed indicators of land-cover change for multitemporal change-vector analysis. *Int. J. Remote Sens.* Vol. 15 (10): 2099-2119. [2]

E.F. Lambin. 1996. Change detection at multiple temporal scales: Seasonal and annual variations in landscape variables. *Photogrammetric Engineering & Remote Sensing,* Vol.62 (8) :931-938. [3]

E.F. Lambin & D. Ehrlich. 1997. Land-cover changes in sub-Saharan Africa (1982-1991): Application of a change index based on remotely-sensed surface temperature and vegetation indices at a continental scale. *Remote Sensing of Environment,* Vol.61(2): 181-200, 1997. [4]

E.F. Lambin & D. Ehrlich. 1997. Identification of tropical deforestation fronts at broad spatial scales. *Int. J. Remote Sens.,* Vol.18(17): 3551-3568. [5]

E.F. Lambin. 1997. Modelling and monitoring land-cover change processes in tropical regions," *Progress in Physical Geography,* Vol.21(3): 375-393. [6]

J.S. Borak, E.F. Lambin & A. Strahler. 2000. Use of temporal metrics for land-cover change detection at coarse spatial scales. Int. J. Remote Sens., Vol.21(6/7): 1415-1432. [7]

F. Lupo, I. Reginster & E.F. Lambin, (in press 2001) Monitoring land-cover changes in West Africa with SPOT VEGETATION: Impact of natural disasters in 1998-1999. Int. J. Remote Sens. [8]

Agro-ecological Zones of the FAO:

http://www.iiasa.ac.at/Research/LUC/GAEZ/index.htm/. [9]

Africa Data Dissemination Service, USGS:

http://edcintl.cr.usgs.gov/adds/adds.html. [10]

World FIRE Web, JRC:

http://ptah.gvm.jrc.it. [11]

FAO, Global Information and early warning System:

http://geoweb.fao.org. [12]

FEWS Famine Early Warning System:

http://www.info.usaid.gov/fews. [13]

United Nations, Contributions for Natural Disasters: http://www.reliefweb.int/fts/. [14]

Vegetation Program:

http://sirius-ci.cst.cnes.fr:8080/index.html. [15]

Observing our environment from Space: New solutions for a new millennium, Bégni (Ed.)
© 2002 Swets & Zeitlinger, Lisse, ISBN 90 5809 254 2

The use of temporal variograms in the analysis of long term global image data

K.R.McCloy
Danish Institute of Agricultural Sciences, Tjele, Denmark

ABSTRACT: The goal of the work discussed in this paper is to develop tools for the display, analysis and interpretation of multi-temporal image data, for the purpose of better understanding the dynamic processes that are captured in the data. It is proposed to do this by means of temporal image transformations to produce temporal images. These are defined as images that collapse a long temporal sequence of image data into the one image, yet retain some or all of the information on dynamic change that is in the source data, in this temporal image. This paper focuses on one aspect of this work, specifically the role of temporal variograms in the production of temporal image transformations and temporal images. For this work the global NDVI data provided by NASA is used to evaluate the algorithms and techniques developed.

1 INTRODUCTION

The existence of various forms of global change has been long recognised. Recent events have caused a recognition that there are many more forms of active global change, and that their effects on the bio-geo-chemical processes on the planet can be significant. This recognition has led to a number of initiatives, including the International Geosphere – Biosphere Programme (IGBP) at http://www.igbp.kva.se/, the International Human Dimensions Programme on Global Environmental Change (IHDP) at http://www.uni-bonn.de/ihdp/, and the World Climate Research Programme (WCRP) at http://www.wmo.ch/web/wcrp/wcrp-home.html.

There are two important components to all three of these programmes that are relevant to this paper. The first is the use of models to investigate, better understand, and ultimately to be able to predict the response of processes to specific changes. The second is the use of image data to provide critical information to these models, and ultimately to test the validity of the model results.

Tucker et. al. (1983) showed that remotely sensed data can be used to assess vegetation greenup across extensive areas. He derived temporal NDVI data from AVHRR data for Africa and later for the globe to show that this NDVI data could depict the dynamic changes that occur in global vegetation. This work formed the basis of the operational application of NDVI data derived from AVHRR image data by the Food and Agricultural Organisation (FAO) at http://www.fao.org/WAICENT/faoinfo/economic/giews/english/giewse.htm. It also formed the basis of the extensive database of AVHRR products that are held by NASA for access by the global research community. A number of long term data sets derived from AVHRR data are freely available on the Internet, primarily from the NASA based sources, including;

- The Pathfinder Program of NASA produces long-term data sets that have been processed in a consistent manner for global change research.
  http://daac.gsfc.nasa.gov/DATASET_DOCS/avhrr_dataset.html.

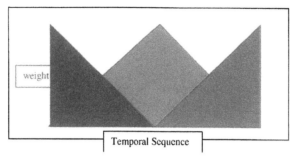

Figure 1. Weighting that could be given to blue green and red for each image through the temporal or hyperspectral sequence. Such a weighting will give blue- cyan- green - yellow - red in the sequence, but no magenta. Other weighting functions could be adopted that include magenta. (Colour plate, see p. 430)

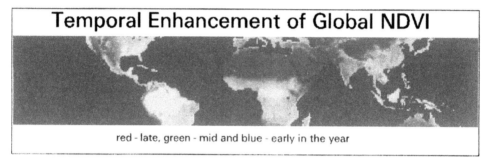

Figure 2. A temporal enhancement of 216 temporal images covering 18 years of data. (Colour plate, see p. 430)

Figure 3. Temporal variograms of the global NDVI data, with red - lag 4, green - lag 3 and blue - lag 2. (Colour plate, see p. 430)

Figure 4. Fourier Amplitude Image of the Temporal variogram dataset. Red - No annual cycles, Green - dominant annual cycles, blue - dominant bi-annual cycles, yellow - Annual with negligible cycles, magenta - Bi-annual with negligible cycles. (Colour plate, see p. 430)

- Sea surface temperature data sets constructed from AVHRR thermal data, available through http://podaac.jpl.nasa.gov/order/order_sstemp.html.

This work was followed by other applications of global NDVI values derived from AVHRR data to derive information useful to the modelling community. DeFries et al(1994) used twelve months of data from this data set to create global landcover maps. Others have used this data to detect changes that are occurring in global vegetation (Los et. al. 2001) and to derive indices of use to the modelling community (DeFries et. al. http://cstars.ucdavis.edu/projects/indexfr.html.).

Long sequences of image data record the expression of temporally dynamic processes in that data. The data should thus be capable of providing insights into these processes, as well as being used to compare models of these processes with actual conditions. Currently the only way to depict such long-term processes is by means of video clips of the data. The ephemeral nature of such displays means that they are not suitable for image analysis. Better tools for depicting the temporally dynamic processes as they are recorded in the image data are required.

The goal of this work discussed in this paper is to develop tools for the display, analysis and interpretation of multi-temporal image data, for the purpose of better understanding the dynamic processes that are captured in the data.

A major component of this goal is to produce temporal image transformations. The purpose of these transformations is to produce temporal images. Temporal images are defined as images that collapse a long temporal sequence of image data into the one image, yet retain some or all of the information on dynamic change that is in the source data, in this temporal image. Such an objective naturally involves surrenders. A significant component of the work in this project is to identify the surrenders that are being made in each form of image being produced.

The author has developed a number of tools to assist in this task, where these tools are described in McCloy (2000). In this paper I wish to focus on some of the results that have been achieved with analysis of these tools to date. Initial work to evaluate these tools is being done using the Global NDVI data set available through NASA at website http://daac.gsfc.nasa.gov/CAMPAIGN_DOCS/LAND_BIO/GLBDST_Data.html. The Global NDVI 1 degree by 1 degree data is extensive, with continuous monthly coverage from April 1981 until July 1994. There was a gap with no acquisition during late 1994, with acquisition being resumed in 1995. It is a relatively small data set, consisting of 360 pixel by 180 line, one byte per pixel images for each month through this period. For this study data was used from July 1981 - June 1994 and from July 1995 to June 2000 to form a 216-image stack consisting of 18 annual cycles of monthly data.

The images have been rectified, calibrated and corrected for some atmospheric effects. The images are calibrated using the Libyan desert as a standard reflectance surface (Rao, 1993) http://psbsgi1.nesdis.noaa.gov:8080/EBB/ml/niccal1.html. The data are then corrected for ozone absorption and Rayleigh scattering (James and Kalluri, 1994). The data have not been corrected for water vapor or aerosol effects. The calibrated data are then used to compute NDVI in the normal way. To create 1 degree by 1 degree pixels, a geometric subset of the pixels in this window are used. The pixel values with the highest NDVI values are then selected to represent that pixel in that month. The highest NDVI values are chosen on the basis that atmospheric effects reduce NDVI values so that the highest values should be the most representative value for actual NDVI.

The imagery was imported into Erdas Imagine, and then an image stack was created consisting of all 216 images in the sequence. Imagine uses the word stack to mean the construction of new images using existing images as layers in these new images. In this paper the word stack is used in a specific way that can be compatible with its usage within Imagine. An image stack is defined as a registered sequence of hyperspectral and hypertemporal images stored within the one file. At present one image format is being used for these images. In this format the data is stored such that the pixels, lines, bands, images change at a decreasing rate. Thus pixels change most rapidly, and images change least rapidly in the format. The construction is identical to Band Sequential (BSQ) if either only one band is stored, or one image, as is the case in this study where only NDVI data are used at each date. Analysis of the image data showed that some of the images contained errors at high latitudes, so the data set was subsetted to contain 360 pixels by 68 lines for the 216 images.

2 THE INITIAL ANALYSES

The initial analyses was conducted using temporal variograms, Fourier transforms of the variograms and the image data, and temporal enhancements (McCloy, 2000). These analyses showed that temporal images could be produced that show more general effects. The temporal enhancement depicted in Figure 2 uses the enhancement scheme shown in Figure 1 to enhance sequences from January through

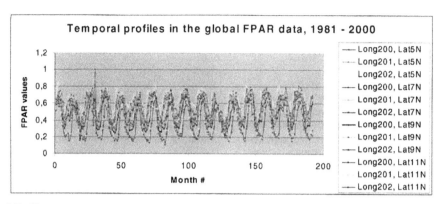

Figure 5. Profiles across global FPAR data at 12 locations in Africa, varying from the Congo Rainforest to the moderately dry Sahel. (Colour plate, see p. 431)

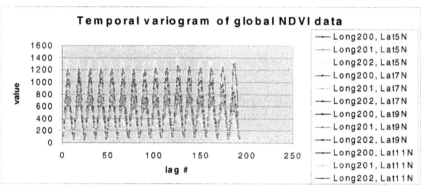

Figure 6. Profiles in the Temporal variograms for 12 locations in Africa, varying from the Congo Rainforest to the moderately dry Sahel. (Colour plate, see p. 431)

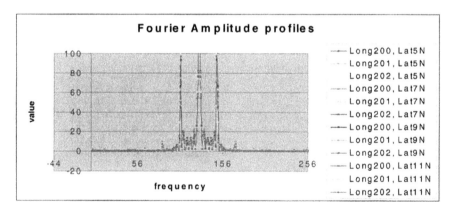

Figure 7. Fourier Amplitude profiles derived from the variogram data for 12 locations in Africa, (a) the whole profiles, (b) at frequency 107 representing annual cycles, and (c) at frequency 85, representing six-monthly cycles.(Colour plate, see p. 431)

340

to December as blue, through green, to red. All of the images in the 18 year stack were used in this enhancement, which thus forms of average over twelve months to construct "typical" response distributions for a twelve-month period.

The image is coloured to depict the period of maximum green vegetation during a typical annual cycle in accordance with Table 1. The image shows as green, the greening up that occurs in northern latitudes in the northern summer period. The areas of monsoonal activity late in the year in India and South East Asia show very clearly as cyan. There are areas of blue in South Africa and South America corresponds to greener vegetation in the southern summer period. Areas of magenta correspond to maximum greenup in the southern autumn months.

The temporal enhancement in Figure 2 indicates the seasonality of the vegetation around the globe. Since the vegetative cover depicted in Figure 2 reflects an integration of the effects of climate, soil and topography on plant growth, and the impact of man and animals on the standing biomass, so these images will not correspond directly with either climate maps or vegetation association maps. The types of vegetation present at each location do affect vegetation greenup. Thus the sharp boundaries that can be seen in Figure 2, most notably in South America, are often be due to differences in vegetative association. Figure 2 thus depicts different information than either climate maps or vegetation association maps. Since it depicts actual vegetation greenup it may be more useful for some modelling work than either climate or vegetation association maps on their own.

Table 1. Colour assignment by month for the temporal enhancement

Month	Primary Colour	Secondary Colour
January	Blue	
February		Blue-magenta
March		Magenta
April		Magenta
May		Magenta-red
June	Red	
July	Green	
August		Green-cyan
September		Cyan
October		Cyan
November		Cyan-blue
December	Blue	

Temporal Variograms were then produced as shown in Figure 3. These variograms are not temporal images in the sense defined in this paper, as they do not collapse the information into a few bands that can be displayed simultaneously. To achieve this goal, the variograms are converted into Fourier amplitude images. Most of the information in the variograms is contained in a few frequencies as can be seen from the profiles depicted in Figure 6. This means that these frequencies will contain most of the information in the Fourier images (Figure 4).

The profiles for the same twelve locations are depicted for the source data in Figure 5, the variograms in Figure 6 and for the Fourier data in Figure 7. The profiles for the source data show a cyclic pattern that is partially masked by quasi-random fluctuations in the data. This means that any techniques developed to show trends in this data will have to contend with this "noise" level. The "noise" in the source data was one reason why the variograms were initially developed. The variogram equation (Cressie, 1993) shows that a variogram should both enhance and smooth the data relative to the source data:

$$\delta(l) = \{\Sigma(x_i - x_{i+1})^2\} / (n - l) \text{ for } l = 1 \text{ to } (n - k) \qquad (1)$$

where n = number of observations to be used, 216 in this case; l = lag interval; k = constant to ensure that an adequate sample is used, usually set to 30.

With cyclic data of the form shown in Figure 5, the variogram will retain information on the amplitude and the phase in the data, but will loose information on the phase shift. Thus variograms created using temporal data that start in December will be the same as variograms that start in January. The variograms are thus independent of the starting date, unlike the source data. However, for some purposes, the phase shift may be significant, and thus for some purposes this may be a disadvantage.

Figure 6 shows that the profiles for the variograms are in fact much more regular than for the source data. The variogram data is also more sinusoidal in shape, so that it will provide a simpler solution in Fourier space than will the source data.

The Fourier amplitude data was produced using the Fast Fourier Transformation (FFT). The FFT requires that the number of bins or frequencies into which the transformed data will be placed must be 2^f where $2^f > n$, where n is the number of observations, 216 in this case. This means that $2F = 2^f$ is to be set to 256 and F = 128. In the transformation, this data is set symmetrically about the central value at F128, so that identical information is shown either side of this value. The central spike is very obvious in Figure 7, as are spikes at F107 and F85. The relationship between frequency (F) and phase (P) in the data is given by the relationship:

$$(F - k/2) * P = k \qquad (2)$$

where k = 256 in this case and F is the x axis or band value in the Fourier data. From this the phase can be calculated using;

$$P = 256 / (F - 128) \qquad (3)$$

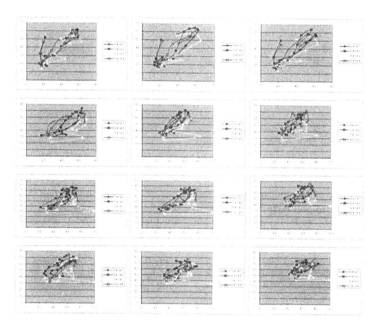

Figure 8. Comparative plots of FPAR for the two AVHRR based datasets. Each plot represents one pixel, at longitudes 200, 201 and 202 degrees, and latitudes 5, 7, 9 and 11 degrees north. (Colour plate, see p. 432)

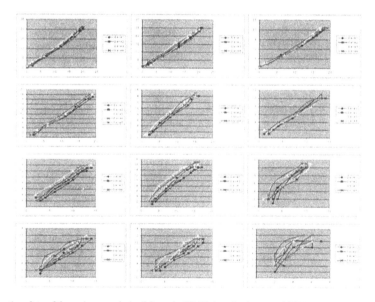

Figure 9. Comparative plots of the variograms derived from the FPAR data for the two AVHRR based datasets. Each plot represents one pixel, at longitudes 200, 201 and 202 degrees, and latitudes 5, 7, 9 and 11 degrees north. (Colour plate, see p. 432)

The major spikes in the data and their related phases are given in Table 2. Note that the relationship above has fractional values whereas the F values are integer. This will lead to rounding off differences.

Table 2. The Phase and Frequency relationship in the data used in this study

F	(F - 128)		Phase Comment
64 and 192	64	4	Three cycles per year
85 and 171	43	6	Bi-annual cycles
107 and 149	21	12	Annual cycles
117 and 139	11	24	Two yearly cycles
128	0	Infinite	Central value

The Fourier profiles in Figure 7 show a clear central spike at F128 for all pixels, with two pairs of spikes distributed about the central spike. The central spike represents no effective cycle to the data. The pairs of spikes set symmetrically about this central spike represent annual and six monthly cycles of the data at F107 and F85 respectively. There are some smaller spikes also in the data. If F128 is depicted as red, F107 as green and F85 as blue, as shown in Figure 6., then the colours in the image have the significance given in Table 3.

Table 3. Significance of the colours in Figure 6.

Blue	Only six monthly cycles
Cyan	Only six and twelve monthly cycles
Green	Only twelve monthly cycles
Yellow	Twelve monthly and no annual cycles
Red	No annual cycles

Figures 6 and 7 show clearly that all pixels have a strong component of non-cyclic or constant greenness effect. Thus the deserts and parts of the Amazon rainforest are all depicted in red. It is also to be expected for data with pixels of the size used for this data set. What are of interest are the large areas of significant six monthly cyclic contributions, in central Africa, the horn of Africa, northern India, southern USA, Bangladesh and Cambodia. Clearly agriculture is a major factor in this effect for some if not all of these areas.

3 CONCLUSIONS DRAWN FROM THIS ANALYSES

This initial work showed that the hypotheses that temporal images can be constructed, and that they will depict dynamics processes is valid. But the images produced to demonstrate this have been only able to depict more general trends. The temporal enhancement shows general seasonal patterns. The Fourier images show the general cyclic patterns within a year. Development of Fourier images from the source data created Fourier amplitude images that contained a much larger scattering of data through many more frequencies than is the case when using the variograms. It was not found to be possible to display simple Fourier images from the source data to depict the information as shown in Figure 4.

This work thus showed that the variograms enhance and smooth the fluctuations in the data, that they retain the general trends that exist in the data, including the cyclic patterns but that they loose phase shift information that exists in the source data. Their capacity to enhance and smooth the data made production of Fourier images that depicted temporal dynamic information straightforward, whereas this was not the case if the source data was used. However both temporal images showed very general information. The focus of the work then shifted to depiction of more detailed information on dynamic change.

Depiction of more detailed information on dynamic trends means that this information needs to be retained in the temporal images, yet it still needs to discriminate this information from the "noise" that is in the data. This means that a closer understanding of what represents noise, or isolated fluctuations in the data, and what represents the more detailed, has to be developed. Since a major use of this information is to analyse how well modelled result compare with actual conditions, it was decided to focus on the development of tools to compare two temporal data sets. As a first step in this process, the data set of 216 images was subsetted into two parts, each of 108 images, (July 1981 - June 1990) and (July 1990 - June 2000 excluding July 1994 - June 1995). These two data sets were then compared, producing a plot at each pixel showing the changes that have occurred at that pixel over time in the two data sets. The results for this are shown in Figures 8 and 9. Figure 8 shows the trajectories for the source data, and Figure 9 for the variograms.

4 THE SUBSEQUENT ANALYSES

Since similar data sets are being compared, it was expected that the comparative images, depicted in Figures 8 and 9, would all display linear features at about 45 degrees to the axes. If two sinusoidal features with the same phase and phase shift are compared in this way, then it can be shown that the comparative data will lie along a line. The orientation of that line is a function of the amplitudes of both data sets. If the data sets have the same amplitude, then the line will lie at 45 degrees to the axes. Consider the two sine waves:

$$X_i = A_0 * \sin(B_0 * t + C_0) \text{ and} \quad (4)$$

$$Y_i = A_1 * \sin(B_0 * t + C_0)$$

Then to remove t from the equation:

$$X_i / A_0 = Y_i / A_1 \qquad (5)$$

so that this is a line of gradient that depends on the two amplitudes. If these amplitudes are the same or nearly so, then the line will be at 45 degrees to both axes.

The comparisons shown in Figures 8 and 9 represent 12 months of comparisons in each colour. The are four colours depicted, so that there are four comparisons, as given in Table 4. The comparisons shown in Figure 8 represent the actual data set. In this figure, two of the pixels at latitude 11 degrees N do shown strong linear shapes, but no other pixels show this shape in any of the data displayed. In the comparisons shown in Figure 9, for the variogram data, most of the northern pixels show linear shapes, but the closeness to linearity decreases as the pixels analysed approach the Congo basin. In both figures, the pixels along one latitude show strong similarity. This similarity is also evident, but to a lesser degree between pixel profiles along different latitudes.

Table 4. The comparisons given in Figures 8 and 9 and their colour codes.

Dark blue	(1981 - 1982)	to	(1990 - 1991)
Magenta	(1982 - 1983)	to	(1991 - 1992)
Yellow	(1983 - 1984)	to	(1992 - 1993)
Cyan	(1984 - 1985)	to	(1993 - 1994)

5 CONCLUSIONS

The work has shown that temporal images can be produced that retain temporal dynamic information as is held in long sequences of temporal image data. The temporal images produced thus far provide very general dynamic process information. The work on investigating how to depict more detailed dynamic process information has shown that the "noise" in the data becomes a serious challenge. Some of this "noise" represents errors in the data, and some of it represents localised, once off events. Neither is of interest to those who wish to derive dynamic process information. Techniques to depict detailed dynamic process information will need to discriminate between the "noise" in the data and this dynamic information.

Temporal data sets are affected by all of the sources of error that affect a single image. They can be affected by other sources of error as well, including;

- Variations in sensor calibration with time, ensuring that sensor degradation follows a continuous curve function and not a step type of function.
- Use of different sensors to acquire different parts of the data sequence, with differences in calibra-

tion, and possible differences in spectral ranges in the sensors.

- Differences in orbit characteristics, between the different sensors, and over time for the one sensor, introducing differences in radiance.
- The type of spatial and temporal resampling used to create the source data.

The existence of these additional sources of error means that hypertemporal image data sets will contain significant noise. Techniques developed will need to deal with this noise.

Temporal images have been shown to contain a wealth of information that is not contained in single images in the source data set. The types of information contained in hypertemporal image data sets include;

- Localised once-off effects. These may be very important in the analysis of individual images, but they constitute noise if the analyst is interested in the dynamic process information that is stored in the data.
- Long term trends. Techniques for deriving long term trend information from hypertemporal data have been developed and shown to yield valuable information. The developed techniques can certainly be improved.
- More localised or shorter term dynamic process information. The work has shown that this type of information is more difficult to discriminate from the "noise" fraction in the data than is the longer-term information.

Variograms have been shown to be a valuable tool in removing the effects of "noise" in the data. They are thus an important tool in the armoury of those who wish to analyse hypertemporal image data.

REFERENCES

Cressie, N.A.C. 1993. Statistics for Spatial Data, Wiley Interscience Publication. John Wiley and Sons, New York.

DeFries, R.S. & Townshend, J.R.G. 1994. NDVI derived landcover classification at a global scale. *International Journal of Remote Sensing,* 15: 3567-3586.

DeFries, R.S., Field, C.B., Fung, I., Justice, C.O., Los, S., Matson, P.A., Matthews, E., Mooney, H.A., Potter, C.S., Prentice, K., Sellers, P.J., Townshend, J.R.G., Tucker, C.J., Ustin, S.L. & Vitousek, P.M. 1996. Mapping the Land Surface for Global Atmosphere- Biosphere Models: Toward Continuous Distributions of Vegetation's Functional Properties. http://cstars.ucdavis.edu/projects/indexfr.html.

James, M.E. & Kalluri, S.N.V. 1994. The pathfinder AVHRR Land dataset - an improved coarse resolution dataset for terrestrial monitoring. *International Journal of Remote Sensing.* Vol. 15(17): 3347-3363.

Los, S.O., G.J. Collatz, L. Bounoua, P.J. Sellers, & Tucker, C.J. 2001. Global interannual variations in sea surface tem-

344

perature and land surface vegetation, air temperature and precipitation. *Journal of Climate* 14: 1535-1549.

McCloy, K.R. 2000. Creation of temporal images from sequences of remotely sensed images, *Proceedings, XXXIIIrd ISPRS Symposium.* Amsterdam, Holland.

Rao, C.R.N. 1993. Degradation of visible and near Infra-red channels, NOAA Tech. Report NESDIS-70, NOAA/NESDIS.

Tucker C.J., Vanpraet, C.L., Boerwinkle, E. & Easton, A. 1983. Satellite remote sensing of total dry matter accumulation in the Senegal Sahel. *Remote Sensing of Environment.* 13: 461-469.

Observing our environment from Space: New solutions for a new millennium, Bégni (Ed.)
© *2002 Swets & Zeitlinger, Lisse, ISBN 90 5809 254 2*

Integrating spectral and textural information for land cover mapping

A.M.Jakomulska & M.N.Stawiecka
Remote Sensing of Environment Laboratory, Faculty of Geography and Regional Studies, University of Warsaw, Poland

ABSTRACT: Experiments aiming at integration of spectral and textural information showed great potential in image classification. The objective of this study was to assess applicability of variogram-derived texture measures in integrated optical and microwave images classification. The study was conducted using Landsat TM and ERS SAR images, covering Kolno Upland, an agricultural region in northeastern Poland. A set of textural parameters was derived within a moving window of an adapting size (determined by range of a variogram). Decision binary trees (DBT) were used to assess the potential of texture derived from variograms, cross variograms and pseudo-cross variograms to discriminate between land cover classes. Accuracy of DBT constructed on both spectral and textural parameters increased by as much as 10%, comparing with the accuracy of standard per-pixel classification. However, as the maximum likelihood decision rule did not perform well, further methods for improvement of the variogram-derived texture classification are discussed.

1 INTRODUCTION

Traditional elements of photointerpretation include characteristics of first order (tone/color), second order (spatial arrangement: size, shape and pattern) and third order (height, shadow). In digital remote sensing third order image characteristics are considered a nuisance, while potentially useful spatial information has been usually ignored, due to lack of methodology and computational limitations. Parallel to advances in spectral image classification techniques, research has been undertaken, albeit by a limited number of researchers, on incorporating texture and contextual information in image classification. Positive results of these studies as well as a growing number of high ground resolution satellite sensors operating or scheduled for launch within the next years provide extra incentives for further research in this direction.

The objective of this study was to assess the potential of variogram-derived texture measures applied to classification of satellite images of medium ground resolution: Landsat TM and ERS-1. The method was applied to an agricultural landscape, characterized by a distinct mosaic of small parcels. The technique employed was a combination of the approaches used to date, with a few improvements. The following section describes variogram applications in classification of remotely sensed images.

1.1 *Variogram applications in image classification*

Geostatistics is currently a well-understood and frequently applied image processing technique. It has been shown (Woodcock 1988a, 1988b) that range is directly related to the texture and/or objects size, while sill is proportional to global object (class) variance, although it is also affected by external factors, such as sensor gain, image noise, atmospheric conditions etc. An increasing interest in applications of variogram-derived texture has been noted in image classification.

Two approaches to texture derivation from the variogram have been suggested and proved to increase image classification accuracy: semivariogram textural classifier algorithm STC (Miranda et al. 1992, 1996) and modeling the variogram (Ramstein & Raffy 1989). In STC, semivariances for the first consecutive lags are used directly as additional input layers to image classification, while in the second technique variogram model coefficients are used. However, both techniques have a significant shortcoming: variogram computation is based on a moving kernel of a fixed size, where window size is chosen experimentally in the pre-processing phase and depends highly on the particular data used in a study.

There is a trade-off between application of too large and too small a window. The first approach leads to straddling class boundaries and encompassing adjacent classes in the same variogram. On the other hand, several authors reported applications, where variograms were derived from small moving windows (e.g. Chica-Olmo & Abarca-Hernández 2000 used only the first lag of the variogram). Although the first few lags may be a good approximation of variogram shape, and hence theoretical variogram model type, their application seems to give an incomplete picture, since, in many cases, the sill is not achieved. Variogram range measures texture coarseness and has been proven to be an important discriminant of forest stands, vegetation types and land cover classes. Range is also directly related to the size of objects: for small patches (in dispersed landscapes) range is shorter. Hence it is inappropriate to measure all the classes and objects with the same measure. Furthermore, Berberoglu et al. (2000) pointed out that variograms for short lags measure field edges, rather than within-field variability. In fact, semivariance at the first lag of a variogram is a strong edge detector (Fig. 1), and semivariances at consecutive lags are gradually smoother edge detectors.

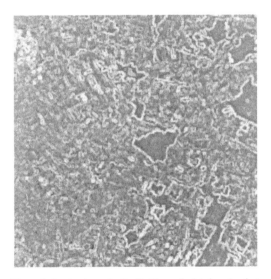

Figure 1. Semivariance for the first lag, calculated on Landsat TM green band, exhibits edge-detecting properties.

Franklin & McDermid (1993) and Franklin et al. (1996) proposed an alternative technique: a geographic window, where dimensions of a moving kernel are customized and are determined by variogram range. The technique proved to be promising and has been applied to derive texture measures like the co-occurrence matrix for image classification as well as first and second order texture and semivariance moment texture (nugget, range, sill, slope, mean semivariance) for LAI estimation (Wulder et al. 1997).

Jakomulska & Clarke (2001) applied variogram-derived texture for large-scale vegetation mapping using high ground resolution aerial images. Experimental variograms were derived for each image pixel from a moving geographic window, the size of which was determined locally by a range. Next, a set of parameters was calculated from the variogram and their discriminative power was assessed using decision binary trees (DBT). The final image classification was based on the original spectral bands and a few texture parameters, which best exploited the differences between classes tested. Significant increase of classification accuracy has been noted for the highly textured vegetation canopies (chaparral, oak woodlands).

In this study we apply variogram-derived measures for medium resolution satellite images with the land cover mapping application. We assess the applicability of textural classification for:
– optical imagery (Landsat TM),
– microwave imagery (ERS SAR), and
– integrated optical and microwave dataset.

To explore wide capabilities of variogram-derived measures we propose application of univariate and multivariate estimators of the variogram function. Since techniques described in this paper depend on the variogram, it is introduced here briefly; for a more exhaustive explanation and illustration the reader is referred to a geostatistical literature.

Variogram, a function of variance with distance, describes spatial dependence in data: the likelihood that observations close in space are more alike than those further apart. Mathematical expression of a variogram is given by Equation 1 (Cressie 1993):

$$\gamma_k(\mathbf{h}) = \frac{1}{2n(\mathbf{h})} \sum_{i=1}^{n(\mathbf{h})} \{dn_k(x_i) - dn_k(x_i + \mathbf{h})\}^2 \quad (1)$$

where $n(h)$ = the number of pairs in lag h; $dn\{\}$ = the DN values at location x_i and $x_i + h$; and k = a sensor band.

The multivariate variograms (Journel & Huijbregts 1978, Wackernagel 1995) measure the joint spatial variability between two bands (the cross variogram, Equation 2):

$$\gamma_{jk}(\mathbf{h}) = \frac{1}{2n(\mathbf{h})} \sum_{i=1}^{n(\mathbf{h})} \{[dn_j(x_i) - dn_j(x_i + \mathbf{h})] \cdot$$
$$[dn_k(x_i) - dn_k(x_i + \mathbf{h})]\} \quad (2)$$

and the variance of cross differences of two bands (the pseudo-cross variogram, Equation 3):

$$\gamma_{jk}(\mathbf{h}) = \frac{1}{2n(\mathbf{h})} \sum_{i=1}^{n(\mathbf{h})} \{dn_j(x_i) - dn_k(x_i + \mathbf{h})\}^2 \quad (3)$$

where j and k indicate two radiometric bands (in this study two principal components).

2 STUDY AREA AND DATA

The researched area, Kolno Upland is located in northeastern Poland (between 21°70' and 22°60' E and 53°20' and 53°60' N). The landscape is agricultural, with arable lands prevailing on the upland and meadows along valleys. The arable lands are characterized by a complicated pattern of ownership resulting in a mosaic of small parcels (on average 5 ha), cultivated in a varied way. Due to the postglacial origin of the upland and unfertile soils, the main crops in the area are rye and potatoes. Coniferous and deciduous forests cover a few percent of the area. Along valley bottoms sparsely timbered deciduous forests are common. Urbanized areas are rare, but dispersed households and little villages are frequent in the region.

The analysis was performed on Landsat Thematic Mapper acquired on 25.08.87 (Fig. 2) and ERS-1 acquired on 31.08.94 (Fig. 3). Exploratory data analysis and parameter assessment were based on the study area covering the central part of the Kolno Upland. Color aerial images in 1 : 26 000 scale were used for derivation of training samples and accuracy assessment.

The following land cover classes were analyzed:
- built-up areas,
- arable lands,
- coniferous forests,
- deciduous forests and
- meadows.

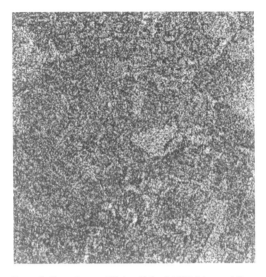

Figure 3. Central part of Kolno Upland: ERS-1 image. Microwave imagery allows for discrimination between rough, high objects (light) and humid ones (dark).

In spite of few land cover classes present within the researched region, the area is highly varied with respect to spectral response and spatial arrangement of objects, due to:
- complex pattern of arable lands,
- heterogeneous urban areas,
- dispersed dwellings, villages and sparsely timbered deciduous forests.

Furthermore, objects are differentiated due to roughness and humidity:
- high versus low objects (e.g. deciduous forests vs. meadows),
- increased soil humidity in the valleys.

3 METHODS

3.1 *Computation of variogram-derived texture*

Preprocessing of the data was limited to geometrical correction and filtering of the microwave image. To reduce speckle present in the microwave image a Gamma-Map filter (3x3 window) was applied. Both Landsat TM and ERS-1 images were rectified to UTM zone 34 N using nearest neighbor resampling method. ERS-1 image was resampled to the pixel size of Landsat and compressed to 8-bit radiometric resolution.

Experimental variograms were calculated for each of the six TM bands (excluding thermal) and ERS-1 image. Multivariate variograms (cross variograms and pseudo-cross variograms) were calculated between pairs of the first three principal components derived from the six TM bands (explaining 98.7% of the variance). The computation was carried out, us-

Figure 2. Central part of Kolno Upland: red band of Landsat TM. Complex mosaic of arable lands is characteristic for the entire region. Meadows and forests are comparatively homogenous. The image covers about 100 km².

ing a program written in the C language. To determine variogram range, variances were obtained for
up to 34 lags (1 km) on a per-pixel basis. A bounded
semivariogram model is assumed: if a range is not
achieved during the calculation, it is set at the
maximum allowable lag (lag=34). In spite of a high
limit of a range, the maximum value was rarely
reached, with a mean value of the range remaining
below 200 m (Table 1).

Table 1. Range values (in pixels) obtained in particular bands.

Band	Mean	Standard deviation
TM1	5.9	4.7
TM2	5.5	4.7
TM3	6.6	5.2
TM4	7.1	5.7
TM5	6.7	5.4
TM7	6.8	5.2
ERS-1	2.9	7.0

Comparing ranges for the optical and microwave
data it can be noted that mean ranges derived from
ERS-1 image are shorter by half than those derived
from Landsat bands, but standard deviation is
higher. This is a result of a high variance achieved at
very short distances due to high variance of the microwave image (in spite of the fact that the image
has been filtered). Ranges calculated on infrared images are slightly longer than those derived from visual bands.

Figure 4 illustrates the concept of a moving kernel of an adapting size. Range remains short for homogenous objects (e.g. meadows or forests) and increases for textured objects. Range is also low at
object borders. In this case variance increases very
rapidly due to the fact that the window straddles adjacent classes, and then stabilizes. This allows the
moving window to remain small when the edge of
the object is approached (notice low range values at
the woodland patch edges).

A set of indices was calculated for each central
pixel within a moving window of a changeable size
(Table 2). Altogether 46 textural variables were derived from the six optical and one microwave images.

Table 2. Textural parameters derived from Landsat and ERS
images.

Parameters derived from:	Vario-gram	Cross vario-gram	Pseudo-cross variogram
At a range:			
Range	X	X	
Sill	X	X	X
At lag = 0:			
Semivariance			X
Global (calculated from consecutive lags up to a range):			
Sum of semivariances	X	X	
Mean of semivariances	X	X	

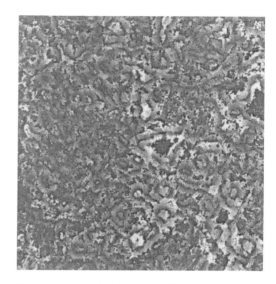

Figure 4. Range derived from the red band of Landsat TM.
Low values of range are depicted in dark colors, high values of
range in light colors.

3.2 Assessment of texture in image classification

Due to a large set of derived textural variables, the
parameters had to be screened out. To evaluate the
discriminative power of texture indices, the parameters have been assessed using decision binary
trees. Firstly, training samples were extracted from
the images. 200 pixels for each class were chosen at
random from a land cover layer produced through
photointerpretation of aerial photos. Secondly trees
were constructed using: a) original spectral bands, b)
both spectral bands and a set of textural parameters.
The final image classification (integrated bands of
Landsat TM, ERS-1 and textural parameters) was
performed using the maximum likelihood decision
rule. Classification accuracy was evaluated through
crosstabulation of a land cover layer achieved
through photointerpretation of aerial images and the
results of classification.

4 RESULTS

4.1 Spectral and textural properties of the studied land cover classes

Landsat TM sensor has been specifically designed
for land cover and vegetation mapping applications.
Furthermore, additional information from the active
radar sensor should facilitate proper identification of
object characterized by varied roughness and/or humidity. However, even a visual analysis of spectral
characteristics of the researched land cover classes
shows some difficulties in proper class identification.

Mean values of the spectral responses of the classes are very close to each other, especially for the built-up areas and arable lands classes (Fig. 5). Also the minimum-maximum ranges of spectral responses often overlap. This is mainly a result of the high within class variance, which is the case of the urban class. Similarly arable lands, highly varied due to a mosaic of small parcels, exhibit high standard deviation from the mean values.

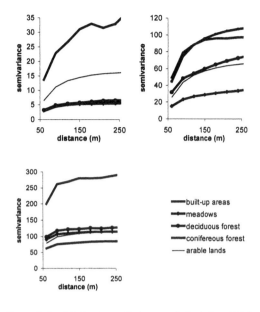

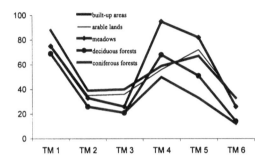

Figure 5. Spectral signatures.

Figure 6. Spatial signatures. Variograms derived from TM 3 (upper left), TM 4 (upper right) and ERS-1 (lower left).

Spatial information is of particular benefit for mapping classes characterized by complex spatial pattern represented by heterogeneous spectral response. Variograms derived from training samples show better differentiation for the selected classes than spectral signatures, especially if within-class variance is higher than interclass variance (Fig. 6). Built-up areas can be easily distinguished from arable lands using sills of variograms derived from the red band or microwave image. Meadows are distinct from all the other classes due to a very low variance in the infrared portion of electromagnetic spectrum (Fig. 7). Dense and uniform canopies of coniferous forests result in low sills of the variogram derived from a microwave image.

Although Ramstein & Raffy (1989) showed that land cover classes can be well differentiated by range only (assuming an exponential variogram model), other reports (St-Onge & Cavayas, 1995; Wallace et al., 2000; Jakomulska & Clarke 2001) show, that both sill and range are distinctive for different vegetation communities.

In this study, range has been rarely used in the final classification and shows high dependence on the object size. It preserves buffer characteristics and significantly decreases at object edges and along linear structures. Furthermore, for a particular class, range seems to be more stable for all bands, while sill varies, allowing additionally multivariate analysis (Fig. 6). Four examples of variogram-derived texture are shown on Figure 7.

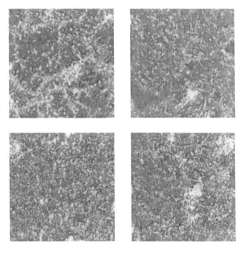

Figure 7. Variogram-derived texture parameters. Upper left: sill of a TM 3 variogram, upper right: sill of a TM 4 variogram, lower left: sill of a ERS-1 variogram, lower right: sill of a cross variogram calculated between the first two principal components.

4.2 Evaluation of variogram-derived measures

Decision binary trees were constructed on samples of 200 pixels chosen randomly for each class from a land cover layer produced through photointerpretation of aerial images. Decision rules were complicated: the final decision tree, constructed from both textural and spectral variables, constituted of 70 final nodes. A simplified, non-overlapping decision binary tree (constructed on all available parameters: radiometric Landsat and ERS-1 bands as well as derived texture from optical and microwave images) is shown on Figure 8: three of the four decision split rules are based on spatial information.

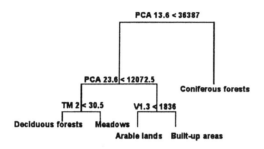

Figure 8. A simplified non-overlapping decision binary tree.

Coniferous forests are distinguished from all the other classes by a high sill of a pseudo-cross variogram calculated between the first and the third principal component. A low sill of a pseudo-cross variogram calculated between the second and the third principal component distinguishes vegetated areas (deciduous forest and meadows) from barren regions (arable lands and built-up areas). Arable lands are characterized by lower mean of variances (calculated up to a range) than built-up areas. Finally, only meadows are distinguished from deciduous forests by a spectral characteristic: higher reflectance in green band.

The following parameters proved to have the highest discriminative power, and were chosen for the final classification of the entire image:
– sills of variograms and cross variograms,
– mean semivariances and cross semivariances,
– sum of cross variances (up to a range),
– pseudo-cross variances at lag 0 and at a range.

Variogram derivatives were calculated for six TM bands and microwave image. Multivariate parameters were derived from the combinations of the first three principal components acquired from the TM image. Altogether 28 texture measures, six spectral TM bands and ERS-1 band were used in final classification.

With the exception of pseudo-cross variances at lag=0, all texture indices are derived at a range or involve computation based on consecutive variances up to a range. Means and sums of variances and cross variances calculated up to a range incorporate both sill and range information and approximate the shape of the variogram.

4.3 Image classification: assessment of variogram-derived measures performance

The accuracy of decision binary trees constructed using both optical and microwave bands outperformed classification based on optical data only by 1.5%. Although ERS-1 image increased accuracy of discrimination between arable lands and meadows (due to increased humidity of meadows), the overall accuracy remained similar. However, addition of textural information improved per-pixel classification by as much as 10%. Per-class accuracy is shown in Table 3.

Table 3. Comparison of accuracy assessment (in %) of decision binary tree and maximum likelihood classification.

Classification method	Decision binary tree			Maximum likelihood
	TM	TM+ERS	TM+ERS +texture	TM
Built-up areas	86.8	86.0	93.7	34.0
Arable lands	73.6	77.8	86.1	87.0
Meadows	73.6	77.2	88.7	66.0
Deciduous forests	76.8	77.5	82.4	47.5
Coniferous forests	81.2	81.3	89.3	75.5
Total	78.4	79.8	88.0	74.0

Incorporating texture into the decision binary tree classification significantly decreased class confusion of all the researched classes. Analysis of the users and producers accuracy revealed that classification improved due to:
– decreased omission error for homogenous classes (e.g. pixels classified as deciduous forest shifted to meadows class),
– decreased commission error for more varied classes (e.g. pixels classified as built-up areas shifted to arable lands class).

Textural variables actually used in tree construction were used as input in the final image classification using the maximum likelihood decision rule. The overall accuracy of the maximum likelihood, per-pixel classification based on optical data was unsatisfactory and slightly lower than the accuracy of the DBT. Furthermore, neither addition of microwave nor textural information layers improved the classification. In fact, addition of all 28 texture parameters reduced the accuracy. Low accuracy results from the non-linear relation of variogram derived texture measures and objects registered on remotely sensed data. Moreover, neither spectral, microwave nor textural information used in this study had normal distribution. However, a minimum distance decision rule also produced very poor results. Poor performances of the standard classification decision rules suggest that they are not suitable for the textural classification. Parameters used in the construction of decision binary trees show that few classes can be properly identified by spectral responses only while some require addition of textural information. Since both minimum distance and maximum likelihood decision rules use all the input parameters in the classification, it can be concluded, that for the discrimination of some classes (or pixels), texture may actually deteriorate classification accuracy.

Finally, Table 4 presents total accuracy assessment of decision binary trees classification performed on different sets of information data. Classification of the original, radiometric data was generally worse than classification with texture. Although results of classification based on ERS only are unsatisfactory, addition of texture derived from ERS only increased accuracy by as much as 15%. Lower, but still significant increase in accuracy was noted for the optical data. The results suggest that texture is of particular interest when only microwave data is available.

Table 4. Accuracy of classification (in %) performed on raw spectral data and textural information (decision binary tree decision rule).

Bands used	Accuracy	Difference
TM	78.4	9.3
TM+texture derived from TM	87.7	
ERS	58.0	15.0
ERS+texture derived from ERS	43.0	

5 DISCUSSION

The results of textural analysis are encouraging: variogram-based measures proved to increase overall and class-specific accuracy of decision binary trees classification. The technique does have the potential for providing improved discrimination of land cover classes, especially between pairs of classes, which have similar spectral response but contrasting spatial structure (e.g. arable lands and built-up areas). Although variogram-derived texture is still a computationally intensive technique, its incorporation in the image classification produces good results and can be recommended even for classification of medium resolution imagery. While texture analysis has been proved to be useful in the analysis of visible/infrared image data, it is showing even greater applicability to microwave imagery.

However some confusion between classes is still present. For example, high variance is characteristic for both:
- heterogeneous objects (like urban areas), and
- objects of high reflectance (e.g. arable lands in visual bands).

It is expected, that improved results could be achieved if variances were normalized by a reflectance of the object.

Further abatement of error could be obtained if other classification techniques were employed. Decision binary tree classification performed on training samples has shown that some classes were discriminated using solely spectral information, some just spatial information, and others both. It is expected then, that knowledge-based classifiers would perform better than parametric ones, since decision rules can be designed to include only textural or only spectral information. A better solution could be application of artificial neural networks, since they do not require a priori knowledge of statistical distribution of data. The decision rules are not fixed by a deterministic rule applied to the training signatures (as opposed to employment of hard rules by expert systems), but are determined iteratively by minimizing classification error. Other advantages of neural networks, like the possibility of including a large number of diverse variables, have been widely emphasized in the literature. This hypothesis will be tested in the future.

6 ACKNOWLEDGEMENTS

The authors acknowledge the European Space Agency for granting ERS-1 image within the Third Announcement of Opportunity offered to conduct research and application development projects using data from the first and second European Remote Sensing satellites, ERS-1 and ERS-2.

REFERENCES

Berberoglu S., Lloyd C. D., Atkinson P. M., Curran P. J., 2000, The integration of spectral and textural information using neural networks for land cover mapping in the Mediterranean, Computers & Geosciences, 26, 385-396

Carr J. R., 1998, The semivariogram in comparison of the co-occurrence matrix for classification of image texture, IEEE Transactions on Geoscience and Remote Sensing, 36 (6)

Chica-Olmo M. & Abarca-Hernández F., 2000, Computing geostatistical image texture for remotely sensed data classification, Computers & Geosciences, 26, 373-383

Cressie N. A. C., 1993, Statistics for Spatial Data, John Wiley & Sons, New York, 900 pp.

Franklin S. E. & McDermid G. J., 1993, Empirical relations between digital SPOT HRV and CASI spectral response and lodgepole pine (Pinus contorta) forest stand parameters, International Journal of Remote Sensing, Vol. 14, No. 12, 2331-2348

Franklin S. E., Wulder M. A., Lavigne M. B., 1996, Automated derivation of geographic window sizes for use in remote sensing digital image texture analysis, Computers & Geosciences, Vol. 22, No. 6, 665-673

Jakomulska A. & Clarke K. C., 2001, in press, Variogram-derived measures of textural image classification. Application to large-scale vegetation mapping. In: Geostatistics and Quantitative Geology, Kluwer Academic Publishers

Journel A. G. & Huijbregts C. J., 1978, Mining Geostatistics, Academic Press, London 600 pp.

Miranda F. P., Macdonald J. A., Carr J. R., 1992, Application of the semivariogram textural classifier (STC) for vegetation discrimination using SIR-B data of Borneo, *International Journal of Remote Sensing*, Vol. 13, No. 12, 2349-2354

Miranda F. P., Fonseca L. E. N., Carr J. R., Raranik J. V., 1996, Analysis of JERS-1 (Fuyo-1) SAR data for vegetation discrimination in northwestern Brazil using the semivariogram textural classifier (STC), *International Journal of Remote Sensing*, Vol. 17, No. 17, 3523-3529

Ramstein G. & Raffy M., 1989, Analysis of the structure of radiometric remotely-sensed images, *International Journal of Remote Sensing*, 10, 1049-1073

St-Onge B. A. & Cavayas F., 1995, Estimating Forest stand Structure from High Resolution Imagery using the Directional variogram, *International Journal of Remote Sensing*, Vol. 16, No. 11, 1999-2021

Wackernagel H., 1995, Multivariate geostatistics. Springer-Verlag, Berlin, 256 pp.

Wallace C. S. A., Watts J. M., Yool S. R., 2000, Characterizing the spatial structure of vegetation communities in the Mojave Desert using geostatistical techniques, *Computers & Geosciences*, 26 (2000) 397-410

Woodcock C. E. et al., 1988a, The use of variograms in remote sensing: I real digital images, *Remote Sensing of Environment*, 25:323-348

Woodcock C. E. et al., 1988b, The use of variograms in remote sensing: II real digital images, *Remote Sensing of Environment*, 25:349-379

Wulder M. A., Lavigne M. B., LeDrew E. F., Franklin S. E., 1997, Comparison of texture algorithms in the statistical estimation of LAI: first-order, second-order, and semivariance moment texture (SMT), *Canadian Remote Sensing Society Annual Conference, GER'97, Geomatics in the Era of Radarsat*, May 24-30, 1997, Ottawa, Canada

The estimation of the herbaceous biomass in the Sahelian pastoral zone using a GIS

T.De Filippis, A.Di Vecchia & P.Vignaroli
CeSIA-Accademia dei Georgofili, Florence, Italy

B.Djaby & B.Koné
Agrhymet Regional Center, Niamey, Niger

ABSTRACT: In the Sahel region the pastorals resources are strongly linked to the fluctuation of the biomass production In the present study, an integrated approach of multi-source, multi-type and multi-scales data analysis in the pastoral zone in Senegal, Mali, Burkina Faso, Mauritania, Niger and Chad has been developed using Geographic Information System potentialities for the purpose of Sahelian rangeland production estimation and as well as the breeding systems, which are based on the transhumance and the nomadism as well.

1 INTRODUCTION

Depending of the fragility of the environmental and the socio-economical context in the Sahelian zone, the breeding sector plays an important role in the mitigation or the worsing of food crisis.

For hence, the availability of information on pastoral resources is a very important component for food security and a right land management. Those information are often used to understand better pastoral crisis and contribute to increase livestock farmers income. The integration of pastoral monitoring in famines and early warning systems started 10 years ago. The experience developed in the present paper has been carried out by the Early Warning and Agricultural Productions Forecast Project (Pj. AP3A) hosted by the AGRHYMET Regional Centre in Niamey, Niger, funded by Italian Co-operation and sponsored by the World Meteorological Organisation.

The objective is mainly to give an adequate information about rangelands such as inventory fixtures during the main production period, the raining season in the sahelian region, and the end of green period, period where qualitative and quantitative state of rangelands have an effect on animals driving. In addition to fields census, data approaches has been developed with remote sensing such as NOAA – AVHRR to model rangeland production with normalized difference vegetation index (Justice and al. 1986). This model based on NDVI has a challenge to estimate a dry matter yield at regional scale over many ecosystems. Although the use of field data is justified, the process takes a considerable amount of time and money especially as on a large scale. In the

Biomass modelling rank, a progress has been done with the P.P.S project (the Sahelian Rangeland Production Project) in Mali, which developed the method based on taking into account the water and nutrients balance for rangeland production estimation. One of the interesting points of the methodology is the global assessment using data over a large working scale surface. This non-destructive model at this scale is necessary to give an accurate information about vegetal production assessment.

Within the PJ. AP3A objectives, many data have been collected, tabular data as well as numeric coverage data, particularly in the pastoral domain where additional coverages about the Sahelian pastoral unit have been digitised using PC-Arc/Info™. The main purpose is to integrate all the data to produce pastoral map risk, which is a real need for regional environment monitoring and national early warning services. The thematic approach permits a circumstantial risk assessment that indicate the level of current year production, advanced knowledge of seasonal movement and pressure on natural resources.

The whole process is built over a territorial system analysis (SAT) and circumstantial system analysis (SAC) within a GIS-based and database management system environment.

The SAT is a complete GIS database which contains an historical and structural data about nine countries. Domains covered by SAT are biophysical data, socio-economic data and thematic maps. A vulnerability assessment is a first objective of SAT.

The SAC contain an actual GIS database with current year monitoring data such a satellite image analysis, and biomass provided by the actual model.

2 STUDY AREA

The permanent Interstate Committee for Drought Control in the Sahel (CILSS) covers nine (9) West African countries, which are the Cape Verde Islands, Senegal, Guinea-Bissau, the Gambia, Mauritania, Mali, Burkina Faso, Niger, Chad (Figure 1). The study area extends from 12° to 18° N and 18° W to 24° E, and covers about 1.050.000 km² over six (6) countries in West Africa. This area corresponds to the pastoral zone in the Sahelian region (CIRAD-IEMVT-CTA 1989) which is delimited by the 150 mm to 600 mm (Hiernaux 1983). More aspects on this zone are important for pastoral use like quality of herbaceous and lack of certain diseases (trypanosomiasis) and a low density of trees. The vegetation of the area is composed of semi-desert grassland and grassland with trees and shrubs typical for the Sahelian Zone (Justice et al. 1986). The rainfall distribution is seasonal, occurring between May and October, with a maximum of precipitation in August and a high spatial and temporal variability from one year to another. The production of annual grasses and the floristic composition depend on this variability, which has serious effects on the livelihoods of local pastoral community in the drought years.

Figure 1. – Pastoral zone of CILSS countries.

3 GIS APPLICATION

Using GIS to assess rangelands production in Sahelian present a great challenge about spatial area of climatic variability and spatial output usable by another systems.

The Biomass production model (figure 2) was developed by the AP3A project in order to give a structural information about different climate conditions and predict as well the potential carrying capacity. The model is now operational and is able to simulate historical data on a yearly basis, assuming that the rainfall data is available.

The model biomass yield computation is based on a simplified water infiltration and nitrogen balance, which are also based on surface runoff. The model-input coverages are the maps of pastoral units (Djaby et al. 1996) and rainfall data. The major output components are the dry matter production map at

scale 1km*1km or 5km*5km and the biomass quality based on nitrogen content. The results is published in Agrhymet annual bulletin and the analysis can be made by comparison with vegetation front line.

4 METHODOLOGY

4.1 Model Definition

In the Sahelian region, the rangeland production depends primarily on the rainfall below 250 mm and the nutrient components as well above this limit. The availability of water in the soil depends on streaming, infiltration and soils texture. The study region is first divided according to the climatic limit, and an mathematical function found by P.P.S project is applied to compute the biomass yield.

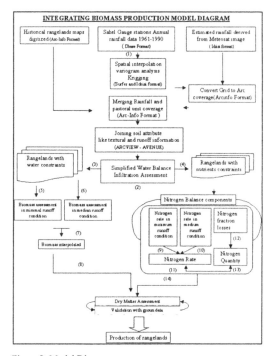

Figure 2. Model Diagram

The model requires various data to simulate long-term or annual biomass. Hence, the important data used by the model derives from pastoral units and as well as rainfall data maps. As far as rainfall is concerned, we used either synoptic and secondary national networks raingauge stations data or Meteosat full-resolution images (AGRHYMET 1994).

4.2 GIS Layers

Two PC-Arc/Info™ coverages are required to run the model with an Avenue™ procedure. These are the rainfall and pastoral units maps.

Compiling the rainfall map using meteorological stations data, a variogram analysis has been carried out in order to better interpolate this data value over a region. This spatial interpolation is necessary for those points data and his low density in a Sahel region. Indeed, the study shows that the Sahelian rainfall is correlated to the latitude. The softwares used for this analysis are Variowin™, Surfer™ and PC-Arc/Info™. Variowin software provides parameters such as kriging sill and range. Those parameters are used as input to Surfer™ software to produce an Erdas™ Grid File with Idrisi™ software. The conversion of rainfall grid to a polygon coverage is accomplished using GRIDPOLY command which converts the grid layer into a coverage containing polygons with 1 km x 1km resolution. Using rainfall derived from Meteosat follows the same step to produce PC-Arc/Info™ coverage.

The pastoral units are digitized from ATLAS-IEMVT maps at 1:500.000 scale (De Filippis et al. 1996). Each unit of map has his texture derived to morphological information. Hence, a table containing for each unit, the texture, the historical data for biomass yield and the geo-morphological characteristics is created.

Those two layers, rainfall and pastoral units maps are used to create new arc coverage by using PC-Arc/Info™ command INTERSECT. Each pastoral unit is cut into segments with rainfall polygons and encoded with respective rainfall covers ids.

An avenue macro has been written and uses as input file soil, rainfall and pastoral maps.

For structural studies in regard with the project objectives, analysis of wet and dry years is carried out to assess pastoral conditions.

4.2.1 Link soils parameters

In Arc-View™, in addition to the new coverage created in the last step, soils parameters, derived from pastoral maps are linked to the coverage with pastoral unit ids for each country map.

The source of soil information derives from pastoral maps. An assimilation with six types of soils which have a runoff coefficient under different climatic zones allows to compute each sub-pastoral units infiltration (Breman and Nico de Ridder, 1991). For each pastoral unit, soils characteristics are assumed uniformly distributed. Three components are identified, sandy soils, shallow detritic soils and fluviatile soils. Sandy soils are deep, homogeneous and of eolic origin. Detritic soils are developed on laterite or sandstone. Fluviatile soils are deep, clayey soils, or loamy-clayey subsoils covered with sandy-loamy topsoil (Penning de Vries, 1986).

4.2.2 Assessment of water balance

Water balance is a complex phenomena which depend on runoff, infiltration, intensity and rainfall duration, soils characteristics and vegetation cover.

Those items vary considerably from one area to another. With the global context of this assessment, the water balance is resumed by infiltration. The infiltrated quantity is done by corrected rainfall with water stream near the surface. The infiltrated quantity is equals to the amount of rainfall.

4.3 Assessment of the biomass yield below 250 mm rainfall

Between the biomass and then rainfall, the relation is not direct, but depends on soil characteristics. Two relations are defined to compute these parameters.

First, the biomass yield is computed from conditions where runoff is very low, and second, under the medium condition. Interpolations give the result in a condition of combined pastoral and rainfall unit.

4.4 Assessment of the biomass yield above 250 mm rainfall

The process here is based on nitrogen assessment. The Nitrogen Balance assumption is that the actual amount of N in Vegetation, plus inorganic N in the soil during the growing season, reflects the sum of all processes, including uptake (Penning de Vries et M. Djitèye, 1982). Considering the equilibrium situation, the amount of N, which enters the system, is equal to the amount that leaves it. In the model, annual fraction of N lost by the system is computed. It is a function of rainfall. Also, the fraction of the N contained in the biomass at flowering is computed with the rate of N in the biomass at different condition of runoff.

4.5 Assessment of the biomass in flood zone

In flood zones, the cycle of vegetation depends of water supplied by river. Information does not exist any more in this zone. Within a pastoral study undertaken by IEMVT, historical data can be found. Considering that those zones have the same type of soils, biomass are computed like other zones and corrected with a coefficient derived from historical data.

357

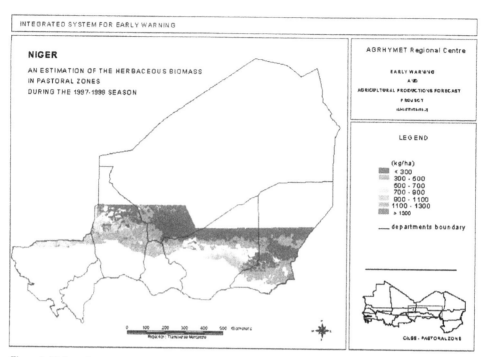

Figure 3. (Colour plate, see p. 433)

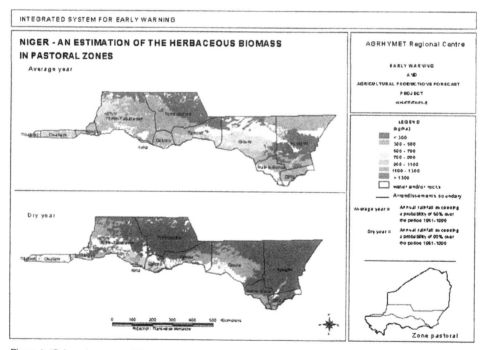

Figure 4. (Colour plate, see p. 433)

5 RESULTS

The main results obtained are the biomass yield in each component, pastoral unit crossed by rainfall data and forage quality (Figs.3-4). The data could cover many or one year. With the rainfall historical data, model could produce a long-term reference. The extend of the result, although limited can be developed for administrative level or bio-climatic levels.

The comparison with historical data or ground data provided by national services shows a good accuracy of dry matter. So the method results in an output closed to the assessment using the NDVI. The difference occurs in the precision due to soil differences, which are taken into an account in the later methodology.

Automating the computation with ArcView™ is a good approach to disseminate the methodology, hence users could change parameters, such as rainfall and runoff.

CONCLUSIONS

Using GIS potentialities allows implementing a rough assessment of dry matter for early warning purpose. This approach produces results that can contribute to identifying a vulnerable zone where carrying capacity is very low during one year, or in extreme drought conditions. Another important issue is the possibility to use grid data such as estimated rainfall data provided by Meteosat to substitute for ground data, for the biomass assessment in October, a month over which data are not gathered.

However, the approach has a limit that does not take into account the floristic diversity and the landuse. So, the challenge of the future is to implement a brief period assessment that uses decadal rainfall data as well as a more accurate topographic zone.

REFERENCES

AGRHYMET 1994. Actes de l'Atelier sur l'estimation des pluies par satellite, Niamey 2-4 décembre 1993. Niamey: Centre Régional AGRHYMET.

Breman Henk & Nico de Ridder 1991. Manuel sur les pâturages des Pays sahéliens. CABO-DLO, ACCT-CTA-KARTHALA.

CIRAD-IEMVT, CTA 1989. Atlas Elevage et potentialités pastorales sahéliennes - Synthèses Thématiques. Pay Bas: Maison-Alfort.

De Filippis Tiziana. & Alio Agoumo 1996. Numérisation des cartes des potentialités pastorales du Sahel. *Note Technique /14/AP3A/12/96, Projet Alerte Précoce et Prévisions des Production Agricoles.* Niamey: Centre Régional AGRHYMET.

Djaby, B. & De Filippis, T. & Koné, B. & Vignaroli, P. 1996. Intégration des données de l'Atlas pastoral dans le caractérisation du territoire. 1ère partie: présentation de l'Atlas et choix des cartes. *Note Technique /03/AP3A/02/96. Projet.*

Alerte Précoce et Prévisions des Production Agricoles. Niamey: Centre Régional AGRHYMET.

Hiernaux, P. 1983. Recherce sur les systèmes des zones arides du Mali. CIPEA. *Rapport de Recherche No. 5.* Addis Ababa.

Penning de Vries & Djitèye, M.A. 1982. La productivité des pâturages sahéliens, une étude des sols, des végétations et de l'exploitation de cette ressources naturelle. *Agricolture Research Rep. 918.* Wageningen.

Justice, C.O. & Hiernaux, P. 1986. Monitoring the grasslands of the Sahel using NOAA AVHRR data: Niger 1983. *International Journal of Remote Sensing, 7, 1475.*

Observing our environment from Space: New solutions for a new millennium, Bégni (Ed.)
© 2002 Swets & Zeitlinger, Lisse, ISBN 90 5809 254 2

Radargrammetry helps fight hunger in Ethiopia

T.K.Kanshie
Agricultural Bureau for Gedeo Zone, Dilla, Ethiopia

P.Romeijn
TREEMAIL International Forestry Advisers, Heelsum, Netherlands

E.Nezry & F. Yakam Simen
PRIVATEERS N.V. Private Experts in Remote Sensing, Netherlands Antilles

ABSTRACT: This paper reports the operational implementation of radargrammetry for the production of Digital Elevation Models, or DEMs, to areas of rugged topography. The Southern Ethiopian Highlands east of lake Abaya, with elevations between *ca.* 900 and 4,400 meters, were mapped. Currently available topographical maps are of insufficient quality to assist a study of the area's unique land use system, which is arguably the oldest and most durably sustained land use system of the planet. Without external inputs or terracing, the land use system maintains soil fertility and staves-off hunger. It has been doing so during the past 30 years of unrest and civil war, in one of the most crowded regions of Ethiopia. However, the central role of the staple crop enset within the land use system and its production cycles has hardly been the subject of scientific study. Understanding of this system is most likely to be relevant to enhancement of health and productivity in many regions of the world. Upon the request of the Agricultural Bureau for Gedeo Zone, geocoded and georeferenced topographical maps with accuracy of 20 meters (x, y and z) were made by PRIVATEERS N.V. on the basis of RADARSAT multi-incidence (S2/S7) images. These maps are now incorporated as the basic layer within the Bureau's GIS system. Map production techniques proved to be cost effective and relevant; especially for mountainous areas with poor accessability where correct geographic information is not available. The ease of orientation proved of invaluable help to rationalize execution and planning of cost-effective environmental field work and reporting.

1 INTRODUCTION AND SCOPE OF THE RESEARCH

1.1 *The operational site*

The Gedeo live in the escarpments of the South East Ethiopian Highlands, around Dilla overlooking Lake Abaya in the Rift valley, between *ca.* 900 and 3000 meters above sea level (Kippie 1994).

Their country is located in the humid southeastern Ethiopian highlands, approximately between 5° and 7° North latitude and between 38° and 40° West longitude, 360 km southeast of Addis Ababa. Dating back from neolithic times (Simoons *ex.* Westphal 1975), Gedeo land use systems are among the oldest agricultural systems in the world.

With more than 420 persons per square kilometre, the Gedeo highlands are one of the most densely populated regions in the country (Cent. Stat. Office 1992).

Currently available topographical maps are of poor precision and reliability, thus rendering them of little use to land use planning in the Gedeo zone.

1.2 *Ensete Ventricosum*

The Gedeo have developed *enset* agriculture which is based on the group-wise use of *enset (Ensete ventricosum)* (see Table 1) (Kippie 2001), the rotation of which is maintained in the context of the larger multiple rotation of the components accompanying *enset*. (see Tables 2 & 3) (Kippie 2001, Kippie & Oldeman 2000). The shorter rotation (*enset*) and the longer one (*Coffea arabica*) are interwined in the mixed population of multi-purpose trees with very long. natural rotations, the latter timing the average ecological turnover. Implicit in the rotation of *enset* was the concept of sustainability, showing that the Gedeo had long been aware of the concept of sustainability (Kippie 1994). This sustainability hence preceded by far the forester Von Carlowitz (1713, from Oldeman *et al.* 1994), the first European to have published the German concept of *Nachhaltigkeit, i.e.,* conservation for posterity. Sustainability since long was seen from three angles, *i.e.,* an ecological, a social and an economical one (*eg.* Oldeman 1986; Swift *et al.* 1993. Otto 1987, Kuper & Massen 1997, Eppel 1999).

Enset sustains the highest population density in Ethiopia (420 persons per km^2, in Gedeo province (Eth. Stat. 1997). *Enset* is not affected by excess rain nor by drought, as has been witnessed during the famines of the 1980s (Brandt *et al.* 1997). Besides, the stability of the SE Ethiopian highlands owes much to the *enset* culture (Amare 1984). Therefore, Brandt *et al.* (1997) rightly called *enset* the tree against hunger (AAAS *et al.*, Brandt *et al.* 1997), while Emperor Menelik's appreciation of the plant saying it was appropriately called *worqe*, literally meaning "my own gold" as it was not affected by the lack of oxen during the Great Famine of 1888-1892 (Pankhurst 1985), is well placed.

The Gedeo numbering around one million belong to the Eastern Cushitic-speaking group (Murdock 1959, Ehret 1979) and are the people most dependent on *enset* (Kippie 1994), where it is believed to have been the ancient food plant (Rossel 1998). The history of *enset* is only fragmentarily known, as when, where and by whom was first domesticated not precisely known (Rossel 1998). Though *enset* occurs in many parts of Africa and Asia, (Taboje 1997, Brandt 1997, Rossel 1998), Ethiopia is the only country known for the cultivation of *enset*, in its Southern Peoples' Region.

The Gedeo have developed their own ways of growing and processing *enset*, their livelihood in the steeply escarpments of the SE Ethiopian highlands dependent on it.

Enset indeed is a very versatile crop worth of further research and development, as it is the most unstudied of all domesticated crops in Africa (Brandt *et al.* 1997); it is also one of the unharnessed agricultural resources for the erosion- and drought-prone Ethiopian highlands, representing perhaps a viable solution to the recurring food crises.

1.3 *Scope of the research*

With the foregoing perspectives, understanding of the land use system is of critical importance for the region. This has recently been confirmed in a study by the American Association for the Advancement of Science (AAAS *et al.*). The study's title depicts enset as the *"tree against hunger"*. According to the study, enset provides a healthy saple food and simultaneously enhances soil fertility. The study thus reaches the conclusion that: "Research and development are needed to address sustainability issues and the place of enset as a major contributor to the food security of Ethiopia" (Brandt *et al.* 1997). Such understanding may well prove to be of fundamental relevance to other densely populated mountainous regions, as was confirmed for the Andean region. If so. the lack of understanding this unique land use system can be seen a hindrance to its beneficial promotion.

Results of the present study will contribute to the better understanding of the land use systems, and will directly benefit an estimated 15 million people dependent on enset cultivation and another 12 million dependent on coffee cultivation, the two major components of the Gedeo land use system. Moreover, the study happens to come at the right moment when general principles from complex land use systems are being considered for possible use in designing stable and sustainable cropping systems fit for the present age (compare Neugebauer *et al.* 1996).

2 RESEARCH PHASES AND TECHNICAL APPROACH

The present study use methods that represent the existing land use systems' complexity as fully as possible. The basis of the methodologies is graphical and employs profile diagrams and maps at the appropriate scales (*cf.* Koop 1989, Kippie 1994, Oldeman 1998). The method is complemented with FAO's criteria of land evaluation (FAO 1976). Data are collected at two scales, *i.e.* landscape and farm, in three consecutive phases. A Digital Elevation Model (DEM) is used to extrapolate the farm system results to the scale of landscape for the Gedeo zone.

Phase 1 covered a preliminary survey, including a pilot study to check the methodology. Phase 1 was initiated in 1998 by the Director and a team of field researchers at the Agricultural Bureau of the Gedeo Zone. The field work consisted of a general survey, resulting in: lists of land marks and limits to be entered on the map; a selection of 14 farms, two per altitudinal zone, to be the object of the case studies; a selection of "good farming" families or households as sources of local information; and a questionnaire. The agents of the Agricultural Bureau for Gedeo Zone were trained in the approach of farmers for research purposes, proper conduct of sound field interviews, recording of responses and distinguishing and recording of land marks. Supervision was provided by the Director.

Phase 2 covers the extrapolation of these results to a landscape scale. The need for a basic geographic reference map was satisfied by the making a spatio-map, established from geo-coded and geo-referenced RADARSAT Synthetic Aperture Radar (SAR) data of the Gedeo province and a DEM, established by Radargrammetry (Leberl 1990). This map was prepared by PRIVATEERS N.V. and TREEMAIL. The map is essential to the correct interpretation and extrapolation of the Phase I field research results, and constitutes a sound basis to all future Gedeo Zone mapping purposes.

Table 1: Complete enumeration and distribution of a population of *enset* (*Ensete Ventricosum* (Welw.) Chessman MUSACEAE) over Gedeo size classes and over the 2.5 ha farmland feeding 23 household members altogether. Gedeo size-classes: subscript 1 represents the oldest phase; K = *Kaassa*; S = *Saxxaa*; G = *Guumee*; B = *Beyaa*; D = *Daggicho* (for meaning see text). Symbols: D_{rc} = diameter above root collar; H = height of pseudostem excluding leaves; LN = leaf number; LL = leaf length; LW = leaf width.

size-class (1)	D_{rc} cm (2)	h m (3)	LN m (4)	LL m (5)	LW m (6)	Number (7)	%total (8)
K_2	10-15	0.5-1.0	2- 3	1.5-2.0	0.1-0.3	885	29.5
K_1	15-19	1.0-1.8	3- 6	2.0-2.5.	0.3-0.5	672	22.4
S_2	19-22	1.8-2.0	6- 8	2.5-3.0	0.5-0.8	486	16.2
S_1	22-39	2.0-2.3	8- 9	3.0-3.5	0.8-0.9	285	9.5
G_2	39-41	2.3-2.5	9-10	3.5-3.6	0.9-1.0	165	5.5
G_1	41-43	2.5-2.7	10-11	3.6-3.8	0.9-1.0	138	4.6
B_2	43-65	2.7-3.5	13-15	4.0-6.0	0.8-0.9	126	4.2
B_1	65-100	2.7-3.5	13-15	4.0-6.0	0.8-0.9	123	4.1
D	90-100	3.5-4.0	15-18	5.0-6.0	0.9-1.0	120	4.0
Total						3000	100

Table 2: The place of *enset* within the multiple rotations of the agro forest.

Species	0-10	10-20	20-30	30-40	40-50	50-60	60-70	70-80	80-120
Ensete,	x								
Ensete, *Milletia*		x							
Ensete, *Milletia*, *Coffea*			x						
Ensete, *Milletia*. *Coffea*. *Pygeum*				x					
Ensete, *Milletia*. *Coffea*. *Pygeum*. *Albizzia*					x				
Ensete, *Milletia*, *Coffea*. *Pygeum*, *Albizzia*, *Celtis*						x			
Ensete. *Milletia*. *Coffea*. *Pygeum*. *Albizia*, *Celtis*, *Aningeria*							x		
Ensete, *Milletia*, *Coffea*, *Pygeum*. *Albizzia*, *Celtis*. *Aningeria*, *Cordia*								x	
Ensete, *Milletia*, *Coffea*, *Pygeum*. *Albizzia*, *Celtis*, *Aningeria*, *Cordia*. *Podocarpus*									x

Table 3: Estimate of the natural rotation time for some perennial components of Gedeo agroforests.

Species	Rotation time (years)	Remark
Ensete ventricosum	10	plant yielding staple food
Coffea arabica	30	cash crop
Milletia ferruginea	50	used for soil enrichment. shade and as source of fuel wood & fodder
Cordia africana	70	used as shade & source of construction wood
Croton macrostachys	50	mainly used as source of fuel wood
Pegeum africanum	70	mainly used as source of fuel wood
Albizzia gummifera	50	mainly used for soil enrichment & also as shade
Erythrina abyssinica	30	mainly used for soil improvement & also as shade
Celtis sp.	70	mainly used as source of construction wood
Aningera friedereci	100	source of timber regarded as one of the best
Podocarpus gracilor	120	source of timber regarded as one of the best

Phase 3 covers the scale of farms and fields. The basic methodology has been applied earlier by Kippie (1994). Its methodologies and procedures were first described by Oldeman (1979) and are meanwhile extensively used, for instance in research on local South-East Asian agroforests (eg. Michon 1983, Sansonnens 1994) or as basic method in the Brazilian programme of research at ESALQ, Piracicaba, Saõ Paulo, on forest fragments of the Atlantic Rain Forest and its management. A simple line transect, proceeding at 400 meters a day, will give the whole farm context with its borders with neighbours. On this transect, plots are pinpointed which represent typical key situations in the sense of land use and/or ecology. These plots are mapped in a more precise way by block transects, including soil and herbarium sampling and the establishment of profile diagrams. Finally, in these plots, key species are to be found, i.e. those with a key rotation for the whole set of multiple rotation ("pacemaker species").

Later this year, T.Kippie plans to report the overall phase 1-3 results in Treebook 5 (http://www.treemail.nl/books) and in a Ph.D dissertation for the Wageningen University in the Netheralnds.

3 THE PHASE 2 MAPPING PROGRAM

To achieve the foregoing, farm design and dynamics together with the associated management practices, having in view the level of ecological and socio-economic efficiency attained, will be researched. The manner in which the diverse farm components, farms and land use versions are interacting to give a balanced productivity and stability will be described in terms of agrosystem dynamics and associated management practices. In parallel, trends in the farmers' adaptive strategies across the three land use versions will be documented as these will help both to reflect on their past land use history and in making predictions.

The main task of the present Phase 2 of project is mapping the three altitude-based Gedeo land use systems. In Phase 3, this will allow to assert whether or not changing biophysical factors (along with altitudinal zones) elicit corresponding changes in land use system design, stability, productivity as well as in the farmers' adaptive strategies.

3.1 *RADARSAT data processing*

Radiometric calibration has been performed, according to the RADARSAT SGC product specifications.

The RADARSAT SAR images have been resampled to a pixel size of 20 meters x 20 meters, which corresponds closely to the spatial resolution of the RADARSAT SGC SAR data products.

Conversion from original data format (16-bits per image pixel) to 8-bits per image pixel has been made, in a similar way for the four calibrated RADARSAT SAR images.

Speckle filtering has been carried out using an adaptive Maximum A Posteriori (MAP) speckle filter (Lopes et al. 1993) especially designed for multi-channel SAR images (Nezry et al. 1996, 1997). In the Gaussian-Gaussian MAP filter for multi-channel SAR images, new structure and texture detectors (Nezry et al. 1995) based on the second order statistics (autocorrelation properties) of both the speckle and the imaged scene are incorporated.

3.2 *DEM generation*

Using the RADARSAT S2/S7 speckle filtered dataset, a stereo-image has been generated.

Using this stereo-image (or anaglyph), a DEM with a sampling rate of 20 meters x 20 meters has been produced by radargrammetry (see Figures 1 and 2), as described in Leberl (1990). This technique has already been used by PRIVATEERS N.V. in the past with satisfactory results (Pénicand et al. 1995, Nezry & Demargne 1998, Nezry & Yakam Simen 1999).

Accuracy of the DEM can be evaluated by analysing the results of the geocoding procedure applied to the RADARSAT images. In the present case, the accuracy is of the order of 20 meters in altitude, which is good enough for the purpose of this project (Though some residual wrong altitude evaluations are still locally present in the DEM).

The spatial coverage of this DEM is 10,000 km^2, englobing the whole of the Gedeo Country.

An elevation contour map, every 100 meters altitude, has also been produced, at a spatial sampling rate of 20 meters x 20 meters (see Figure 3).

In addition, the anaglyph is extremely useful for photo-interpretation, in order to understand the thematic contents of the image.

3.3 *Production of cartographic maps*

The speckle filtered images have been geocoded (i.e. corrected for the geometrical distorsions due to relief), using the DEM generated previously.

Radiometric Correction for terrain (variation of pixel area, with slope and orientation, influence of actual incidence angle) effects have been applied. The physical model used in the correction for the effects of the actual incidence angle is a Lambertian scattering model (Beaudoin et al. 1995).

All processed satellite SAR image and elevation products have been projected to Universal Transverse Mercator (UTM), in order to produce cartographic products.

The georeferencing system is: Universal Transverse Mercator (UTM), Zone 37, Row N. Earth Ellipsoid WGS 1984 (see Figure 3).

Figure 1: Digital Elevation Model (DEM) of Gedeo Zone (Ethiopia).

Figure 2: Three dimentional view of Gedeo Zone (Ethiopia). Filtered RADARSAT-1 SAR image and DEM.

Figure 3: The Digital Elevation Model (DEM) of Gedeo Zone (Ethiopia), in Universal Transverse Mercator (UTM) cartographic projection, with superimposed elevation curves.

4 CONCLUSION

The availability of an accurate DEM enables the study of the rivers coming down from the mountains down to Lake Abaya, and of their watersheds.

Indeed, this continuous elevation information, now available from the source of the local rivers and streams, across all of Gedeo country, and until Lake Abaya, will support in the near future accurate studies of the runoffs, and of the water balance.

It will also support studies aiming at the evaluation of global and local erosion risks in the whole of Gedeo country.

Finally, it will also support studies aiming at the optimization of the management of the water resources or agriculture purposes.

The satellite radar maps, built using speckle filtered RADARSAT SAR images enable the accurate location of the communication network in the Gedeo country. In particular, the only asphalted road running from the Northernmost part of Gedeo country to its Southernmost part is easily indentified.

In addition, the very characteristic response of built-up areas (due to corner effects and/or resonance effects on solid roofs and/or corrugated metal sheet roofs) enables a very easy location of the towns, of the villages, and even of the isolated settlements.

The satellite maps and the DEM, in cartographic WGS 1984 projection, can be used during ground work, with, or even without a GPS (Global Positioning System) device.

5 ACKNOWLEDGEMENTS

The present study was co-funded by PRIVATEERS N.V. and TREEMAIL.

6 REFERENCES

Amare G. 1984. Stability and instability of mountain ecosystems in Ethiopia. In *Mountain Research & Development 4 (1): 39-44.*

American Association for the Advancement of Science with Awassa Agricultural Research Center, Kyoto University, Center for African Area Studies & University of Florida. The "tree against hunger". Enset based agricultural systems in Ethiopia. *On-line publication: URL http://www.aaas.org/international/ssa/enset/index.htm.*

Beaudoin A., Castel T., Deshayes M., Stachs N., Stussi N. & Le Toan T. 1995. Forest biomass retrieval over hilly terrain from spaceborne SAR data. *Proceedings of the Symposium on Retrieval of bio- and geophysical parameters from SAR data for land applications. Toulouse (France). :131-140. 10-13 October 1995. CNES/IEEE Publication.*

Brandt S.A., Spring A., Hiebsch C., McCabe J.T., Taboje E., Diro M., Wolde-Michel G., Yntiso G., Shigeta M. & Tesfaye S. 1997. The "tree against hunger": enset-based agricultural systems in Ethiopia. *American Association for the Advancement of Science with Awassa Agricultural Research Center, Kyoto University Center for African Area studies and University of Florida.*

Ehret C. 1979. On the antiquity of agriculture in Ethiopia. *Journal of African History. 20: 161-177.*

Eppel J. 1999. Sustainable development and environment: a renewed effort in the OECD. *Environment, Development and Sustainability, 1:41-53.*

Ethiopian Statistical Authority. 1997, Statistical abstracts, *Addis Ababa, Ethiopia.*

FAO. 1988. Traditional food plants. *Food and Nutrition, Paper No. 42: 269-273.*

Kippie T.K. 1994. The Gedeo agroforests and biodiversity: architecture and floristics. *Departments of Forestry and Ecological Agriculture, Wageningen Agricultural University, The Netherlands, MSc thesis, 89 pp.*

Kippie T.K., 2001. Production and uses of enset (Ensete ventricosum (Welw.) Cheesman MUSACEAE). *Agricultural Bureau for the Gedeo Zone, P.O.Box, 400, Dilla, Ethiopia. To be published.*

Kippie T.K. & Oldeman R.A.A. 2000. The Gedeo agroforests and biodiversity: Enset (Ensete ventricosum (Welw.) Cheesman MUSACEAE) in multiple rotations. a case study from southern Ethiopia. *Agricultural Bureau for the*

Gedeo Zone, P.O.Box, 400, Dilla, Ethiopia & Wageningen Agricultural University (NL) Wageningen The Netherlands. To be published.

Kuper J.H. & Maessen P.P.T.M. 1997, Sustainability the pro silva way. *Proceedings of the 2nd International Pro Silva Congress. Publisher: The Dutch Pro Silva Congress Foundation, 277 pp.*

Leberl F.W. 1990: Radargrammetric image processing. *Artech House Inc., USA.*

Lopes A., Nezry E., Touzi R. & Laur H. 1993. Structure detection and statistical adaptive speckle filtering in SAR images. *International Journal of Remote Sensing, 14 (9), 1735-1758.*

Murdock G.P. 1959. Africa: Its Peoples and Their Culture History. *McGraw Hill, New York, USA.*

Neugebauer B., Oldeman R.A.A. & Valverde P. 1996. Key principles in ecological silviculture. Fundamentals of Organic Agriculture. *D-66636 Thole-Theley, IFOAM, Okozentrum Imsbach, :153-175. T.V. Ostergaard editors, Netherlands.*

Nezry E. & Demarge L. 1998. Using SPOT and radar data to inventory forests in Sarawak, Malaysia. *Spot Magazine, (29) :28-31.*

Nezry E., Leysen M. & De Grandi G. 1995. Speckle and scene spatial statistical estimators for SAR image filtering and texture analysis: Some applications to agriculture, forestry and point targets detection. *Proceedings of SPIE, 2584 :110-120.*

Nezry E., Yakam Simen F., Zagolski F. & Supit I. 1997. Control systems principles applied to speckle filtering and geophysical information extraction in multi-channel SAR images. *Proceedings of SPIE, 3217 :48-57.*

Nezry E., Zagolski F., Lopes A. & Yakam Simen F. 1996. Bayesian filtering of multi-channel SAR images for detection of thin details and SAR data fusion. *Proceedings of SPIE, 2958 :130-139.*

Oldeman R.A.A. 1979. Quelques aspects quantifiables de l'arborigénèse et de la sylvigénèse. *Oecol Plant, 14 (3): 1-24.*

Oldeman R.A.A., Parvian J. & Stephan K.H. 1994. Les traitements forestiers. *Naturopa, 75 :15-19.*

Pankhurst R. 1985. The history of famine and epidemics in Ethiopia prior to the twentieth century. *Relief and Rehabilitation Commission, Addis Ababa, Ethiopia.*

Pénicand C., Rudant J.P. & Nezry E. 1995. Utilisation opérationnelle des images de télédétection radar pour la cartographie. *Bulletin de la Société Française de Photogrammétrie et de Télédétection, (137) :35-41.*

Rossel G. 1998, Taxonomic-Linguistic study of plantain in Africa. *Doctoral Thesis. Research School CNWS, School of Asian, African and American Studies. Leiden, Netherlands.*

Taboje E. 1997, Morphological characterization of enset (Ensete ventricosum (Welw.) (Cheesman) clones and the association of yield with different traits. *MSc. Thesis University of Agriculture, Alemaya, Ethiopia, 89 pp.*

Westphal E. 1975. Agricultural systems in Ethiopia. *Wageningen Agricultural University, The Netherlands, Alemaya College of Agriculture and Haile Sellassie I University,*

Ethiopia, Centre for Agricultural Publishing and Documentation; Agricultural Research Report 826, 278 pp.

Yakam Simen F., Nezry E. & Ewing J. 1999. The legendary lost city "Ciudad Blanca" found under tropical forest in Honduras, using ERS-2 and JERS-1 SAR imagery. *in: "JERS-1 Science Program '99 - Global Forest Monitoring and SAR Interferometry", Earth Observation Research Center, National Space Development Agency of Japan (NASDA/EORC), :139-143.*

Observing our environment from Space: New solutions for a new millennium, Bégni (Ed.)
© 2002 Swets & Zeitlinger, Lisse, ISBN 90 5809 254 2

Use of multiresolution and multitemporal remote sensing data for sand study around Al-Hassa oasis

R.A.Vaughan & M.S.Al-Rowili
Centre of Remote Sensing and Environmental Monitoring, University of Dundee, UK

ABSTRACT: Al-Hassa Oasis is the largest oasis in the Kingdom of Saudi Arabia and is affected by the migration of the Jafurah Sand Sea, which will bury the oasis in the long term if it is not stopped. With the rapid growth of the Al-Hassa oasis and the major development programmes in progress the problem is very acute with the development of methods for controlling the sand movement having enormous importance. Active sand in the area around Al-Hassa Oasis has been characterised using multi-resolution and multi-temporal remote sensing data (Landsat MSS, TM, SPOT XS, and IRS). These images were compared for their ability in the spatial and spectral regimes for resolution of sand dune features. The study will consider sand dune distribution, sand dune classification, sand dune movement and direction. Sand dune movement can be detected and mapped according to its rate of movement which can be classified into three types, transfer dunes, barchan dunes and dome dunes. To detect the active sand around the area an Active Sand Index (ASI) has been applied to the imagery in order to classify the sand type based on the normalised difference between the blue and mid-infrared spectral values. This formula highlights and maps active sand around the Al-Hassa area. The results of the study have also been presented in a number of image maps showing the different kinds of sand dune movement using ASI, and using visual interpretation to map sand dune distribution and Sabkhas. Image classification techniques such as NDVI have also been used in this study.

1. INTRODUCTION

Saudi Arabia is one of the countries in the world with very extensive sand cover. The Al-Hassa Oasis, the largest oasis in the Kingdom of Saudi Arabia, is I effected strongly by the sand movement, and will be buried by the sand movement of the Jafurah Sand Sea within 600 years if it is not stopped (Al-Abduwahed, 1982). The Al-Hassa is also well known as a great agricultural area, and is very important if Saudi Arabia is to achieve the goal of self - sufficiency in food production, especially in dates and date palm (Al-Taher, 1999).

The problem of sand dune movement and drift by wind has been one of the major concerns in the area of the Al-Hassa Oasis for hundreds of years, in which time many villages and springs have been buried. Sand movement is a very active process that 1 interferes with some of the vital aspects of urbanised, industrial and agricultural areas (Al-Taher, 1999). With the rapid growth of Al-Hassa Oasis and the major development programmes in progress, the problem is very acute and the development of methods to detect active sand for controlling the sand movement has gained enormous importance. To un-derstand and perhaps predict the movement of the sand dunes that are in the North and East of the study area. the research will undertake to study dune movement using remote sensing which will lead to a general study of sand movement in the surrounding region. The ability to discriminate active and inactive sand by remote sensing can help to assess and movement (Lillesand & Kiefer 1994). In addition, remote sensing can be used to monitor and characterise the active sand in the area. Remote sensing, together with Global Positioning Systems CaPS) to take the location of the sampling area where the sand sample was taken, will identify the exact location of the active sand.

Multi-resolution and multi-temporal images (Landsat MSS, TM, SPOT XS, and IRS) have been compared for their ability in the spatial and spectral regimes for resolution of sand dune features, movement, sand detection. Field studies were also used to investigate the potential links between the remotely sensed data and the sand dune size, shape and other characteristics of the active sand. In addition, GPS was used to locate the sample areas and to georeference the satellite imagery in the remote area where there are no mapped land features. A new map for active sand will be produced using satellite data, which may also include geomorphological and Sabkhas zonation mapping. Image classification techniques and principal component analysis have also been used in this study for image analysis.

2 METHODOLOGY AND RESULTS

The methodology applied involved: (i) field work, (ii) geometric correction, (iii) studies of the wind speed and direction (iv) mapping the dune movement hazard, (v) detecting active sand by applying a model, (vi) mapping sand dune areas and sand distribution, and (vii) classifying the sand dunes in the area.

2.1 *Field Work*

The study area was visited in January 2001 to compare the remote sensing data with the ground truth and to take measurement of the sand dune size, shape and height.

2.2 *Image Processing*

2.2.1 *Image Enhancement.*
Image enhancement and contrast manipulations were used to improve the quality of the image and to remove noise.

2.2.2 *Geometric Correction*
A common geometric and geo-reference method was used. To achieve this, four (A 1) maps at 1:50.000 scale were scanned and then mosaiced and geo-referenced together. After this, the image was corrected using the map (image to map correction).

Image to map correction normally enables the exact overlaying of all the images together, but in this study the correction was not sufficiently accurate because there are no landmarks in the desert, especially in the areas away from the oasis. In this case the GPS was used during the fieldwork to correct the images in the field by using prominent stable geological features such as the horns of crescent sand dunes.

2.3 *Active Sand Model.*

To detect active sand, the *Active Sand Index* (ASI) has been developed and applied, making use of the blue band which is conceived as being important for many remote sensing applications (Karnieli 1997).

The index was computed from a new equation, which was developed by trial and error, and applied

to give better accuracy for mapping the active sand area. The new index is as follows:

$$ASI = (MIRED - BLUE) / (MIRED + BLUE)*100$$

Where MIRED is channel 7 and BLUE is channel 1 in the TM images. Channel 7 was chosen rather than channels 4 or 5 simply because it appeared to give better results.

The new equation highlighted well the active sand, coloured red around the oasis (Figure 1). The ASI detects active sand according to its DN values which are higher (brighter) than the average values for inactive sand in Landsat TM channel 7 (Paisley *et al*, 1991). This was consistent with the results of fieldwork and with the interpretation of the original image.

2.4 *Spectral Mapping for Sand Dunes.*

Spectral mapping is a method for comparing image spectra with individual spectra by determining the similarity between them by calculating the spectral angle between them, treating them as vectors in a space with dimensionality equal to the number of bands. This provides a good first cut at spectral

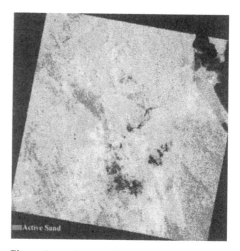

Figure 1. Active Sand Index result

mapping of the predominant spectrally active material present in a pixel. However, natural surfaces are rarely composed of a single uniform material. Spectral mixing occurs when material with different spectral properties are represented by a single image pixel.

Spectral mapping was applied to detect and map the sand area around the oasis. Three different areas were selected as being representative of the sand

Figure 2. Spectral Mapping result

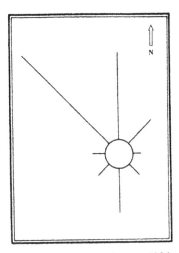

Figure 3. Wind rose diagram 1984 to 1998

dune and vegetated areas. The result of the mapping is shown in Figure 2, where yellow colour is used for active sand, light yellow for sand sheet, and blue for the vegetation and other types of soil. It is seen that both spectral mapping and ASI give similar results. All images show the active sand in the same areas but the spectral image shows the sand sheet better than the ASI. On the other hand, the ASI shows the areas which have only very active sand and which do not have any vegetation.

3. CLASSIFICATION

3.1 *Sand Dune Movement.*

Sand dune movements in the north of the oasis were recorded during the summer season of June, July and August. The highest annual rates of sand drifting occurs in this area in early summer when the shallow water table in the sabkhas is low and when there is little vegetation to stabilise the surface (Al-Hinai 1989 and Al-Taher, 1999). The barchan dune has the highest rate of sand movement followed by the dome. On other hand, the parabolic dune has the lowest rate of dune movement and the sand sheets in the area have a very low annual rate of sand drift (Bader, 1994, Al-Taher, 1999).

To detect the direction of the dune movement, and to give early warning to the area, the wind direction and dune shapes were compared and mapped according to a wind rose diagram (Figure 3).

The result of classifying the wind direction and speed shows that most of the wind is blowing from the north-west (Figure 4), with drifting of the sand about 10m every year (AbolKhair 1981 and Al-

Taher, 1996). The sand which is blowing from these areas affects some areas that have not been protected by the stabilisation project or any other protection method more than others.

That the major hazard comes from the north-west and north of the oasis is also evident from the shape of the dunes whose slope face gives the direction of movement of the dune. The major hazard comes from areas numbered 3, 4, and 5 in Figure 4. The affect of sand movement is easily recognisable by observing the position and direction of the dunes. Area number 7 is already protected by a stabilisation project and is monitored by the ministry of agricultural and waters so will not be considered further. Movements in areas number 4 and 5 (Figure 4) need to be stopped by applying stabilisation projects. This area is, at least in the short term, more important than area 3.

3.2 *Mapping of Sand Dune Distribution and Sand Dune Classification.*

The major sand dunes in Al-Hassa were mapped and classified using Landsat MSS, TM, SPOT XS, and IRS data (Figure 5). Comparing the differences between images, it was observed that, while MSS images clearly identified large dune areas (north of the lakeside area 2, Figure 4) it gives very poor identification within these areas. This is because most of the sand dunes are smaller than 200m, and so cannot be discriminated on MSS images (see Table 1). TM and SPOT XS images give betters results and dunes that are smaller the 200m can be identified and mapped. Sabkhas, however, could be identified more easily in the MSS images. SPOT PAN, with a

Table 1. Comparison of multi-resolution images for their ability in the spatial and spectral regimes for recognising sand dune features.

Sensor	Sensor Resolution	Identified Dune Size				
		5-10m	10-50m	50-100m	100-200m	>200m
Landsat MSS	79m	X	X	X	X	✓
Landsat TM	30m	X	X	X	✓	✓
SPOT XS	20m	X	X	X	✓	✓
SPOT PAN	10m	X	X	✓	✓	✓
IRS PAN	5.6m	X	✓	✓	✓	✓

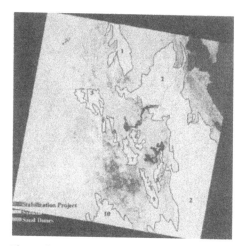

Figure 4. Major sand dunes areas around AL-Hassa Oasis.

resolution of 10m, and IRS PAN, with resolution 5.6m, gave very good results by discriminating dune sizes between 10 to 100m.

Using the TM 1999 image, the major dune area was divided into 10 smaller areas (Figure 4). The largest sand dune area was in the East/Southeast (number 2). There are also other areas in the north (3, 4, 5, 6 and 7). The major risk comes from areas numbered 4 and 5 that affect the city of Al-Uyun and other agricultural areas. These areas need suitable stabilisation systems to be applied to stop the movement of these dunes. From a comparison of imagery from different dates, it can also be seen that the effect of the stabilisation project in changing the area from active to inactive can be easily identified by noting the change in colour of the sand and the increase in area of vegetation.

There are many varieties of sand dunes in the area, each type being created by interaction of the sand and the atmospheric boundary. The distribution and the classification of the dunes were obtained mainly from the high-resolution IRS image. An edge detection filter and a variance 25 * 25 spatial filter (Figure 6) were applied to detect the slope face of the dunes. Classification of the dune types in the study area found three types of dune using field measurement and remote sensing. The first type was the transverse dune which tends to form in areas where there is a large supply of sand and little vegetation. The shape of these dunes is wavelike. They have a height between 5 to 14 metres and width about six times their height. Transverse sand dunes are located in most of the areas (Figure 4). Another type of sand dune was the crescent shaped barchan sand dune which tends to form in areas where there is a limited amount of sand and little vegetation. Barchan dunes in the area have heights of 4 to 15 metres (Table 2). Most of the barchan dunes are in the north of the oasis, especially in areas 3, 4, 5, 6, and 7. The last types of sand dune were the star and dome, which occur in just small numbers in to the north of the stabilisation project associated with Uruq Al-Thawwajivah (area7).

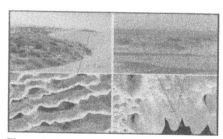

Figure 5. Comparing the dune pattern using IRS image and the field check.

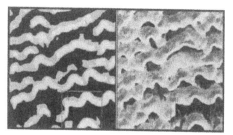

Figure 6. (a) Variance 25 * 25 spatial filter and (b) IRS original image.

372

Table (2). The sand dune classes in the study area

Area	Dune type	Dune Size Average			Total Area Size (km²)	Total Perimeter (km)
		Length (m)	Width (m)	Height (m)		
1	Transverse and barchan	80	50	7	211	121
2	Transverse and barchan	500	100	16	2,325	496
3	Transverse and barchan	50	12	5	122	100
4	Transverse and barchan	55	14	5	15	22
5	Transverse and barchan	53	14	5	25	41
6	Transverse and some barchan	55	15	5	42	53
7	barchan	100	63	9	63	49
8	Transverse and barchan	70	30	8	55	78
9	Transverse and barchan	60	30	7	73	72
10	Transverse and barchan	80	28	6	206	107
11	Stabilisation Project barchan	40	15	3	43	83

Figure 7. The NDVI for Landsat TM 87 image.

Figure 8. The NDVI for Landsat TM 99 image.

3.3 *Normalised Difference Vegetation Index*

The Normalised Difference Vegetation Index (NDVI) has been applied to detect the oasis vegetation in both the TM 87 and the TM99 image (Figures 7 and 8).

To study the changes between the two images, over a 12-year period, the maps of vegetation were compared (Figures 7 and 8). In the area south of the oasis, some new development has taken place and some new land has been reclaimed and used for farming, especially around the Qatar Road. Also, for the area, which is in the Stabilisation Project, some reclaimed land can be seen which is used for preserving the main agricultural area and to act as the first buffer to stop the sand movement. The effect of the project is to reduce or stop the sand movement and to turn the area into sand sheet rather than remain as an inactive sandy area.

A newly reclaimed area can be identified easily from the five circles of green land in the north (Figure 8, area 3). There are two reasons why such an area is situated this far from the oasis. Firstly the area has sufficient water and secondly it is very close to the Riyadh Damamm highway. Another newly reclaimed area lies to the north of the main agricultural area (Area 2 and 4 in figure 8). This development was among the sand dunes in the north of Al-Uyun City and north of stabilisation project. The last to be developed was in the south of oasis around Qatar road (Area 1 in figure 8).

4. CONCLUSION

Monitoring of active sand using multi-temporal and multi-resolution data (Landsat MSS, TM, SPOT XS, and IRS) has proved to be a useful tool. In addition GPS was found useful in the correction and geo-referencing of the satellite imagery in the field when linked to geological features. This study has focussed on the development and testing of a new index, the Active Sand Index. Characterising the sand dunes and comparing multiresolution and multitemporal remote sensing data demonstrated the ability in the spectral and spatial rimes for the resolution of sand dune features.

The result of this work has been presented in the form of spectral and ASI maps. Other maps illustrated sand dune distribution, sand dune patterns and classification, particularly for the transverse and Barchan, together with some parabolic sand dunes. Final spectral maps were illustrated with the detection of vegetation in the in the oasis using NDVI.

5. REFERENCES

AbolKhair Y, M. 1981. Sand encroachment by wind in El-Hasa of Saudi Arabia, PhD. Geography Department, Indiana University.

Al-Abduwahed, Y. 1982. Stabilisation Project in Al-Hassa. Department of Pastures and Forest, Saudi Arabian Ministry of Agriculture and Water.

Al-Hinai, Khatab. 1989. Evaluation of Remote Sensing Data for Sand Studies in Saudi Arabia, Workshop on Desert Studies in the Kingdome of Saudi Arabia, 21 to 23-11-1989, pp 67-90.

Al-Taher, A. A. 1996. The Effect of Sand and Dust Storm in the Farms Soil at Al-Ahassa Oasis, Kingdom of Saudi Arabia. Geographical Research, Saudi Geographical Society, King Saud University. No 24.

Al-Taher, A. A. 1999. Al-Ahassa Geographical Study. King Saud University College of Art, Geography Department. Riyadh.

Bader. T. 1989. Scientific means and studies used to stabilise dunes in Eastern Province. Workshop on Desert Studies in the Kingdome of Saudi Arabia, 21-23 Nov., pp. 45-66.

Bader. T. 1994. Hazard of sand drifting in Saudi Arabia. Symposium on Desert Studies in the Kingdome of Saudi Arabia "Extant & Implementation". Nov 3, pp 115-129.

Campbell, J. B. 1996. Introduction to Remote Sensing. Taylor & Francis. London

Karnieli, A. 1997. Development and implementation of spectral crust index over dune sand. Int. J. Remote Sensing. Vol. 18, No. 6, pp 1207-1220.

Lancaster, N.1995. Geomorphology of Desert Dunes. Routledge London & New York.

Lillesand. & Kiefer. 1994. Remote Sensing and Image Interpretation third edition. John Wiley & Sons, Inc. New York.

Paisley, E. C. I., Lancaster. N,. Gaddis, L, R & Greeley, R. 1991. Discrimination of Active Sand from Remote Sensing : Kelso Duns, Mojave Desert California. Remote Sensing Environment Nov 73, pp 153-166.

Observing our environment from Space: New solutions for a new millennium, Bégni (Ed.)
© 2002 Swets & Zeitlinger, Lisse, ISBN 90 5809 254 2

Satellite map of the Narew river national park

St.Lewiński & B.Zagajewski
Remote Sensing of Environment Laboratory, Faculty of Geography and RS, University of Warsaw, Poland

ABSTRACT: The increasing volume of satellite data encourages the search for the increasingly advanced and better methods of processing and amalgamation of information originating from various sources, with the aim of obtaining the detailed image of the Earth's surface. The paper presents the methodology of elaboration of the satellite map of a protected area featuring a particular natural value. The area in question is the Narew River National Park (Polish acronym: NPN) along with the surrounding protective zone, located in the northeast of Poland (near Białystok). The source materials used include satellite (Landsat TM 5) and aerial data, which served to obtain the colour image on the scale of 1:25,000.

1 INTRODUCTION

The valley of the Narew river is characterised by the dense network of river beds, which occupy the entire valley over its numerous stretches. Insignificant inclination (by 0.15‰ on the average), slow water current, as well as flat, peaty terrain, cause that the flow of the surface and ground waters is hampered and held back, and the whole valley is within the reach of the long-lasting flooding by the river waters, transported across it by the branching-out network of the old river channels. In the period of the Spring and Summer high discharges, water flows over the whole width of the valley (2-3 km). The consequence is the water-and-swamp setting of the natural environment, with the specific ecological conditioning, different from those prevailing in other river valleys. This leads to a rich mosaic of the ecosystems having arisen from the aquatic, aquatic-and-meadow, land-and-swamp, as well as land environments (Banaszuk, H. 1996). This area, so valuable from the point of view of nature assets, plays a significant role in the international system of migrations of living organisms. The area corresponding to the NPN was considered an international core area (25/M) in the European programme ECONET (Liro, A. (ed.) 1995). Yet, quite an intensive anthropogenic pressure of the 1970s and 1980s, followed by the advancement of the re-naturalisation works, coupled with the economic crisis, entailed an important transformation of the Narew river catchment area. Currently, after the National Park has been established, there has arisen the institutional capacity of conducting the proper environmental monitoring.

The Remote Sensing of Environment Laboratory of the Faculty of Geography and Regional Studies of the Warsaw University, having conducted for years studies related to this area and the Narew River Landscape Park, gathered a significant volume of research output, used to determine the directions of changes taking place in the environment of the Narew River National Park.

Elaboration of the report gave rise to the necessity of solving a number of methodological questions and the nature-related interpretation of the images processed. Thus, in order to obtain the image of the colours close to the natural ones, an original method was applied of amalgamating colour compositions of the satellite images. The Landsat TM 5 imagery was transformed in such a manner as to have, after the addition of the colour-infrared photomap, the possibility of obtaining the detailed colour image on the scale of 1:25,000, which would be a sum of information on vegetation, soil, built-up area and the remaining elements of surface cover.

The satellite map of the Narew River National Park was elaborated for the purpose of obtaining a detailed image of land cover and the information on the condition of the plant communities. This information was provided by the Landsat TM 5 imagery and the colour-infrared photomap with the field resolution of 0.5 m. The satellite and aerial photography were taken in the same period of vegetation growth.

2 METHODOLOGY

The schedule of the project was as follows:
- Analysis and selection of the appropriate RGB colour compositions of the Landsat TM colour photographs for the particular classes of land cover and the surroundings of the Park.
- Implementation of the Tasselled Cap Transformation of Landsat TM image.
- Empirical determination of the values of the "greennness" channel, related to the selected land cover forms, expressed with the RGB colour compositions (3,2,1) and (2,4,3) of the Landsat TM imagery.
- Association of the (3,2,1) and (2,4,3) compositions with application of the masks of the values (0,1), determined on the basis of the "greenness" channel.
- Application of the IHS transformation to the amalgamation of the satellite colour composition with the PC1 obtained on the basis of aerial spectrozoual photograph.
- Establishment of the vector information layers: boundaries of the NPN, roads, railways, and main settlement centres.
- Establishment of the legend of the map containing such items as: damaged-, coniferous forest, damaged-, deciduous forest, thicket, meadow, pasture, arable land, arable land with a vegetation cover, built-up area, river and drain.
- Printout of the map on a high resolution printer.

The detailed algorithm of image processing is shown in Figure 1.

3 RESULTS

Owing to the application of a number of transformations, including the original method of associating various colour compositions, the fundamental objective of the project was attained, namely elaboration of the colour *Satellite Map of the Narew National Park on the scale of 1:25,000* (Figure 2. colour plate), allowing to discern the areas damaged or threatened with degradation. The specific character of the protected area required application of the algorithm that would make it possible to present the condition of the vegetation. This particular information was mainly taken from the spectral renditions of the satellite imagery. The statistical analyses carried out confirm high correlation between the NDVI map formed on the basis of the satellite data and the product of the RGB (2,4,3) composition and the (0,1) mask resulting from the "greenness" channel (correlation coefficient = 0.93, and R^2 = 0.86). For the green areas from the composition [RGB (2,4,3) x mask (0,1)] the correlation coefficient ranges from 0.73 to 0.84, while the error R^2 – between 0.55 and 0.64. When analysing the map one should remember

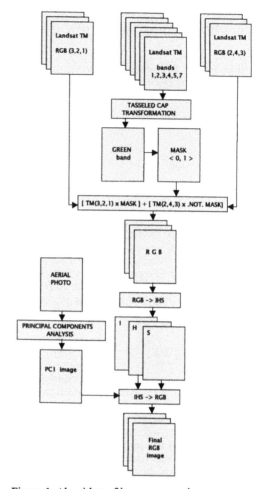

Figure 1. Algorithm of image processing

that there exists a dependence between the intensity of the green colour on the map and the condition state of the vegetation, and also between the humidity of the meadows and pastures, and the green-blue-violet colour on the map. The increase of humidity of the uncovered arable lands is reflected through the darker tones of the brown colour. It is also interesting that there is a possibility of observing the segetal plants, which appear on the arable land.

The map presented is a spectacular instance of the possibility of associating the satellite and aerial data for the purpose of obtaining a detailed image of the Earth's surface. Full interpretation of the image obtained will be elaborated in the consecutive research work stages, related to the monitoring of the environment of the NPN. This work will be conducted in the framework of the doctoral dissertation being prepared by B. Zagajewski, as well as in the framework of the Protection Plan of the Park, currently under elaboration by the Board of the NPN.

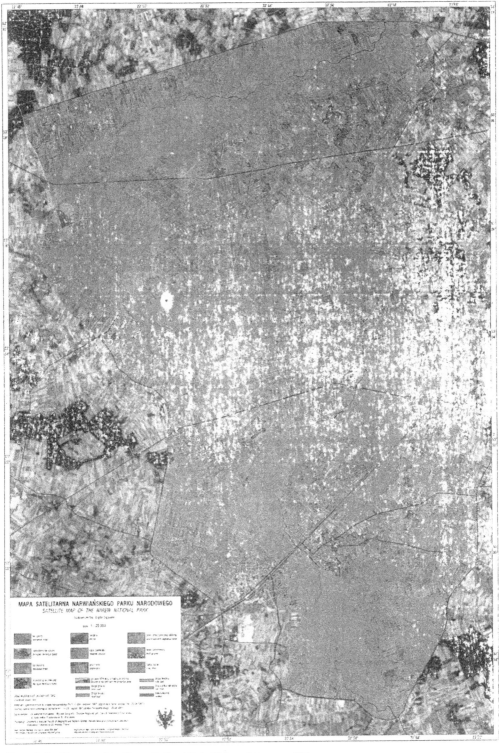

Figure 2. Satellite map of the Narew National Park. (Colour plate, see p. 434)

4 REFERENCES

Banaszuk, H. 1996. Paleogeografia Naturalne i Antropogeniczne Przekształcenia Doliny Górnej Narwi. Wydawnictwo Ekonomia i Środowisko, Białystok.

Carper, W.J., T.M. Lillesand & R.W. Kiefer, 1990. The use of intensity-hue-saturation transformations for merging SPOT Panchromatic and multispectral image data. Photogrammetric Engineering & Remote Sensing, vol. 56, no. 4, pp. 459-467.

Chavez, P.S. Jr., C.S. Stuart & J.A. Anderson, 1991. Comparison of three different methods to merge multiresolution and multispectral data: Landsat TM and SPOT Panchromatic. Photogrammetric Engineering & Remote Sensing, vol. 57, no. 3, pp. 295-303.

Lewinski, St. 2000. The satellite maps of Poland elaborated on the basis of Landsat MSS, TM and IRS-1C images. Proceedings of 28th International Symposium on Remote Sensing of Environment, Cape Town, RPA, March 27-31.

Liro, A. (red.) 1995. Koncepcja krajowej sieci ekologicznej ECONET – POLSKA. Foundation IUCN Poland. Warszawa

Satellite maps:

St. Lewinski, Z. Goljaszewski. Satellite map of Nowy Dwór and Legionowo district, in colours similar to natural. Scale 1:50 000, IRS-1C / LISS + PAN. Institute of Geodesy and Cartography. Warsaw 1999.

St. Lewinski, Z. Goljaszewski. Satellite map of Warszawa. Scale 1:50 000, IRS-1C / PAN + Landsat TM. Institute of Geodesy and Cartography. Warsaw 2000.

St. Lewinski, Z. Goljaszewski. Satellite map of Warszawa - Centrum/Ursynów. Scale 1:25 000, IRS-1C/PAN + Landsat TM. Institute of Geodesy and Cartography. Warsaw 2000.

Observing our environment from Space: New solutions for a new millennium, Bégni (Ed.)
© 2002 Swets & Zeitlinger, Lisse, ISBN 90 5809 254 2

Examining the land-use classes by means of digital elevation model and remotely sensed data

S.Kaya, N.Musaoglu & C.Goksel
ITU, Civil Engineering Faculty, Department of Geodesy and Photogrammetry, Remote Sensing Division, Maslak-Istanbul, Turkey

C.Gazioglu
University of Istanbul, Institute of Marine Sciences & Management, Vefa-Istanbul, Turkey

ABSTRACT: Digital Elevation Model makes it possible to display and model all the analyses concerning the surface and topography. Thanks to available information on elevation, it is possible to create new data, carry out evaluations about the study area that would not otherwise be possible through two-dimensional image and to determine the structure of the area. Therefore, it has found a vast scope of application today. Data sources of Digital Elevation Model are the ground measurements, photogrammetry, remote sensing and available topographic data which are digitised for this purpose. In this study, digital elevation model that has been obtained after having digitised available topographic maps has been merged with remote sensing data and a topographic map of 1/5000 scale. In the areas where the land-use tends to change fast, evaluation of the three-dimensional land classes have been made on the multitemporal data. Test areas used in the classification were determined by means of aerial photographs. Consequently, advantages and disadvantages of the process where the land-use classes are evaluated through digital elevation model created by having merged, with the remote sensing data, the land-use classes that had been previously assessed two-dimensionally.

1 INTRODUCTION

In many studies carried out around the world, remote sensing data are being used in the process of planning the land-use, monitoring the changes in the land-use and preparing revision maps (Lillesand and Kiefer, 1987, Ehlers et. al, 1990, Green et. al., 1994). Determining the land-use and its control is rather difficult especially in the cities, which are developing and surrounded by a rapidly changing environment. In order to solve a number of the problems faced by the cities with dynamism, it is essential to have updated information (Jensen, 1996). It may sometimes be impossible for many public institutions and authorities to get access to such data. Moreover, it is necessary to compare data of different times so as to have some statistical conclusions about urban development. It is indeed difficult to make such a comparison by means of ground data obtained through classical methods. Using high-resolution remote sensing data today, urban areas, especially where the land-use has been changing rapidly can be monitored. Comparison of urban development of a given period with that of a previous date can be made by using remote sensing data obtained within certain time intervals as a result of which quick access to data is possible.

In this study, Istanbul – Ikitelli region, an area which developed rapidly from the year 1985 onwards has been analysed by using remote sensing data (Kaya,1993). Making the digital elevation model of the area, remote sensing data have been merged with topographic map. The purpose of this merging process is to enhance, on three-dimensional data, the interpretability of land-use classes, which are difficult to be interpreted by means of two-dimensional evaluation. Land-use changes of the study area have been analysed structurally to yield conclusions. It has been observed that some of the classes in remote sensing data obtained as a result of classification were compatible with the topographic map while some were not.

2 METHOD

2.1 Data Used

Landsat 5 TM data of the years 12 August 1984, 06 September 1992 and 18 July 1997, Spot Panchromatic 13 June 1993 data and the map revised in 1990 – 1994 were used in the study. Furthermore, contours of the 1/25 000 scale topographic maps have been digitised to compose the digital elevation model.

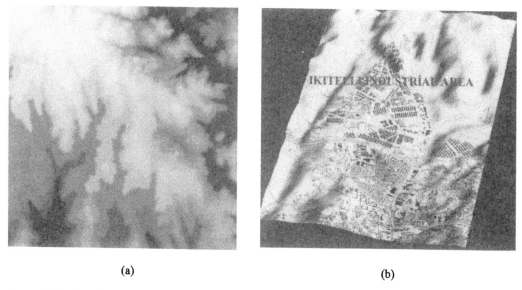

(a)   (b)

Figure 1. Ikitelli Region a) Digital Elevation Model, (b) Digital Elevation Model overlaid with Map.

2.2 *Rectification and Classification*

Different data groups must have same geometric features and be defined on the same co-ordinate system when they are to be used together (Sunar and Kaya ,1997). Moreover, registration is necessary to eliminate some distortions in satellite images and to sharpen the details in the images (Welch and Ehlres, 1987). In this study , co-ordinate system of the 1/ 25 000 scale topographic maps on the UTM (Universal Transversal Mercator) projection plane which is the national co-ordinate system of topographic maps, has been used. All of the data have been resample on this system at 0.5 RSE accuracy.

Map of 1 / 5 000 scale has been scanned and transferred to computer media and resampling was made to obtain geometric resolution of 3 meters.

Landsat 5 TM data of the years 1984 , 1992 and 1997 have been classified to determine the land classes of the study area. Six main classes , i.e. settlement areas , industrial areas, forests, empty areas, roads and rocky areas have been determined. By using ISODATA method (ERDAS ver. 8.2), 15 classes have been determined and same ones have been combined. In defining the land-use classes , available terrestrial data, Spot panchromatic data and Landsat 5 TM data viewed through different band combination have been utilised. Also the image which displays the three-dimensional land classes merged with topographic map has served as an important source used in determining the land-use classes.

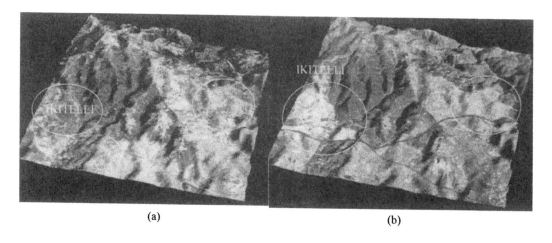

(a)   (b)

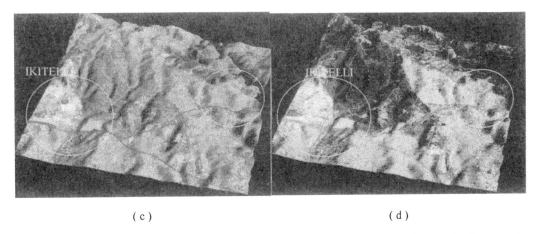

(c)   (d)

Figure 2. Ikitelli Region, (a) Digital Elevation Model overlaid with 1984 Landsat image(3,2,1 canal), (b) Digital Elevation Model overlaid with 1992 Landsat image(7,4,1 canal), (c) Digital Elevation Model overlaid with 1997 Landsat image (7,4,1 canal), (d) Digital Elevation Model overlaid with Spot P.

2.3 *Joint Evaluation of Digital Elevation Model with Remotely Sensed Data*

In general sense , digital elevation model is the digital and three-dimensional expression of the surface of the land . By means of elevation information , it is possible to create new data , carry out evaluations about the study area that would not otherwise be possible through two-dimensional analysis and find out the topography of the area. In the study made , contours have been digitised at every 20 m through the map of 1 / 25 000 scale. Then digital elevation model was obtained from these data and converted into raster data form. A lot of informations concerning the surface of the land (land-use, surface morphology, size of the land classes etc.) do not exist on these data groups. Therefore , not much can be said about the surface of the land by using only the digital elevation model. However, when combined with data groups which define the information on the land surface , which have the possibility of evaluating information elicited at different dates , which define the topographic features of the surface of the land and provide fast flow of data ; it is then possible to obtain multi-discipline interpretable data. Joint use of digital elevation model with remote sensing data today sheds light to a host of studies. Because , remote sensing data offer significant advantages in knowing the morphological features of the surface of earth , in determining the land-use classes and in obtaining the actual co-ordinates on the earth's surface and the area size of these classes (Lillesand and Kiefer, 1987, Kaya, 1999) .

In this study , first of all digital elevation model was merged with Landsat 5 TM data which have viewing possibility in different band combinations. Morphological features of the study area were stud-

ied through these data. Images pertaining to the years 1984, 1992 and 1997 as well as the classified data obtained from these images were individually merged with digital elevation model. Analyses from both the two-dimensional and three-dimensional images have been made to determine the land-use , urban development areas and the reasons of growing or decreasing the number of the established land classes. 12-year development of the study area was evaluated according to the land classes. According to these results , it has been ascertained that the area attracted massive human settlement since 1984 and the land-use around the industrial areas has changed rapidly. Most of these settlement areas are unplanned ones. Analysis of especially the image obtained from the remote sensing data merged with digital elevation model contributed greatly to structural evaluation of the settlement areas.

3 CONCLUSIONS AND SUGGESTIONS

In this study, benefits of integrating remote sensing data with digital elevation model in monitoring the land use have been explained. It has been ascertained that use of the satellite images whose integration is made at a certain accuracy can produce up-to-date information about the issues of land-use and directions of urban development . On the other hand, it has been concluded that structural evaluation of the area could be made through classified and enhanced satellite images. Advantages of evaluation through three-dimensional model are to be able to analyse structurally the land-use classes which are impossible to be evaluated and interpreted two-dimensionally and to monitor the temporal changes three-dimensionally. As for the disadvantages , it can be

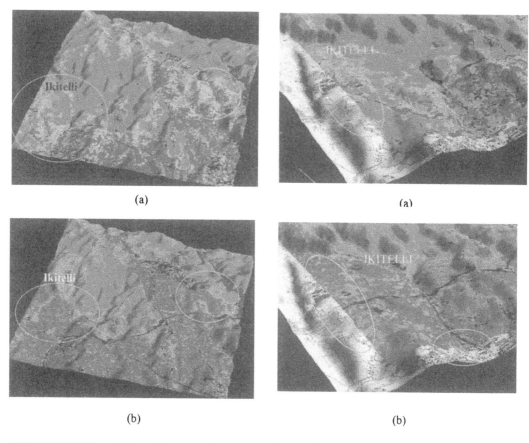

(a)   (a)

(b)   (b)

Figure 4. Ikitelli Region (a) Digital Elevation Model overlaid with map and Classified image of 1984, (b) Digital Elevation Model overlaid with map and Classified image of 1997. (Colour plate, see p. 435)

(c)

●   Road
●   Urban Area
      Bare Area
●   Forest and Green Area
●   Water
●   Industrial Area

Figure 3. Classified images, (a) Classified image of 1984, (b) Classified image of 1992, (c) Classified image of 1997. (Colour plate, see p. 435)

said that , although digital elevation model was not made at a certain accuracy (digitising the contours at every 20 m , 10 m , 5 m or 1 m) , especially the linear contours could not match on the model and relatively flat areas are not visible. Therefore , in composing the digital elevation model , it is necessary to digitise the contours with higher frequency .

The fact that remote sensing data are of iterative nature , they brought significant innovations to the field of analysing dynamic surfaces in terms of both global and local areas. In case one is engaged in studies with high-resolution satellite data today, such evaluations will be made in more details. In the future, making the geometric conversions of high-resolution satellite images by means of GPS co-ordinates will enhance the accuracy of the integration and will ensure the detailed description of topographic features of the surface of earth and more sensitive determination of the land-use classes. Furthermore, important information about the development of the city will be obtained through three-dimensional model .

4 REFERENCES

Ehlers, M., Jadkawski, A.M., Howard, R.R., Brostven, D., 1990. Application of Spot Data for Regional Growth Analysis and Local Planing, *Photogrammetric Engineering and Remote Sensing*, Vol.56, No:2, 175-180.

Green, K., Kempka, D. and Lackey, L. 1994. Using Remote Sensing to Detect and Monitor Land-cover and Land-use Change, *Photogrammetric Engineering and Remote Sensing*, 60 (3), pp 331-337.

Jensen, J. R. 1996. Introductory Digital Image Processing A Remote Sensing Perspective, Second Edition, *Prentice-Hall,Inc., USA.*

Kaya, S. 1993. The Metropolitan Analyses by Using the Remote Sensing Data Within the Example of Istanbul – Ikitelli, Master Thesis, *ITU Institute of Science and Technology,* Istanbul.

Kaya, S. 1999. Study of Geomorphological and Geological Characteristic Along the Northern Strand of the North Anatolian Fault Between Gallipolly and Isiklar Mountain by Using Remote Sensing Data and Digital Elevation Model, Ph.D. Thesis, *ITU Institute of Science and Technology.* Istanbul.

Lillesand, T.M., Kiefer, R.W., 1987. Remote Sensing and Image Interpretation, *Chichester: John-Wiley and Sons.,* Canada.

Sunar, F., and Kaya, S. 1997. An Assessment of the Geometric Accuracy of Remotely Sensed Images. *Int.J.Remote Sensing.* Vol. 18. No:14. pp 3069-3074.

Welch, R., and Ehlers, M. 1987. Merging Multiresolution SPOT HRV and Landsat Data, *Photogrammetric Engineering and Remote Sensing*, 53, pp 301-303.

Observing our environment from Space: New solutions for a new millennium, Bégni (Ed.)

Geomorphological map (Tykocin sheet). Methodology

W.Wołk-Musiał, St.Lewiński & B.Zagajewski
Remote Sensing of Environment Laboratory, Faculty of Geography and RS, University of Warsaw, Poland

ABSTRACT: The study reported aimed at the analysis of the surface relief within the central part of Tykocin Basin in the northeast of Poland and at presentation of this relief on a detailed geomorphological map of the scale of 1:25000. Implementation of the project required elaboration of the methodology, which would make use of the integrated research techniques from the domains of remote sensing, physical geography, and GIS. An attempt was undertaken of developing a new cartographic presentation of the map being elaborated using a number of digital transformations, incorporating the photographic map, into the geomorphological contents.

1 INTRODUCTION

The gemorphological map, representing a detailed image of the relief, the primary element of the natural environment, constitutes the basis for the rational use of geographical space. It shows the locations and the classification of all relief forms, accounting for the morphological and morphographic features, origins (on the basis of the spatial distribution of relief forms, and the lithofacial analysis of the sediments), as well as the age of forms. In order to optimise the study of relief the methods were made use of from the domains of remote sensing, physical geography – with special emphasis on geomorphological analysis, and from the Geographical Information Systems (Wołk-Musiał, E. 1999, Wołk-Musiał, E. & B. Zagajewski 1999, 2000) It is in particular the remote sensing materials, the aerial photography, that provide the possibility of perceiving the area from the bird's eye view and of encompassing with one glimpse all the primary features of relief, identification of its separate elements, as well as analysis of their mutual relations in the spatial setting. This allows for the determination of the geological structure of the area considered, and therefore also to determine its origins. The study of relief with the remote sensing materials is possible first of all owing to the broad use of relations existing among all the landscape components. This kind of approach to the geomorphological photointerpretation is referred to as the landscape method of analysing aerial photography. The method consists in the analysis of interdependencies between, on the one hand, the directly cognisable components of the geographic environment, such as the types and

forms of relief, the character of the hydrographic network, the vegetation, and the elements related to human activity, and, on the other hand, the geological structure, soils, and underground waters.

It is therefore essential in the analysis of the remote sensing input to determine, within the boundaries of a "landscape" or of its morphological parts, the directly cognisable components of the geographical environment. They yield ultimately a definite structure of the photographic image, referred to as the photomorphic unit. The study of the thus conceived photomorphic units constitutes the basis for the geomorphological interpretation of aerial photography. The main advantage from application of aerial photography in the study of relief is constituted by the faithfulness of the image. In comparison with topographical maps, on which numerous forms, especially of fluvial relief, are not designated at all, the image of relief on the remote sensing materials is much more detailed. This concerns first of all the areas of glacial origins.

Likewise, the dynamic development of computer applications caused that many domains of human activity gained a new dimension. One of the spheres, in which the greatest changes took place, is constituted by the sciences associated with space. The development of the GIS made possible a new approach to the analysis and processing of data describing space, as well as new solutions in visualisation of spatial data, also in cartographic form. Introduction of the GIS as a method of study of relief gives the opportunity of tracking the relations between the data of various kinds and originating from different sources, the data, which can be unequivocally located in any arbitrary coordinate system. This con-

Table 1. Morphometric types of relief for the Tykocin sheet on the scale of 1:25,000

Symbol	Relief form	Elevation difference in metres	Element's basis in metres	Slopes
1	Plains in the bottoms of valleys and basins	Up to 2.5		<2%, <1%
1a	Higher elevations within plains in the bottoms of valleys and basins			
2	Plains outside basins	Up to 2.5		<2%
3	Low hillocks of small perimeter	< 15	Up to 200	10-20%
4	Low hillocks of large perimeter	< 20	200-500	2-10%, rarely less
5	High hillocks	> 20	200-500	around 20%
6	Low hills	10-15, maximum up to 20	> 500	up to 5%
7	High hills	> 20	> 500	5-10%
8	Flat-bottom (small) valleys			
9	Concave slopes and valleys			
10	Slope flats			< 1°
11	Slopes of low inclination			1° – 3°
12	Slopes of medium inclination			3° – 9°
13	Slopes of high inclination			> 9°
14	Anthropogenic landforms (road canyons, embankments, gravel-pits, waste heaps, terraces, etc.)			
15	Small basins without outlets		< 100	.
16	Water surfaces			

sequently leads to the precise location and classification of all the forms of relief with due account of the morphometric and morphographic features, the origin, and the age of forms.

2 RESEARCH ALGORITHM

The study here reported was aimed at the analysis of the relief of the central part of Tykocin Bowl, to the South of the locality of Długołęka, with application of the integrated research methods.

The effects obtained are shown on the detailed geomorphological map of the scale of 1:25,000, Tykocin sheet (coordinate system 1942). An attempt was undertaken of developing a new cartographic presentation of the elaborated geomorphological map, by showing it against the background of the ortho-photographic map, made of the panchromatic aerial photographs of the scale of 1:25,000.

Hence, remote sensing plays a double role in the work presented. It both provides the source material for the acquisition of information on the surface relief forms, and the "background" for the presentation of the results obtained from the geomorphological analysis.

The methodology of the study encompassed the following stages:

1. The geomorphological cognition of the study area on the basis of the available cartographic materials (topographic and thematic maps), and the analysis of the panchromatic aerial photographs on the scale of 1:25,000. Consequently, a map was developed of the photomorphic units, constituting the basis for the genetic analysis of the surface relief. Within the framework of the very same units the attempt was made of determining the forms having emerged due to the deposition by inland ice, as well as due to glacifluvial deposition and erosion. One of the most characteristic elements of the area considered is constituted by valleys. Their shape depends upon the volume of water flowing and upon the quality of the material being modelled. The area studied is covered by a thick layer of the glacial sediments, with many valleys having undergone far-reaching transformations in the periglacial conditions. They are now visible in the relief in the form of elongated troughs with relatively small elevation differences between the bottoms and the watershed surfaces. Yet, owing to the sensitivity of the stereoscopic model all these depressions, hardly visible in the field, become well seen.

In the photo-interpretative analysis of the quaternary glacial surface relief of the lowland areas, side by side with the direct recognition features, which

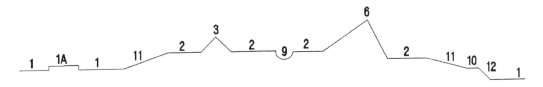

Figure 1. The model sequence of relief forms (numerical symbols conform to Table 1).

can be read out from the stereoscopic model, an essential significance is ascribed the landscape method. This method accounts for the relations existing between surface relief, lithology, soils, vegetation, and various objects associated with human activity. Analysis of these relations allows for the proceeding with the lithological-genetic identification of the surface relief forms.

2. Elaboration of the morphometric classification of the glacial relief of the study area on the basis of the topographic maps of the scale of 1:25,000. The relief types were obtained taking into account the elevation differences, the diameter of the basis of the form, and the dominating inclination of the area surface (see Table 1).

Analysis of the contents of the map indicated that the landscape of the river valleys is decidedly dominated by the plains in the bottoms of valleys and depressions (1), and the higher elevations within them (1a). The hillocks of small and large perimeter (3, 4) are sporadically encountered. Within the uplands the landscape is dominated by weakly inclined slopes (11), in some places turning into moderately inclined slopes (12) and plains outside the depressions (2). Uplands are cut through by the secondary river valleys (1) and the dense network of small concave valleys (9) of peri-glacial origin. The convex forms are dominated by the low hillocks of small and large perimeter (3, 4). Sporadically, low hills (6) and high hills (7), are also encountered. The slope flats distinguished are poorly visible.

A typical sequence of relief forms – from the valley to the upland – is shown in Figure 1. The setting of the elements along the cross-section is locally modified.

The morphometric classification of relief was processed in the digital form and constitutes an independent information layer in the GIS.

3. The lithological identification of the study area was carried out on the basis of the map of soil types on the scale of 1:25,000, elaborated on the foundation of the soil-and-agricultural maps on the scale of 1:5,000 and own field studies. Elaboration of the map in the ArcInfo system made it possible to assign to each basic spatial unit (polygon) the attributes containing three information pieces coded separately: the surface form, its thickness, and the bedding. This yielded the detailed identification of lithology at the depths of 0.5, 1.0, and 1.5 metres.

The mineral bedding of the valley bottom of Narew river lies at the depth of about 1 metre. It is only over a small area, directly to the North of Tykocin, that the bedding falls down somewhat, reaching the depth of 1.5 metre. It is made on the entire area of loose sands. In the northern part of the valley they are covered by peat, whose upper layer is subject to the rotting process. Solely at the confluence of Nereśla and Narew rivers the bottom of the valley is covered by the one-metre-thick layer of the loose sands lying on the light clayey sands (to the West of Nereśla). The lithology of the valley of Narew river to the North-East of Tykocin looks somewhat differently. The share of the sandy forms decreases here to the advantage of peats.

The areas to the South of the Narew river valley are composed, as a rule, of the light boulder clays, covered by the clayey sands. The thickness of the latter amounts to 1 metre, and only in the direct vicinity of the Narew river valley it drops down to approximately half a metre.

The northern part of the study area has an entirely different lithological structure. Boulder clay covered by clayey sands appears on only a small fragment between the localities of Góra and Morusy, where it covers with the half-metre thick layer the loamy dusts. The remaining part of the area is covered by sands and cut through by the valley of Nereśla, filled by fluvial sands, with two patches of peat filling up the local depressions in the valley.

4. Determination of the morphogenetic types of relief through application of the GIS-driven spatial analyses to the previously elaborated information layers: the photomorphic layers, the types of relief, and the lithological structure of the study area for three depths (0.5, 1.0, and 1.5 metres). The genetic types of relief are the effect of the activity of the morphogenetic factors:
   - deposition by inland ice,
   - initial melting of the inland ice surface,
   - final melting of the inland ice surface,
   - glacifluvial and glacilacustrine deposition and erosion,
   - erosion and denudation,
   - fluvial deposition and erosion,
   - aeolian deposition and erosion,
   - biogenic deposition,
   - anthropogenic activity.

5. The detailed geomorphological analysis for the particular types of relief, carried out on the basis of the precisely defined catalogue of the genetic landforms. An essential role is played here by the DTM, which allows for identification of the relief (absolute elevations, relative elevations of landforms, slopes, inclinations). Yet, it first of all makes it possible to perform the analysis of the hypsometric level assignment to landforms, classified into the morphogenetic relief types, with determination of the sequence of their emergence. The effect was the geomorphological map of the study area on the scale of 1:25,000, digitally elaborated. The legend to the map was put together in the ERDAS Imagine software, where the sequences of colours corresponds to the genetic types of relief, while hachure reflects the genetic types of glacial and post-glacial relief (ground moraines, hills, ramparts and plateaux resulting from thawing of inland ice, kame hillocks, hills, and terraces, glacio-lacustrine plains, as well as the domain of fluvial and aeolian forms).

6. The final stage is constituted by the presentation of the geomorphological map against the background of the panchromatic aerial photograph with the purpose of showing the detailed location of landforms and the land use capacities.

Due to a number of transformations the image was obtained, on which, owing to the colour distinctions of the geomorphological classes, the outline of the land surface can be seen.

In the work related to image processing black-and-white aerial photographs and the colour classification image were used. Aerial photography was first amalgamated into one image file, which was thereafter subject to two kinds of transformations. The first one consisted in application of the edge filter in order to obtain the image of the edges of landforms. The second kind of transformation consisted in such an elongation of the histogram as to make more pronounced the image of forested areas while simultaneously the image of the remaining part of the image was weakened and fuzzyfied.

The resulting black-and-white images were associated with the classification colour image. This was achieved through application of simple operations of addition and subtraction. The image after filtration was added to the colour composition of the classifying image, and then the second image was subtracted from it. On the resulting image, within the framework of individual classes of geomorphological distinctions, the lighter edges of the land cover forms are visible along with the darker forest surfaces.

The new cartographic approach makes it possible to assess the land use within the particular geomorphological units. On the valley areas the hay-growing meadows have a small-block structure. The increasing humidity of these meadows, designated on the photographic map with a dark-grey phototone, appears through a darker colour tone on the geomorphological map. On the other hand, the fragments of the valley bottom where land improvement had been carried out, and which are used as farming land, show a striped structure on the map.

On the uplands, side by side with the setting of fields, designated on the background of the map by the lighter edges, the areas, which are more humid or are under root crops, are characterised by a darker colour of the geomorphological units. Best seen are the forested areas, as well as in-field groves and swamp forests, since in the background of the landform indicated the fine-grained structure and the darker distinguishing colour are well seen.

3 RESULTS

The study area, incorporating the valley of Narew river along the stretch Tykocin – Góra, together with the adjacent uplands, was encompassed by the middle-Polish glaciation, the North-Masovian stage (Mojski, J.E. 1972). In the photomorphic division of Poland (Olędzki, J. 1992), performed on the basis of the satellite images, the area analysed is located at the border of two units: the Biebrza River Basin and the High-Masovian Upland.

The hypsometric differentiation, which is observed on this area, is significant for an old glacial relief. The lowest fragments appear in the valley of Narew river to the North of Tykocin with the elevation of 104 m a.s.l., while the culmination of the area is constituted by the elevation in the southern part of the area – 166 m a.s.l. Hence, the elevation difference exceeds 60 metres.

The dominating form of the study area is the valley of Narew, primarily filled by the peat plain, cut across by a rich system of old river-beds. Close to the main river bed a part of those landforms play the role of flood channels in conditions of high discharges. Currently, due to the intensive land improvement works in the valley of Narew river, a significant lowering of the groundwater table was brought about. This entailed drying out of a number of old river-beds and intensification of the rotting of peat.

Numerous aeolic forms appear in the valley of Narew river, mainly the parabolic dunes with deflation depressions. A significant complex of such forms stretches between the localities of Dobki and Góra, where it is elevated by 7 metres above the bottom of the valley. The riverside dunes of Tykocin Basin are not well established and are subject to destruction by the wind. It was stated on the basis of the analysis of fossil soils that they emerged in two dune-generating periods: in younger Dryas and in the younger phases of the Holocene. The aeolic activity in the second of these periods took place most probably around the 12th century BC, and then between the 13th-14th and 18th-19th centuries, as well as in the 20th century, and was brought about by the local human activity (Grzybowski, J. 1981).

To the South of the valley of Narew river a ledge stretches of a variable width, rising from the elevation of 115-117 m a.s.l. This particular level is mostly made of sand and loam sediments with an admixture of gravel. Only in the vicinity of Tykocin and Popowlany the boulder clay gets uncovered. These levels were described in the literature as the fluvial deposition and erosion (S. Żurek, 1975; H. Banaszuk, 1980). In the light of the new views concerning the glacial origin of the Biebrza Basin and the glacilacustrine origin of the fragments of Narew river valley, the surfaces mentioned were classified into the meltdown levels. This classification is additionally confirmed by the lack of distinct erosion cuts on the slopes of the Narew river valley in the direction of High-Masovian Upland.

The highest elevation within the study area (166 m a.s.l.), located to the South of Tykocin, is a meltthrough hill. This landform constitutes a hydrographic node, from which river valleys spread out radially. This highest point is surrounded by an inland ice thaw plateau, with kame hillocks and hills sticking above it, having elevation between 147 and 155 m a.s.l. Farther on in the direction of the valley of Narew river, the plateau turns into a clayey ground moraine, cut through by a system of dry valleys having their origin related to the level of the inland ice thawing.

Within the area of the ground moraine an elongated hill can be perceived of the elevation of 124.2 m a.s.l., to the South-East of Tykocin, being a glacio-fluvial and ablation rampart. It is characterised by a high inclination of its slopes towards the valley of Narew river and the low inclination of its slopes towards the plateau, which is an evidence of its appearance in the zone of contact with the dead ice filling out in the final stages of deglaciation the lacustrine stretches of the valley.

To the North of the valley of Narew river, between Morusy and Góra, flat surfaces exist in the form of patches of various dimensions, adjacent to the valley of Narew. This plain of disjunctive character, whose elevation is 106-108 m a.s.l., was also classified into the level of the inland ice thawing.

The domain of the aeolic forms – dune ramparts, parabolic dunes, or fields of aeolian sands, is visible in the northern part of the study area. These landforms reach the elevation of 128 m a.s.l., and their shape and morphology indicate that they developed mainly owing to the winds blowing from the western sector. The fields of aeolian sands, with weakly shaped dune forms, appear primarily in the valley of Nereśla, at the elevations of 108-110 m a.s.l. The average thickness of sand on the area considered is 4.5 metres.

In the north-eastern part of the study area clay appears under a thin sandy cover. Such a situation can be observed in the vicinity of the village of Góra, where the level commented upon reaches the elevation of 106-110 m a.s.l. It is dominated by a 5-6 metre culmination constituting a kame hill. It is made of the intermittent series of gravels and layered sands covered by the unstructured thick-grained gravels. This is an evidence for the in-glacial origin of the kame. The areas described have been classified by H. Banaszuk (1980) as moraine islands. Yet, in the light of the glacial origins of valleys of Biebrza and Narew rivers they can be the inland ice thawing levels made of boulder clay.

The area considered is strongly dismembered by numerous depressions genetically associated with the thawing of the dead ice. They are filled with peats and fluvial sands. In the stretch of Nereśla river close to its confluence with Narew a meridionally oriented belt of dunes appears, surrounded by the fields of aeolian sands. The valley of Nereśla played a significant role in the outflow system of the disappearing inland ice, since it channelled out water and supplied the glaciofluvial material to the Tykocin Basin. During the analysis of the hypsometric situation within the mouth-adjacent stretch of Nereśla river a vast depression without an outflow was identified, which should have been filled out in accordance with the theory mentioned. Yet, the origin of this depression is related to the later dune forming processes, which caused the piling of a rampart in the form of irregular dunes closing the valley of Nereśla from the South. This could constitute one of the causes of appearance of a depression in the bottom of the valley.

Landforms identified within the study area indicate the surfacial manner of the deglaciation of inland ice cover dependent upon the relief of the bedding.

4 CONCLUSIONS

The establishment of the geomorphological identification of the lowland areas featuring glacial relief requires application of complementary research methods from the domains of remote sensing and physical geography. Cognition of the character, origin, age of forms, as well as determination of the morphogenetic processes is possible owing to the application of the GIS system, providing the capacity of gathering information from various sources in a definite system of co-ordinates and of tracking relations among them. Integration of the detailed geomorphological map and the panchromatic aerial photograph allows for the correct spatial location of landforms. Thereby, the source material is provided for the study of rational development of space, carried out through presentation of the land use against the background of the geomorphological units synthesising the basic components of the geographical environment.

5 REFERENCES

Banaszuk, H. 1980. Geomorfologia południowej części Kotliny Biebrzańskiej. Prace i Studia WGiSR UW, Warszawa

Grzybowski, J. 1981. Rozwój wydm w południowo-wschodniej części Kotliny Biebrzańskiej. Prace Geograficzne IG PAN, z.4.

Mojski, J. 1972. Nizina Podlaska. In: Geomorfologia Polski. PWN. Warszawa

Olędzki, J. 1992. Geograficzne uwarunkowania zróżnicowania obrazu satelitarnego Polski i jego podziału na jednostki fotomorficzne. Rozprawy UW. Warszawa

Wołk-Musiał, E. 1999. Komplementarność metod badawczych w kartowaniu rzeźby. w: Geografia na przełomie wieków – jedność w różnorodności. WGiSR UW. Warszawa.

Wołk-Musiał, E. & B. Zagajewski 1999. Geoinformation as en effective toll of environmental protection and land use optimisation. In: Operational Remote Sensing for Sustainable Development. Balkema. Rotterdam.

Wołk-Musiał, E. & B. Zagajewski 2000. Lithological and geomorphological large-scale mapping using remotely sensed data, GIS and terrain analysis. In: Remote Sensing in the 21st Century Economic and Environmental Application. A. A. Balkema. Rotterdam /Brookfield.

Observing our environment from Space: New solutions for a new millennium, Bégni (Ed.)
© *2002 Swets & Zeitlinger, Lisse, ISBN 90 5809 254 2*

Remote sensing imagery in monitoring spatial pattern changes in forest landscape

M. Kunz
Institute of Geography, Nicholas Copernicus University, Torun, Poland

A.Nienartowicz
Institute of Ecology and Environment Protection, Nicholas Copernicus University, Torun, Poland

ABSTRACT: GIS and remote sensing technology have been used to compare Normalized Difference Vegetation Index (NDVI), diversity and fragmentation of two forest complexes different each from the other in the structure of land use in the 19th century and intensively of timber exploitation in the recent several years. Both complexes have been located in Zabory Landscape Park in Pomerania Province. The first site M has been comprised secondary forests restored by introduction of pine seedlings on former poor seep pastures and arable lands. After 100 years, due to intensive timber exploitation a big part of those area has been occupied by clear-cuttings and few-year-old pine plantations apart from the oldest pine forests. The second site P has been forest complex where the timber exploitation was significantly reduced since the middle of eighties. The structure of landscape in the first site has been assessed on the basis of five sample plots about 150 or 500 hectares. Four plots of similar size have been chosen in the site P. The Landsat TM imagery from 28 July 1990 and Idrisi-32 software have been used in analysis. It results from the calculations that landscape with intensively exploited forests on former agricultural soils has been characterised by lower values of NDVI and higher values of diversity and fragmentation in comparison to those with sustainable forest management. Significant differences in the structure of landscapes have also been clearly expressed by the methods of numerical classification.

1. INTRODUCTION

Production intensity and wood harvests had big influence on the landscape structure. Among all technologies used in forestry to harvest the main product clear cuttings created the biggest changes in the landscape structure. As a result of their application and then restoration of forest ecosystems the increase of tree stands age diversity took place among forest subsections, and considerable decrease of this parameter in each separate section. Because till not long ago only one type of tree was used on clear cuttings to restore a forest, which was Scotch pine, at the same time the decrease of species diversity took place as well in each separate section as in the whole forest complex. Moreover, on forest areas clear cuttings caused landscape fragmentation increase, and as a result of a pine introduction there was the decrease of habitat quality.

During recent years many papers were published concerning landscape change structures as the result of man's economical activities. Such a comparisons were made as well in time as in space. The objects of comparisons were very often forest areas, and the goal of those papers was the estimation of the influence of different economical activities such as the introduction of section lines, forest roads, clear cuttings, and also natural factors such as wind or fire on the diversity and heterogeneity of the landscape. The estimation of habitat quality and landscape structure was usually prepared from aerial and satellite level.

This paper presented the application of remote sensing methods to define the influence of economical activities which took place in forestry on a habitat quality and a landscape spatial pattern. The influence of man's activity was defined by the comparison of two forest areas on which clear cuttings took place with different intensity. One was the area of intensive forest economy and the second was the area of „Bory Tucholskie" National Park where a few years before this form of preservation clear cuttings took place. The comparison of habitat quality and landscape structure was made on the basis of NDVI values, diversity, and fragmentation. Analysis are realised for the grant of Polish Research Committee (No P04F 01313).

2. STUDY AREA

The research was made mainly in Zabory Landscape Park, situated in southern part of Pomerania

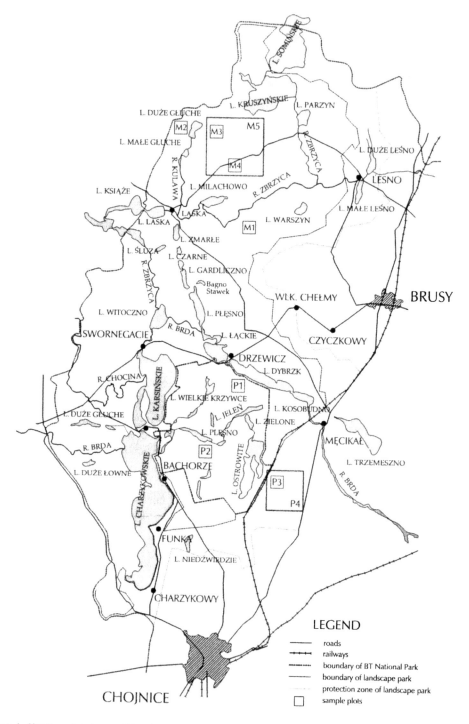

Figure 1. Situation of study area (M – site of intensive timber exploitation, P – site of „Bory Tucholskie" National Park).

Province, on west from Brusy town (Figure 1). In 1996 in southern part of Zabory Landscape Park, which belonged to Chojnice district, „Bory Tucholskie" National Park was created. National Park area is not big, because councils of other districts did not agree to cover their areas with this form of nature protection.

Vast forest complexes in the area under study are part of Przymuszewo Forest Inspectorate (Przymuszewo and Laska forest districts) and Rytel Forest Inspectorate (Klosnowo forest district), which is subordinate to the Regional State Forest Management in Toruń.

The research took place in two areas: in Przymuszewo Forest Inspectorate, in northern part of Zabory Landscape Park (site M) and in the area which some time ago was created a part of Bory Tucholskie National Park (site P) (see: Figure 1). On the first site there were dry and fresh pine forests (Sokołowski, 1965). They were created as the result of clearings reforestation with mono-cultures of pine. Those clearings were very often more fertile forest habitats and agricultural lands. Much area was covered by *Pinus-Calluna* communities, which appeared after reforestation of pastures and other grassland used in the past to sheep-grazing (Boiński, 1988). The rest of deciduous forests – dry-ground forests, beech woods, oak-hormbeam forests, and alder bog forests were preserved on small areas in valleys of many lakes which exist there. Big area was covered by degeneration forms of dry-ground forests and beech woods – *Pinus sylvestris - Deschampsia flexuosa* (Boiński, 1988). In the area of the national park a typical fresh pine forest was a dominating forest type. In the past this area was not used agriculturally. There were many lakes on both sites near which on small areas there were also wet pine forests (*Pinus-Molinia* communities) and swampy pine forests (*Vaccinio uliginosi - Pinetum*).

The first site was the area of intensive wood production. At the beginning of 90s wood was gained with the clear cutting method. During last few years after the introduction of new methods for forest breeding such clearings took place on a smaller scale, and if they took place usually clumps of seed trees were left on the area.

On P forest site, on which in 1996 National Park was created, just from the middle of 80s wood harvests were limited. Because of that there were no clearing areas, and the age of pine mono-cultures was higher than on M site.

Detailed analyses of NDVI, diversity, and fragmentation on both sites were made on smaller analysed areas, each of them covered six forest sections (about 150 ha each) and on bigger areas which were created by 21 forest sections (about 500 ha). Analysed areas were on high plains, on which there were no water ecosystems.

On M site four small and one big analysed plots were created. They were situated in the following places:

- M1 plot - near Rolbik village, where land properties were nationalised, reforested and joined to national forests after the Second World War,
- M2 plot – the furthest put forward from all analysed areas, situated near agricultural village Zapceń, on sander plain, between valleys of Kłonecznica and Kulawa rivers,
- M3 plot – situated on a highland, eastward from the Duże Głuche lake valley, and flowing through it Kulawa River,
- M4 plot – situated northward from the Milachowo lake, on the area of Polish land property, which in 1890 was sold to Prussian forest administration, and included to Zwangshof Forest Inspectorate,
- M5 plot (about 500 ha) covering the area of above mentioned land property and areas nearby it.

On P site four plots were marked, three smaller and one big. Their localisation was the following:

- P1 plot - situated in the area of today's National Park, near its border, covering fire belts, and because of partial clearings with its structure it was similar to areas situated on M site,
- P2 plot – situated south-eastward from the Płęsno lake, and near the Gacno Wielkie lake,
- P3 plot (about 150 ha) and P4 (about 500 ha) – situated eastward from the Ostrowite lake, near Dębowa Góra forester's lodge, in direct neighbourhood of the National Park.

3. METHODS

Spatial change of landscape pattern was analysed on the basis of Landsat TM satellite imagery made on 28 July 1990 with 30 meter resolution. Idrisi-32 software was used in image analysis. For each of nine sample plots the normalised difference vegetation index of separate pixel was counted according to the formula NDVI $= [IR - RED] / [IR + RED]$, where RED and IR are reflectance values equivalent to Thematic Mapper (TM) Bands 3 (RED, 630-690 nm) and 4 (IR, 760-900 nm).

Mean value of NDVI, range of its changes, and standard deviations were also defined. Total theoretic range of change NDVI from -1 to $+1$ was graduated into 256 classes which had numbers from 1 to 255. Frequency and percentage of pixels in each class were defined, histograms were prepared. Analysed plots were compared taking into consideration NDVI changes, with numerical taxonomy methods. The classification was made with the average linkage clustering method (Orlóci,1978) using Euclidean distance as the measure of differences between plots. Frequency was used as comparing features of objects

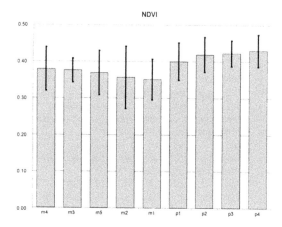

Figure 2. Mean value and standard error of sites M and P.

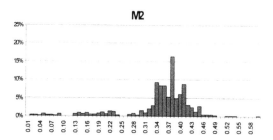

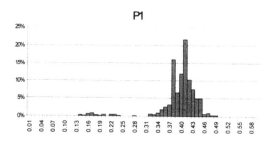

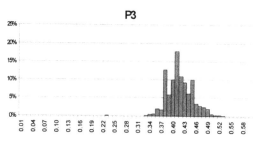

Figure 3. Distribution of NDVI frequency for three sample plots.

(percentage) in 256 classes of NDVI value. MVSP program was used in calculations (Kovach, 1993).

Diversity and fragmentation of sample plots were calculated by marking in their areas squares covering different number of pixels. Three sizes of squares comparing 3x3, 5x5 and 7x7 pixels were used in analysis.

Diversity was calculated using Shannon's formula:

$$H = -\sum_{i=1}^{n} p_i \ln p_i,$$

where p_i is share of i – th is class pixel (i = 1, 2, 3, ..., n), n – number of pixel categories in a square.

Fragmentation was calculated according to the formula:

$$F = (n - 1)/(c - 1),$$

where n states, as in the previous formula, number of pixel categories, and c – number of pixels in a square (9, 25 or 49).

Each area was characterised by average H and F values. This was arithmetic average from values calculated for a square group the middle area of which was one by one each pixel of the sample plot. Working according to this schema for pixels at the borders of analysed plot chosen squares 3x3, 5x5, 7x7 covered also pixels out of the sample plot. To preserve the same calculation formula for all squares of sample plots in diversity and fragmentation analysis, three additional ranges of pixels being out of each border of analysed area were taken into consideration.

Such activity was according to suggestions concerning the above method, which were stated by Upton and Fingleton (1985), Hauser and Mucina (1991).

4. RESULTS

On the basis of comparisons, it was stated that areas localised in national park were characterised by higher NDVI values than areas covering intensively exploited forests (Figure 2).

In the area of higher preservation statutes, NDVI exceeded a little 0.4 value. In all areas covering secondary pine forests in which intensive wood production took place, average NDVI value did not exceed that value. Moreover, areas localised out of a national park were characterised by changes in range of this factor.

On Figure 3 the distribution of NDVI frequencies was marked on M2 sample plot, with characteristic, the biggest changes among nine compared objects and P3 plot, which NDVI covered the smallest number of classes. There was also marked the histogram of P1sample plot on this picture.

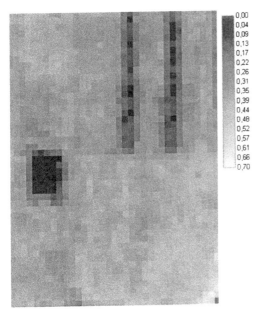

Figure 4. Spatial changes of NDVI on sample plot M1.

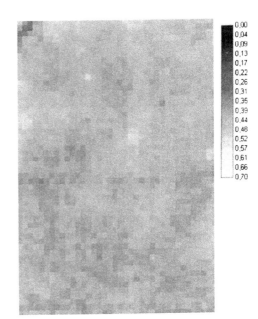

Figure 6. Spatial changes of NDVI on sample plot P3.

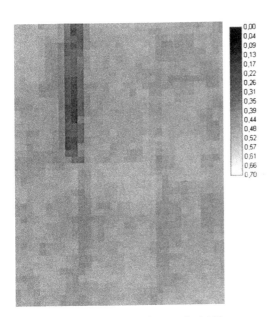

Figure 5. Spatial changes of NDVI on sample plot P1.

It had also an indirect character, and although it was situated in the area of the National Park its NDVI spatial change was rather substantial. This area was situated on the border of today's National Park and covered fire belts. Because trees were all the time cut from the belts, there were considerable differences of

biomass states, and in the consequence also NDVI, between separate parts of P1 sample plot. Spatial distribution of normalised vegetation index for three described areas is presented on Figures 4-6.

NDVI values and the ranges of its changes were very similar within each subgroup of sample plots. On the dendrogram created with MVSP program considerable differences appeared between areas marked in the National Park and ones situated out of its range, and in the consequence the elements of both subgroups created separate concentrations (Figure 7). In the concentration of sample plots in the National Park there was a considerable separateness of P1 plot covering fire belts.

Because the areas marked in intensively used forests covered as well cuttings as older stand of

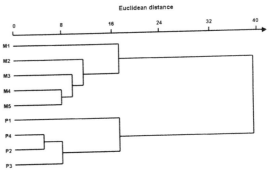

Figure 7. Dissimilarities between sites M and P.

395

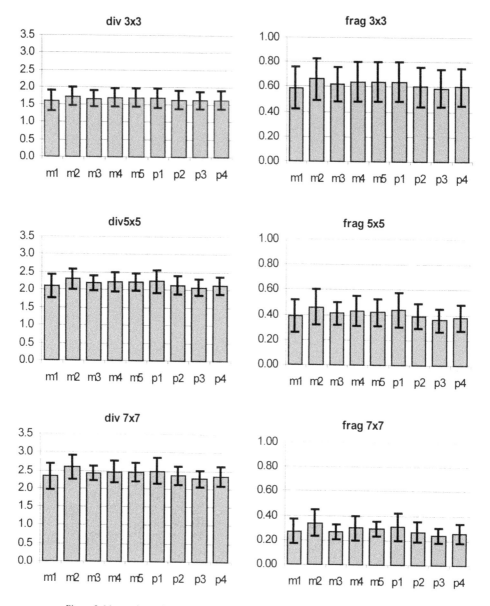

Figure 8. Mean value and standard error of diversity and fragmentation indices of site M and P.

trees of different age class, their diversity, and fragmentation were a little bigger than those of analysed plots in the National Park (Figure 8).

On the presented histograms there was again dissimilarity of P1 site from the rest of areas in the National Park, and at the same time its considerable similarity to sample plots covering intensively used forests. Among all areas which belonged to the last group of objects the biggest similarity, concerning diversity and fragmentation, to sample plots in the National Park was characteristic for M1 site.

From the analyses of histograms it was also concluded that when the area of a square was enlarged (one after the other 9, 25 and 49 pixels) there was the increase of diversity and decrease of fragmentation.

5. CONCLUSION

From the comparisons of ranges and average index values of vegetation on separate areas and historical cartographic materials it was concluded that NDVI corresponded with the way the land was used. On the forest areas situated in northern part of Zabory Landscape Park where M1-5 sites were defined, there was more pine forests restored on post-agricultural areas. Considering species composition they related to dry forests. This similarity resulted mainly from smaller share of dwarf shrubs, and high frequency and cover of overground lichens. Substantial area of this part of the landscape park was covered by *Pinus-Calluna* community marked by Boiński (1988). Heath forests were created mainly because of reforestations of poor pastures between XIX and XX ages. In this community besides common heather, moss and lichen were very important. Other forest groups dominated in southern part of the analysed area, which was the place where the National Park was created. Because this area was not agricultural in the past, and degeneration of forest communities was connected mainly with the introduction of pine monocultures for clear cutting, today fresh pine forest were dominating here, and they were classified as *Leucobryo-Pinetum* As it was presented by Barcikowski (1992, 1996), those forests were characterised by higher level of green biomass and chlorophyll content than dry forests. It was connected not only with mature treestands, but also with all analogical development phases in secondary forest succession in those habitats. Because of lower green biomass and lower chlorophyll content in one area unit, noticed in forests similar to dry forests, NDVI in formerly arable complexes, in which such phytocenoses were dominating, was lower. It was established with remote sensing methods and described in separate paper by Kunz et al. (2000).

Achieved average NDVI values from 0.35 to 0.45 are approximate to data given for plantations of other pine species in different parts of the world. Values gained for Zabory Landscape Park are approximate to 0.3 – 0.5 values, which were presented for *Pinus densiflora* plantations in Japanese mountain regions by Lee and Nakane (1997). Gholtz et al. (1997) analysing spectral answer of *Pinus elliotti* var. *elliotti* in Florida, USA, gained higher values 0.55-0.66. So high NDVI values in Zabory Landscape Park are reached only by forests with older treestands, and big density in fresh pine forest habitat.

From data presented by Barcikowski (1992, 1996) it can be concluded that presence of clear cuttings and pine cultivation of several years were factors that mostly influenced lower NDVI values in M sites, covering intensively exploited forests. Biomass and chlorophyll index in such ecological patterns was very low. In phytocenosis changes of both parameters depending from treestands age were low for about 20 years. That was why in the area of today's national park, where age diversity concerned first of all III-VI classes, there is present not only higher NDVI value but also lower diversity and fragmentation of landscape. In sites covering intensively exploited forests because of the presence of many clear cutting fields and plantations of several years, NDVI decreased, but diversity and fragmentation increased. Changes direction of last two parameters is according to calculations, which were made on theoretical models of forest landscape structure with different share and distribution of clear cuttings by Franklin and Forman (1987).

In comparison of landscape, made in this paper, a satellite image from 1990 was used. State existing over five years before Tuchola Forest National Park was established, which was the period when clear cuttings still took place. After national park establishment wood cuttings were limited, and heavy clear cuttings did not take place at all. This is why the differences between that part of analysed forest complex and its northern part became bigger, although in the areas of intensive forest economy clear cuttings were limited, and if they took place then biogroups of trees, which were seed producers, were left in the areas. If the national park establishment decreased diversity and fragmentation of forest landscape in this part of Tuchola Forest, and what were the results of new forest technologies could be defined by the comparison of satellite images from 1990 and 2000.

REFERENCES

Barcikowski A., 1992. Differentiation in the structure and energy flow in phytocenoses with primary and secondary succession. In: R. Bohr, A. Nienartowicz, J. Wilkoń-Michalska (eds.), *Some ecological processes of biological systems in North Poland.* N. Copernicus University Press, Toruń.

Barcikowski A., 1996. Biomass and chlorophyll of photo-synthesizing organs of plant communities in secondary succession in pine forest habitat. *Photosynthetica* 32: 63-76.

Boiński M., 1988. Roślinność rzeki Kulawej. *Acta Univ. N. Copernici, Biologia* 32, *Nauki Mat.-Przyr.* 69: 73-95.

Franklin J.F., Forman R.T.T., 1987. Creating landscape patterns by forest cutting: Ecological consequences and principles. *Landscape Ecology* 1(1): 5-18.

Gholz H.L., Curran P.J., Kupiec J.A., Smith G.M., 1997. Assessing leaf area and canopy biochemistry of Florida pine plantations using remote sensing. In: Gholz H.L., Nakane K., Shimoda H. (eds.), *The use of remote sensing in the modelling of forest productivity,* pp 3-22. Kluwer Academic Publishers, Dodrecht Boston London.

Hauser M., Mucina L., 1991. Spatial interpolation methods for interpretation of ordination diagrams. In: E. Feoli and L. Orloci (eds.), *Computer Assisted Vegetation Analysis,* pp. 299-316. Kluwer Academic Publishers, Dordrecht/Boston/London.

Kovach W.L., 1993. *MVSP- A Multi Variate statistical Package for IBM PC's, version 2.1.* Kovach Computing Services. Pentraeth. Wales, U.K.

Kunz M., Nienartowicz A., Deptuła M., 2000. The use of satellite remote sensing imagery for detection of secondary forests on post-agricultural soils: A case study of Tuchola Forest, Northern Poland. In: J.L. Casanova (ed.), *Remote Sensing in the 21st Century. Economic and Environmental Applications*, pp. 61-66. Proceedings of the 19th EARSeL Symposium on Remote Sensing in the 21st Century/Valladolid/Spain/ 31 May – 2 June 1999. A.A. Balkema/Rotterdam/Blookfield/2000.

Lee N.J., Nakane K., 1997. Forest vegetation classification and biomass estimation based on Landsat TM data in a Mountain Region of West Japan. In: Gholz H.L., Nakane K., Shimoda H. (eds.), *The use of remote sensing in the modelling of forest productivity*, pp. 159-172. Kluwer Academic Publishers, Dodrecht Boston London.

Orlóci L., 1978. *Multivariate Analysis of Vegetatio Research.* Dr W. Junk Publishers, The Hague Boston.

Sokołowski A., 1965. Zespoły leśne Nadleśnictwa Laska w Borach Tucholskich. *Fragmenta Floristica et Geobotanica* 11: 96-119.

Wilkoń-Michalska J., Nienartowicz A., Kunz M., Deptuła M., 1999. Old land-use maps as a basis for Interpreting of the contemporary structure of forest communities in Zabory Landscape Park. *Phytocoenosis* 11: 139-154.

398

Miscellaneous

List of participants

ABDUL-Qadar Amal Mrs.
University of Valladolid
LATUV - Dpto. Fisica Aplicada I
47005 Valladolid, SPAIN
Fax: +34 983 423 952
jois@latuv.uva.es

AHOKAS Eero
Finnish Geodetic Institute
Geodeetinrinne 2
02431 Masala, FINLAND
Fax: +358 9 295 55200
Eero.ahokas@fgi.fi

ALIASGHARI Robert
RESEARCH SYSTEMS
105 rue Marcel Dassault
92774 Boulogne-Billancourt Cedex, France
Fax: +33 01 46 94 92 09
france@rsinc.com

AL-RAWI Kamal
University of Valladolid
LATUV - Dpto. Fisica Aplicada I
47005 Valladolid, SPAIN
Fax: +34 983 423 952
kamal@latuv.uva.es

ANTOHE Carmen Ms.
Universitatea Valahia
B-dul Regele Carol I nr.2
0200 Targoviste, ROMANIA
Fax: +40 45 2176 92
vloghin@valahia.ro

ARINO Olivier Dr.
ESA/ESRIN
Via Galileo Galilei
00044 Frascati (Rome), ITALY
Fax: +39 06 94180552
Olivier.arino@esa.int

BARACHE Damien
RESEARCH SYSTEMS
105 rue Marcel Dassault
92774 Boulogne-Billancourt Cedex, France
Fax: +33 01 46 94 92 09
france@rsinc.com

BARTLETTE Roberta A.
USDA, Forest Service
Rocky Mountain Research Station
Fire Sciences Laboratory, POB 8089
Missoula, Montana 59807, U.S.A.
Fax: +1.406 329 4820
rbartlette@fs.fed.us

BEGNI Gérard Dr.
MEDIAS-France
Bpi 2102
18 avenue E. Belin
31401 Toulouse Cedex 4, FRANCE
+33 (0) 05 61 28 2905
begni@medias.cnes.fr

BENES Tomas Dr.
UHUL Forest Management Institute
Nabrezni 1326
250 44 Brandys-nad-Labem, CZECH REPUBLIC
Fax: +420 202 803371
benes@uhul.cz

BIELECKA Elzbieta
Institute of Geodesy & Cartography (IGIK)
Ul. Jasna 2/4
00-950 Warsaw, POLAND
Fax: +48 22 827 03 28
elab@igik.edu.pl

BRIGGS Stephen Dr.
ESA/ESRIN
Via Galileo Galilei
00044 Frascati (Roma), ITALY
sbriggs@esrin.esa.it

BRUZZI Stefano Dr.
ESA, EO Applications Programmes
8-10 rue Mario Nikis
75738 Paris Cedex 15 France
Fax: +33 (0)1 5369 7674
sbruzzi@esa.int

BUCHROITHNER Manfred Prof.
Institute of Cartography,
Technical University Dresden
Mommsenstr. 13
D-01062 Dresden, GERMANY
Fax: +49 351 463 7028
buc@karst9.geo.tu-dresden.de

BUSCH Andreas Dr.
BKG Federal Agency for Cartography and Geodesy
Richard-Strauss Allee 11
60598 Frankfurt-am-Main, GERMANY
Fax: +49 69 6333 441
busch@ifag.de

CANNIZZARO Giovanni
TELESPAZIO
Via Tiburtina 965
I- 00156 Rome, ITALY
Fax: +39 064 0793638
Giovanni_cannizzaro@telespazio.it

CASANOVA José-Luis Prof.
University of Valladolid
LATUV, Dpto. Fisica Aplicada I
47005 Valladolid, SPAIN
Fax: +34 983 423 952
jois@latuv.uva.es

COOLS June Ms.
VITO
Boeretang 200
Mol, BELGIUM
Fax: +32 14 33 55 97
June.cools@vito.be

CORREA Aderbal Dr.
University of Missouri-Columbia
Civil & Environmental Engineering Dept., E2509 EBE
Columbia, Missouri 65211, USA
Fax: +1 573 882 4784
correaA@missouri.edu

COSKUN Gonca Dr.
T. U. Istanbul
Civil Engineering Faculty
Remote Sensing Division
80626 Maslak/Istanbul, TURKEY
Fax: +90 212 285 6587
gonca@itu.edu.tr

COSTA Nina Dr.
Joint Research Centre EC
SAI Institute, SSSA Unit
Building. 26.A
I-21020 Ispra (Va), ITALY
Fax: +39 0332 78 5461
Nina.costa@jrc.it

De FILIPPIS T.
CeSIA/IATA-CNR
Via Caproni 8
I-50145 Firenze, ITALY
Fax: +39 055 308910
filippis@iata.fi.cnr.it

DENEGRE Jean Director,
Ecole Nationale des Sciences Géographiques
6-8 ave. Blaise Pascal
Cité Descartes, Champs sur Marne
77455 Marne-la-Vallée, FRANCE
Fax: +33 01 64 15 3107

DJABY Bakary
AGRHYMET Regional Centre
B.P. 11011
Niamey, NIGER
Fax:+227 732 435
djaby@sahel.agrhymet.ne

DUPERET Alain
Institut Géographique National
6-8 avenue Blaise Pascal
Cité Descartes, Champs sur Marne
77455 Marne-la-Vallée, FRANCE
Fax: +33 (0)1 64 15 31 07

EGELS Yves.
Institut Géographique National
6-8 avenue Blaise Pascal
Cité Descartes, Champs sur Marne
77455 Marne-la-Vallée, FRANCE
Fax: +33 (0)1 64 15 31 07

ERASMI Stefan Dr.
University of Goettingen
Cartography, GIS & RS
Goldschmidststr. 5 D
37077 Goettingen, GERMANY
Fax: +49 551 39 8071
serasmi@gwdg.de

FELLOUS Jean-Louis Dr.
Centre National d'Etudes Spatiales
2 Place Maurice Quentin
75039 PARIS Cedex 01, FRANCE
+33 1 44 76 76 76

FICHAUX Nicolas
Groupe Télédétection et Modélisation
Ecole des Mines de Paris
BP 207
06904 Sophia Antipolis, FRANCE
Fax: +44 (0)4 9395 7535
Nicolas.fichaux@ensmp.fr

FLATI Antonio
ESA/ESRIN
Via Galileo Galilei
00044 Frascati, ITALY
Fax: +39 06 94180 632
Antonio.flati@esa.int

GIRAUD Anne
GEOMEDITERRANEE
80, route des Lucioles
06560 Valbonne, FRANCE
Fax: +33 (0)4 93 00 40 01
geomed@geomediterranee.com

GOEMAERE Mr.
VITO
Boeretang 200
Mol, BELGIUM
Fax: +32 14 33 55 97
Stijn.geomaere@vito.be

GOITA Kalifa Dr.
Université de Moncton   Campus d'Edmundston
Ecole de Sciences Forestières
165 Boulevard Hébert, Edmundston
New Brunswick, CANADA E3V 2S8
Fax: +1.506 737 5373
kgoita@cuslm.ca

GOKSEL Cigdem Dr.
Istanbul Technical University
Civil Engineering Faculty
Remote Sensing Division
80626 Maslak/Istanbul, TURKEY
Fax: +90 212 285 6587
goksel@itu.edu.tr

GOOSSENS Rudi Prof.
Dept. Geography, University of Gent
Krijslaan 281 (S8-A1)
B-9000 Gent, BELGIUM
Fax: +32 9 264 4985
Rudi.goossens@rug.ac.be

GRUJARD I.
Société Française de Photogrammétrie et de
Télédétection, c/o IGN
6-8 ave. Blaise Pascal
Cité Descartes, Champs sur Marne
77455 Marne-la-Vallée, FRANCE
Fax: +33 (0)1 64 15 32 85
sfpt@ensg.ign.fr

GUDMANDSEN Preben Prof. (em.)
Technical University Denmark
Building 348
DK-2800 Lyngby, DENMARK
Fax: +45 45 931634
pg@oersted.dtu.dk

HALLIKAINEN M. Prof.
Helsinki University of Technology
Laboratory of Space Technology
P.O. Box 3000
02015 HUT, FINLAND
Fax: +358 9 4512898
Martti.hallikainen@hut.fi

HARBI Mohamed
Consultant ITU
B.P. 820
Place des Nations
CH-1211 Geneva 20, SWITZERLAND
Fax: +41 22 7883602
harbi@itu.int

HENNINGS Ingo Dr.
GEOMAR Forschungszentrum
Wischhofstr. 1-3
D-24148 Kiel, GERMANY
Fax: +49 431 600 2926
ihennings@geomar.de

HOFMANN Alexandra D.
Inst. of Photogrammetry & Remote Sensing
Dresden University of Technology
Mommsenstr. 13
D-01062 Dresden, GERMANY
Fax: +49 351 463 7266
Alexandra.hofmann@mailbox.tu-dresden.de

IBRAHIM Hussein Dr., Director General
General Organization of Remote Sensing
P.O. Box 12586
Damascus, SYRIA
Fax: 963 11 3910700

ICHOKU Charles Dr.
Science Systems & Applications, Inc.
Building 33, Room E306
NASA/GSFC (Code 913)
Greenbelt, MD 20771, USA
Fax: +1.301 614 6307
ichoku@climate.gsfc.nasa.gov

JAKOMULSKA Anna Dr.
University of Warsaw
Remote Sensing of Environment Laboratory
Faculty of Geography & Regional Studies
Ul. Krakowskie Przedmiescie 30
00-927 Warsaw, POLAND
Fax: +48 22 55 21 521
Anna.jakomulska@mercury.ci.uw.edu.pl

KAYA Sinasi Dr.
Istanbul Technical University
Civil Engineering Faculty, Remote Sensing Division
80626 Maslak/Istanbul, TURKEY
+90 212 285 6587
skaya@srv.ins.itu.edu.tr

KING Christine Dr.
BRGM – ARN
3 avenue Claude Guillemin
BP 5009
45060 Orléans Cedex 2, FRANCE
Fax: +33 (0)2 38 64 33 99
c.king@brgm.fr

KISSIYAR Ouns
University of Bonn
Remote Sensing Research Group
Meckenheimer Allee 166
53115 Bonn, GERMANY
Fax: +49 228 73 48 63
ouns@rsrg.uni-bonn.de

KONECNY Gottfried Prof.(em.) Dr.
University of Hannover
Inst. f. Photogrammetry & Engineering Surveys
Nienburgerstr. 1
D-30167 Hannover, GERMANY
Fax: +49 511 762 2483
gko@ipi.uni-hannover.de

KRISHNAMOORTHY R. Dr.
University of Madras
Dept. of Applied Geology
School of Earth & Atmospheric Sciences
A.C. College Buildings, Guindy Campus
Chennai 600025, INDIA
Fax: +91 44 235 1870
Krish_r_46@hotmail.com

KUDASHEV Efim Prof.
Space Research Institute
84/32 Profsouznaya str.
117810 Moscow, RUSSIA
Fax: ++7 095 1056
eco@iki.rssi.ru

KUNZ Mieczyslaw Dr.
Nicolas Copernicus University
Institute of Geography
Laboratory of Remote Sensing and Cartography
Gagarina 5
87-100 Torun, POLAND
Fax: +48 56 62 273 08
met@cc.uni.torun.pl

LACAUX Jean-Pierre Dr.
MEDIAS-France
Bpi 2102
18 avenue E. Belin
31401 Toulouse Cedex 4, France
Fax: +33 (0) 05 61 28 2905
begni@medias.cnes.fr

LAMBERTI Fiorella Ms.
DATAMAT SpA
Via Laurentina 760
I-00143 Rome, ITALY
Fax: +39 06 5927 4500
flambert@datamat.it

LANDRY Robert Dr.
Canada Centre for Remote Sensing
588 Booth Street
Ottawa, ON K1A OY7, CANADA
Fax: +1 613 947 1385
Robert.landry@ccrs.nrcan.gc.ca

LANEVE Giovanni
University of Rome
Via Salaria 851
Rome, ITALY
Fax: +39 06 8106351
laneve@psur.uniroma1.it

LASAPONARA Rosa
IMAAA-CNR/DIFA-
Un. Della Basilicata
C/da S. Loja
Tito Scalo
Potenza, ITALY
Fax: +39 0971 427271+39
lasaponara@imaaa.pz.cnr.it

LAURORE Catherine
GEOIMAGE
80, route des Lucioles
06560 Valbonne, France
+33 (0)4 93 00 40 01
com@geoimage.fr

LOGHIN Vasile Mr.
Universitatea Valahia
B-dul Regele Carol I nr.2
0200 Targoviste
ROMANIA
Fax: +40 45 2176 92
vloghin@valahia.ro

LUPO Fréderick
Laboratoire de Télédétection
Dept. Geography, U. C. Louvain-la-Neuve
Bâtiment Mercator
Place Louis Pasteur, 3
B-1348 Louvain-la-Neuve, BELGIUM
Fax: +32 010 47 27 77
lupo@geog.ucl.ac.be

MARINI Alberto Prof.
University Cagliari, Lab. TeleGIS
Via Trentino, 51
09127 Cagliari, ITALY
Fax: +39 070 675 7735 07
marini@unica.it

MASSE Bernard
WEU Satellite Centre
Apdo. de Correos 511
E-28850 Torrejon de Ardoz, SPAIN
Fax: +34 91 678 6006
info@weusc.es

MATHER Paul Prof.
Department of Geography
University of Nottingham
Nottingham NG7 2RD, England, UK.
Fax: +44 115 951 5249
Paul.mather@ntlworld.com

MBOW Cheikh
University Cheikh Anta Diop de Dakar
Institut des Sciences de l'Environnement
Faculté des Sciences et Techniques
Dakar, SENEGAL
Fax: +221 825 4821
enrecada@telecomplus.sn

McCLOY Keith Dr.
Danish Institute of Agricultural Sciences
P.O. Box 50
8830 Tjele, DENMARK
Fax: +45 8999 1819
Keith.McCloy@agrisci.dk

MENDAS Abdelkader Mr.
Laboratoire de Géomatique, CNTS
BP 13 - Arzew
13200 Oran, ALGERIA
Fax: +312 41 47 34 54
amendas@usa.net

MENZ Gunter Prof.Dr.
University of Bonn
Remote Sensing Group
Meckenheimer Allee 166
D-53115 Bonn, GERMANY
+49 1 228 739702
menz@rsrg.uni-bonn.de

MUSAOGLU Nebiye Dr.
Istanbul Technical University Civil Engineering Faculty
Remote Sensing Division
80626 Maslak/Istanbul, TURKEY
Fax: +90 212 285 6587
nmusaaoglu@srv.ins.itu.edu.tr

OLUIC Marinko Dr.
GEOSAT
Poljana B. Hanzekovica 31
Zagreb, CROATIA
Fax: +385 1 383 3910
Geo-sat@zg.tel.hr

PAGANINI Marc Dr.
ESA/ESRIN
Via Galileo Galilei
00044 Frascati (Rome), ITALY
Fax: +39 06 94180552
Marc.paganini@esa.int

PARLOW Eberhard Prof. Dr.
MCR Laboratory
Institute of Geography, University of Basel
145 Spalenring
CH-4055 Basel, SWITZERLAND
Fax: +41 61 272 6923
Eberhard.parlow@unibas.ch

PAVO LOPEZ Marcos Francisco
Instituto Geografico Nacional
Remote Sensing Unit
C/General Ibanez de Ibero 3
28003 Madrid, SPAIN
Fax: +34 91 5979770
aarozarena@mfom.es

PETERI Renaud
Groupe Télédétection et Modélisation
Ecole des Mines de Paris
BP 207
06904 Sophia Antipolis, France
Fax: 33 (0)4 9395 7535
Renaud.peteri@ensmp.fr

PIKEROEN Bernard
THALES ISR
3 rue Ampère
91349 Massy Cedex, France
Fax: +33 (0)1 69 76 21 95

PIPPI Ivan Dr.
CNR-IROE
Via Panciatichi 64
50127 Florence, ITALY
+39 055 410893
pippi@iroe.fi.cnr.it

POETE Peter
University of Bonn
Remote Sensing Group
Meckenheimer Allee 166
D-53115 Bonn, GERMANY
Fax: +49 1 228 739702
poete@rsrg.uni-bonn.de

POGGESI Marco Dr.
CNR-IROE
Via Panciatichi 64
50127 Florence, ITALY
Fax: +39 055 410893
pippi@iroe.fi.cnr.it

POGLIO Thierry
Groupe Télédétection et Modélisation
Ecole des Mines de Paris
BP 207
06904 Sophia Antipolis, France
+33 (0)4 9395 7535
Thierry.poglio@ensmp.fr

POULIT M., Director General
Institut Géographique National
136bis, rue de Grenelle
75700 PARIS 07 SP, France

POZO Theilen-del Jutta
Space Applications Institute
JRC-EC-TP 263
Via E. Fermi
I-21020 Ispra (Va),   ITALY
Fax: +39 0332 785500
Jutta.thielen@jrc.it

PREUX Daniel
Ecole Nationale des Sciences Géographiques
6-8 ave. Blaise Pascal
Cité Descartes, Chalps sur Marne
77455 Marne-la-Vallée, France
+33 01 64 15 3107

RABAUTE Thierry
SCOT Conseil
8-10, rue Hermès
31526 Ramonville Cedex, France
Fax: +33 (0)5 61 39 46 10
Theirry.rabaute@svot.cnes.fr

REUTER Rainer Dr.
University Oldenburg
Fach. Physik
D-26111 Oldenburg, GERMANY
Fax: +49 441 798 3201
r.reuter@las.physik.uni-oldenburg.de

RICHTERS Jochen
University of Bonn
Remote Sensing Group
Meckenheimer Allee 166
D-53115 Bonn, GERMANY
Fax: +49 1 228 739702
richters@rsrg.uni-bonn.de

ROBIN Cécile Ms.
RESEARCH SYSTEMS
105 rue Marcel Dassault
92774 Boulogne-Billancourt Cedex, France
Fax: +33 01 46 94 92 09
france@rsinc.com

ROCHON Gilbert Dr.
Sustainable Environments Branch
U.S. EPA
A.W. Breidenbach Environmental Research Center
26 W. Martin Luther King drive, Mail Stop 497
Cincinnati, Ohio 45268, USA
Fax: +1.513 569 7111
Rochon.gilbert@epamail.epa.gov

ROSSI Federico
DATAMAT
Via Laurentina, 760
00143 Rome, ITALY
Fax: +39 06 5027 4500
fedro@datamat.it

ROTT Helmut Dr.
Inst. f. Meteorologie & Geophysik
University Innsbruck
Innrain 52
A-6020 Innsbruck
AUSTRIA
Fax: +43 512 507 2924
Helmut.rott@uibk.ac.at

SABATIER Philippe Dr.
INRA – Ecole Vetérinaire de Lyon
1 Avenue Bourgelat
69280 Marcy L'Etoile, FRANCE
Fax: +33 04 78 87 26 67
sabatier@clermont.inra.fr

SACAU-CUADRADO Ma.. del MAR
University of Vigo
Lagoas-Marcosende
36200 Vigo, SPAIN
Fax: +34 986812556
marsacau@uvigo.es

SCHAEPMAN Michael.Dr.
Remote Sensing Laboratories
Dept. of Geography, University of Zurich
Winterthurerstr. 190
CH-8057 Zurich, SWITZERLAND
+41 1 635 6846
schaep@geo.unizh.ch

SCHMIDT Michael
University of Bonn, Remote Sensing Group
Meckenheimer Allee 166
D-53115 Bonn, GERMANY
Fax: +49 1 228 739702
Michael@rsrg.uni-bonn.de

SCHMITT Ursula Dr.
Joanneum Research
Institute of Digital Image Processing
Wastiangasse 6
A-8010 Graz, AUSTRIA
Fax: +43 316 876 1720
ursula.schmitt@joanneum.ac.at

SCHWARZ M.
Swiss Federal Research Institute WSL
Zürcherstrasse 111
CH-8903 Birmensdorf, SWITZERLAND
Fax: +41 1 739 22 15
Markus.Schwarz@wsl.ch

SEIDEL Klaus Dr.
Institute for Communications Techniques
ETH Zurich
Gloriastr. 35
CH-8092 Zurich, SWITZERLAND
Fax: +41 1 632 1251
seidel@vision.ec.eth.ch

SERRE Roger
Ecole Nationale des Sciences Géographiques
6-8 ave. Blaise Pascal
Cité Descartes, Champs sur Marne
77455 Marne-la-Vallée, France
Fax: +33 01 64 15 3107

SHAROV Aleksey Dr.
Institute of Digital Image Processing
Joanneum Research
Wastiangasse 6
A-8010 Graz, AUSTRIA
Fax: +43 316 876 1720
Aleksey.sharov@joanneum.ac.at

SIMONIELLO Tiziana
IMAAA-CNR
Infor-Com-Un. Studi di Roma 'La Sapienza"
C/da S. Loja, Tito Scalo
Potenza, ITALY
Fax: +39 0971 427271
simoniello@imaaa.pz.cnr.it

SMAHI Zakaria
National Centre for Space Techniques (CNTS)
BP 13
Arzew 31200, ALGERIA
Fax: +213 41 47 34 54
smahiz@usa.net

SMITH Alistair
Kings College London
Department of Geography – Room 113N
London WC2R 2LS
England, UK
Fax: +44 207 848 2287
Alistair_smith_kcl@yahoo.co.uk

STAWIECKA Magdalena Ms.
Remote Sensing & Environment Laboratory
University of Warsaw
Ul. Krakowskie Przedmiescie 30
00-927 Warsaw, POLAND
Fax: +48 22 55 21 521
szkocja@mercury.ci.uw.edu.pl

STEELE Caiti Dr.
University Southampton
Department of Geography
Southampton SO17 1BJ
England, UK
Fax:+44 23 8059 3295
Caiti.steele@soton.ac.uk

SUBA Zsuzsanna Ms.
FÖMI Remote Sensing Centre
Bosnyak tér 5
H-1149 Budapest, HUNGARY
Fax: +361 252 8282
Zsuba@rsc.fomi.hu

TALBI M. Dr.
Institut des Régions Arides
Remote Sensing and Geo-Information Laboratory
Médenine 4119, TUNISIA
Fax: +216 5 633 006
Mohamed.talbi@ira.rnrt.tn

TALBI Okacha
National Centre of Spatial Techniques (CNTS)
BP 13
Arzew 21300, ALGERIA
Fax: +213 41 47 34 54
otalbi@usa.net

THAMM Hans-Peter Dr.
University of Bonn, Remote Sensing Research Group
Meckenheimer Allee 166
53115 Bonn, GERMANY
Fax: +49 228 73 48 64
thamm@rsrg.uni-bonn.de

TORRES-PALENZUELA M. Jesus
University of Vigo
Lagoas-Marcosende
36200 Vigo, SPAIN
Fax: +34 986812556
jesu@uvigo.es

UNG Anthony
Groupe Télédétection et Modélisation
Ecole des Mines de Paris, BP 207
06904 Sophia Antipolis, FRANCE
Fax: +44 (0)4 9395 7535
Anthony.ung@ensmp.fr

Van LEEUWEN W.
METEO-FRANCE
42, ave. G. Coriolis
31057 Toulouse, FRANCE
Fax: +33 5 61 07 96 26
Willem.vanLeeuwen@cnrm.meteo.fr

VANDERSTRAET Tony
Department of. Geography
University of Gent
Krijslaan 281 (S8-A1)
B-9000 Gent, BELGIUM
Fax: +32 9 264 4985
Tony.vanderstraete@rug.ac.be

VASILEISKI Alexandre
University of Valladolid
LATUV - Dpto. Fisica Aplicada I
47005 Valladolid, SPAIN
Fax: +34 983 423 952
asvas@wildcat.i&i.rssi..ru

VASILIEV Leonid
Institute of Geography
Russian Academy of Sciences
Staromonetny 29
Moscow 109017, RUSSIA
Fax: +7 095 959 0033
vasiliev@igras.geonet.ru

VASQUES DE LA CUEVA Antonio Dr.
CIFOR-INIA
Crta. De la Coruna km 7
28040 Madrid, SPAIN
Fax: +34 91 3572293
vazquez@inia.es

VAUGHAN Robin Dr.
Centre for RS and Environmental Monitoring
University of Dundee
Dundee DD1 4HN, Scotland, U.K.
Fax: +44 1382 345415
r.a.vaughan@dundee.ac.uk

VESCOVI Fabio Dr.
University of Bonn
Remote Sensing Research Group
Meckenheimer Allee 166
D-53115 Bonn, GERMANY
Fax: +49 228 731889
Fabio@rsrg.uni-bonn.de

WALD L. Dr.
Ecole des Mines de Paris
Groupe Télédétection et Modélisation
BP 207
06904 Sophia Antipolis Cedex, FRANCE
Fax +33 (0)4 9395 7535
Lucien.wald@ensmp.fr

WIDEN Nina
Finnish Geodetic Institute
Geodeetinnrinne 2
02430 Masala, FINLAND
Fax: +358 9 295555200
Nina.widen@fgi/fi

WINKLER Peter Dr.
FÖMI Remote Sensing Centre
Bosnyak tér 5
H-1149 Budapest, HUNGARY
Fax: +361 252 8282
Peter.winkler@rsc.fomi.hu

WIRNHARDT Csaba Mr.
FÖMI Remote Sensing Centre
Bosnyak tér 5
H-1149 Budapest, HUNGARY
Fax: +361 252 8282
Cs.Wirnhardt@rsc.fomi.hu

WOLK-MUSIAL Elzbieta Dr.
University of Warsaw
Remote Sensing of Environment Laboratory
Faculty of Geography & Regional Studies
Ul. Krakowskie Przedmiescie 30
00-927 Warsaw, POLAND
Fax: +48 22 55 21 521
Anna.jakomulska@mercury.ci.uw.edu.pl

YAKAM SIMEN Francis
PRIVATEERS N.V.
8 rue Delayrac
31000 Toulouse, FRANCE
Fax: +33 (0)5 61 99 1724
106341.2602@compuserve.com

ZAGAJEWSKI Bogdan
University of Warsaw
Remote Sensing of Environment Laboratory
Faculty of Geography & Regional Studies
Ul. Krakowskie Przedmiescie 30
00-927 Warsaw, POLAND
Fax: +48 22 55 21 521
bzag@wgsr.uw.edu.pl

ZEGRAR Ahmed
Centre National de Techniques Spatiales
1, avenue de la Palestine, BP 13
31200 Arzew, ALGERIA
Fax: +213 6 41 473 454
zeg-ahm@usa.net

ZOMER R. Dr.
International Centre for Research in Agroforestry
P.O. Box 30677
United Nations Avenue, Gigiri
Nairobi, KENYA
Fax: +254 2 524 001
z.zomer@cgiar.org

ZOUMAS Athanassios
Department of Geography
King's College London
Strand, London WC2R 2LS, England, UK
Fax: +44 207 848 2287
Athanassios.zoumas@kcl.ac.uk

Author index

Colour plates

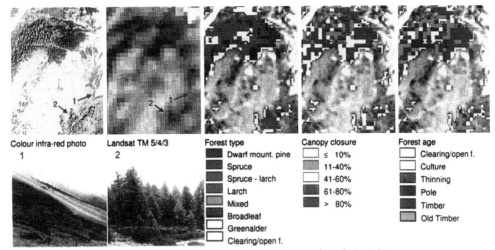

Colour infra-red photo
1

Landsat TM 5/4/3
2

Forest type
- Dwarf mount. pine
- Spruce
- Spruce - larch
- Larch
- Mixed
- Broadleaf
- Greenalder
- Clearing/open f.

Canopy closure
- ≤ 10%
- 11-40%
- 41-60%
- 61-80%
- > 80%

Forest age
- Clearing/open f.
- Culture
- Thinning
- Pole
- Timber
- Old Timber

Plate 1. Dwarf mountain pine (1) and larch (2) in a selected area at the south face of the Dachstein massif and forest classification results superimposed to panchromatic SPOT image. (Schmitt et al., see p. 43)

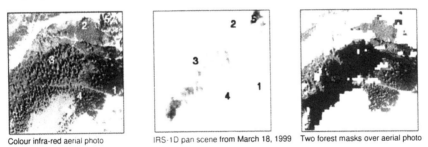

Colour infra-red aerial photo   IRS-1D pan scene from March 18, 1999   Two forest masks over aerial photo

Plate 2. Forest border based on summer Landsat TM and SPOT data (all coloured areas in right image) compared to forest border derived from combined winter IRS panchromatic and summer SPOT data (green in right image): 1 = dwarf mountain pine, 2 = greenalder, 3 = Larch stand with bushes, 4 = larch with very low crown closure + dwarf mountain pine, 5 = rock. (Schmitt et al., see p. 43)

Plate 3. Distribution and crown closure of forest on slopes steeper than 30%, superimposed to SPOT image: Crown closures 11-30% = red, 31-60% = yellow, >60% = green. (Schmitt et al., see p. 43)

413

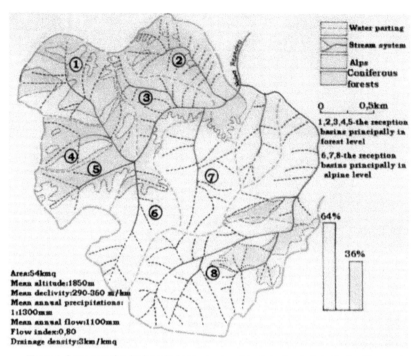

Figure 4. Some reception basins principally developed in alpine level of the Rodna Mountamis. Torrential drainage is responsible for the intense erosion and for the flood generated by the rivers from the Maramures Depression (Viseu, Iza), Tisa's tributaries. (Loghin & Antohe, see p. 50)

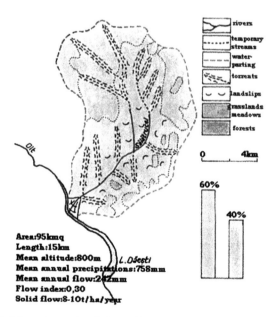

Figure 6. A deforested secondary hydrographic basin – Salatrucel. The geornorphologic consequences: mass movements (landslips, land fall) torrential erosion, alluvial transfer in collector river (Olt), colmatage of the retention lakes. (Loghin & Antohe, see p. 51)

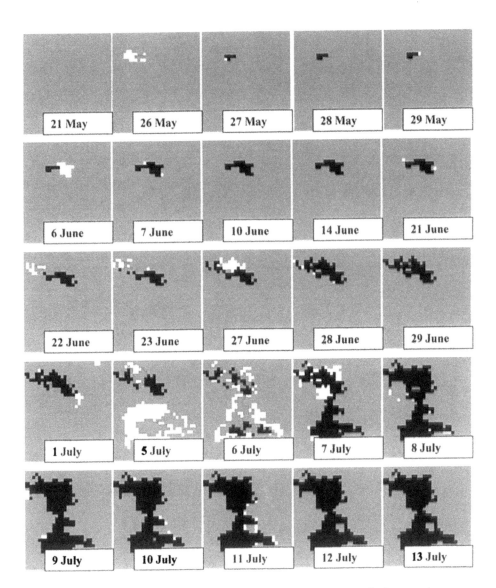

Figure 4. Fire monitoring using IFEMS. Red, black, white, and yellow represent fire front, area that burned completely, at the time of the current image, area that burned between the time of the previous image and the current image, and area that are located beneath the flames, respectively. (Al-Rawi et al., see p. 99)

Figure 1. True-colour (Red: 11th channel, Green: 6th channel, Blue: 1st channel) image acquired by the MIVIS over the Alps (Italy) and showing a natural fire affecting a large wood. The fire-front and the burned areas may be identified even in this visible picture, however the smoke spread around heavily dims the visibility and sometimes hinders the fire.(Barducci et al., see p. 104)

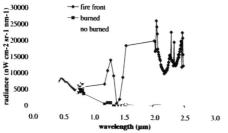

Figure 6. MIVIS spectra collected by the MIVIS over the fire-front, the burned area and the unaffected soil in the wavelength range 0.4 μm – 2.5 μm. The main difference between these spectra is due to the strong black-body emission starting from about 900 nm of wavelength. This strong emission is however modulated by the noticeable extinction produced by the above cool atmospheric (inversion) layer.(Barducci et al., see p. 106)

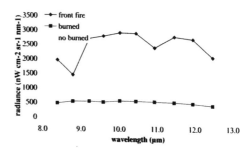

Figure 7. MIVIS spectra collected by the MIVIS over the fire-front, the burned area and the unaffected soil in the thermal infrared spectral range. The main difference between these spectra is due to the strong black-body emission produced by the fire that, however, is only partially appreciated at these wavelengths. This emission seems to be modulated by two absorption features that could be produced by O_3. (Barducci et al., see p. 106)

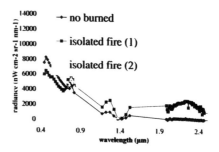

Figure 9. Spectra collected by the MIVIS over the two isolated fire spot indicated in the previous Figure8 as spot (1) and spot (2), in comparison with spectra of an unaffected soil in the visible and near infrared spectral range.(Barducci et al., see p. 107)

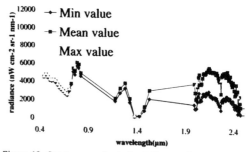

Figure 12. Spectra over the isolated fire spot indicated in the previous Figure8 as spot (1) and averaged upon four pixels around the spot fire. Statistical elaboration is also showed. (Barducci et al., see p. 108)

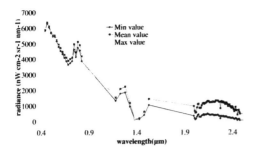 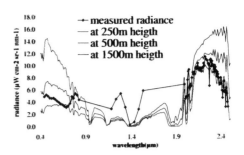

Figure 13. Spectra over the isolated fire spot indicated in the previous Figure8 as spot (2) and averaged upon four pixels around the spot fire. Statistical elaboration is also showed. The solar radiance results more intense than the radiance emitted from fire. (Barducci et al., see p. 108)

Figure 14. Spectra simulated with MODTRAN 4 of radiance emerging from a 800K fire at different heights. In the graphic it is also depicted the measured radiance from fire spot(1). (Barducci et al., see p. 109)

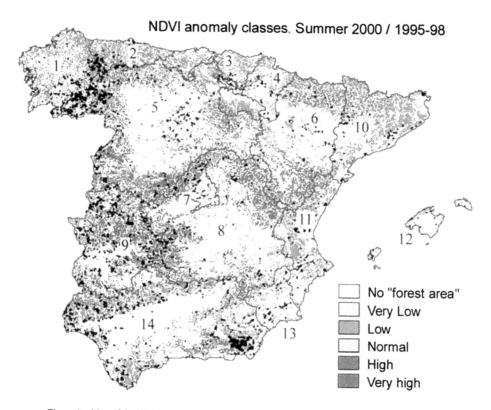

NDVI anomaly classes. Summer 2000 / 1995-98

Legend:
No "forest area"
Very Low
Low
Normal
High
Very high

Figure 1. - Map of the NDVI anomalies for the 2000 summer in relation to the mean value for the 1995-98 (June to September) reference period based on NOAA-AVHRR data distributed by DLR. Red colors indicate negative anomalies and green one positives. Black points represent the pixels with "hot spots" based on the WFW. (Vázquez et al., see p. 112)

Figure 1. Burned area from Kanlisirt. (Musaoglu et al., see p. 118)

1992

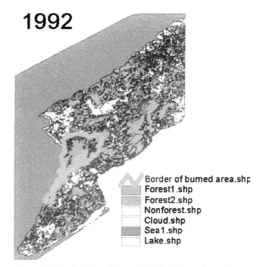

Border of burned area.shp
Forest1.shp
Forest2.shp
Nonforest.shp
Cloud.shp
Sea1.shp
Lake.shp

Figure 4. Distribution of layers (1992). (Musaoglu et al., see p. 119)

Figure 2. Classification result (1992). (Musaoglu et al., see p. 119)

1998

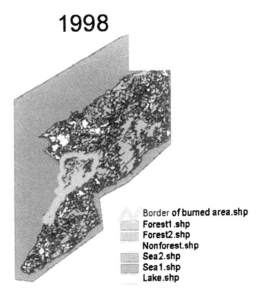

Border of burned area.shp
Forest1.shp
Forest2.shp
Nonforest.shp
Sea2.shp
Sea1.shp
Lake.shp

Figure 3. Classification result (1998). (Musaoglu et al., see p. 119)

Figure 5. Distribution of layers (1998). (Musaoglu et al., see p. 120)

1992

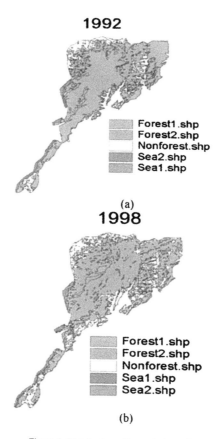

Forest1.shp
Forest2.shp
Nonforest.shp
Sea2.shp
Sea1.shp

(a)

1998

Forest1.shp
Forest2.shp
Nonforest.shp
Sea1.shp
Sea2.shp

(b)

Figure 6. Distribution of layers in burned area
a) 1992 b) 1998. (Musaoglu et al., see p. 120)

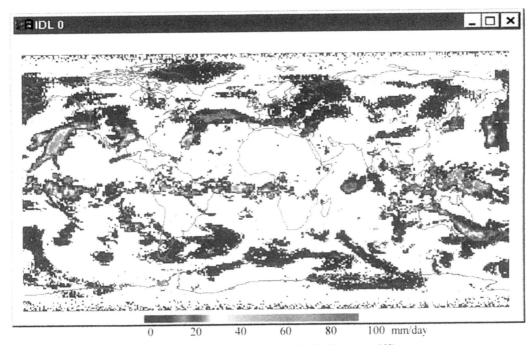

Figure 1. Daily 1 x 1 degree image of global precipitation on 21 May 1997. (Vasiliev, see p. 160)

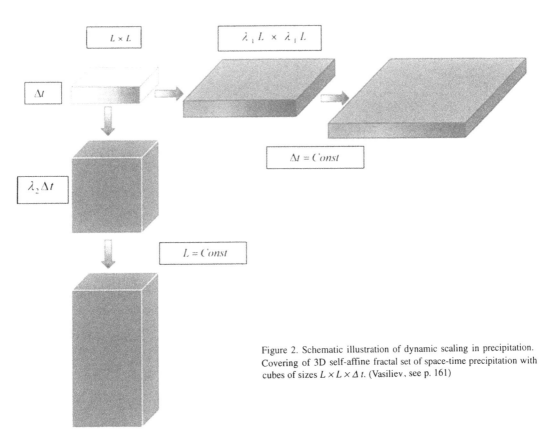

Figure 2. Schematic illustration of dynamic scaling in precipitation. Covering of 3D self-affine fractal set of space-time precipitation with cubes of sizes $L \times L \times \Delta t$. (Vasiliev, see p. 161)

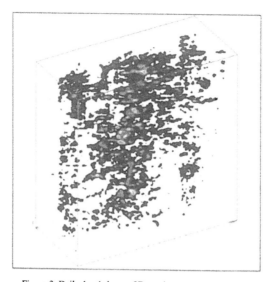

Figure 3. Daily 1 × 1 degree 3D spatio-temporal pattern of precipitation balls on a global scale during 1997. (Vasiliev, see p. 162)

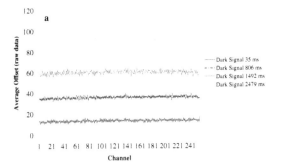

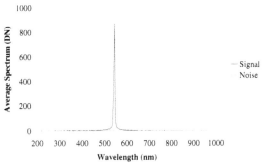

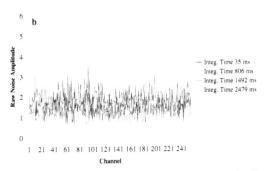

Figure 3. Spectrum of the averaged (ten trials) instrument's offset and noise for different integration times. a: offset, b: noise. Let us note that the offset as well as the experimental noise are spectrally flat; i.e.: they do not change with changing wavelength.(Barducci et al., see p. 199)

Figure 7. Spectrum of a coloured He-Ne laser source acquired by the spectro-irradiometer. The plotted data were averaged over ten independent measures and wavelength calibrated using the results previously discussed. (Barducci et al., see p. 201)

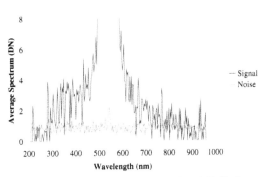

Figure 8. Wings of the spectrum of a coloured He-Ne laser source acquired by the spectro-irradiometer. The plotted data were averaged over ten independent measures and wavelength calibrated using the results previously discussed.(Barducci et al., see p. 201)

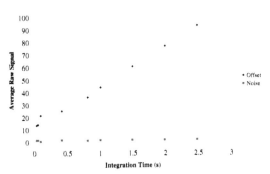

Figure 4. Average instrument's offset and noise amplitude versus the imposed integration time for the high gain. For any point in the plot ten measurements were acquired and the noise amplitude was computed as the signal standard deviation for the single measure. The plotted data have been averaged over repeated trials and over wavelength (different channels) for both the offset and the noise amplitude. (Barducci et al., see p. 199)

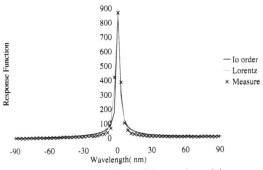

Figure 9. Plot of the fit result and of the experimental data versus the wavelength. It is evident that the Lorentz's profile (red line) is the better approximation to the experimental data (crosses). The response function is not normalized. (Barducci et al., see p. 201)

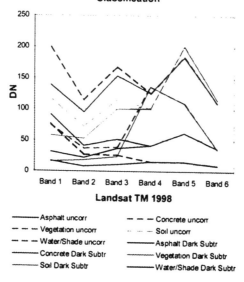

Figure 2. Spectral signatures of the image endmember.
(Hofmann, see p. 205)

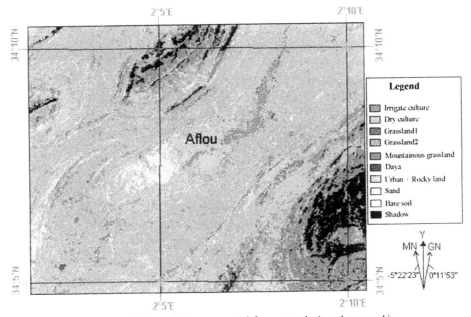

Figure 2. Extracted window of the result image corrected from atmospheric and topographic effects. (Smahi & Bensaid, see p. 218)

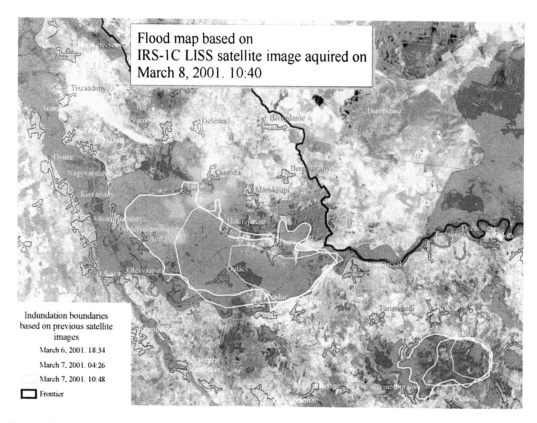

Figure 6. This high resolution flood map shows the actually flooded areas on March 8, 2001, as well as the boundaries of flooded areas based on previous high-, medium- and low resolution IRS LISS, IRS WiFS and NOAA AVHRR data. (Lelkes et al., see p. 263)

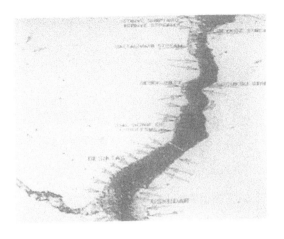

(a)

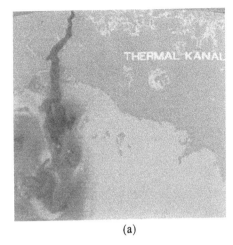

(a)

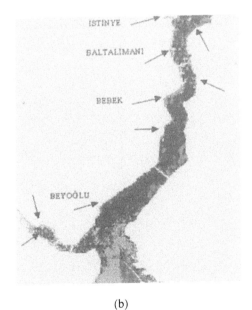

(b)

Figure 3. Turbidity distribution map of Bosphorus derived from Landsat-5 TM of 1986 (a) and 1997 (b), formed by controlled classification method. (Ganca Coşkun et al., see p. 277)

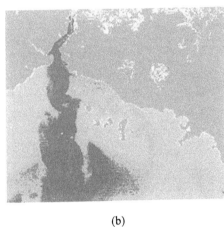

(b)

Figure 4. Thermal images are formed by the thermal bands of Landsat-5 TM dated 1984 (a) and 1997 (b) of the Bosphorus and the Marmara Sea. (Ganca Coşkun et al., see p. 278)

Figure 1. Terkos Water basin Borders. (Goksel et al., see p. 300)

Figure 3a. Classified 1996 LISS image. (Goksel et al., see p. 302)

Figure2. Merge Image (IRS IC + LISS 111). (Goksel et al., see p. 301)

Figure 3b. Classified 2000 LISS image. (Goksel et al., see p. 302)

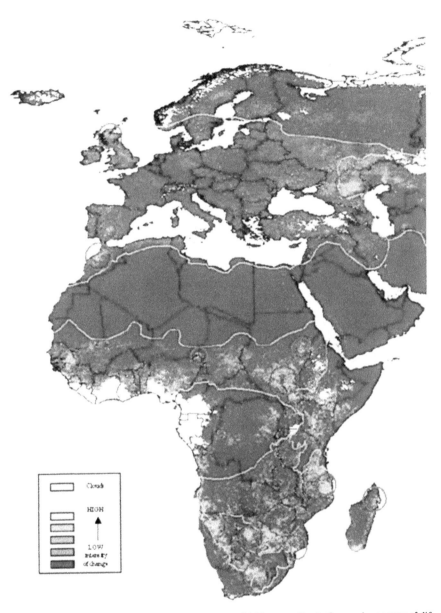

Fig. 1. Map of land-cover changes in Europe and Africa according to the growing seasons of different agro-ecological regions from May 1998 to April 2000. The change intensity is measured by the change vector magnitude. The clouds are represented in white (mask) and the red circles delimitate area where a detailed validation has been done (see http://www.geo.ucl.ac-be/Disasters.htm). (Lupo et al., see p. 335)

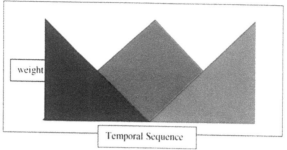

Figure 1. Weighting that could be given to blue green and red for each image through the temporal or hyperspectral sequence. Such a weighting will give blue- cyan - green - yellow - red in the sequence, but no magenta. Other weighting functions could be adopted that include magenta. (McCloy, see p. 338)

Temporal Enhancement of Global NDVI

red - late, green - mid and blue - early in the year

Figure 2. A temporal enhancement of 216 temporal images covering 18 years of data. (McCloy, see p. 338)

Figure 3. Temporal variograms of the global NDVI data, with red - lag 4, green - lag 3 and blue - lag 2. (McCloy, see p. 338)

Figure 4. Fourier Amplitude Image of the Temporal variogram dataset. Red - No annual cycles, Green - dominant annual cycles, blue - dominant bi-annual cycles, yellow - Annual with negligible cycles, magenta - Bi-annual with negligible cycles. (McCloy, see p. 338)

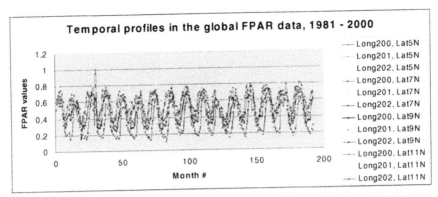

Figure 5. Profiles across global FPAR data at 12 locations in Africa, varying from the Congo Rainforest to the moderately dry Sahel. (McCloy, see p. 340)

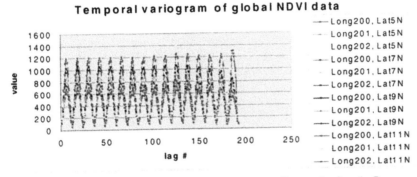

Figure 6. Profiles in the Temporal variograms for 12 locations in Africa, varying from the Congo Rainforest to the moderately dry Sahel. (McCloy, see p. 340)

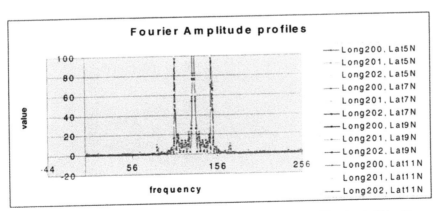

Figure 7. Fourier Amplitude profiles derived from the variogram data for 12 locations in Africa, (a) the whole profiles, (b) at frequency 107 representing annual cycles, and (c) at frequency 85, representing six-monthly cycles. (McCloy, see p. 340)

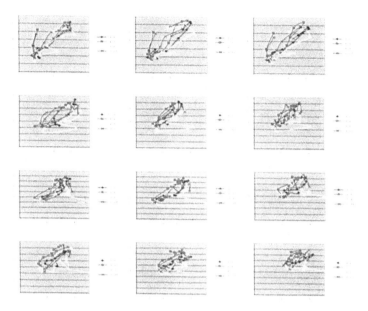

Figure 8. Comparative plots of FPAR for the two AVHRR based datasets. Each plot represents one pixel, at longitudes 200, 201 and 202 degrees, and latitudes 5, 7, 9 and 11 degrees north. (McCloy, see p. 342)

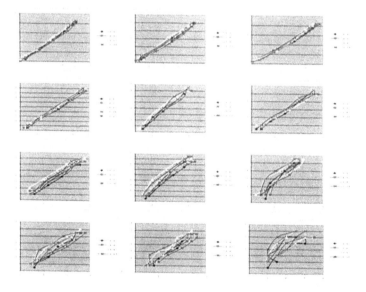

Figure 9. Comparative plots of the variograms derived from the FPAR data for the two AVHRR based datasets. Each plot represents one pixel, at longitudes 200, 201 and 202 degrees, and latitudes 5, 7, 9 and 11 degrees north. (McCloy, see p. 342)

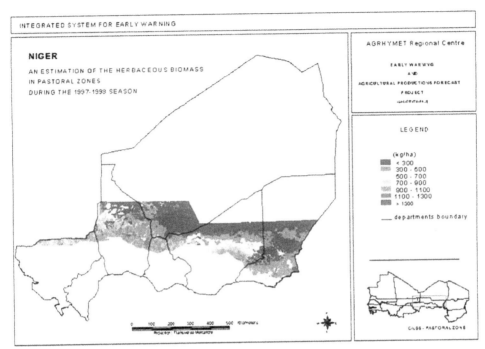

Figure 3. (DeFilippis et al., see p. 358)

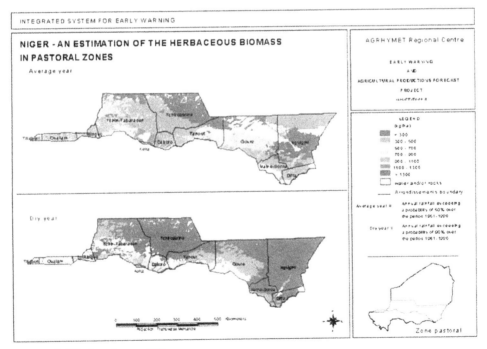

Figure 4. (DeFilippis et al., see p. 358)

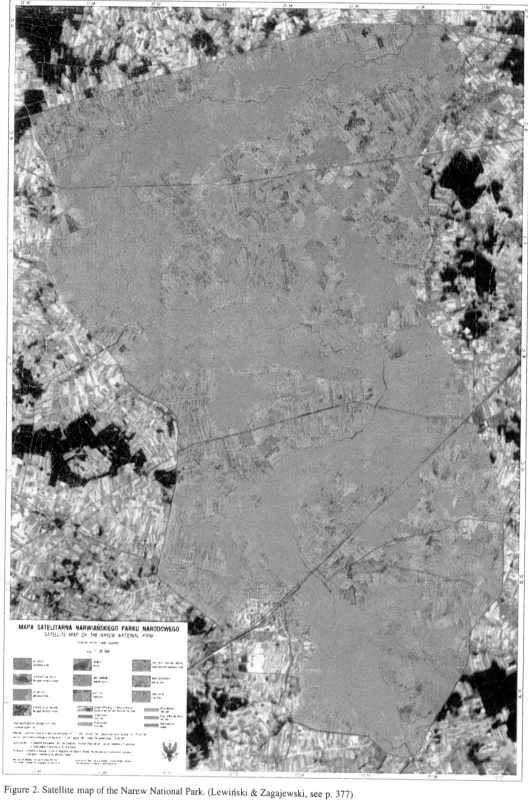

Figure 2. Satellite map of the Narew National Park. (Lewiński & Zagajewski, see p. 377)

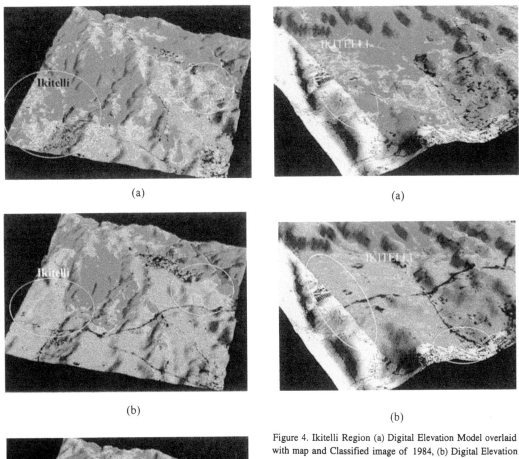

(a)   (a)

(b)   (b)

Figure 4. Ikitelli Region (a) Digital Elevation Model overlaid with map and Classified image of 1984, (b) Digital Elevation Model overlaid with map and Classified image of 1997. (**Kaya** et al., see p. 382)

(c)

- ● Road
- ● Urban Area
-   Bare Area
- ● Forest and Green Area
- ● Water
- ● Industrial Area

Figure 3. Classified images, (a) Classified image of 1984, (b) Classified image of 1992, (c) Classified image of 1997. (Kaya et al., see p. 382)